Productivity Growth in Agriculture

An International Perspective

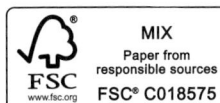

This book is dedicated to the memory of Dr Vernon W. Ruttan (1924–2008) whose work and character continue to inspire generations of students.

Productivity Growth in Agriculture

An International Perspective

Edited by

Keith O. Fuglie

Sun Ling Wang

and

V. Eldon Ball

Economic Research Service, US Department of Agriculture

(dbj
www.cabi.org

CABI is a trading name of CAB International

CABI	CABI
Nosworthy Way	875 Massachusetts Avenue
Wallingford	7th Floor
Oxfordshire OX10 8DE	Cambridge, MA 02139
UK	USA
Tel: +44 (0)1491 832111	Tel: +1 800 552 3083 (toll free)
Fax: +44 (0)1491 833508	Tel: +1 (0)617 395 4051
E-mail: info@cabi.org	E-mail: cabi-nao@cabi.org
Website: www.cabi.org	

A catalogue record for this book is available from the British Library, London, UK.

Library of Congress Cataloging-in-Publication Data

Productivity growth in agriculture : an international perspective / edited by Keith O. Fuglie, Sun Ling Wang, and V. Eldon Ball.
    p. cm.
  ISBN 978-1-84593-921-2 (hardback)
  1. Agricultural productivity. I. Fuglie, Keith Owen. II. Wang, Sun Ling. III. Ball, V. Eldon. IV. C.A.B. International.

  HD1415.P697 2012
  338.1′6--dc23

                       2012015289

ISBN-13: 978 1 84593 921 2

Commissioning editor: Claire Parfitt
Editorial assistant: Chris Shire
Production editor: Simon Hill

Typeset by SPi, Pondicherry, India
Printed and bound in the UK by CPI Group (UK) Ltd, Croydon, CR0 4YY.

Contents

Contributors

_____

Mirian Rumenos Piedade Bacchi, Assistant Professor, Center for Advanced Studies on Applied Economics (CEPEA), University of São Paulo, Av. Padua Dias, no 11, PO Box 132, Piracicaba, SP 13400-970, Brazil. E-mail: mrpbacch@esalq.usp.br

V. Eldon Ball, Economic Research Service, US Department of Agriculture, 355 E Street, SW, Washington, DC 20024. E-mail: eball@ers.usda.gov

Eliana Teles Bastos, Ministry of Agriculture of Brazil, Esplanada dos Ministerios Bloco D, Annex B, PO Box 02432, Brasilia, DF, Brazil. E-mail: eliana.bastos@agricultura.gov.br

Hans P. Binswanger-Mkhize, Adjunct Professor, College of Economics and Management, China Agricultural University, No. 17 Qinghua East Road, Beijing PR China. E-mail: binswangerh@gmail.com

Rita Butzer, Belmont, Maine. Dr Butzer's contribution to this volume was part of her PhD dissertation research at the University of Chicago. E-mail: ritabutzer@gmail.com

Sean A. Cahill, Research and Analysis Directorate, Agriculture and Agri-Food Canada, 1341 Baseline Road, Tower 7, 4th Floor, Room 330, Ottawa, ON K1A 0C5, Canada. E-mail: sean.cahill@agr.gc.ca

Alwin d'Souza, Room 16, Jhelum Hostel, Jawaharlal Nehru University, New Delhi 67, India. E-mail: alwdsouza@gmail.com

Keith O. Fuglie, Branch Chief for Resource, Environmental, and Science Policy, Economic Research Service, US Department of Agriculture, 355 E Street, SW, Washington, DC 20024. E-mail: kfuglie@ers.usda.gov

Lilyan E. Fulginiti, Professor, Department of Agricultural Economics, University of Nebraska at Lincoln, 307 Filley Hall, Lincoln, NE 68583-0922. E-mail: lfulginiti1@unl.edu

José Garcia Gasques, Institute for Applied Economic Research (IPEA) and Ministry of Agriculture of Brazil, Brasilia, DF, Brazil. E-mail: jose.gasques@agricultura.gov.br

Emily M. Gray, Australian Bureau of Agricultural and Resource Economics and Sciences (ABARES), GPO Box 1563, Canberra, ACT 2601, Australia. E-mail: emily.gray@abares.gov.au

Donald F. Larson, Senior Research Economist, Development Research Group, World Bank, 1818 H Street, NW, Washington, DC 20433. E-mail: dlarson@worldbank.org

Frikkie Liebenberg, Department of Agricultural Economics, Extension and Rural Development, University of Pretoria, Private Bag X20 Hatfield, Pretoria, 0028, South Africa. E-mail: frik.liebenberg@gmail.com

Yair Mundlak, Ruth Hochberg Professor (Emeritus) of Agricultural Economics, Hebrew University of Jerusalem, P.O. Box 12, Rehovot, 76100, Israel. E-mail: mundlak@agri.huji.ac.il

Alejandro Nin-Pratt, International Food Policy Research Institute, 2033 K Street NW, Washington, DC 20006-1002. E-mail: a.ninpratt@cgiar.org

Alejandro Plastina, International Cotton Advisory Committee, 1629 K Street NW, Suite 702, Washington, DC 20006-1636. E-mail: alejandro@icac.org

Nicholas E. Rada, Economic Research Service, U.S. Department of Agriculture, 355 E Street, SW, Washington, DC 20024. E-mail: nrada@ers.usda.gov

Tabitha Rich, Research and Analysis Directorate, Agriculture and Agri-Food Canada, 1341 Baseline Road, Tower 7, 4th Floor, Room 330, Ottawa, ON K1A 0C5, Canada. E-mail: Tabitha.rich@agr.gc.ca.

David Schimmelpfennig, Economic Research Service, U.S. Department of Agriculture, 355 E Street, SW, Washington, DC 20024. E-mail: des@ers.usda.gov

Juan P. Sesmero, Assistant Professor, Department of Agricultural Economics, Purdue University, Krannert Building, Room 591A, 403 West State Street, West Lafayette, IN 47907-4773. E-mail: jsesmero@purdue.edu

Yu Sheng, Australian Bureau of Agricultural and Resource Economics and Sciences (ABARES), GPO Box 1563, Canberra, ACT 2601 Australia. E-mail: yu.sheng@abares.gov.au

Waleerat Suphannachart, Lecturer, Department of Agricultural and Resource Economics, Kasetsart University, 50 Phahonyothin Road, Jatujuk Bangkok 10900, Thailand. E-mail: waleerat.sup@gmail.com

Johan Swinnen, Professor and Director, Centre for Institutions and Economic Performance (LICOS), Katholieke Universiteit Leuven, Kantoorgebouw Waaistraat, Waaistraat 6, 3000 Leuven, Belgium. E-mail: jo.swinnen@econ.kuleuven.be

Haizhi Tong, Dublin, Ohio. Dr Tong's contribution to this volume was part of her graduated studies at the University of Nebraska at Lincoln. E-mail: haizhitong@yahoo.com

Constanza Valdes, Economic Research Service, US Department of Agriculture, 355 E Street, SW, Washington, DC 20024. E-mail: cvaldes@ers.usda.gov

Kristine Van Herck, PhD student, Centre for Institutions and Economic Performance (LICOS), Katholieke Universiteit Leuven, Kantoorgebouw Waaistraat, Waaistraat 6, 3000 Leuven, Belgium. E-mail: Kristine.vanherck@econ.kuleuven.be

Liesbet Vranken, Assistant Professor, Division of Agricultural and Food Economics, Department of Earth and Environmental Sciences, Katholieke Universiteit Leuven, Celestijnenlaan 200 E, Box 2411, B-3001 Leuven, Belgium. E-mail: Liesbet.Vranken@ees.kuleuven.be

Sun Ling Wang, Economic Research Service, US Department of Agriculture, 355 E Street, SW, Washington, DC 20024. E-mail: slwang@ers.usda.gov

Peter Warr, John Crawford Professor of Agricultural Economics and Head of Department, Arndt-Corden Department of Economics, Crawford School of Economics and Government, College of Asia and the Pacific, Australian National University, Lennox Crossing, Building #132, Canberra, ACT 0200, Australia. E-mail: peter.warr@anu.edu.au

Bingxin Yu, International Food Policy Research Institute, 2033 K Street NW, Washington DC 20006-1002. E-mail: b.yu@cgiar.org

Shiji Zhao, Productivity Commission, GPO Box 1428, Canberra, ACT 2601, Australia. Shiji Zhao was with ABARES when the work presented in this volume was carried out. E-mail: shiji.zhao@pc.gov.au

Foreword

By the middle of this century there will be more than 9 billion people to feed, clothe and shelter on our planet. Meeting these basic human needs, and doing so in a sustainable way that maintains resources for the future, is the most fundamental challenge facing our society and our agriculture today.

Increased agricultural productivity is critical to meeting this challenge. However, our measures of productivity growth and our understanding of its causes are limited. Continuing efforts to improve productivity measures and research into the forces that drive it are necessary for agricultural economists to provide policy makers with the insights to develop the policies and incentives to meet the challenges of a growing world.

Unfortunately, the economics of productivity and the forces that drive it are often misunderstood, even by many economists. As a result the seemingly uncontroversial proposition, that increased productivity is key to meeting human needs in a growing world, has become an increasing source of controversy in the debate over what constitutes a sustainable agriculture. Even among agricultural economists studying productivity there is no consensus on whether the rate of growth in agricultural productivity is slowing.

Fortunately, there is a vibrant research community, actively engaged in expanding our knowledge of the economics of agricultural productivity. Researchers are working on the theoretical basis for measurement, data issues, cross-sectional and time series analyses of the forces driving productivity. They are also beginning to address the consequences of productivity growth for economic welfare, poverty reduction, food security and natural resource conservation. This growing body of research is global, and spans both industrial and developing economies around the world.

This volume grew out of a May 2010 international conference on agricultural productivity growth sponsored by USDA's Economic Research Service and Farm Foundation NFP, with support from the Global Harvest Initiative. It will not answer all of the questions we have about agricultural productivity growth but it brings together a rich collection of the most current research on this critical issue. Although it is targeted at the community of agricultural economists working on productivity-related research, the insights in this collection deserve the attention of all economists and serious students of public policy who are concerned about meeting the challenges of feeding a growing world.

Neil Conklin
President, Farm Foundation NFP

Acknowledgements and Disclaimer

The authors would like to thank the Economic Research Service, Farm Foundation and the Global Harvest Initiative, Neil Conklin and Bill Lesher in particular, for their encouragement and support for the research on agricultural productivity presented in this volume. The authors would also like to thank the participants of an ERS-Farm Foundation conference on the *Causes and Consequences of Global Agricultural Productivity Growth*, held in Washington, DC, in May 2010, where early drafts of the chapters in this volume were discussed and critiqued. The views expressed in the volume are the authors' alone, and no endorsement of the US Department of Agriculture, Farm Foundation, Global Harvest Initiative or any of the contributors' affiliated institutions should be inferred.

1 Introduction to Productivity Growth in Agriculture

Keith O. Fuglie, Sun Ling Wang and V. Eldon Ball
*Economic Research Service, US Department of Agriculture,
Washington, DC*

What this Volume is About

Improving agricultural productivity has been the world's primary defence against a Malthusian crisis – the idea that food demand from a rising population will confront limits to natural resources and lead to famine. In fact, throughout the 20th century real (inflation-adjusted) agricultural prices fell (Giovanni, 2005), implying supply was growing faster than demand, in spite of a global population increase of 3.7 times (United Nations, 2004). Hayami and Ruttan (1971, 1985) showed that the agricultural success story of the 20th century was increasingly about raising the productivity of agricultural resources, rather than expanding the resource base. Figure 1.1 updates a graphical depiction of long-term trends in agricultural land and labour productivity that was popularized in the texts of Hayami and Ruttan (1971, 1985). The graph shows the progression in output per worker and output per area for major global regions during the past 50 years. Generally, industrialized nations have defined, and steady pushed out, the 'technology frontier', or the highest land- and labour-productivity combinations. Currently, SE Asia, China and Latin America are approaching the productivity levels that today's industrialized nations were at in the 1960s.

For the past decade, however, agricultural commodity prices have changed course. According to the International Monetary Fund's (IMF's) food commodity price index, world prices of agricultural commodities rose by about 125% between 2000 and 2011 (or by about 63% in inflation-adjusted US dollars). This decade-long rise in agricultural prices, coupled with the new demands on agriculture from energy markets and concerns about climate change, have renewed concerns about the limits to agricultural growth.

Of particular concern is whether the kinds of gains in agricultural productivity that characterized the 20th century can be sustained in the 21st. Moreover, it is increasingly recognized that future productivity gains in agriculture need to save not only land (by raising yield on existing agricultural land) but also a wider array of natural resources (water quantity and quality, soil quality, biodiversity, etc.) so as to avoid negative impacts to the environment from agricultural intensification. And finally, the issue is not only about raising productivity per se but raising productivity in such a way that it better serves the needs of vulnerable populations – the poor and malnourished – which, as most of the world's poor and malnourished are themselves farmers, means raising agricultural productivity in poor areas.

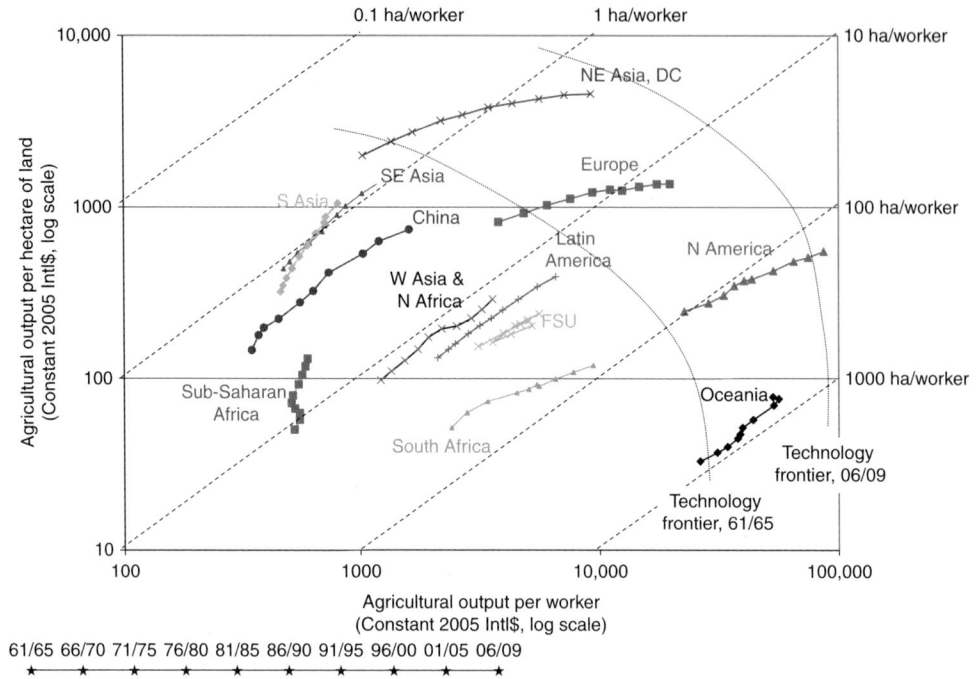

Fig. 1.1. Trends in agricultural land and labour productivity in major world regions. The points on the curves show the progression in output/worker and output/land combinations (averaged over 5-year periods) between 1961/1965 and 2006/2009. Output is gross crop and livestock output measured in constant 2005 international dollars. The number of workers is the number of economically active adults in agriculture. Land is the sum of total cropland and permanent pasture in hectares. The dashed diagonal lines show the average land–labour ratio in hectares per worker. The 'technology frontiers' denote the highest land and labour productivities achieved in these regions in 1961/1965 and 2006/2009. In most regions, labour productivity grew faster than land productivity as average area per worker rose (the exceptions are South Asia and sub-Saharan Africa). Countries of the former Soviet Union (FSU) suffered a reversal in land and labour productivity in the 1990s following the transition from centrally planned to market economies, but had recovered most of these losses by 2006/2009. NE Asia, DC includes Japan, South Korea and Taiwan. Oceania includes Australia and New Zealand. N America includes Canada and the USA. Europe includes all European states except those in the former Soviet Union (FSU). Derived from FAO data.

The first step in being able to assess prospects for future productivity growth is to understand the past. Although much work has been done on analysing trends in agricultural productivity, our understanding of this issue remains far from complete. Recently, Alston, Babcock and Pardey (2010) compiled a number of studies on trends in agricultural productivity in various regions of the world. Their conclusion was that 'agricultural productivity has slowed, especially in the world's richest countries'. But they recognized that the evidence was mixed, and, given the importance of the issue, that it needed further investigation. In many ways the present volume builds upon this earlier work. This volume extends their coverage to new regions of the world, to more recent years and, in some cases, with improved methods and data. In particular, the present volume devotes an entire section to agricultural productivity in sub-Saharan Africa, a region of particular concern given the concentration of the world's poor and malnourished living there (and where half of the world's population increase this century is expected to occur). Another new contribution in the present volume is that it goes beyond national productivity aggregates. For five of the largest global agricultural producing nations (Australia, Brazil, China, Indonesia and the USA), it reports trends in agricultural productivity growth at the provincial or state level.

The present volume is written primarily for agricultural economists doing research on productivity. It includes fairly detailed discussions of the theoretically underpinnings of productivity measurement (see Chapter 4 by Zhao, Sheng and Gray, in particular) as well as the many practical considerations that go into translating this theory into actual measures of aggregated outputs and inputs (Chapter 3 by Cahill and Rich contains considerable discussion on data issues in describing their new productivity accounts for Canada). The unifying concept of agricultural productivity used across the chapters of this volume is aggregate total factor productivity (TFP) of the sector. TFP is a broad measure encompassing the average productivity of all inputs with market value (land, labour, capital and materials) employed in the production of all crop and livestock commodities. It excludes, however, 'non-market' inputs and outputs, such as changes in environmental services resulting from agricultural activities.

The volume also contains detailed analysis of the underlying *causes* of agricultural productivity growth. In the case studies on the USA, Indonesia, Thailand and sub-Saharan Africa, the authors develop econometric models to explain different productivity growth patterns across countries, states or provinces, and/or over time. All of these chapters find a central role for public investments in agricultural research and development (R&D), both local and external, in stimulating productivity growth. The chapter on the transition economies of the former Soviet Union and Eastern Europe uses a more heuristic approach to explain the widely varying output and productivity performance of these nations following their transition from centrally planned to market-based economies. The pace of institutional and policy reform is an important element in explaining these differences.

The volume does not generally delve into *consequences* of agricultural productivity growth, such as for economic welfare, trade competitiveness, poverty alleviation and the environment, leaving this for future work. Exceptions, however, are Binswanger-Mkhize and d'Souza (Chapter 9) who examine the implications of agricultural growth for structural transformation, employment generation and poverty reduction in the Indian economy, and Nin-Pratt and Yu (Chapter 13) who explore patterns of agricultural productivity and poverty rates in sub-Saharan Africa. Both studies find that agricultural productivity growth is associated with poverty reduction in these regions.

Synopsis of the Volume Chapters

The chapters in **Part I** examine agricultural productivity in high-income and transition countries. Chapter 2 by Wang, Ball, Fulginiti and Plastina uses 1980–2004 state panel data to assess the impact of local public goods – public agricultural R&D, extension and road infrastructure – on US agricultural productivity growth. The authors found that the impact of own-state R&D was greatly enhanced by R&D investments in other states and regions through knowledge spill-ins. Returns to local R&D were also higher when it was coupled with greater local extension activity and road infrastructure. The internal rate of return to public R&D is taken as a measure of the contribution of research expenditures to productivity growth. The estimated local rate of return (taking into account only in-state benefits) was around 13% per year. The social rate of return (which takes into account not only benefits to the state where the R&D was conducted but also benefits of spillovers to other states) was found to be around 45%. Moreover, the levels of R&D spill-ins, extension and road infrastructure were found to play a significant role in explaining the differences in productivity growth rates among the states and regions.

The production accounts developed by the US Department of Agriculture (USDA) underpin the estimates of productivity growth in the US states. In Chapter 3, Cahill and Rich from Agriculture and Agri-Food Canada (AAFC) describe in some detail the methods and data used by AAFC to construct the production and productivity accounts for Canadian agriculture. Like the USDA, they use gross output rather than value added (which subtracts the value of intermediate inputs from gross output) as the measure of agricultural output. As a result, inputs of intermediate goods, obviously crucial to agricultural production, are treated symmetrically with capital and labour inputs. The alternative approach using value added as a measure of agricultural output is clearly inferior to gross output in modelling the behaviour of agricultural producers. Not surprisingly, most empirical research on agricultural productivity has come to rely on gross output, rather than value added (including all of the chapters in this volume except Chapter 11 on Thailand, which uses value added). The authors estimate that gross output expanded at a 2.3% average annual rate over the 1961–2006 period, while aggregate input growth averaged just 0.7% annually. Thus productivity growth, at 1.6% per year, was the principal source of economic growth of the Canadian farm sector, and its contribution has remained fairly steady over time.

Chapter 4 focuses on the economic performance of Australian agriculture. Since the mid-1990s, the Australian Bureau of Agricultural and Resource Economics and Sciences (ABARES) has compiled and published annual estimates of productivity growth for the Australian broadacre and dairy industries and have extended these estimates back to the late 1970s. These estimates are widely used by government, industry groups and the wider research community for purposes of informing decision makers of the likely impacts of agricultural policy. Although the statistics on productivity growth have a wide audience, the absence of documentation of methods and data makes it difficult for users to understand and interpret the statistics.

This lack of documentation also makes it difficult for ABARES to engage the research community in an effort to improve measurement of productivity growth. In this chapter, Zhao, Sheng and Gray attempt to bridge this divide by reviewing the concept of productivity growth and providing the technical details underlying ABARES productivity statistics.

There have been dramatic changes in productivity over the past 15 years in the transition countries. In Chapter 6, Swinnen, van Herck and Vranken look at the impact of recent economic and institutional reforms on agricultural productivity growth in Central and Eastern European countries and the former Soviet Union. They provide a conceptual framework for addressing the evolution of productivity gains and relate these gains to initial factor endowments, the pace of reforms and the overall level of economic development. In general, they observe a U-shaped pattern of agricultural growth. Virtually all countries witnessed an initial decline in output and productivity, following transition to market economies and virtually all countries are currently experiencing increases in productivity. In several transition countries, the rate of productivity growth over the past five years has been quite spectacular. However, the depth and length of the initial decline differed greatly among countries. The authors argue that productivity changes were related to the manner in which reforms were implemented. In the most advanced countries (mostly in Central Europe), the decline in productivity was relatively mild and recovery started soon after the onset of reforms. In the Baltic countries, recovery started relatively quickly, reflecting the fast pace of reforms, but the initial fall in productivity was much deeper than in Central Europe, possibly reflecting institutional impediments to growth in these former Soviet Union countries. In several of the other former Soviet Union countries as well, the decline in productivity was quite dramatic and lasted for much of the 1990s. It is clear that the early reformers, such as the Central European countries, witnessed technological adoption and productivity growth much earlier. However, the initial level of development and economic structure also made a difference. The institutional and human capital hurdles to creating a market economy were higher in the east, so it is no surprise that productivity declines were deeper and longer in those countries.

Chapter 5 assesses the prospect of slowing agricultural productivity growth in Western Europe. As previously mentioned, Alston, Babcock and Pardey (2010) concluded that agricultural productivity growth had slowed, at least in the world's richest countries. But, apart from the UK, they were not able to look in any detail at Western Europe. In this chapter, Wang, Schimmelpfennig and Fuglie employ data compiled by Eldon Ball and his colleagues to construct measures of productivity growth in 11 European Union countries over the period from 1973 to 2002 (Ball et al., 2010). They apply statistical techniques to individual country data to determine whether any of them experienced a slowing of productivity growth. Although the authors observed stagnant output growth, this was attributed to the withdrawal of resources from the agricultural sector. They did not find evidence of a productivity slowdown.

Part II examines agricultural productivity growth and its driving forces in five important agricultural producers in Asia and Latin America – Brazil, China, India, Indonesia and Thailand. Chapter 7, by Gasques, Bastos, Valdes and Bacchi, analyses the TFP growth and structural transformation in Brazil for agricultural census years between 1970 and 2006. Censuses contain the most complete account of all the factors employed in a country's agriculture and therefore form a particularly strong base from which to evaluate productivity. On the basis of their estimates, productivity growth was the major driver of growth in Brazilian agriculture. They estimate a 2.1% per annum TFP growth rate for the whole period, with accelerating TFP growth over time, although they conclude that this is probably low because of incomplete information on agricultural outputs in the 2006 Census. Nonetheless, they found gradual diversification in the

composition of agricultural output over time both at the national level and for most states. They assert that the diversification can have positive effects on employment and income as farmers allocate more resources to higher valued products.

In Chapter 8, Tong, Fulginiti and Sesmero evaluate agricultural productivity growth in 30 of China's provinces during 1993–2005. This period is particularly interesting for China because the uncertainty over land tenure security rose as long-term leases were set to expire in 1998. They contrast results from two widely used methodologies for measuring productivity, the Malmquist Index method and the stochastic production frontier method. They find that China achieved high rates of agricultural productivity growth throughout the whole period, averaging around 4% per year. They also find significant regional disparities in productivity performance, with eastern coastal provinces outperforming the central and western provinces. This is consistent with the pattern of regional economic growth in China. Agriculture therefore might be contributing to the growing regional disparities in China's economy. They also find differences in productivity growth estimates from the two methods, and attribute this to the assumptions imposed by each method. Whereas the Malmquist index demonstrates large variability in productivity growth from year to year, the stochastic frontier econometric method imposes uniformity and smoothness in growth estimates over time.

Chapter 9, by Binswanger-Mkhize and d'Souza, analyses the structural transformation of the Indian economy and its agriculture between 1961 and 2009. Although the agricultural share of gross domestic product (GDP) fell dramatically between the early 1960s and late 2000s, it remained the main employment sector throughout the period. This lack of convergence in the agricultural GDP and labour force shares of the economy is mirrored in the widening and accelerating divergence in average labour productivity between agricultural and non-agricultural sectors. The authors see the turning point in India's structural transformation, in which this productivity differential begins to narrow, as still in the distant future. Agricultural growth, and productivity growth in particular, will need to accelerate for India to generate sufficient new employment for its burgeoning rural population. Agricultural employment is especially important for India's poor and least educated workers, who are less able to migrate to non-farm sectors. The authors identify productivity improvement in the livestock sector as an underappreciated success story in India's agriculture.

In Chapter 10, Rada and Fuglie examine provincial level agricultural TFP growth in Indonesia based on Fisher-ideal quantity index estimates. They find that Indonesian agriculture exhibited divergent patterns of regional growth in the 1985–2005 post-Green Revolution period. The results show that Java and Bali, regions heavily reliant on irrigated rice production, experienced the slowest agricultural production or productivity growth of any region. In the western islands of Sumatra and Kalimantan, however, and in parts of Sulawesi, agricultural output grew about twice as fast as that of Java and Bali. Both significant agricultural land expansion and productivity growth were taking place in these 'outer' islands of Indonesia. They find that an important element behind this agricultural growth success was Indonesia's market and trade liberalization policies, supported by public investment in research and other agricultural services. Research spending by the quasi-independent plantation crop institutes seems to earn a higher rate of return than research spending by government food crop and livestock research institutes. One reason could be stronger institutional and financial support in the plantation crop research system. They also find international technology transfer in food crops to be an important contributor to Indonesian agricultural TFP.

Chapter 11, by Suphannachart and Warr, looks at the TFP measurement and determinants in Thai agriculture during 1970–2006. The findings show that TFP made an important contribution to both crop and livestock output growth during the study period, but has stagnated or regressed

since the mid-1990s. In general, TFP accounts for 21% of crop output growth and 17% of livestock output growth. The contribution of TFP to output growth is relatively low compared with other studies from this volume, as well as Fuglie's estimates of TFP growth for Thailand reported in Chapter 16. One reason for the lower values might be differences in measurement. For output, the authors of Chapter 11 use value-added instead of gross output, which is generally viewed as a less desirable way of constructing productivity accounts. On the input side, they assign a relatively high cost share to capital services (which they do not measure directly but as a residual after accounting for labour and land costs). If agricultural terms of trade declined (reducing measured output growth) or if capital costs were overestimated (exaggerating measuring input growth), it would probably lead to an underestimate in the growth in TFP. However, a strength of their approach is the quality adjustments they make for land and labour inputs. They analyse the TFP determinants by fitting an error correction model. They find that drivers of agricultural TFP include not only domestic agricultural research, but also international technology transfer, domestic investments in extension and infrastructure, weather and disease shocks, and international commodity prices. One of the innovations in both Chapters 10 and 11 is an attempt to capture the contributions of foreign or international R&D to national TFP growth. Both studies find this to be an important source of productivity growth, with national R&D being a complement to it.

Low rates of agricultural growth are thought to underlie the acute poverty and food insecurity that characterizes much of sub-Saharan Africa (SSA). The chapters in **Part III** focus on measuring and identifying constraints to agricultural productivity growth in this region. Assessing productivity growth in SSA is a particular challenge because of a lack of detailed information on agricultural inputs and their costs, except for South Africa. This precludes the kind of detailed growth accounting procedures that many of the other country case studies in this volume have followed. Other approaches are required. In Chapter 12, Fuglie and Rada econometrically estimate a production function for SSA agriculture and from this derive TFP indexes for each country from 1961 to 2008. They then examine what might be driving or constraining TFP. They find national and international agricultural research investments have had a positive and significant effect on productivity growth. But the rate of return to research seems to be much higher in larger countries of the region, which they attribute to economies of size in research systems. Economic reforms that reduced agricultural taxation and improved the terms of trade for the sector were also positively associated with more rapid TFP growth. Civil war, the spread of HIV/AIDS and low levels of education were significant constraints to productivity growth.

In Chapter 13, Nin-Pratt and Yu use a different method, a non-parametric Malmquist index, to measure agricultural TFP growth in sub-Saharan Africa. Their estimates are quite similar to Fuglie's and Rada's (Chapter 12), showing productivity flat or declining from the 1960s to the early 1980s followed by moderate TFP growth since then. They attribute this recovery in agricultural productivity growth to the structural adjustment reforms that were implemented in a number of SSA countries beginning around this time and continuing into the 1990s.

In Chapter 14, Liebenberg reassessed agricultural productivity growth in South Africa, the African country that has been most studied and which has by far the most comprehensive data of its agricultural sector. What Liebenberg finds, however, is that the statistical reporting of South African agriculture was heavily influenced by the country's past racial policies: agricultural data largely focused on white-owned commercial farms and gave inconsistent coverage of semi-subsistence black-owned farms (which were restricted to certain areas of the country). Thus, past measures of productivity performance mostly refer to the commercial sector, and could give a distorted view of the country's agriculture as a whole. Liebenberg also finds evidence of stagnation in South African agricultural TFP, which he attributes to reduced public investment in the sector, particularly in agricultural research. The consequence has been a prolonged period of slow output growth and reduced net exports, especially in field crops.

Part IV of the volume contains two chapters that give a global perspective on agricultural productivity. One of the greatest challenges in measuring agricultural productivity (at the national or international level) is constructing consistent measures of capital stock and its service value. One reason is that, in agriculture, capital stock arises from two distinct sources: from purchases of fixed capital (machinery and structures) from outside the sector and from capital of agricultural origin (animal breeding stock, draft power, orchards and other tree crops). Previous studies have tended to rely on simple counts of farm machinery in use to measure fixed capital and size-adjusted counts of animals on farms to measure capital of agricultural origin with rarely any consideration of treestock. In Chapter 15, Butzer, Mundlak and Larson discuss a new approach for measuring agricultural capital stocks in a consistent fashion for the purpose of international comparisons. For 30 countries they now have consistent estimates of agricultural fixed capital, livestock capital and tree stock capital for 1967–2000 (and 57 countries for 1967–1992). In their approach, fixed capital is measured from past annual investments in machines and structures, appropriately discounted for depreciation. Their approach for measuring capital of agricultural origin (livestock and tree capital) can be extended to include any country in the FAOSTAT dataset. With these new measures of capital they then reassess the contribution of capital accumulation to agricultural growth. Their results confirm a strong role for capital in raising labour productivity and output in agriculture. Another important implication of their results is that a simple count of machinery in use is likely to understate capital accumulation in agriculture. This bias is likely to be larger in higher income countries where fixed capital plays a relatively larger role than capital of agricultural origin. The implication for productivity is that any underestimation in the growth of agricultural capital is likely to imply an overestimation in the measured rate of TFP growth.

In the volume's final chapter, Fuglie makes an ambitious effort to derive estimates of agricultural TFP growth by country, region and for the world as a whole for the period 1961–2009. To estimate TFP, he uses a consistent growth accounting framework based primarily on FAO input and output data, and draws upon country-level case studies for estimates of input cost shares needed for aggregating inputs into a total. The results show accelerating productivity for the world as a whole over time, mostly resulting from improved productivity performance in developing countries. There is wide variation among countries, however. To help explain these differences, Fuglie examines the correlation between country-level indexes of 'technology capital' and their long-run agricultural TFP growth rates. He finds evidence that national capacities in agricultural and industrial research are significantly correlated with agricultural TFP growth. National capacities in agricultural extension and farmer schooling are also important, but, at the margin, seem to be less essential than research capacity in releasing constraints to productivity.

Synthesis of Findings: Where Agricultural Productivity is Growing and Why

One general conclusion from the chapters in this volume is that in recent decades agricultural productivity growth in some developing countries has accelerated but remains extremely uneven across countries. Another conclusion is that there has been a significant recovery of agricultural productivity in Eastern Europe and the former Soviet Union, but this could be temporary once the efficiency gains from moving toward a market economy have been exhausted. The picture is less clear on productivity trends in developed countries, although this volume

adds important new evidence from Western Europe and Canada.

We bring together the findings from the various chapters on trends in global agricultural TFP growth in Fig. 1.2. The data are drawn from the country and regional case studies and then supplemented for the rest of the world using the less precise method Fuglie describes in Chapter 16. The figure shows where TFP growth has been high, moderate and low since the mid-1990s. A unique feature of Fig. 1.2 is that it shows sub-national TFP growth rates for five large agricultural producers: Australia, Brazil, China, Indonesia and the USA. The sub-national estimates reveal considerable within-country variation in addition to the cross-country variation in productivity growth. In China, TFP growth has been very strong in coastal provinces but slackens off as one moves into the interior, with low TFP growth in some agriculturally important provinces such as Sichuan. The coastal provinces of Brazil have also experienced robust agricultural productivity growth (and notably the historically poor north-east region) but unlike China high TFP growth is also evident in some parts of the interior – such as Mato Grosso in the Cerrado, now the main soybean and cotton producing state in the country. In Indonesia, productivity growth seems to be concentrated in western regions of the country, such as Sumatra and Kalimantan, regions where export commodities like oil palm and cocoa have been booming. In contrast, TFP growth has been low or stagnant in much of Java and the eastern provinces. This seems to be in sharp departure from the 'Green Revolution' decades of the 1970s and 1980s that disproportionately benefited irrigated rice production, which is especially important in Java and Bali. In the USA, productivity growth has been moderately strong in agricultural important areas such as the Corn Belt and Lake States but low in Plain States, Appalachia and major horticultural producers such as California and Florida. Although Australian 'broadacre' (dryland) agricultural TFP has been stagnant nationally, this has primarily affected eastern and southern portions of the country. Figure 1.2 also indicates a good

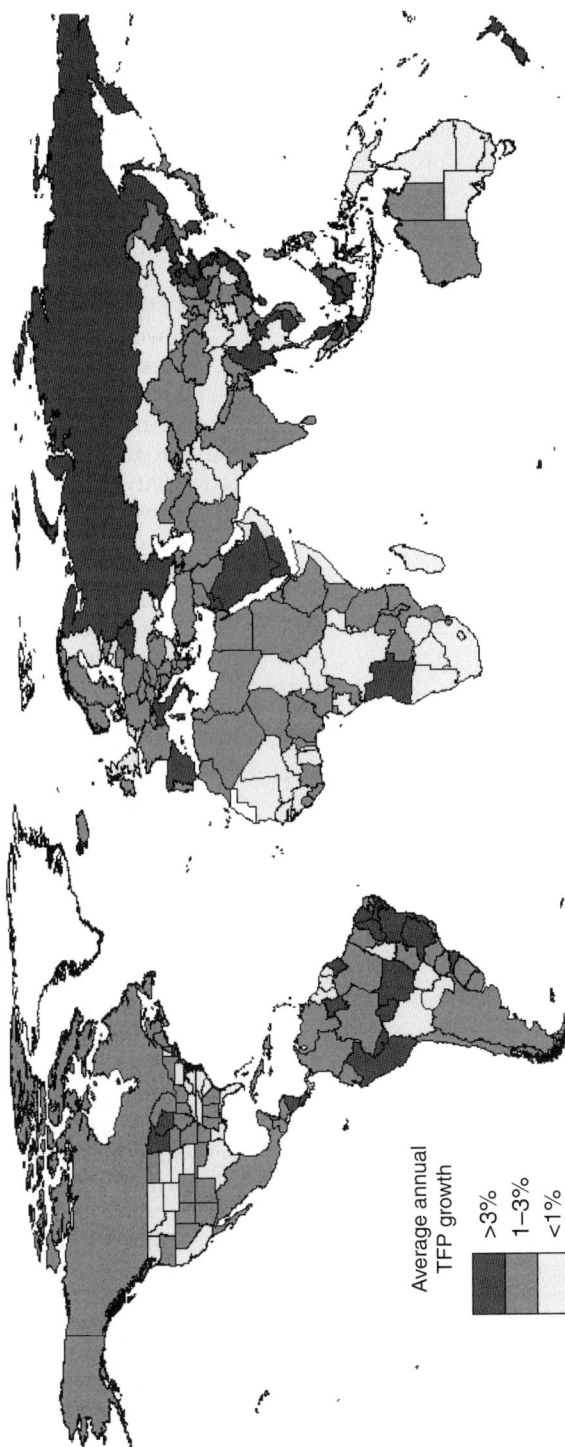

Fig. 1.2. Total factor productivity (TFP) growth in global agriculture since the mid-1990s. Agricultural total factor productivity (TFP) growth rates are taken from the chapters in this volume. The period of coverage is from the mid-1990s to the mid-to-late 2000s (actually years depend on data availability). US states, 1993–2004 (Wang et al., Chapter 2); Brazil states, 1995 and 2006 (Gasques et al., Chapter 7); China provinces, 1993–2005, Malmquist method (Tong et al., Chapter 8); Indonesia provinces, 1993–2005 (Rada and Fuglie, Chapter 10); Australian states, 1993–2009, for broadacre agriculture except Northern Territories, which is for beef only (Zhao et al., Chapter 4; note that state level TFP indexes are not reported in Chapter 4 but are available from the authors); Western European countries, 1993–2002 (Wang et al., Chapter 5); Transition countries of the former Soviet Union and Eastern Europe, 1995–2007, except Kazakhstan, which is 1995–2004 (Swinnen et al., Chapter 6); sub-Saharan Africa, 1993–2008 (Fuglie and Rada, Chapter 12); Thailand, 1993–2006 (Suphannachart and Warr, Chapter 11); Canada, 1993–2006 (Cahill and Rich, Chapter 3); South Africa, 1993–2010 (Liebenberg, Chapter 14); All other countries:, 1993–2009 (Fuglie, Chapter 16).

productivity growth performance in a number of sub-Saharan African countries although this can be somewhat misleading, as some of this is simply recovery of declining TFP in earlier years. Sub-Saharan Africa remains perhaps the biggest challenge in achieving sustained, long-term productivity growth in its agricultural sector.

Perhaps the single most important factor separating countries of the world that have successfully sustained long-term productivity growth in agriculture from those that have not is their national capacity in agricultural research. Countries that have built national research systems capable of developing and adapting a continuous stream of new technologies suitable for local farming systems are generally the ones that achieved higher rates of long-term agricultural TFP growth. In this volume, several of the chapters present statistical models that test this proposition, and they all confirm the importance of public investments in agricultural R&D as a strong determinant of agricultural productivity growth. An important role of local (national or provincial) R&D is technology adaptation: 'knowledge spillovers' from research conducted elsewhere are important determinants of local productivity growth, but local R&D is often crucial for translating these spillovers into usable technologies for local farming systems. Being actively engaged with external R&D institutions and knowledge flows significantly raises returns to local investments in agricultural research.

But these models find that a number of other factors, which can broadly be characterized as the 'enabling environment' for the dissemination of new technologies and practices, also exert a significant influence on the rate of agricultural productivity growth. Policies that strengthen economic incentives for producers, provide agricultural educational and extension services, and improve access to markets are positively correlated with agricultural productivity growth. At the same time, economically disruptive 'shocks', such as disease epidemics and civil unrest, can seriously depress agricultural productivity growth. Having a more favourable enabling environment complements but does not substitute for research. By themselves, these factors are likely to exert only a short-term influence on productivity growth. But having a more favourable enabling environment significantly raises the payoff from public investments in agricultural R&D.

Dedication

This volume is dedicated to the memory of Dr Vernon W. Ruttan, a pioneer in the analysis of international agricultural development, agricultural productivity and the role of public investments in releasing resource constraints to agricultural production. Ruttan, together with Yujiro Hayami, was among the first to carefully measure and explain differences in agricultural productivity across countries and over time and their texts (Hayami and Ruttan, 1971, 1985) remain essential reading for today's practitioners in the field. Ruttan was generally optimistic about prospects for raising agricultural productivity to meet the needs of a growing population. In fact, in one of his first scientific papers, he strongly refuted the pessimistic projections of US agricultural productivity that were common in the early 1950s (Ruttan, 1956). However, in one of his last major discussions of this topic, Ruttan expressed concern that it might become increasingly difficult to maintain historical rates of agricultural productivity growth (Ruttan, 2002). He saw the co-evolution of new pests and diseases, climate change, and diminishing returns to research on raising agricultural productivity as posing significant threats to future global food supply. Thus, there is renewed urgency in furthering our understanding of trends and determinants of agricultural productivity. Ruttan set extremely high standards of quality in his research and writings, and in this volume we aspire to maintain his example. His life and work continue to inspire generations of students.

References

Alston, J., Babcock, B. and Pardey, P. (eds) (2010) *The Shifting Patterns of Agricultural Production and Productivity Worldwide*. Midwest Agribusiness Trade and Research Information Center, Iowa State University, Ames, IA.

Ball, V.E., Butault, J., Mesonada, C. and Mora, R. (2010) Productivity and international competitiveness of agriculture in the European Union and the United States. *Agricultural Economics* 41, 611–627.

Giovanni, F. (2005) *Feeding the World: An Economic History of Agriculture, 1800–2000*. Princeton University Press, Princeton, NJ.

Hayami, Y. and Ruttan, V.W. (1971, 1985) *Agricultural Development: An International Perspective*. 1st edn, 1971; 2nd edn, 1985. Johns Hopkins University Press, Baltimore, MD.

Ruttan, V.W. (1956) The contribution of technological progress to farm output: 1950–1975. *Review of Economics and Statistics* 38, 61–69.

Ruttan, V.W. (2002) Productivity growth in world agriculture: Sources and constraints. *Journal of Economic Perspectives* 16, 161–184.

United Nations (2004) *World Population to 2300*. Population Division, Department of Economic and Social Affairs, United Nations Secretariat, New York, NY.

2 Accounting for the Impact of Local and Spill-in Public Research, Extension and Roads on US Regional Agricultural Productivity, 1980–2004

Sun Ling Wang,[1] V. Eldon Ball,[1] Lilyan E. Fulginiti[2] and Alejandro Plastina[3]
[1]*Economic Research Service, US Department of Agriculture, Washington, DC;*
[2]*University of Nebraska, Lincoln;* [3]*International Cotton Advisory Committee,*
Washington, DC

2.1 Introduction

The study of the contribution of research investment to farm production and agricultural productivity was pioneered by Griliches (1958, 1964), Evenson (1967) and their associates. The benefit of public research in agricultural productivity growth has been documented in numerous studies since their seminal work. On the basis of literature surveys by Evenson (2001), Alston *et al.* (2000), Huffman and Evenson (2006), and Fuglie and Heisey (2007), the rate-of-return estimates for agricultural research are high. In general, the rates of return to federal-state investment in agricultural research are in the range of 20–60% (Fuglie and Heisey, 2007).

The high rate of return to research spending is partly attributed to spillover effects, i.e. the adoption of technologies developed in one region or institution by producers in another region or institution (Evenson 1989; Griliches, 1998). Griliches (1992), referring to research and development (R&D) spillovers between firms, addressed the contribution of R&D spillovers in productivity growth indicating that '...where R&D returns can account for up to half of the growth in output-per-man and about three-quarters of the measured TFP growth, most of the explanatory effect coming from the spillover component, which is large, in part, because it is the source of increasing returns...'. In the context of public agricultural R&D, studies of the US agricultural sector have investigated R&D spillovers between states, because state universities and agricultural experiment stations have historically been a crucial source of local agricultural innovation.[1] Huffman *et al.* (2002), Yee *et al.* (2002), Alston *et al.* (2010) and Plastina and Fulginiti (2012) have reported high social rates of return to state agricultural research, where 'social' returns include the economic impact of interstate R&D spillovers, as opposed to 'local' returns that consider only the in-state benefits from that research. However, these studies are not able to explain why states in similar regions, benefitting from research spill-ins from neighbouring states, might none the less experience significantly different rates of productivity growth. Nor are they able to say much about the main channels for dissemination of technology and technical information.

According to the USDA Economic Research Service (USDA–ERS) production accounts, during 1980–2004, average annual agricultural productivity growth rates ranged from a low of 0.95% in Washington State to a high of 3.18% in Massachusetts. Among USDA production regions (see Fig. 2.1), rates ranged from 1.53% in the Appalachian region to 2.60% per annum in the Corn Belt. Among the ten production regions, five had average annual growth rates above 2% (the Corn Belt, the North-east, Northern Plains, Delta and Lake States). The performance of individual states within regions could, however, be heterogeneous or relatively uniform. Figure 2.2 shows that Appalachian states have widely different productivity growth trends, whereas Lake States have a more uniform trend. This observation leads us to hypothesize that, although there might be important differences in R&D spending across states, variations in other public goods might be intimately related and could affect how

local R&D affects productivity growth, such as investments in agricultural extension services and road infrastructure.

Evenson (2001) reviews a number of studies on the impact of research and extension programmes and concludes that extension activities also play an important role in promoting agricultural productivity growth. Antle (1983) reinforced the importance of public transportation infrastructure in enhancing agricultural productivity growth in an international comparisons study. He asserts that '... the extent to which farmers are able to use new technologies to their advantage depends on the costs and benefits of learning and using them. These costs and benefits are hypothesized to be a function of the country's stock of infrastructure capital and the resulting costs of infrastructural service.' In a study of US agriculture, Paul *et al.* (2001) also found that the transportation network enhanced US agricultural productivity growth.

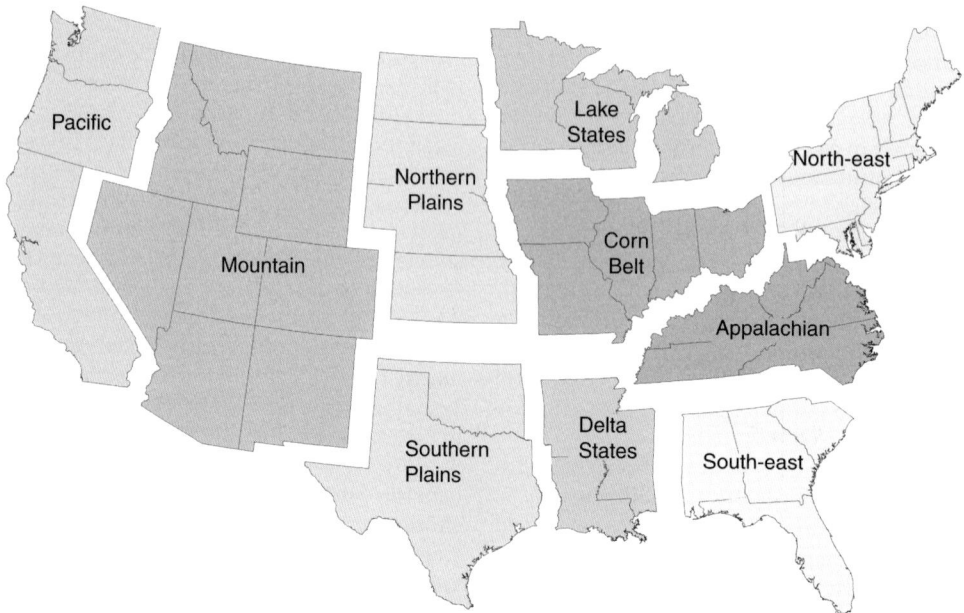

Fig. 2.1. USDA Production regions. States according to region: North-east: NH, PA, ME, MD, RI, MA, DE, CT, VT, NY, NJ; Lake States: MN, MI, WI; Corn Belt: OH, IA, MO, IN, IL; Appalachian: WV, TN, NC, VA, KY; South-east: SC, AL, GA, FL; Delta: LA, AR, MS; Northern Plains: ND, SD, KS, NE; Southern Plains: TX, OK; Mountain: CO, UT, AZ, NM, WY, NV, ID, MT; Pacific: OR, CA, WA. Source: USDA–ERS (http://www.ers. usda.gov/data/farmincome/USDA-Production-regions.htm)

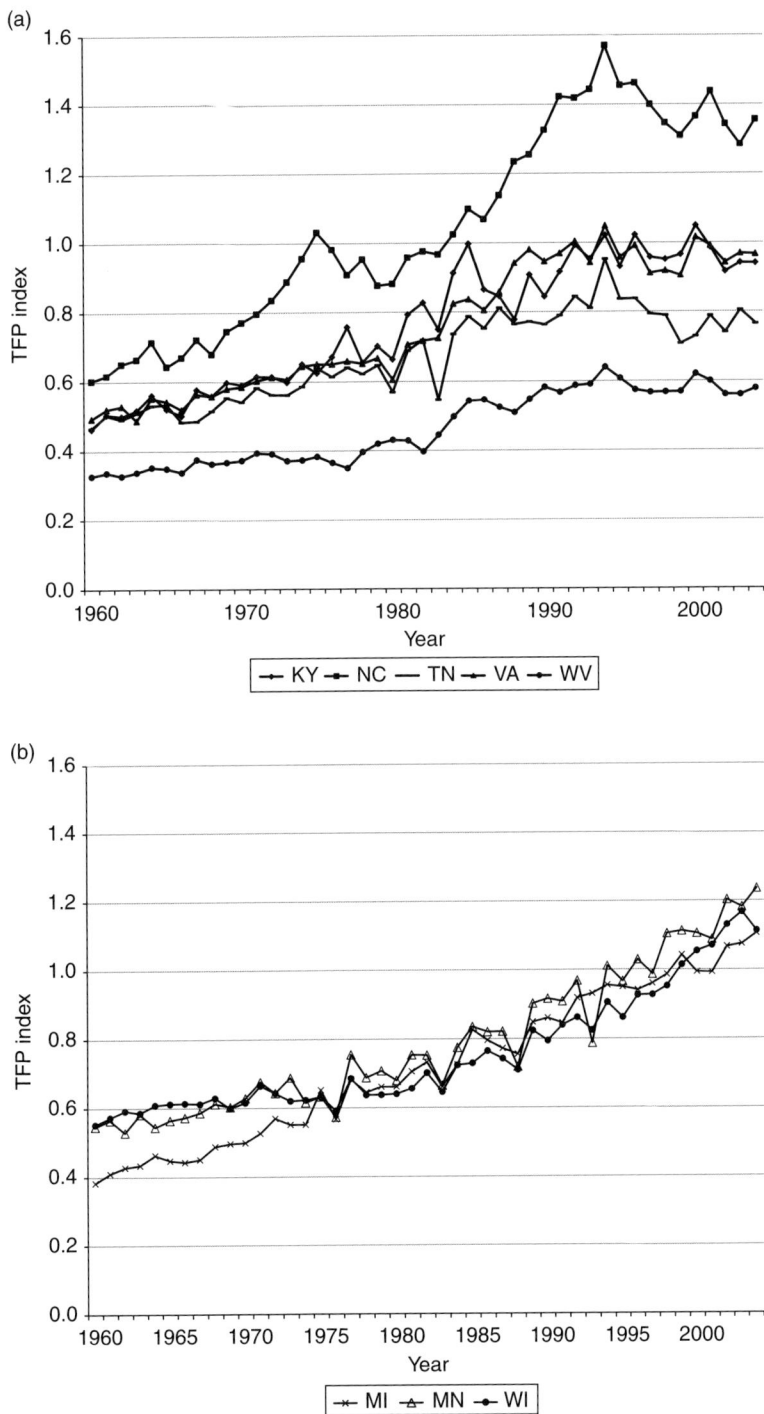

Fig. 2.2. Heterogeneity in agricultural productivity growth among US states. Total factor productivity indexes, Alabama in 1996 = 1.0. (a) TFP indexes for states in the Appalachian region. (b) TFP indexes for states in the Lake States region. Source: The productivity indices are from USDA–ERS.

In this study we examine the role of public R&D expenditures, R&D spill-ins, extension activities, and road infrastructure in US agricultural productivity growth. We hypothesize that a convenient transportation network can provide farmers with an easier way to acquire new technology as it lowers the costs of obtaining inputs that embody new technology, including information. Extension activities may strengthen the dissemination and absorption of technical information. Finally, research spill-ins from other states within the same geo-climatic region could amplify the impact of local R&D on productivity growth. In this way, extension activities, road infrastructure, and R&D spill-ins could act as a catalyst in stimulating local technology development and diffusion as well as utilization of new technical information.

The objective of this chapter is to investigate how local public goods such as extension activities, R&D spill-ins, and road infrastructure can help explain the heterogeneity in agricultural productivity growth among US states and USDA production regions. To do so, we first estimate the impact of public research investments on US agricultural productivity growth using a dual cost function and state-by-year panel data. Second, we evaluate the differential impacts of extension activities and road infrastructure on R&D's contribution to productivity growth. We examine the interaction between local R&D stock and research spill-ins, extension services, and road infrastructure across regions and states. We compare results from models with and without the extension and infrastructure variables in order to assess the impact of these variables on estimates of the rate of return to investment in research.

2.2 Model

We estimate a variable cost function to evaluate the benefit of R&D investments on the cost of production. A reduction in the cost of producing a given level of output can be interpreted as a productivity improvement. To explain costs, we construct variables representing own R&D stocks, spill-in R&D stock, extension activities (ET), and road infrastructure (RO), and estimate their impact on variable cost. We allow for interactions among these variables in order to capture their potential enhancing effect on the diffusion of technical information. We fit a translog variable cost function using state-by-year panel data to estimate productivity growth in US agriculture by state. We assume that each state produces three outputs, livestock (V), crops (C) and other farm-related goods and services (O) using four variable inputs including land (A), labour (L), materials (M) and capital (K), and one fixed input, own agricultural R&D stock (RD). We include extension activities (ET), road density (RO), and R&D spill-in (SR) variables, which we refer to as efficiency variables (E), to examine their interaction with local R&D. The total variable cost (TVC) function is (see equation 2.1 at bottom of page):

where w_i are input prices, y_l are output quantities, D_n are regional dummy variables[2], W is a weather variable, and the α's, β's, γ's, δ's, θ's, ξ's, ρ's and ϕ's are parameters to be estimated.

$$\ln TVC = \alpha_0 + \sum_{n=1}^{10}\sum_{i=1}^{4}\eta_{ni}D_n \ln w_i + \sum_{l=1}^{3}\beta_l \ln y_l + \gamma_{RD}\ln RD + \frac{1}{2}\gamma_{RDRD}\ln RD\ln RD$$

$$+ \frac{1}{2}\sum_{i=1}^{4}\sum_{j=1}^{4}\alpha_{ij}\ln w_i \ln w_j + \frac{1}{2}\sum_{l=1}^{3}\sum_{k=1}^{3}\beta_{lk}\ln y_l \ln y_k + \sum_{i=1}^{4}\sum_{l=1}^{3}\delta_{il}\ln w_i \ln y_l$$

$$+ \sum_{i=1}^{4}\theta_{iRD}\ln w_i \ln RD + \sum_{l=1}^{3}\phi_{lRD}\ln y_l \ln RD + \sum_{h=1}^{3}\xi_{hRD}\ln E_h \ln RD$$

$$+ \sum_{h=1}^{3}\sum_{i=1}^{4}\rho_{ih}\ln E_h \ln w_i + \sum_{i=1}^{4}\rho_{iW}\ln W \ln w_i$$

$$(2.1)$$

We impose symmetry and linear homogeneity in prices in the estimation.

Using Shephard's lemma, the cost share for input i is:

$$S_i = \sum_{n=1}^{10} \eta_{ni} D_n + \sum_{j=1}^{4} \alpha_{ij} \ln w_j$$

$$+ \sum_{l=1}^{3} \delta_{il} \ln y_l + \theta_{iRD} \ln RD$$

$$+ \sum_{h=1}^{3} \rho_{ih} \ln E_h + \rho_{iW} \ln W \qquad (2.2)$$

The estimated system of equations includes the total variable cost equation (Eqn 2.1) and three input cost share equations (Eqn 2.2). The parameters are estimated using Iterative Seemingly Unrelated Regression (ITSUR). To test for the effect of extension activities and roads on productivity growth we fit the system of equations twice, once with these variables included and once without. Model 2 includes all variables of interest, while Model 1 is a special case that sets all parameters on extension and roads to zero.

To account for the impacts of local public goods we calculate the cost elasticity of own R&D and other efficiency variables E_h based on parameter estimates from the above models. A negative elasticity indicates a cost-saving effect. These are:

$$\varepsilon_{RD} \equiv \frac{\partial \ln TVC}{\partial \ln RD} = \sum_{i=1}^{4} \theta_{iRD} \ln w_i$$

$$+ \sum_{l=1}^{3} \phi_{lRD} \ln y_l + \sum_{h=1}^{3} \xi_{E_h RD} \ln E_h \qquad (2.3)$$

$$\varepsilon_{E_h} \equiv \frac{\partial \ln TVC}{\partial \ln E_h} = \xi_{E_h RD} \ln RD + \sum_{i=1}^{4} \rho_{iE_h} \ln w_i \qquad (2.4)$$

The elasticity ε_{RD} measures the percentage cost reduction from a 1% increase in local R&D stock. The elasticity ε_{E_h} measures the percentage cost reduction from a 1% increase in the efficiency variables E_h. If the sign of ε_{RD}, or ε_{Eh} is negative, then an increase in own R&D stock or in the efficiency variables E_h reduces total variable cost.

The marginal effect of the efficiency variables on R&D is given by

$$\frac{\partial \varepsilon_{RD}}{\partial \ln E_h} = \xi_{E_h RD} \qquad (2.5)$$

If the sign of $\xi_{E_h RD}$ is negative, it implies that an increase in extension services or road density enhances the impact of local R&D by enabling it to reduce costs even further.

Based on these estimates and following Wang *et al.* (2009) we calculate internal rates of return to agricultural research investments for Models 1 and 2. The local internal rate of return to research r_1 reflects the benefit from R&D investment by a state to its own agricultural sector. It is the discount rate that equates one dollar of investment today with the present value of all future production cost savings resulting from that research. It is found by solving the following equation (see appendix for the derivation) (see equation 2.6 at bottom of page):

where s is the maximum number of years research investments affect future production and the ω_τ's are the weights used to construct the R&D knowledge capital stocks from annual R&D expenditures (see appendix).

Taking into account the benefits of R&D done in one state to other states (R&D spillovers), the social internal rate of return can be derived by solving for r_2 in the following equation (see appendix for the derivation) (see equation 2.7 at bottom of page):

$$1 = - \left[\sum_{i=1}^{4} \theta_{iRD} \ln w_i + \sum_{l=1}^{3} \phi_{lRD} \ln y_l + \sum_{h=1}^{4} \xi_{E_h RD} \ln E_h \right] * \frac{TVC}{RD} * \sum_{\tau-1}^{s} \frac{\omega_\tau}{(1+r_1)^\tau} \qquad (2.6)$$

$$1 = - \left[\sum_{i=1}^{4} \theta_{iRD} \ln w_i + \sum_{l=1}^{3} \phi_{lRD} \ln y_l + \sum_{h=1}^{4} \xi_{E_h RD} \ln E_h \right] * \frac{TVC}{RD} * \sum_{\tau-1}^{s} \frac{\omega_\tau}{(1+r_2)^\tau}$$

$$- \sum_{j=1}^{n-1} \sum_{\tau=0}^{s} (\xi_{SRRD} \ln RD_j + \sum_{i=1}^{4} \rho_{iSR} \ln w_{ij}) * \frac{TVC_j}{SR_j} * \frac{\omega_\tau}{(1+r_2)^\tau} \qquad (2.7)$$

In this study we calculate local rates of return (r_1) and social rates of return (r_2) based on the parameters and fitted values from two models – one including only the local and spill-in R&D variables (Model 1) and one also incorporating agricultural extension spending and road density (Model 2).

2.3 Data Sources and Description

To estimate the system of Eqns 2.1 and 2.2 we need information on input cost shares, input prices, output quantities, research expenditure, extension activities, and road density. We describe our data and discuss trends in these variables over the study period in this section.

2.3.1 Input shares, input prices and output quantities

We use annual data for the 48 contiguous states from 1980 to 2004 for our analysis.

The agricultural production data were drawn from the state agricultural productivity accounts at the US Department of Agriculture, Economic Research Service (USDA–ERS). The output data were constructed as the nominal output value deflated by the relative price index between the individual state and the base state. Multilateral price indexes were computed following Caves, Christensen and Diewert (1982) and using detailed data described in Ball *et al.* (1999). Figure 2.3 shows average output by region. Among the ten regions, the Corn Belt region has the highest crops and livestock productions. Pacific and Northern Plains regions rank second and third in crop production, whereas Northern Plains and Lake States rank second and third in livestock production. North-east and Delta regions are two smallest regions among the ten in aggregate output during the 1980–2004 period.

As to the input shares, intermediate materials accounts for most of the variable cost with an average cost share of 50% between 1980 and 2004. The labour cost

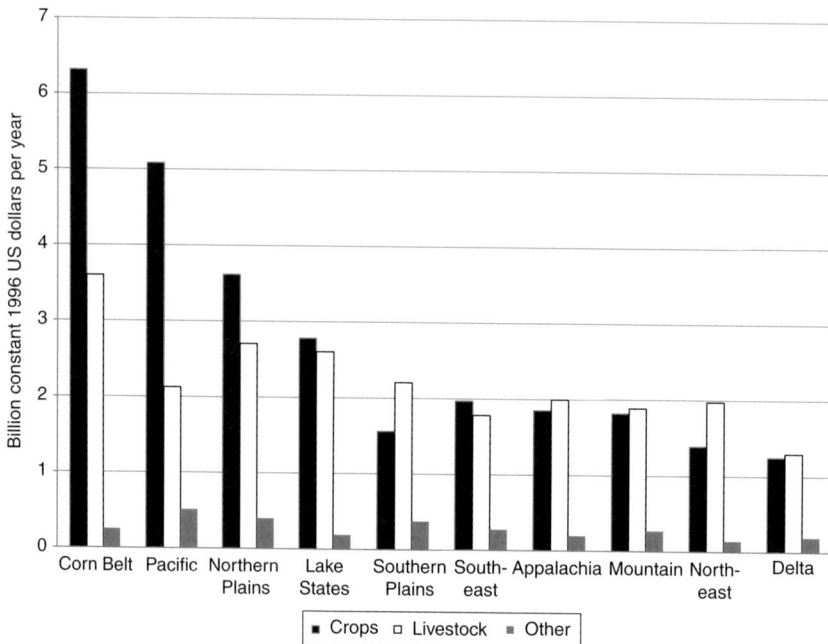

Fig. 2.3. Agricultural output for US regions (average over 1980–2004). Data sources: Authors' calculations from state agricultural productivity accounts available at USDA–ERS.

share is 23%, followed by capital and land with 17% and 15%, respectively. However, there is considerable variation among states in cost shares. The highest materials share is almost 80% of the total variable cost, whereas the smallest land share is 3% of the total variable cost. Among the four input prices, land prices varied the most among observations as they reflect differences in land quality across states.

2.3.2 R&D stocks and R&D spill-ins

In this study we used a trapezoidal-weight pattern proposed by Huffman and Evenson (2006) to construct R&D stocks from R&D expenditures. The annual agricultural research expenditure data and the research price index used to deflate expenditures are provided by Huffman. Huffman (2009) reported that the public research expenditure data

are drawn mainly from the USDA Current Research Information System (CRIS), which contains information on all research projects implemented by USDA's research institutions including the Agricultural Research Service (ARS) and the ERS and the state agricultural experiment stations (SAESs) and the veterinary schools/colleges of the land-grant universities. To focus on productivity-oriented projects, only Huffman (2009) excluded the research expenditures that do not contribute directly to agricultural productivity. Figure 2.4 shows real R&D expenditures and R&D stocks. Although real R&D expenditures have been flat in recent years, R&D stocks were still increasing as they reflect previous years' research expenditures.

We construct R&D spill-ins stocks on the basis of USDA production regions. We assume that states within the same USDA production region could benefit from each

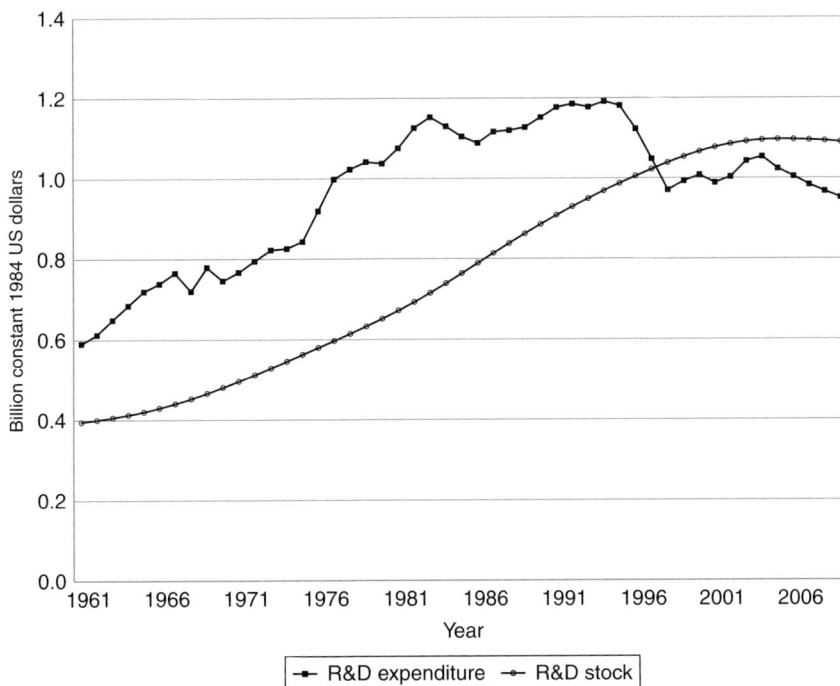

Fig. 2.4. Public R&D expenditures and R&D stocks in US agriculture. Source: Huffman (2009). Public agricultural R&D expenditures include only spending on productivity-oriented research by land grant universities, state agricultural experiment stations, colleges of veterinary medicine and relative institutions. See Huffman (2009) for further explanation.

other's research. The R&D spill-ins are measured as follows:

$$SR_i = \sum_{i \neq j} \Omega_{ij} RD.$$ (2.8)

where SR_i is the spill-in *R&D* stock for state i, Ω_{ij} is the weight used to adjust for the contribution of the *jth* state's innovations to the ith state, and RD_j is own R&D stock generated by state j. In this study we assume $\Omega_{ij} = 1$ for each state other than the own-state within the same production region.

2.3.3 Extension service, roads and weather

We use total full-time equivalent (FTE) extension staff at the state level to construct the extension (ET) capacity indexes for each state. The extension capacity index uses total FTEs as the numerator and the number of farms as the denominator to represent the capacity of the extension service to disseminate technical information. Data on FTEs by state were drawn from the Salary Analysis of the Cooperative Extension Service from the Human Resource Division at USDA. Real extension expenditures, as well as FTEs, have declined over the period of analysis (Fig. 2.5).

As to road infrastructure, we construct a road density index to examine its impact on dissemination of local R&D. The road density index was constructed using total road miles excluding local (e.g. city street) miles for each state divided by total land area. We hypothesize that in states with higher road density the cost of disseminating technical information will be reduced and, therefore, the impact of public R&D on productivity will be enhanced. The information was drawn from the Highway Statistics Publication.

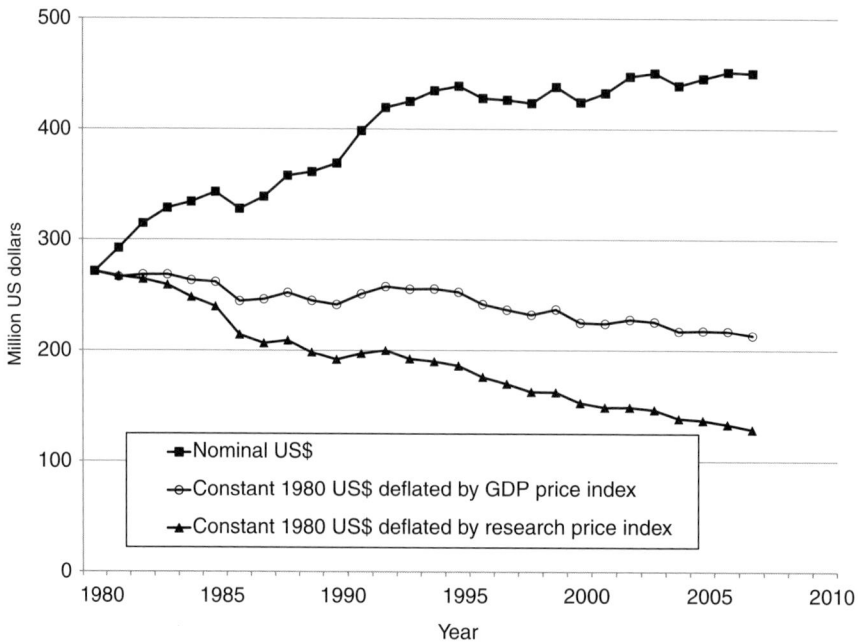

Fig. 2.5. Federal government expenditures for agricultural extension in the USA. Sources: Extension expenditures from USDA, the Office of Budget and Program Analysis; research price deflator from USDA–ERS; implicit GDP price deflator from the US Department of Commerce, Bureau of Economic Analysis.

Weather is treated as a control variable in this model. We use total precipitation in inches from March to November (Schlenker and Roberts, 2006) to capture the short term shocks caused by the rainfall variation.

2.4 Econometric Results and Returns to R&D, Spillovers, Extension and Roads

Two models were estimated: Model 1 includes only the local and spill-in R&D stock variables, whereas Model 2 also incorporates agricultural extension and road density in the estimation. We fit the total variable cost equation (Eqn 2.1) and three input cost share equations (Eqn 2.2) subject to symmetry and linear homogeneity in prices using the Iterative Seemingly Unrelated (ITSUR) approach. Parameter estimates and other statistics from the estimation are given in Appendix Table A2.1, excluding the regional dummies. Most coefficients are significant at the 5% level. The curvature condition, checked after estimation, was satisfied locally but not globally. The joint null hypothesis that the extension and road density related parameters are zero was rejected at the 1% level of confidence (see Wald X^2 statistics in Table A2.1). The likelihood ratio test, also presented in Table A2.1, gives similar results. We conclude that Model 2 is the preferred model.

Using the estimated parameters and equations (Eqns 2.3–2.5) we calculate the impacts of changes in own-state R&D and other efficiency variables, including R&D spill-ins, extension services and road density, on productivity growth, as well as the impacts of the efficiency variables on R&D's cost-saving effect. When other variables are held constant, a 1% reduction in cost caused by the change of the variables of interest can be treated as a 1% growth in productivity. Table 2.1 shows that the cost elasticities are all negative, indicating that an increase in own state R&D stock, or R&D spill-ins stock reduces costs and, therefore, increases productivity. When the extension and road variables are excluded (Model 1), own R&D and spill-in R&D have the same mean elasticity, but when these variables are included (Model 2), the own R&D elasticity declines and the spill-in R&D elasticity increases. The effect of extension is particular strong. This implies that part of the cost-saving effect of own R&D expenditures is generated through its interaction with extension services and road density; excluding these variables from the model biases the impact of R&D on local productivity upward.

Across regions, estimates of the R&D elasticities vary considerably and again are usually lower when extension and road variables are included (Table 2.2). Among the ten regions, investment in public research has a higher productivity impact in the North-east, Mountain, Appalachian, South-east, and Delta regions compared to others. Table 2.3

Table 2.1. Cost elasticities of local R&D, extension, roads and R&D spill-ins from other states.

Public good	Elasticity	Model 1: R&D variables only		Model 2: Extension and roads included	
		Mean	SD	Mean	SD
Local R&D	ξ_{RD}	−0.158	0.119	−0.129	0.090
R&D spill-ins	ξ_{SRD}	−0.158	0.012	−0.164	0.010
Extension	ξ_{ET}	–	–	−0.248	0.021
Roads	ξ_{RO}	–	–	−0.036	0.004

SD, standard deviation.
The elasticities measure the percentage change in agricultural production cost resulting from a 1% increase in the public good. The negative sign reflects a reduction in production cost.

Table 2.2. Cost elasticity of local R&D by region.

Region	Model 1: R&D variables only		Model 2: Extension and roads included	
	Mean	SD	Mean	SD
1. Corn Belt	−0.077	0.036	−0.081	0.035
2. Pacific	−0.036	0.095	−0.028	0.064
3. Northern Plains	−0.029	0.039	−0.022	0.042
4. Lake States	−0.078	0.035	−0.080	0.036
5. Southern Plains	−0.045	0.070	−0.045	0.056
6. South-east	−0.145	0.068	−0.120	0.065
7. Appalachia	−0.183	0.086	−0.154	0.059
8. Mountain	−0.193	0.078	−0.145	0.067
9. North-east	−0.290	0.107	−0.232	0.068
10. Delta States	−0.138	0.034	−0.107	0.029

SD, standard deviation.
The elasticities measure the percentage change in agricultural production cost resulting from a 1% increase in local R&D stock. The negative sign reflects a reduction in production cost.

Table 2.3. Marginal effects of efficiency variables on the elasticity of local R&D.

Sources	Model 1: R&D variables only		Model 2: Extension and roads included	
	Marginal effect on local R&D	t ratio	Marginal effect on local R&D	t ratio
R&D spill-ins	−0.009	−10.28	−0.009	−12.45
Extension	–	–	−0.015	−18.82
Road density	–	–	−0.002	−4.38

The marginal effect is the effect of a 1% increase in the efficiency variable on the cost elasticity of local R&D. The negative value indicates that greater R&D spill-ins, extension services, and roads increase the rate at which local R&D reduces agricultural production costs.

presents the efficiency variables' marginal effects on the local R&D elasticity. It shows that the cost-saving effect of own R&D is enhanced by spillover R&D, extension services and roads. Among the three marginal effects, extension services are the most important for enhancing impact of local R&D.

Based on Eqns 2.6 and 2.7 and using the predicted values from Models 1 and 2, we calculate the local internal rate of return as well as the social rate of return for both models at each observation. We report these estimates by region in Table 2.4. The output-share weighted average local internal rate of return, r_1, is 13.0% for Model 1 and 12.5% for Model 2. The weighted average social rate of return is 43.7% for Model 1 and 45.1% for Model 2. Although the difference in the average numbers between models is small, we find that the estimates in Model 2 show more

dispersion than the estimates in Model 1. Average differences among regions mask the contributions of Model 2 to our understanding of the heterogeneity across regions and across states. We use the coefficient of variation (COV) to show the heterogeneity of the internal rate of return estimates. The COV is measured as the standard deviation divided by the mean. According to the COV criteria, we can see that Model 2 allows better identification of the differential returns among states within the same region. Even when states seem to benefit similarly from R&D spill-ins, this model is able to identify differences in responses to their specific extension activities and road infrastructure.

In general, four regions experienced social rates of returns above 50%. They are the Lake States, the Corn Belt, the Northern Plains and the Southern Plains. These

Table 2.4. Local and social internal rates of return (ROR) to public agricultural research by region (%).

| Region | Model 1: R&D variables only | | | | | | Model 2: Extension and roads included | | | | | |
| | r_1: Local ROR | | | r_2: Social ROR | | | r_1: Local ROR | | | r_2: Social ROR | | |
	Mean	SD	COV	Mean	SD	COV	Mean	SD	COV	Mean	SD	COV
1. Corn Belt	11.6	4.2	36.2	53.8	11.2	20.8	13.0	6.0	46.2	55.2	11.5	20.8
2. Pacific	12.3	2.3	18.7	34.8	13.3	38.2	9.3	3.2	34.4	37.0	16.5	44.6
3. Northern Plains	11.9	9.3	78.2	48.7	13.7	28.1	16.2	13.9	85.8	53.0	16.1	30.4
4. Lake States	11.0	4.6	41.8	50.1	7.8	15.6	11.3	5.3	46.9	51.1	8.5	16.6
5. Southern Plains	20.8	8.1	38.9	52.1	16.6	31.9	20.8	6.6	31.7	54.5	18.4	33.8
6. South-east	10.6	2.6	24.5	36.4	10.3	28.3	7.3	5.1	69.9	34.7	9.4	27.1
7. Appalachia	12.8	5.2	40.6	26.4	10.2	38.6	11.6	5.5	47.4	26.1	9.7	37.2
8. Mountain	11.9	4.6	38.7	39.1	13.9	35.5	9.3	5.2	55.9	40.1	14.3	35.7
9. North-east	11.9	9.7	81.5	32.3	20.3	62.8	9.9	10.1	102.0	31.1	19.9	64.0
10. Delta States	10.2	2.4	23.5	35.7	5.7	16.0	7.6	3.3	43.4	34.9	5.9	16.9
All	11.9	6.6	55.5	39.1	16.6	42.5	10.8	8.0	74.1	39.6	17.6	44.4
Weighted average	13.0			43.7			12.5			45.1		

SD, standard deviation; COV, coefficient of variation (standard deviation divided by the mean in %).

regions are also among the major agricultural production regions in the USA. It seems that states within those regions have benefitted more from R&D spillovers than states in other regions. The Appalachian and the North-east regions have experienced lower social rates of returns. They also play a relatively smaller role in US agricultural production. Figure 2.6 shows the frequency distributions of the local and social internal rate of return across 48 states for both models. Except for the estimated social rates of return in Model 1, the rates of return estimates are not normally distributed and differ across models. Figure 2.6 also shows that most local rates of return are below 15% and most social rates of return are above 30%. This implies that when addressing the benefits from investment in research a common rate of return applied to every state and every time period might not be appropriate.

Figures 2.7 and 2.8 provide additional information on the distribution of internal rates of return among models and states. The median, mean, interquartile range percentile values, as well as the maximum and minimum value, are indicated in the Box–Whiskers plots in Fig. 2.7 for the local and social rates of return estimated by both models. The median local rate of return, r_1, is 12% in Model 1 and 10% in Model 2, whereas the median social rate is much higher, 38% in Model 1 and 37% in Model 2. We also find that the variability is larger in Model 2, as the local rate ranges from 0.2% to 31.4%, and the social rate ranges from 2.3% to 75.0%, a wider interval than for Model 1. We group the 48 states into ten regions and present the estimates by region in Fig. 2.8, panels a–d. Figure 2.8 shows that the benefits from public research differed from region to region. It also

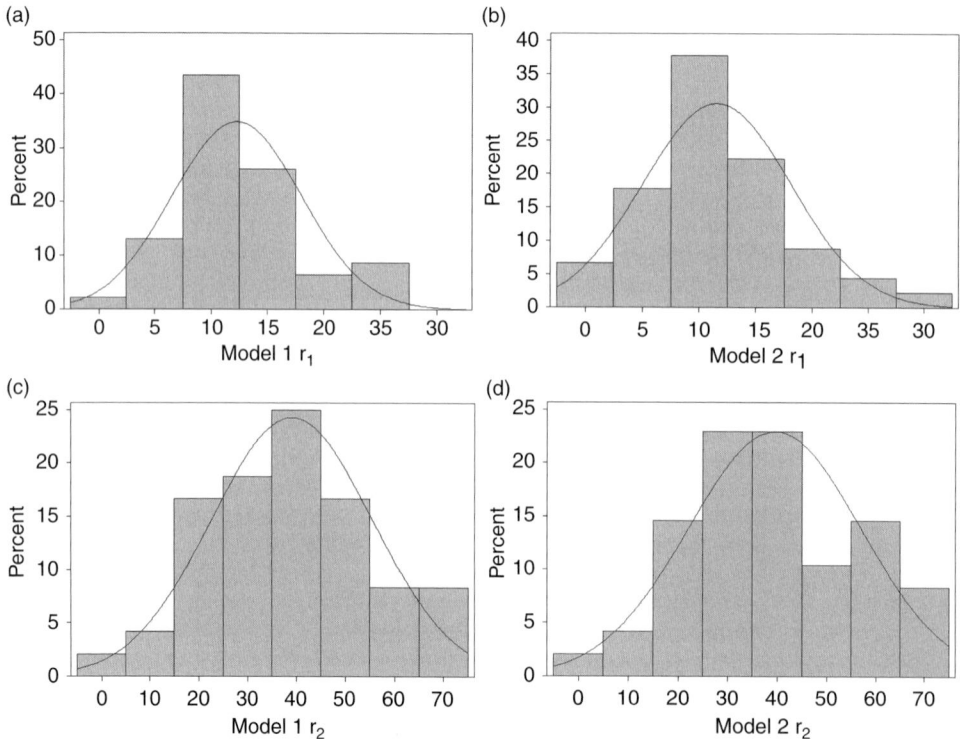

Fig. 2.6. Distribution of the rate of return to agricultural research across US states. (a) Local internal rates of return – Model 1 (R&D variables only). (b) Local internal rates of return – Model 2 (extension & roads included). (c) Social rate of return – Model 1 (R&D variables only). (d) Social rate of return – Model 2 (Extension & roads included). Source: Authors' estimation.

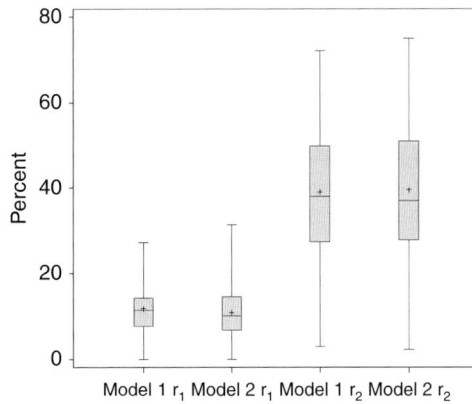

Fig. 2.7. Distribution of rates of return to research by models (Box–Whiskers plots). The Box–Whiskers plot shows the distribution of the rate of return estimates. The horizontal lines from bottom to top show the minimum, lower quartile, median, upper quartile and maximum observation. The star shows the sample mean. Source: Authors' estimation.

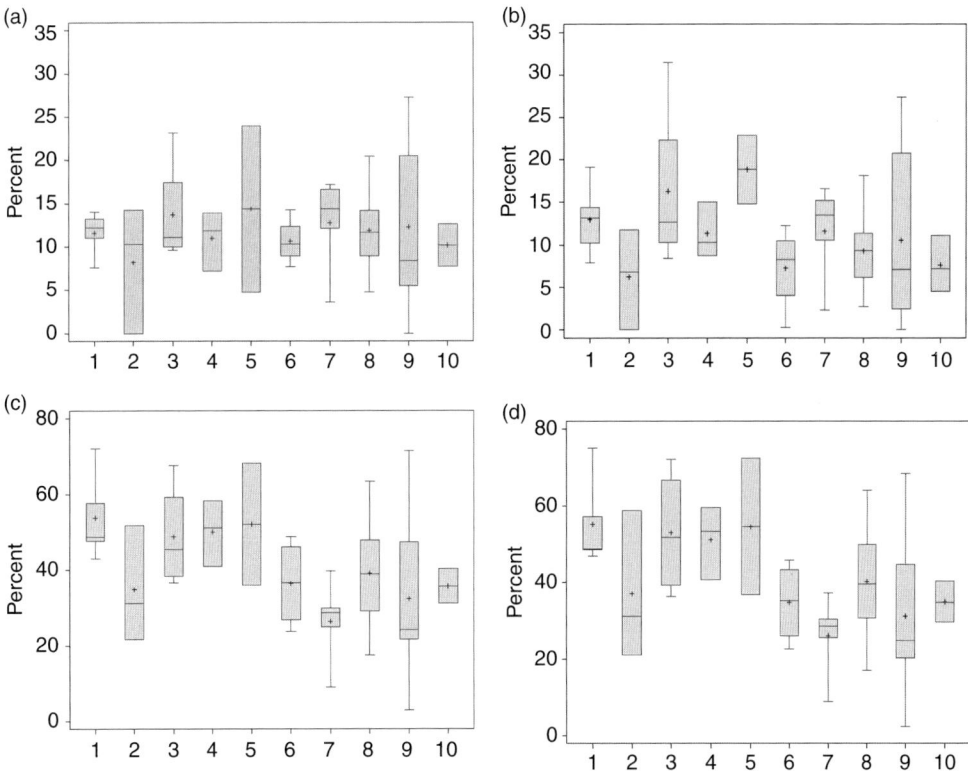

Fig. 2.8. Distribution of rates of return to research among regions (Box–Whiskers plots). (a) Local rates of return by region – Model 1. (b) Local rates of return by region – Model 2. (c) Social rates of return by regions – Model 1. (d) Social rates of return by regions – Model 2. The Box–Whiskers plot shows the distribution of the rate of return estimates. The horizontal lines from bottom to top show the minimum, lower quartile, median, upper quartile and maximum observation. The star shows the sample mean. Regions 1: Corn Belt; 2: Pacific; 3: North Plains; 4: Lake States; 5: Southern Plains; 6: South-east; 7: Appalachian; 8: Mountain; 9: North-east; 10: Delta States. Source: Authors' estimation.

demonstrates a wider interquartile range for the rates of return in Model 2 in most regions, especially for the local rates of return. It shows that returns to research are not only conditional on each state's natural resource endowments, but also could be affected by its public infrastructure, its extension activities and its neighbour's research performance.

2.5   Summary and Conclusions

This chapter uses state-by-year panel data to estimate the contributions of public research, extension activities and the transportation network to state agricultural productivity growth rates during the 1980–2004 period. We estimate alternative models, with and without accounting for extension activities and roads, to examine whether these factors influence how local R&D contributes to a state's productivity growth. The statistical significance of the R&D interactive terms with extension activities, road density and R&D spill-ins indicate that these 'efficiency' variables play an important role in enhancing the utilization and dissemination of local R&D. The negative marginal effects of these efficiency variables show that they amplify the benefits of public research expenditures. The distributions of the rates of return show a wider interquartile range when the impacts of the efficiency variables are included. This suggests that returns to research are not only conditional on each state's natural resource endowment, but also might be affected by its public infrastructure, its extension activities, and its neighbour's research investment. The estimated average internal rates of return mask regional and state heterogeneity. When the extension and roads variables are included in the analysis we can better understand differences in productivity growth rates among states and regions.

The internal rate of return to public R&D is utilized to evaluate the contribution of research expenditures to productivity growth, as well as derive local and social rates of return. The estimated local rates of return are around 13%, whereas the social rates of return are around 45% across regions and models. Use of the model with extension activities and road density enhances our understanding of the heterogenity of R & D performance across states.

Notes

[1] The US public agricultural research system is a Federal–State partnership dating back to the late 19th century. Through a combination of Federal, State and non-government funds, state institutions account for 60–70% of the research expenditures of this system, with USDA intramural research accounting for the rest. State institutions focus primarily on the agricultural concerns of their home state, but the potential for significant inter-state and inter-regional R&D spillovers has important implications for science policy in such decentralized R&D systems.

[2] To conserve degrees of freedom, we only introduce the region dummies in the first-order terms to allow the cost shares to differ among production regions.

References

Alston, J., Chan-Kang, C., Marra, M., Pardey, P. and Wyatt, T. (2000) *A Meta-Analysis of Rates of Return to Agricultural R&D: Ex Pede Herculem?* Research Report No. 113, International Food Policy Research Institute, Washington, DC.

Alston, J., Anderson, M.A., James, J. and Pardey, P. (2010) *Persistence Pays: U.S. Agricultural Productivity Growth and the Benefits from Public R&D Spending.* Springer, New York.

Antle, J. (1983) Infrastructure and aggregate agricultural productivity: International evidence. *Economic Development and Cultural Change* 31, 609–619.

Ball, V.E., Gollop, F., Kelly-Hawke, A. and Swinand, G. (1999) Patterns of productivity growth in the U.S. farm sector: Linking State and aggregate models. *American Journal of Agricultural Economics* 81, 164–179.

Caves, D., Christensen, L. and Diewert, W. (1982) Multilateral comparisons of output, input, and productivity using superlative index numbers. *Economic Journal*, 92, 73–86.

Evenson, R. (1967) The contribution of agricultural research to production. *Journal of Farm Economics* 49, 1415–1425.

Evenson, R. (1989) Spillover benefits of agricultural research: Evidence from U.S. experience. *American Journal of Agricultural Economics* 71, 447–452.

Evenson, R. (2001) Economic impacts of agricultural research and extension. In: Gardner, B. and Rausser, C. (eds) *Handbook of Agricultural Economics, Volume 1, Part A,* Elsevier Science, New York, pp. 573–628.

Fuglie, K. and Heisey, P. (2007) Economic returns to public agricultural research. *Economic Brief No. 10,* Economic Research Service, U.S. Department of Agriculture, Washington, DC.

Griliches, Z. (1958) Research costs and social returns: Hybrid corn and related innovations. *Journal of Political Economy* 66, 419–431.

Griliches, Z. (1964) Research expenditures, education and the aggregate agricultural production function. *American Economic Review* 54, 6, 961–974.

Griliches, Z. (1992) The search for R&D spillovers. *Scandinavian Journal of Economics* 94, Supplement, 29–47.

Griliches, Z. (1998) *R&D and Productivity: The Econometric Evidence*. University of Chicago Press, Chicago, IL.

Huffman, W. (2009) Measuring public agricultural research capital and its contribution to State agricultural productivity. Working Paper No. 09077, Department of Economics, Iowa State University, Ames, IA.

Huffman, W. and Evenson, R. (2006) *Science for Agriculture: A Long-Term Perspective,* 2nd edn. Blackwell Publishing, New York.

Huffman, W., Ball, V.E., Gopinath, M. and Somwaru, A. (2002) Public R&D and infrastructure policies: Effects on cost of Midwestern agriculture. In: Ball, V.E. and Norton, G. (eds) *Agricultural Productivity: Measurement and Sources of Growth*. Kluwer Academic Publishers, Boston, MA, pp. 167–184.

Paul, C., Ball, V.E., Felthoven, R. and Nehring, R. (2001) Public infrastructure impacts on U.S. agricultural production: Panel analysis of costs and netput composition. *Public Finance and Management* 1, 183–213.

Plastina, A. and Fulginiti, L. (2012) Rates of return to public agricultural research in 48 U.S. States. *Journal of Productivity Analysis,* 37, 95–113.

Schlenker, W. and Roberts, M. (2006) Nonlinear effects of weather on corn yields. *Applied Economic Perspectives and Policy* 28, 391–398.

US Department of Agriculture, Economic Research Service. *Agricultural Productivity in the United States.* http://ers.usda.gov/Data/AgProductivity/ (Accessed 7 March 2010).

US Department of Agriculture, National Agricultural Statistics Service. *Data and Statistics.* http://www.nass.usda.gov/Data_and_Statistics/index.asp (Accessed 10 March 2010).

US Department of Agriculture, National Institute of Food and Agriculture. *Salary Analyses of State Extension Service Positions.* http://www.csrees.usda.gov/about/human_res/report.html (Accessed 10 January 2010).

US Department of Transportation, Federal Highway Administration. *Highway Statistical Series.* http://www.fhwa.dot.gov/policyinformation/statistics.cfm (Accessed 15 January 2010).

Wang, S., Ball, V.E., Fulginiti, L. and Plastina, A. (2009) Impact of local public goods on agricultural productivity growth in the U.S. Selected Paper, Agricultural and Applied Economics Association Annual Meeting, Milwaukee, WI.

Yee, J., Huffman, W., Ahearn, M. and Newton, D. (2002) Sources of agricultural productivity growth at the State level, 1960–1993. In: Ball, V.E. and Norton, G. (eds) *Agricultural Productivity: Measurement and Sources of Growth*. Kluwer Academic Publishers, Boston, MA, pp. 185–209.

Appendix 2.1. Internal Rate of Return to Agricultural Research Investment

When ignoring the social effect, the benefit from one dollar of R&D investment to its own state is obtained as the discounted value of all future cost savings. The internal rate of return, r_1, can be obtained by solving the following equation (see equation A2.1 at bottom of page):

where s is the maximum period for the contribution of the research investment in the R&D stock construction, R_t is the own-state research investment at time t and $RD_{t+\tau}$ is the own-state R&D stock at time period $t+\tau$. $\frac{\Delta RD_{t+\tau}}{\Delta R_t}$ is the R&D stock at time period $t+\tau$ generated by one dollar own state research investment at time t. In this study, $\frac{\Delta RD_{t+\tau}}{\Delta R_t}$ is the weight at each period to construct the R&D stock from the research investment at time period t.

$$\frac{\Delta RD_{t+\tau}}{\Delta R_t} = \omega_\tau \qquad (A2.2)$$

The average impact on total variable cost from a one-dollar increase in a state's agricultural research stock can be expressed as:

$$\frac{\Delta TVC}{\Delta RD} = \frac{\Delta \ln TVC}{\Delta \ln RD} \frac{\overline{TVC}}{\overline{RD}} = \varepsilon_{RD} * \frac{\overline{TVC}}{\overline{RD}} \qquad (A2.3)$$

The TVC is the predicted TVC based on the model estimates. The internal rate of return at the sample mean can be obtained by substituting Eqns 2.3, A2.2 and A2.3 into A2.1, and solving for r_1 (see equation A2.4 & A2.5 at bottom of page):

Taking into account the benefits through R&D spillover effects to other states, the internal rate of return can be derived by solving for r_2 in the following equation (see equation A2.6 at bottom of page):

where n is the number of the states within a region or group. Therefore, the number of states benefiting from state i's spillover effect is n−1. The first part of Eqn A2.6 is the own-state benefit discussed in Eqn 2.1. The second part of the equation is the social benefits generated by the ith state's research and can be expressed as follows (see equation A2.7 at bottom of page):

The spill-ins generated by one dollar of investment in state i for state j at time period t is the same weight as in eqn. A 2.2.

$$\frac{\Delta SR_{jt+\tau}}{\Delta R_t} = \omega_\tau \qquad (A2.8)$$

The average impact on one state's total variable cost from its research spill-ins can be expressed as:

$$\frac{\Delta TVC}{\Delta SR} = \frac{\Delta \ln TVC}{\Delta \ln SR} \frac{\overline{TVC}}{\overline{SR}} = \varepsilon_{SR} * \frac{\overline{TVC}}{\overline{SR}} \qquad (A2.9)$$

$$1 = \sum_{\tau=0}^{s} \frac{\Delta TVC_{t+\tau}}{\Delta R_t} \cdot \frac{1}{(1+r_1)^\tau} = \sum_{\tau=0}^{s} \frac{\Delta TVC_{t+\tau}}{\Delta RD_{t+\tau}} \cdot \frac{\Delta RD_{t+\tau}}{\Delta R_t} \cdot \frac{1}{(1+r_1)^\tau} \qquad (A2.1)$$

$$1 = -\sum_{\tau=0}^{s} \left[\sum_{i=1}^{4} \theta_{iRD} \overline{\ln w_i} + \sum_{l=1}^{3} \phi_{lRD} \overline{\ln y_l} + \sum_{h=1}^{4} \xi_{hRD} \ln \overline{E}_h \right] * \frac{\overline{TVC}}{\overline{RD}} * \frac{\omega_\tau}{(1+r_1)^\tau} \qquad (A2.4)$$

or

$$1 = -\left[\sum_{i=1}^{4} \theta_{iRD} \overline{\ln w_i} + \sum_{l=1}^{3} \phi_{lRD} \overline{\ln y_l} + \sum_{h=1}^{4} \xi_{hRD} \ln \overline{E}_h \right] * \frac{\overline{TVC}}{\overline{RD}} * \sum_{\tau-1}^{s} \frac{\omega_\tau}{(1+r_1)^\tau} \qquad (A2.5)$$

$$1 = \sum_{\tau=0}^{s} \frac{-\Delta TVC_{it+\tau}}{\Delta R_t} * \frac{1}{(1+r_2)^\tau} + \sum_{j=1}^{n-1} \sum_{\tau=0}^{s} \frac{-\Delta TVC_{jt+\tau}}{\Delta R_{it}} * \frac{1}{(1+r_2)^\tau} \qquad (A2.6)$$

$$\sum_{j=1}^{n-1} \sum_{\tau=0}^{s} \frac{\Delta TVC_{jt+\tau}}{\Delta R_{it}} * \frac{1}{(1+r_2)^\tau} = \sum_{j=1}^{n-1} \sum_{\tau=0}^{s} \frac{\Delta TVC_{jt+\tau}}{\Delta SRD_{jt+\tau}} * \frac{\Delta SR_{jt+\tau}}{\Delta R_{it}} * \frac{1}{(1+r_2)^\tau} \qquad (A2.7)$$

The internal rate of return including the social benefits at the sample mean can be obtained by substituting Eqns 2.4, A2.5, A2.7–A2.9 into A2.6, and solving for r_2 (see equation A2.10 at bottom of page):

While the internal rate of return can be estimated at a specific point, such as at the mean, it can also be evaluated at each observation.

$$1 = -\left[\sum_{i=1}^{4}\theta_{iRD}\overline{\ln w_i} + \sum_{l=1}^{3}\phi_{lRD}\overline{\ln y_l} + \sum_{h=1}^{4}\xi_{hRD}\ln\overline{E}_h\right]*\frac{\overline{TVC}}{\overline{RD}}*\sum_{\tau-1}^{s}\frac{\omega_\tau}{(1+r_2)^\tau}$$

$$+\sum_{j=1}^{n-1}\sum_{\tau=0}^{s}(\xi_{SRRD}\overline{\ln RD_j} + \sum_{i=1}^{4}\rho_{iSR}\overline{\ln w_{ij}})*\frac{\overline{TVC}_j}{\overline{SR}_j}*\frac{\omega_\tau}{(1+r_2)^\tau} \qquad (A2.10)$$

Table A2.1. Variable cost function parameter estimates from a US state panel, 1980–2004.

Parameters	Model 1 coef.	Model 1 t ratio		Model 2 coef.	Model 2 t ratio	
β						
β_V	1.1167	4.01	***	1.472	5.46	***
β_C	-1.2364	-4.72	***	-0.631	-2.59	***
β_O	-0.0442	-0.15		-0.505	-1.81	*
β_{VV}	0.0242	1.03		0.040	1.80	*
β_{VC}	-0.0168	-0.88		-0.014	-0.77	
β_{VO}	-0.0551	-3.13	***	-0.060	-3.54	***
β_{CC}	0.1672	8.3	***	0.121	6.19	***
β_{CO}	-0.1465	-8.55	***	-0.117	-7.16	***
β_{OO}	0.2073	7.80	***	0.183	7.28	***
γ						
γ_{RD}	-0.6938	-1.52		-0.487	-1.16	
$\gamma_{RD\,RD}$	-0.0255	-1.16		-0.010	-0.50	
α						
α_{AA}	0.0457	16.73	***	0.049	17.96	***
α_{AM}	-0.0316	-9.62	***	-0.035	-10.57	***
α_{AK}	-0.0029	-1.59		-0.003	-1.61	
α_{AL}	-0.0112	-5.14	***	-0.011	-4.87	***
α_{MM}	0.1650	22.32	***	0.161	21.39	***
α_{MK}	-0.0832	-16.17	***	-0.076	-14.89	***
α_{ML}	-0.0501	-12.33	***	-0.050	-11.77	***
α_{KK}	0.1432	25.98	***	0.135	24.60	***
α_{KL}	-0.0571	-29.29	***	-0.056	-28.77	***
α_{LL}	0.1184	33.28	***	0.117	31.58	***
δ						
δ_{AV}	-0.0282	-14.77	***	-0.029	-15.17	***
δ_{AC}	0.0137	6.05	***	0.012	5.29	***
δ_{AO}	0.0045	1.90	**	0.003	1.35	
δ_{MV}	0.0640	16.55	***	0.071	19.37	***
δ_{MC}	-0.0792	-19.12	***	-0.067	-16.92	***
δ_{MO}	0.0356	7.94	***	0.039	9.25	***

Parameters	Model 1 coef.	Model 1 t ratio		Model 2 coef.	Model 2 t ratio	
θ						
$\theta_{A\,RD}$	-0.0159	-5.86	***	-0.0165	-6.06	***
$\theta_{M\,RD}$	-0.0008	-0.16		-0.0044	-0.94	
$\theta_{K\,RD}$	-0.0127	-6.55	***	-0.0129	-6.88	***
$\theta_{L\,RD}$	0.0294	8.22	***	0.0338	9.58	***
φ						
$\varphi_{V\,RD}$	-0.0087	-0.55		-0.0416	-2.73	***
$\varphi_{C\,RD}$	0.0719	5.01	***	0.0498	3.7	***
$\varphi_{O\,RD}$	0.0242	1.49		0.0453	2.95	***
ξ						
$\xi_{ET\,RD}$	–	–		-0.0154	-18.82	***
$\xi_{RO\,RD}$	–	–		-0.0021	-4.38	***
$\xi_{SR\,RD}$	-0.0088	-10.28	***	-0.0091	-12.45	***
ρ						
$\rho_{ET\,A}$	–	–		-0.0106	-4.11	***
$\rho_{RO\,A}$	–	–		-0.0063	-3.71	***
$\rho_{SR\,A}$	-0.0192	-4.21	***	-0.0159	-3.45	***
$\rho_{ET\,M}$	–	–		0.0636	13.04	***
$\rho_{RO\,M}$	–	–		0.0096	3.15	***
$\rho_{SR\,M}$	0.0113	1.54		0.0112	1.52	
$\rho_{ET\,K}$	–	–		-0.0210	-11.5	***
$\rho_{RO\,K}$	–	–		-0.0056	-4.72	***
$\rho_{SR\,K}$	0.0008	0.25		-0.0006	-0.17	
$\rho_{ET\,L}$	–	–		-0.0320	-9.04	***
$\rho_{RO\,L}$	–	–		0.0023	1.04	
$\rho_{SR\,L}$	0.0070	1.21		0.0051	0.87	
ρ_{WA}	-0.0003	-0.51		-0.0002	-0.39	
ρ_{WM}	0.0007	1.11		0.0007	1.05	
ρ_{WK}	-0.0014	-4.3	***	-0.0014	-4.24	***
ρ_{WL}	0.0009	1.77	.	0.0009	1.55	

	Model 1			Model 2		
δ_{KV}	-0.0056	-4.02	***	-0.009	-6.90	***
δ_{KC}	0.0250	15.17	***	0.021	13.49	***
δ_{KO}	-0.0272	-15.26	***	-0.027	-15.99	***
δ_{LV}	-0.0301	-11.14	***	-0.033	-12.31	***
δ_{LC}	0.0406	13.68	***	0.034	11.77	***
δ_{LO}	-0.0129	-3.99	***	-0.015	-4.87	***

Equations	R^2	adj, R^2	R^2	adj, R^2	
LnTVC	0.97	0.97	0.98	0.98	
S_M	0.39	0.38	0.49	0.48	
S_K	0.7	0.7	0.74	0.74	
S_L	0.62	0.62	0.66	0.66	
Wald X^2 Test				486.68	***
Likelihood Ratio Test				487.01	***

Model 1 includes R&D variables; Model 2 also includes extension and road variables.

To understand the subscripts to the parameters, note that V stands for livestock, C for crops, O for other farm-related goods and services, A for land, L for labour, M for materials, K for capital, RD for own agricultural R&D stock, ET for extension, RO for road density, SR for R&D spill-ins.

***, **, and ˙ indicate significance at 1%, 5% and 10% levels, respectively.

3 Measurement of Canadian Agricultural Productivity Growth*

Sean A. Cahill and Tabitha Rich
Agriculture and Agri-Food Canada, Ottawa

3.1 Introduction

In any area of economic research, long time series are valuable because they give a much more complete picture of a trend than shorter series. A difference between the growth rate of one short time series and that of another might indicate a fundamental change or it might simply be due to random deviations from trend. Statistical tests and econometrics can help in determining the significance of the difference, but analysis with a longer time series around these points will always provide more convincing evidence. For productivity growth, where differences in average rates of growth between periods are closely scrutinized and labelled as gaps, slowdowns and so on, there is clear value in a time series that will span these periods and help determine whether or not fundamental changes have taken place.

Over the past 50 years, there have been three studies that together now offer more than 80 years of data on productivity growth in Canadian agriculture. Each of these studies has generated estimates that have been based on detailed production accounts for agriculture and for periods that overlap sufficiently to make a continuous time series. This chapter reports on the third and most recent contribution, a gross output based production account for Canadian agriculture that has been developed at Agriculture and Agri-Food Canada (AAFC). This account is subsequently referred to as the 'AAFC production account'.

The aims of this chapter are twofold. The first is to outline the methodology and data used to construct the AAFC production account and to relate the total factor productivity index estimated with these data to the index data reported in the two previous studies.

The second aim is to highlight some of the measurement challenges faced during the development of the account. Because some of these challenges are likely to be universal, the approaches taken to resolve them in the Canadian context might be

* We would like to thank Eldon Ball (USDA, ERS) for his advice and assistance, Nathalie Troubat, who provided the database architecture and first set of estimates for the account, others who worked on the account at Agriculture and Agri-Food Canada (Tim Colwill, Polina Hristeva, Frederic Lessard, Bruce Phillips and Roger Martini), staff at the Canadian Agriculture Library in Ottawa and the various Statistics Canada and ERS staff who provided both data and advice. Any policy views, whether explicitly stated, inferred or interpreted from the contents of this chapter, should not be represented as reflecting the views of Agriculture and Agri-Food Canada.

useful to others who are in the process of constructing their own production accounts for agriculture.

The chapter is organized as follows. The next section elaborates on the evolution of production accounts for Canadian agriculture and provides discussion around the first two production accounts referred to above as well as several other related studies. The third section is devoted to measurement, covering the methods and describing the data that are used to construct both the output and input sides of the AAFC production account. The fourth section presents the gross output, total input and total factor productivity indexes and a discussion around these results. The splicing of total factor productivity index data over the three studies and the long-run pattern of these data during 1926–2006 is also discussed here. The final section provides a summary and some abbreviated remarks on the particularly challenging measurement issues encountered during the development of the account.

3.2   A Brief History of Total Factor Productivity Measurement for Canadian Agriculture

Extant studies of total factor productivity growth for Canadian agriculture fall into two groups. In the first group, the studies involve the development of comprehensive production accounts. In other words, these studies construct an index of total output using detailed crop and livestock output data. Similarly, an index of total input is constructed by measuring, as completely as possible, all of the inputs used by Canadian agriculture. Treatment of within-sector use of feed, seed and other inputs produced by the sector varies.

Studies in the second group use a much less data-intensive approach. On the output side of the account, either readily available output price indexes are used to deflate cash receipts data for broad aggregates, such as 'crops' and 'livestock' or output indexes are taken directly from statistical publications. These studies provide little detail,

therefore, on the output side of the account because many of the intricacies of measurement have been avoided by using 'off the shelf' data. On the input side, these studies typically report a quite comprehensive accounting of inputs, with the exception of intermediate inputs produced and used within the sector. Within-sector use is never addressed, because the output side of the account is never separated into components that will allow estimates of this input component. For this reason these types of studies are referred to as being 'pragmatic'.

The next two sections provide a brief but comprehensive review of the known studies within each group, focusing on the reported total factor productivity growth estimates.[1]

3.2.1   The production account approach

It would appear that the first comprehensive production account for Canadian agriculture was constructed by Lok (1961). Lok developed a database for the years 1926–1957, measuring output of most major crops and livestock by deflating cash income with price indexes of various types, depending on the commodity. Other outputs such as forest products were incorporated – see Lok (1961, pp. 65–68). Output was measured only in terms of sales, however, and therefore did not give a complete estimate of gross output.

Lok covered an equally comprehensive set of inputs, including purchased inputs, labour and capital – see Lok (1961, p. 49). Labour was estimated using agricultural labour force data. On both sides of the account, Lok was particularly concerned with appropriate aggregation procedures, obtaining constant dollar implicit quantities by deflating current dollar values (all dollars in this chapter are Canadian unless stated otherwise). Both output and input aggregates were estimated by Lok using chain-linked indexes, as well as various fixed base indexes. Average total factor productivity growth between 1926 and 1957 was estimated to be 2% annually – see Lok (1961, p. 76).

The second production account constructed for Canadian agriculture is Danielson (1975). Danielson's production account covered the period 1946–1970 and, like Lok's, comprised a comprehensive commodity-by-commodity treatment of output, supported by detailed tables and documentation. Unlike Lok, Danielson did not rely on cash receipts data to measure output, instead working from production and other data to construct gross output for each commodity. The inputs covered were capital, land, operator labour, hired labour and intermediate inputs. For within-sector use, Danielson included only some seed produced by the sector; other types of within-sector use were excluded.[2] For the 1946–1970 period, Danielson estimated that average annual total factor productivity growth was 2% annually (Danielson, 1975, pp. 185–186).

Statistics Canada started a productivity measurement programme in the 1990s that constructed industry production accounts on the basis of the input–output tables from the National Accounts (Baldwin and Harchaoui, 2002). The programme is an ongoing one, and these data are updated regularly.

In the Statistics Canada production account, gross output is estimated using the range of commodities in the input–output tables, which, while quite comprehensive, can be very aggregated (such as 'vegetables'). The input side of the account covers capital (an aggregate of machinery/equipment and buildings), land, labour (an aggregate of hired and operator/unpaid family labour) and intermediate inputs (energy, materials, services). Data for 1961–2006 are publicly available on CANSIM.[3] According to estimates with these data, average total factor productivity growth in Canadian agriculture was 1% annually over the period 1961–2006.

3.2.2 Pragmatic approaches

It seems that the earliest study of the pragmatic type was Furniss (1964), who used an index of farm production that was available from the Dominion Bureau of Statistics,

while the index of farm inputs was taken from Lok – see Furniss (1964, p. 42).[4] For 1935–1960, Furniss estimated average total factor productivity growth at 2.3% annually (Furniss, 1964). In a continuation of this research, Furniss (1970) reported on productivity growth in Canadian agriculture between 1950 and 1969. The average growth rate estimated over this period was 1.9% per year – see Furniss (1970, p.18, Table 1).

Islam (1982) measured both labour and non-labour inputs on the input side of his account for 1961–1978. In the latter category, Islam included a range of purchased and capital inputs. On the output side, he followed Furniss, using indexes of farm production computed first by the Dominion Bureau of Statistics and, later, by Statistics Canada. The average total factor productivity growth rate calculated with these data for the 1961–1978 period was 1.8% (Islam, 1982, p. 135).

The only other study that fits this category of production accounts is Brinkman and Prentice (1983), who developed a quite detailed input side to their production account. For the output side of the account they followed the approach used by Furniss, i.e. they did not measure the output side directly, instead relying on data from Statistics Canada's index of farm production series. Their estimate of total factor productivity growth for 1961–1980 was 1.8% annually.

3.2.3 Discussion

Up until now, the Statistics Canada database has been the only available source of more up-to-date information on total factor productivity growth in Canadian agriculture. The Statistics Canada data are important in showing how agriculture matches up with other sectors of the Canadian economy, within a consistent framework. However, although quite comprehensive in its coverage of gross output and individual inputs, the Statistics Canada approach has some limitations. In particular, treatment of outputs and within-sector use of output as intermediate input is limited to the commodities and commodity groups (e.g. 'vegetables') that are explicit in the input–output tables.

The operator and unpaid family labour input series in the Statistics Canada database is estimated, for 1961–1975, with occupation data from the Canadian Census of Population as a starting point and with interpolation to the first Labour Force Survey estimate of hours worked in 1976. For 1976–2006, hours data from the Labour Force Survey are used. A drawback to this approach is that the Labour Force Survey, by its design, tends to underestimate labour input to agriculture. This occurs because the Survey allocates hours of work to industries based on the 'main job' of respondents. Hours contributed to the farm operation by operators and unpaid family members with off-farm work are not recorded if they work more hours off farm than on farm.

Finally, commodity-specific payments based on production are not incorporated in the Statistics Canada account.

The AAFC production account offers a database that is specific to Canadian agriculture and that addresses each of the issues identified above. A commodity-by-commodity approach along the lines of that used by Lok and Danielson is followed. As a result, the account also allows a closer link to the earlier studies than the Statistics Canada data and thereby makes possible the creation of a series of production account data that spans the years 1926–2006.

3.3 The AAFC Production Account

The methodology used to construct the AAFC production account closely follows Ball *et al.* (1997). One area where the AAFC approach deviates from Ball *et al.* is with respect to quality adjustment. No inputs or outputs are adjusted for quality in the AAFC account. Although there are compelling arguments for the control of quality effects, for example in the measurement of labour, this production account provides an adjustment-free benchmark.

Some general principles are followed in the measurement of both output and input in the AAFC production account:

- The account is made as complete as possible – actual quantities (by volume, weight, etc.) and prices for each output and input are collected where the data exist.
- The approach is strictly an accounting one; the only exception is the measurement of capital services input, where the price and quantity of these services cannot be estimated without an underlying behavioural model. The model used to derive formulas for the estimation of capital service prices and input is not laid out here; it is similar in most regards to that presented in Ball *et al.* (2008, pp. 1258–1259) for the special case of geometric decay.
- Policies that made commodity-specific payments to producers per unit of output sold are incorporated in prices, the idea being that all of the value of output for each commodity should be attributed to that commodity, so that it is given appropriate weight when aggregated with other commodities to estimate total gross output.
- The total value of gross output must equal the total value of inputs used to produced it – this means that the price for one input (the 'residual claimant') must be endogenous, adjusting to ensure that this equality holds. In the AAFC production account, operators and those family members that work on the farm without pay are the 'residual claimants', i.e. the return to their labour is net farm income, which is the difference between the value of gross output and the cost of all other inputs but operator and unpaid family labour.
- Imputations, interpolations and extrapolations are avoided, but given the importance of Canada's quinquennial Census of Agriculture as a data source, interpolation is needed to develop several key data series (especially operator labour and unpaid family labour and paid labour).

3.3.1 Gross output

Methodology

In the AAFC production account, the concept of gross output is defined, for any commodity, as total production less waste. The term 'gross' refers to the fact that output is not measured simply as the quantity marketed, which nets out various within-sector uses of the commodity within the agricultual sector (such as harvested grain used for seed).

For any commodity and for any year t, gross output is denoted as Y_t and is measured directly as:

$$Y_t = Q_t - L_t \qquad (3.1)$$

using data on production, Q_t and on waste, dockage and loss in handling, L_t.

Where data on production are not available and/or where there is within-sector use, the supply–demand identity is used to estimate gross output indirectly and/or to estimate within-sector use. This identity is, for any commodity in year t:

$$S_{t-1} + Q_t \equiv M_t + D_t + L_t + S_t \qquad (3.2)$$

where S_{t-1} denotes total opening stocks (both on and off farm), M_t denotes marketings to processors, consumers, exporters and to farm households as income in kind, D_t denotes within-sector use (for seed, feed, etc.) and S_t denotes total closing stocks.[5] The left-hand side of the identity (Eqn 3.2) is total supply and the right-hand side is total demand.

Where gross output must be estimated indirectly, isolate $Q_t - L_t$ on the left-hand side of Eqn 3.2 and use definition Eqn 3.1 to get:

$$Y_t = M_t + D_t + S_t - S_{t-1} \qquad (3.3)$$

where $S_t - S_{t-1}$ is the change in stocks. Gross output can then be estimated using data for each of the elements on the right-hand side of Eqn 3.3. This approach is used for several of the livestock commodities in the account.

For most grain and oilseed crops, both Statistics Canada and AAFC provide crop year data in the form of supply and disposition tables that give quantities for each of the elements of Eqn 3.3, with S_{t-1} being carry-over stocks from crop year $t-1$ and the remaining elements being production, marketings etc. for crop year t. Data for seed and feed – the elements that comprise within-sector use D_t – can be estimated using data taken directly from these tables.

For those livestock commodities where gross output must be estimated indirectly, data for each element of Eqn 3.3 is collected from a variety of sources because supply and disposition tables of the type described for grains and oilseeds are not available for livestock.

Crops

Data for most grain and oilseeds are reported on a crop year basis, whereas data for all other crops are reported on a calendar year basis.[6] As most of the data in the AAFC production account are for the calendar year, data reported for the crop year must be converted to calendar year quantities.

For calendar year t, production is the same as that for crop year t and within-sector use of seed is the same as that for crop year $t-1$ (because seed for planting in year t is drawn from the previous year's crop). Within-sector feed use, then, is the only element for which conversion from crop year to calendar year is non-trivial.

It is reasonable to assume that feed use will be fairly constant over the year, varying only as the total number of animals being fed increases or decreases. For any commodity, these variations will be reflected in the data reported in the supply and disposition tables for within-sector feed use, which will change from one crop year to another as livestock numbers change. On the basis of this premise, within-sector use of feed for calendar year t is estimated as a weighted average of the quantity of within-sector use of feed in crop year $t-1$ and that in crop year t. The weighting is simply the proportion of the months of calendar year t covered by within-sector feed from crop year $t-1$ (8/12) and the proportion of the months covered by within-sector feed use from crop year t (4/12); these weights are then applied to quantities of within-sector feed use in crop years $t–1$ and t respectively to estimate within-sector feed use for calendar year t.[7]

Grains, oilseed and other crops included in the AAFC production account are listed in Table 3.1. There are 53 crop commodities separated into five groupings (this is reflected by the sub-headings in the table). In addition to the 8 major grain and oilseed crops, the account measures gross output for 12 types of fruit, 17 types of vegetables, 17 'other' crops – a heterogeneous group of commodities ranging from buckwheat to potatoes – and 4 commodities under the heading 'miscellaneous horticulture'.

Production and other quantity data for all crops other than fruits and vegetables are taken from Statistics Canada CANSIM tables and/or from supply and disposition tables provided by AAFC's Market Analysis Group. Fruit and vegetable data are taken from the Statistics Canada publication 'Fruit and Vegetable Production', from a database developed by Statistics Canada for providing horticulture data up to 1985 and from CANSIM tables.

Livestock

Commodity coverage for livestock is given in Table 3.1; there is a total of 20 livestock commodities included in the output side of the account. For most of these commodities, data on production are not available and must be estimated using Eqn 3.3, i.e. gross output is estimated indirectly as the sum of marketings, within-sector use and stock change.

Livestock gross output is measured on a live weight basis rather than the number of head for cattle, hogs and sheep, which is consistent with Statistics Canada's method of calculating farm cash receipts. Gross output of milk and eggs is also converted to a weight basis. Poultry marketings are reported on a carcass basis and converted to live weight using a uniform conversion factor. Home consumption (of farm output by farm households) is included in marketings for the broad groups of livestock.[8]

Beef cattle, calf and dairy production are interdependent. The time dimension associated with cattle production in particular poses challenges for measurement of both outputs and inputs. Calves are used as inputs in other parts of the sector and may generate output as much as 18 months into the future. Using estimates of livestock supply and disposition, the difference between births of calves in the current year and the number marketed in the current year is the number of calves that will be used internally, either as additions to the breeding herd or fed further for final beef demand.

Gross output of calves in a given year is the sum of marketings in the current year (slaughter of domestically sold and exported live animals, minus imports), home consumption, herd replacement and those sold as intermediate inputs (feeders). Most of the births in a given year are kept within the sector to be fed and are treated as within-sector use, which is estimated as gross calf production, less marketings and replacement heifers. This method implicitly includes farm-to-farm sales within the sector but it does not attempt to model the flows within any given year, and thus does not explicitly capture the timing of feeder cattle marketings.

Cattle gross output is defined as marketings plus stock change. Marketings are defined as domestic slaughter (including home consumption) plus exports minus imports. Canadian cattle slaughter data are available by type of animal over a long period of time. Therefore marketings reflect the variation in quality and type of animals produced over time. Carcass weight and yield figures for each cattle type (steers, heifers, cows, bulls and calves) obtained from marketing reports are also available over time and provide another way that marketings can reflect yield enhancing changes in production. The ratio of carcass weight to yield gives an estimate for the live weight of a slaughtered animal.[9] Following this, stock change must also be measured on a live weight basis.

For example, measurement of the change in cattle stocks comprises two elements: the change in the number of animals, given the average yield (the ratio of carcass weight to carcass yield) and the change in

Table 3.1. The AAFC production account for Canadian agriculture: output and input coverage.

Outputs			Inputs		
I. Crops	II. Livestock	III Other	I. Capital	II. Labour	III. Intermediate inputs
Grains and oilseeds	Bulls	Custom work	Building, farm machinery & vehicle services	Operator/unpaid family labour/hired labour	Within-sector use (Dt): feed
Barley	Calves	Forest products	Farm buildings		Barley
Canola	Calves for cattle production	Land services	Farm machinery		Corn
Corn	Chickens		Passenger vehicles		Flaxseed
Flaxseed	Cows		Commercial vehicles		Fodder corn
Oats	Eggs		Livestock services		Hay
Rye	Eggs used for hatching		Beef cows		Oats
Soybeans	Fowl		Beef heifers		Rye
Wheat	Hatchery chick – broiler		Boars		Wheat
Fruits	Hatchery chick – egg		Bulls		Within-sector use (Dt): seed
Apples	Hatchery chick – turkey		Ewes		Barley
Blueberries	Heifers		Milk cows		Canola
Cherries, sour	Hogs		Milk heifers		Corn
Cherries, sweet	Honey		Rams		Flaxseed
Cranberries	Milk		Sows		Oats
Grapes	Milk fed to livestock		Land services		Rye
Loganberries	Sheep and lambs				Soybeans
Peaches	Steers				Wheat
Pears	Turkeys				Within-sector use (Dt): other
Plums/prunes	Wool				Calves
Raspberries					Eggs used for hatching
Strawberries					Hatchery chicks – broiler
Vegetables					Hatchery chicks – egg
Asparagus	Crops, Cont'd				Hatchery chicks – turkey
Beans	Misc. horticulture				Milk fed to livestock
Beets	Greenhouse cucumbers				Purchased inputs
Cabbage	Greenhouse tomatoes				Commercial feed
Carrots	Mushrooms				Commercial seed
Cauliflower	Potatoes				Electricity
Celery	Other crops				Fertilizer and lime
Cucumbers	Buckwheat				*Continued*

Table 3.1. Continued.

	Outputs			Inputs		
I. Crops	II. Livestock	III Other		I. Capital	II. Labour	III. Intermediate inputs
Lettuce	Canary seed					Fuel for machinery and vehicles
Onions, dry	Chick peas					Heating fuel
Parsnips	Dry beans					Pesticides
Peas	Dry peas					Telephone
Rutabagas & turnips	Fodder corn					Other inputs
Spinach	Hay					Custom work
Sweet corn	Lentils					Irrigation (expenses only)
Tomatoes	Maple products					Artificial insemination/ veterinary (expenses only)
	Mustard seed					Twine, wire and containers (expenses only)
	Sugar beets					
	Sunflower seed					
	Tobacco					

the average yield, given the number of animals.[10] This can be expressed as

$$S_t^v - S_{t-1}^v \equiv \left(S_t^n - S_{t-1}^n\right)\left({}^{x_t}/y_t\right)$$
$$+ S_t^n \left({}^{x_t}/y_t - {}^{x_{t-1}}/y_{t-1}\right) \quad (3.4)$$

where S_t^v is the stock on live weight basis, S_t^n is the stock (in number of head), x_t is carcass weight and y_t is carcass yield. This approach makes it possible to capture the change in average animal size when calculating the change in stocks, thus measuring the effect on the change in stocks of cattle held in inventory to achieve higher finished weights.

Hog gross output is treated essentially as one commodity, that being finished hogs. As with cattle, hog gross output is the sum of home consumption, domestic slaughter and exports, less imports. Conversion to live weight is computed using the warm carcass weight per head of finished hogs and an assumption for carcass yield, which is used for the entire time period because of data limitations. Stock change is calculated according to the method described above for cattle.

Gross output for sheep and lambs is combined because the only data consistently reported by Statistics Canada are for the combination of the two. Sheep and lamb marketings include home consumption, domestic slaughter and exports, and exclude imports. The carcass yield for sheep and lambs is constant over the time period.[11] Stock change is calculated with the same procedure used for cattle and hogs.

Poultry gross output is defined as marketings plus home consumption, where both are reported on an eviscerated weight basis. Marketings are defined as registered slaughter plus live exports minus live imports, all reported on a carcass basis. The yield percentages for chicken, turkey and fowl (hens) are each assumed constant over the period, and represent an average of the yield across the different weight classes. Net trade is incorporated in the slaughter statistics available from Statistics Canada; this differs from the approach taken for other livestock because poultry are not imported for finishing by poultry producers, but rather for

immediate slaughter by processors. Net imports are deducted from the slaughter amount by Statistics Canada to arrive at domestic marketings. Also included in output is hatchery chick production destined for meat production. Output is defined as placements of chicks on broiler, laying hen and turkey farms. Prices are obtained from the Farm Input price index for turkey and broiler chicken. The price for egg type chicks is not as widely available, and must be estimated on the basis of its share of placements for production.

Egg gross output comprises both sales to final demand and within-sector use. Gross output is estimated as marketings of 'table' eggs plus home consumption less waste (there is no stock). Egg marketings reflect production from both registered and small flock producers. Both the laying flocks and eggs destined for hatching are recorded as outputs and as intermediate inputs.

Gross output of milk comprises milk shipments to purchasers of milk for both fluid and processing (industrial), milk fed to livestock and milk consumed by the farm household.[12] Milk fed to livestock represents an intermediate input, so it is also included on the input side of the account. There is no stockholding of milk, so there is no stock change component in the measurement of milk gross output.

The remaining livestock commodities in the account are honey and wool. Production data for these commodities are available and are used to estimate gross output directly.

Other components of gross output and commodities not in the AAFC production account

OTHER COMPONENTS. There are three other components of gross output in addition to crop and livestock output. These are: (i) land services provided to other farms and non-farmers from rental of agricultural land and buildings; (ii) custom work; and (iii) forest products from agricultural land (such as logs taken from woodlots).

The value of land services output is based on rental income from the lease of farmland to farms or individuals outside the agricultural sector, which is measured by Statistics Canada as part of Agricultural Value-added account from 1981 onwards. Rent paid by farms is reported in CANSIM Farm Operating Expenses tables for the whole 1961–2006 period; although not used directly in the account, these rental data are a useful source of information for imputing rental income between 1961 and 1980.[13] Between 1981 and 2006, rent paid and rental income (to farms and individuals outside the sector) were similar, although the expense data were about 4% higher than rental income. The proportion of rent received by farms relative to total rent received over this period was also quite stable, at about 6% in the early 1980s. Rental income for 1961–1980 is therefore imputed by applying the two proportions to the rent paid series. The imputed/observed 1961–2006 series of estimated current dollar rental income is then deflated using the price of land services (defined below in the Capital Services section) to obtain a constant dollar implicit quantity of land services output.

Custom work receipts are also available from the Agricultural Value-added account and start in 1981. When compared with data on custom work expenses for the early 1980s, receipts seem to have been about 70% higher than expenses. This proportion is used to extrapolate back to 1961 using custom work expense data.[14] The constant dollar implicit quantity of custom work output is then estimated by deflating imputed/observed custom work receipts by the custom work deflator that is a component of the Farm Input Price Index, which is available for the full 1961–2006 period. This gives an estimate of the constant dollar implicit quantity of custom work services output.

Forest products revenue for 1961–2006 is taken from CANSIM Farm Cash Receipts tables. This source of revenue primarily applies to farms in Ontario, Quebec and the Maritimes. The continuous revenue series is therefore deflated by a price index for hardwood ties and lumber. The deflated value gives the constant dollar implicit quantity of forest products output.

COMMODITIES NOT IN THE AAFC PRODUCTION ACCOUNT. There are many commodities produced by farms that are not included in the AAFC production account (i.e. that are not listed in Table 3.1). There are three reasons for this. The first is continuity; a survey might have collected data for a commodity for only part of the 1961–2006 period. This type of data is not usable, even though it exists for some years, because there is not sufficient information to estimate output and prices for the commodity in those years for which data are missing.

The second reason is coverage; there are commodities that have not been measured in terms of output or price at any point in time because they were deemed to be of only marginal importance by Statistics Canada (which collects almost all Canadian agricultural statistics). It is not possible to find data for any of these commodities in extant statistical publications or databases. Their value could, however, be captured by farm cash receipts but be not listed as separate items there, i.e. included only in a generic category such as 'other' or 'miscellaneous'.

The third reason is confidentiality. If the population of producers of a particular commodity is so small that confidentiality becomes an issue, any collected data is suppressed in Statistics Canada reports and databases. The issue here is similar to the first reason given above – information for those years where data were not suppressed is not usable because there is no way to estimate the missing data.

The result of these commodity exclusions is that total gross output for Canadian agriculture is somewhat underestimated in the AAFC production account. This underestimate is calculated to be in the range of 1–4%, using cash receipts data for miscellaneous and other outputs not identified here.

Crop and livestock prices and direct payments

OUTPUT PRICES. Prices for crops are collected from sources similar to those used for measuring gross output. These are defined as average farm prices or as farm gate prices, i.e. these are prices that the producer receives. Up to and including 1984, Statistics Canada reported farm prices for almost all crops and some livestock commodities. The farm price series ended in 1984 and, from that point on, information sources are more varied. Most grains and oilseed prices from 1985 on are taken from the AAFC Market Analysis Group supply and disposition tables; the prices reported in these tables are typically averages across all sales and are not necessarily farm gate prices. Several of the prices reported in these tables are for specific grades (e.g. for wheat it is average No.1 CWAD 11.5) that are representative of the average quality sold. All prices that are market prices rather than farm returns are discounted by 15% to take into account transportation and related charges incurred getting the product to selling point. This margin is based on comparisons with farm gate prices for selected commodities that have been submitted by AAFC to the Organisation of Economic Co-operation and Development (OECD) for the calculation of producer support estimates (PSEs) for Canada.

For other crops except fruits and vegetables, more diverse sources are used, with sometimes more than one source drawn upon to have a complete series. These other sources include:

- Farm cash receipts (for calculating unit values with marketings where these can be measured).
- Provincial government agriculture ministries (for hay and fodder corn data) and industry sources (sugar beets data).
- FAO (which in some cases is more easily accessible than Canadian sources).
- Price indexes computed directly (such as for land services output) or taken from CANSIM tables (e.g. custom work).

Few imputations are needed, but where there are short gaps in the data (1–2 years) linear interpolation is used to fill these.

For fruits and vegetables, the sources used for gross output data provide farm gate value for all years between 1961 and 2006. Unit farm gate prices are derived by dividing these values by production.

Prices for livestock differ between commodities and are summarized as follows:

- Prices for cattle are reported on a live weight basis. Hog prices are reported in dressed weight and must be converted to live weight, which is done using the same carcass yield as for hog production.[15] Market prices for both cattle and hogs are each weighted on a regional basis to estimate a value of output consistent with Statistics Canada's data for cash receipts at the national level.
- Production data available for sheep and lambs indicate that most of the meat produced (more than 80%) is from slaughtered lambs. Moreover, sheep price data are also quite inconsistent over time. The lamb price is therefore used to value sheep and lamb gross output, although this will overstate the value of output for sheep and lambs together.
- Unit prices for poultry are calculated using value of production data supplied by Statistics Canada.
- The price of milk is constructed as a unit value using farm cash receipts (including dairy subsidies – see Policies section below) and fluid milk shipments. This is effectively an average price for milk and is a simplified way to measure the value of raw milk compared to a more complicated approach that would reflect the industry pricing structure for milk components.
- Sources and definitions of prices for other livestock are similar to those given earlier in the description of gross output.

POLICIES AND PROGRAMME PAYMENTS. Two principles are used when incorporating payments made to producers under certain Canadian and provincial government policies and programmes. First, only policies or programmes that made direct payments to production (e.g. a deficiency payment per tonne) are incorporated. Second, an accrual rather than cash accounting approach is used. Thus, the value of gross output in any year reflects the returns that accrued to it, even if payments were made several years later.

There are four policies that provided direct payments to producers based on production between 1961 and 2006.[16] The first of these is the Agricultural Stabilization Act, which created the Agricultural Stabilization Board (ASB). The ASB provided supplementary payments per unit produced on a wide range of commodities between 1961 and 1990. Often, a payment was announced after some or all of the commodity had been marketed, so that the ASB reports frequently refer to payments made for commodities produced one, two or three years ago. In these cases, payments to that commodity are added together over all years and then applied to the appropriate production year.

The second policy provided supplementary payments to dairy producers per unit of milk sold. These payments were originally administered by the ASB, but administration was turned over to the Canadian Dairy Commission in 1967. Programme payments were phased out completely in 2002.

The third policy is the Western Grains Stabilization Act (WGSA). The WGSA programme, which began in 1976 and ended in 1991, was jointly funded by contributions made by producers (through levies) and the federal government. Interest was earned (or costs accumulated) on the balance. Payments to producers of eligible crops were usually paid for the previous year out of current year balances. The programme initially was based on a calendar year, but in 1984 this was changed to a crop year basis. This change allowed payments to be made earlier in the year.

WGSA payments are incorporated in the AAFC production account as follows:

- Payments to producers are treated as being net of accumulated levies and interest on these accumulated levies. Levies in any year are treated as savings by producers after marketings rather than taxes on marketings. These savings are then drawn upon in subsequent years.
- Net payments are incorporated in the value of gross output according to the year to which they were applied, not the year in which the payments were actually made.

- Interest cost on negative balances (which occurred after 1984) is not treated as part of any net payment to producers, but rather as an administrative cost.
- 'Write downs', to cover a negative balance at the end of the programme, are not treated as net payments to producers – the negative balance is the result of accumulated net payments to producers and so this value is already reflected in the value of gross output.
- Net payments to producers are allocated across eligible crops according to the share of that crop in total marketings made by programme participants (not all grain and oilseed producers participated in the programme).

The fourth and final set of programmes comprises provincial-level stabilization programmes. In the late 1970s, some provincial governments introduced a variety of policies to supplement federal programmes (ASA and WGSA) and to address perceived gaps in support provided by the various alternatives that replaced them. These programmes are referred to here generically as 'provincial stabilization programmes', following the terminology used by the OECD in its PSE database.

The provincial stabilization programmes have covered a range of commodities but cattle, hogs and corn averaged around 81% of payments to producers between 1986 and 2000. Barley, canola, soybeans and wheat together amounted to a further 10% on average.

Payments from one or more of these policies made to a particular commodity in any year are added together and converted to unit payments by dividing the total payments to that commodity, G_t by gross output, i.e. the unit payment is $g_t = G_t / Y_t$. The price for gross output is then $P_t + g_t$.

3.3.2 Inputs

Capital services

Capital services are one of the most difficult inputs to quantify and the large body of literature on the subject attests to this.[17] Estimation of the level of capital services is done in several stages. First, the capital stock is estimated. Then a formula based on prices of new investment, the rate of physical decay and an appropriate rate of return (interest rate) is used to estimate the price of capital services. The cost of capital services is calculated next, using the price of capital services and constant dollar capital stock data. Finally, the constant dollar implicit quantity of capital services is estimated by deflating the cost of capital services by the price of capital services.

There are three categories of capital services included in the AAFC production account. The first category comprises services from physical capital for the following asset groupings: farm buildings, farm machinery, commercial vehicles and passenger vehicles used for farm work.[18] The second category comprises services from livestock capital. The third category comprises land services. Estimation of the services input and price is described for each category in turn.

CAPITAL SERVICES FROM FARM BUILDINGS, FARM MACHINERY, COMMERCIAL VEHICLES AND PASSENGER VEHICLES. The measurement of services for these asset groupings begins with the estimation of the capital stock. The stocks are computed by applying the perpetual inventory method (with geometric decay) to investment data for 1926–2006.

Some imputations are needed both at the beginning and end of the investment series to separate the nominal value of machinery and equipment into farm equipment and vehicles components. For 1926–1935, current dollar investment for farm machinery and the two vehicle types are extrapolated backwards using investment shares for 1936–1945; similarly, the disaggregation of machinery and equipment investment data for the 1997–2006 period are extrapolated forwards using investment weights from Statistics Canada's Farm Financial Survey (taken every other year; inter-survey year data are obtained using linear interpolation).[19] Investment deflators

are constructed for the 1926–1935 period using import unit values for representative farm machinery and vehicles taken from Trade of Canada publications. For all other years, disaggregated deflators are available for all four asset groupings. Current dollar investment data are converted to constant 1992 dollars by dividing current dollar investment by the investment deflator.

The perpetual inventory formula used to estimate the capital stock for each asset grouping is

$$K^R_{1926+\tau} = K^R_{1926} (1-\delta)^t + \sum_{v=1}^{\tau} I^R_{1926*v}(1-\delta)^{\tau-v} \tag{3.5}$$

where K^R_{1926} is the constant dollar implicit capital stock (evaluated at the investment price in the base year b, τ years after 1926, K^R_{1926} is the seed stock (calculated using 1926 investment and the assumption that $K^R_{1926}=K^R_{1925}$); I^R_{1926+v} is the implicit constant dollar level of investment in year 1926 + τ and δ is the physical decay rate. The decay rates used are 0.04, 0.18, 0.17 and 0.18 for buildings, farm machinery, commercial vehicles and passenger vehicles respectively; these are based on Patry (2007, p. 22).

The quantity of capital services can be estimated from the capital stock data only through the use of a model that compares the flow of returns from an investment in physical capital to the returns from an alternative investment over an infinite planning horizon. Several conditions must be fulfilled in order to obtain a solution for the level of services input from this model.

The most important of these conditions is that the quantity of capital services be proportionate to the size of the capital stock and that this proportion be constant over time. Griliches and Jorgenson (1962, p. 51) state that 'the quantity of a particular type of capital as an asset is proportional to its quantity as a service'. Harper, Berndt and Wood (1989, p. 341) observe that 'we follow tradition here and make the assumption that capital services ... are a constant proportion of capital stocks'. OECD (2001, p. 84) is more vague, relating a change in capital stock to a flow and thereby a change in output; because services 'flow' from capital, there seems to be a proportionality argument here as well.

There are several other conditions that must be met, including:[20]

- Static expectations – producers have static expectations of the price of aggregate gross output, i.e. when forming a production plan they assume that output prices will remain unchanged over the infinite horizon.
- The alternative investment is an annuity – this defines the opportunity cost of capital.
- The marginal product of capital services is constant over the infinite horizon.

Estimation of the price of capital services relies on the fact that present value of rent to the owner of an incremental investment in farm capital equals the cost of financing that investment. The price of services is then estimated from this equality since the rent paid by the operator to the owner is the valuation of the productive services that the incremental investment provides.[21]

The expression defining the price of capital services is:

$$w_t = q_t \,(\rho_t + \delta)/\theta \qquad (3.6)$$

where q_t is the unobservable aggregate price for all investment within the asset group in year t, ρ_t is the rate of return from an annuity purchased in year t and θ is the proportionality constant that relates capital services to the capital stock.

The price defined by Eqn 3.6 is not observable because, although an aggregate price q_t for investments within the asset group exists in theory, it cannot be measured in practice. The same holds true for the constant θ. To have an observable price, Eqn 6 must be expressed in relative terms, i.e. as the current price of capital services relative to the price in any base year chosen for the calculations. This is specified as:

$$w_t^s = q_t^s \, \frac{(\rho_t + \delta)}{(\rho_b + \delta)} \qquad (3.7)$$

where q_t^s is the investment deflator (with base year b) in year t, $q_t^s = q_t/q_b$ where q_b is the unobservable investment price in the

base year, ρ_t is the year t real rate of return and ρ_b is the year b real rate of return.

The current dollar total cost of capital services for any asset grouping is, like all inputs, equal to price times quantity. In the case of capital, this is the product of the price of capital services w_t and the quantity of capital services k_t. It is possible to estimate this total cost using the following expression

$$C_t = w_t k_t \equiv q_t^s (\rho_t + \delta) K_t^R \qquad (3.8)$$

Note that $k_t = \theta K_t$ by the proportionality assumption, so $k_t = \theta K_t^R/q_b$ (as $K_t^R = q_b K_t$ implicitly).

The constant dollar implicit quantity of capital services can then be estimated in the usual manner, noting that this is the unobservable base year price of capital services times the unobservable quantity of capital services, i.e.:

$$k_t^s = w_b k_t \equiv C_t/w_t^{ks} \equiv (\rho_t + \delta) K_t^R \qquad (3.9)$$

In the AAFC production account, ρ_t is set to a constant real rate of 3.5%. This means that, because the decay rate is also constant, Eqn 3.7 simplifies to:

$$w_t^{ks} = q_t^s \qquad (3.10)$$

but Eqn 3.9 remains unchanged (except that ρ_b can be replaced by the constant real rate of return ρ).

The capital service estimates for the four asset groupings are illustrated in Fig. 3.1. Note the large increase in services for farm machinery, commercial vehicles and passenger vehicles between 1973 and 1981, followed by a large decrease between 1982 and 1992; this is a function of the capital stock which displays an identical pattern. Farm machinery capital stock data estimates for several countries are given in Ball et al. (2008, p. 1265); these data also follow a similar course. A cause for this growth and diminution is not sought here, but the pattern observed with Canadian capital stock data for farm machinery (constructed using the 'conventional approach' outlined above) is not reflected in the total number of tractors or trucks data collected by the Census of Agriculture. (Fig. 3.2 provides a

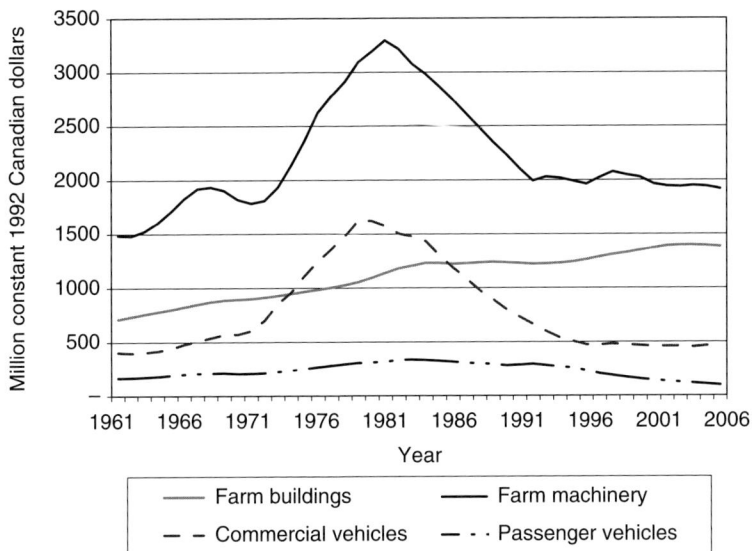

Fig. 3.1. Implicit quantity of capital services in Canadian agriculture, 1961–2006. Source: Authors' estimates.

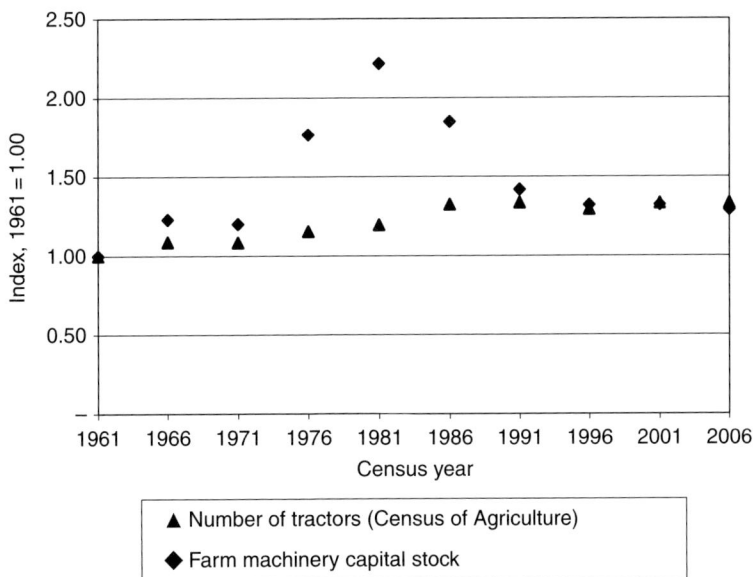

Fig. 3.2. Farm machinery capital stock and number of tractors in use in Canadian agriculture. Indexes, 1961 = 1.00. Source: Canadian Agricultural Censuses (Statistics Canada, 2007).

comparison between the farm machinery stock and the number of tractors; the pattern is much the same for commercial vehicles and trucks.)[22] It is unclear which of the two series gives the most accurate depiction of what happened over this period, but Anderson *et al.* (2009, p. 33) discuss a similar difference between ERS estimates of capital stock and their 'physical' capital data. This suggests that the difference between the two series is not an anomaly but is more reflective of a systematic conflict between the two approaches to measuring capital.

CAPITAL SERVICES FROM LIVESTOCK.   Some livestock provide a flow of services over several years. In particular, breeding animals (bulls, milk cows, beef cows, dairy heifers, beef heifers, boars, sows, rams and ewes) together generate new livestock (some of which become herd replacements themselves).

The stock of these livestock assets is reported in number of head for each type. Prices are in terms of 'values per head' and are reported in a separate price series by Statistics Canada when valuing on farm stocks in the balance sheet for the sector.[23]

The total cost of livestock services can be estimated for each livestock type using Eqn 3.8. In this case, the 'investment deflator' is just the price per head in the current year relative to the base year, and the constant dollar stock is just the number of head times the base year price. As a result, the cost of livestock services, at a constant real rate of return and with $\delta = 0$, simplifies to the current value of the herd times the real rate of return.[24]

CAPITAL SERVICES FROM LAND.   For each province, agricultural land area consists of both improved land (land in crops, summer fallow and pasture/grazing) as well as unimproved land (natural pasture/grazing and 'other', which comprises woodlots etc.). The total area of both of these types of land is used in calculating the input of land services, because it is not possible to separate the portion of land that is actually used from that which is not. It is probably fair to say that all of these land resources are used, to varying degrees both within and across time, to generate agricultural output.

Data on the total value of land and buildings by province are based on Statistics Canada's Value of Farm Capital series, which is part of the Agriculture Economic Accounts and are benchmarked to the Census of Agriculture. These data are supplemented by information from Farm Credit Canada's annual survey of land values, which allows a division of the value of land and buildings into land and buildings components – only the former series is used here because buildings services are calculated as described above. Application of this ratio to the total value of land and buildings gives an estimate of land value for each province that, when divided by area data (in acres) gives an estimate of the price of land per acre.

A price deflator for land services at the national level is estimated by first constructing service prices for each province using Eqn 3.7 and then aggregating across provinces using estimated provincial price of land services and derived provincial land area series. (This deflator is also used in the calculation of land services output – see the Other components of gross output section above.) As it is assumed that $\delta = 0$, the price for land services, simplifies, like livestock, to a price of land relative to the base period (set arbitrarily to 1997). Although no adjustment is made for land quality, this approach, by aggregating across different provincial land prices, does capture some of the heterogeneity of farm land in Canada.

The cost of capital services is the total nominal value of land (summed over provincial values) multiplied by the constant real rate of return (3.5%). The implicit constant dollar quantity of land services is then estimated by dividing the cost of land services by the constructed national deflator.

Labour

OPERATOR AND UNPAID FAMILY LABOUR.   Farm operators and their families supply most of the labour input to Canadian agriculture. Family labour can be paid for directly as a farm expense, but operators usually receive no explicit wage for their labour input. To date, little has been done to estimate either the number of hours devoted to farm work or the value of this work, in spite of the importance of operator and unpaid family labour in agriculture.

As noted earlier, one of the principles followed when building the AAFC production account is that operators and family members that work on the farm without pay are the 'residual claimants' of net farm income (the difference between the value of gross output and the cost of all other inputs but their labour). This return is expressed here as an implicit wage, which is the ratio of net farm income to hours of operator and unpaid family labour input. This wage is referred to as 'implicit' because it is not of the usual type, i.e. a set dollar amount per hour worked. Instead, the implicit wage varies as market prices and output/input quantities change and as the number of hours worked changes. Measurement of the labour input from operators and unpaid family members is therefore important not only because this represents part of total input, but also because the number of hours worked must be known in order to estimate the implicit wage.

The approach taken here begins with the measurement of operator labour – the input of unpaid family members is measured simply as a proportion of operator labour input. The only data source used is the Census of Agriculture; the input quantity is derived from questions on the census about off-farm work and, in more recent censuses, about on-farm work.

This approach is different from those used in the studies by Lok (1961), Danielson (1975) and Brinkman and Prentice (1983), where Labour Force Survey and Census of Canada (population census) data were used to compute either the number of operators and unpaid family workers or to estimate the number of hours worked by them, using some proportionality assumption.

Operator and unpaid family labour input in the AAFC production account is also measured differently from the input in the Statistics Canada production account, which is based primarily on Labour Force Survey estimates of hours worked. It has already been pointed out that this approach will tend to underestimate the labour input from operators and unpaid family members.

The measurement of operator labour input with off-farm work data acknowledges the importance of work off the farm for many Canadian operators. Bollman (1982, p.314) observes that part-time farming 'has always existed in Canada' and cites some historical evidence to support this statement. Census of Agriculture data between 1961 and 1991 show that the participation rate in the off-farm labour market has been quite consistent, with about 30–40% of operators working at least some of the time off farm. This means that employment data could overestimate the level of operator labour input.[25]

Estimation of operator labour input is explained in two separate parts. The first, which covers the period 1961–1991, uses off-farm labour statistics from the Censuses of Agriculture. For the second, data are still taken from the Census, but are in terms of hours worked on farm. The estimates from the two parts are then combined to create an estimate of operator labour input, over the period 1961–2006. Unpaid family labour input estimates are then discussed.

Derivations with Census of Agriculture data, 1961–1991

The importance of off-farm work is reflected in the type of questions that have been asked in the Census of Agriculture questionnaires. Between 1961 and 1991, the questionnaires had similar questions, which were along the lines of: 'how many days did the operator work at non-agricultural work and agricultural work off the farm during the past 12 months?' Two responses were allowed: 'none' or 'number of days', where the latter response would be an integer ranging between 1 and 365. The question: 'how many days (or hours) did you work on the farm?' was not asked in any of the questionnaires for these census years, so no direct observations of on-farm work are available.

For a time series of operator labour data, either days or hours of on-farm work must be the unit of measure, given that the data collected is in one or the other of these units. The common unit of measure used here is hours. This means that off-farm data measured as days, as is the case up to and including 1991, must be converted to hours. To do this, two assumptions are needed, namely that:

A1. The average operator with off-farm employment works a combined total (on and off farm) of 6 days per week, or 313 days per year. This allows for one day of rest each week.

A2. A full work day on the farm is 10 hours, on average.[26]

Assumption A1 makes it possible to estimate the days worked on farm by those operators who worked off the farm (that is, 313 days times the number of operators working off farm less the number of days worked off farm). This, added to the number of days worked on farm by operators who did not have off-farm employment (313 days times the number of operators fitting this description) gives the total number of days worked on farm by all operators. Assumption A2 is used to convert the estimated number of days worked into total number of hours worked on farm by all operators.

An hours-of-work series for operator labour over the period 1961–1986 is estimated using the total number of days of work on farm for Canadian operators derived using the procedures noted above and using A2 to convert total days into hours. Data for intercensal years are estimated using linear interpolation between estimates for census years.

For 1991, a similar approach is used, but an additional aspect to the data must be taken into account. In particular, the 1991 Census of Agriculture allowed for more than operator, so that, although the days worked off farm were reported in the same way as previous censuses, each operator reported days of off-farm work. For consistency with previous censuses, it is necessary to adjust days of off-farm work to remove the effect of additional operators. This is done by using the notion of a 'primary' operator. 'Primary' is used here to mean the operator that best fits the description from the 1986 questionnaire, namely 'the person [most] responsible for the day-to-day decisions made in the operation', where 'most' is added to fit the case of several operators. With this modified description, it should be possible to identify the one individual out of two or more operators that is the most important in making these decisions.

To adjust the total number of operators reported in these censuses, the number of primary operators is taken to be equal to the

number of farms. For 1991, the datum for hours worked off farm is multiplied by the ratio of the total number of primary operators to the total number of all operators (this assumes that all operators have the same distribution of hours of off-farm work), after which calculations made are the same as those for earlier censuses. Intercensal observations between 1986 and 1991 are estimated using linear interpolation.[27]

Derivations with Census of Agriculture data, 1996–2006

The 1996 Census of Agriculture introduced a very different question, which also appeared in the 2001 and 2006 questionnaires. As in 1991, the questionnaire allowed for more than one operator but now, for each of the operators it posed the question: 'What was this person's time contribution to the operation of this agricultural operation?' One of three responses was allowed: (i) on average, more than 40 h/week; (ii) on average, 20–40 h/week; or (iii) on average, less than 20 h/ week. Once again, the questions did not ask how many hours the operator(s) worked on the farm, instead eliciting answers only in terms of these ranges of hours.

Although this is an apparent improvement in relation to earlier Census of Agriculture questionnaires, this question still fails to offer the data needed. In particular, for the 1996, 2001 and 2006 censuses, where on-farm work is reported as a range, it is necessary to have a third assumption to get an estimate of the total number of hours worked on farm by operators. In particular this is:

A3 For operators working:
  (i) <20 h/week on farm, the average operator puts in 19 h of on-farm work.
 (ii) between 20 h/week and 40 h per week on farm, the average operator puts in 39 h of on-farm work.
(iii) >40 h/week, the average operator puts in 59 h of on-farm work.

The number of hours assumed for A3(i), A3(ii) and A3(iii) are the levels that generate a set of estimates that are largely consistent with the trend in estimated total on-farm hours of work in census years prior to 1996. Assumption A3 makes it possible to calculate the total operator input to Canadian agriculture by multiplying the number of primary operators in each category by the average number of hours worked each week.

For the 1996, 2001 and 2006 Census of Agriculture, the number of operators in each range in A3 is multiplied by the ratio of total primary to the total number of all operators. Intercensal observations between 1991 and 2006 are estimated using linear interpolation between estimates for the census years.

When taken together, estimates from the two periods show a substantial decline in operator labour. According to these estimates, estimated operator labour input in 1961 was nearly 1.3 billion hours, or an average 2660 h/farm (taken over all farms). In 2006, this input had fallen to about 512 million hours, which is equivalent to about 2230 h/farm. Total operator labour input to Canadian agriculture thus decreased by 60% between 1961 and 2006.

Unpaid family labour

Unpaid family labour input is estimated using several information sources. The first is the 1961 Census of Agriculture, which collected data on the number of weeks of unpaid family work (this is the only time that this has ever been done). The second is Statistics Canada's Labour Force Survey, which gives the number of unpaid family workers as well as, for 1976–1991, the number of hours of unpaid family and operator labour. The third is the set of estimates of operator labour input described above.

The average weeks worked per unpaid family worker is calculated with the 1961 datum. Then, based on the assumption that unpaid family members worked 5 days per week at 3 hours per worker per day, the estimated average labour input is 761 h/farm in 1961, which is 29% of the hours worked by the average operator per farm.[28] The proportion 0.29 is then applied to the estimated total hours of operator labour input to get the estimate of total 1961 unpaid family member labour input, which is 365.7 million hours.

A ratio of unpaid family labour to operator labour is then estimated. This is initially composed of the 1961 datum and the ratios estimated with the Labour Force Survey hours series for operators and unpaid family members. To interpolate between 1961 and 1976, the trend in employment ratios during 1961–2006 is first estimated using a second-order polynomial. Then the second-order coefficient is adjusted to allow an interpolation between the two points. For 2001–2006, a simple extrapolation is used. The estimated proportions are then applied to the time series of estimated operator labour input to get a full-time series of unpaid family labour input between 1961 and 2006.

According to these estimates, unpaid family labour input fell from 365.7 million hours in 1961 to only 20.5 million hours in 2006; this represents a decrease of 94%. When expressed in terms of hours per farm, unpaid family labour input decreased from 761 h/farm in 1961 to only 90 h/farm in 2006, a decrease of 88%.[29]

Combined operator and unpaid family labour input

On the basis of these estimates, the combined operator and unpaid family labour input decreased from 1.644 billion hours in 1961 to just over 532 million hours in 2006, a decrease of close to 68%. Figure 3.3 provides a comparison between the combined operator and unpaid family labour input per farm and the number of farms for the census years 1961–2006. The reduction in combined labour input is less rapid than the reduction in the number of farms. According to these estimates, combined operator and unpaid family labour input per farm in 1961 was about 3400 h, whereas in 2006 it was about 2300 h, a decrease of 32%. This compares with a decrease of 52% in the number of farms over the same period.

HIRED LABOUR. Each Census of Agriculture between 1961 and 2006 has collected data on expenditures and weeks of paid work. This is the main source of data for hired labour. A secondary source is the Labour

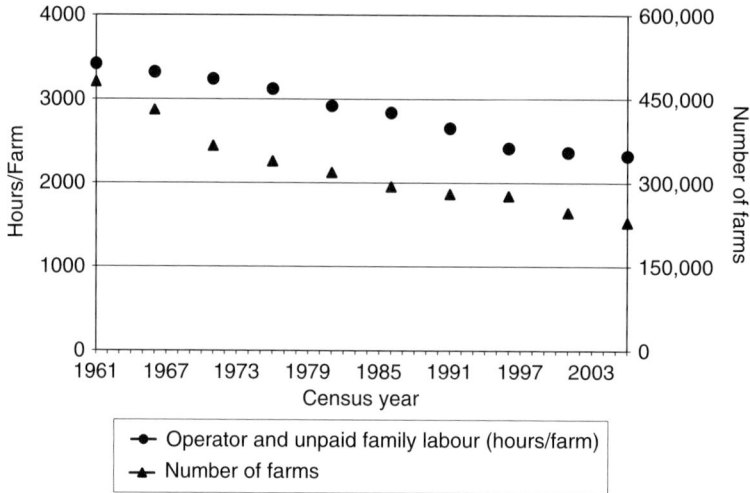

Fig. 3.3. Operator and unpaid family labour per farm and number of farms in Canadian agriculture. Source: Canadian Agricultural Censuses (Statistics Canada, 2007).

Force Survey, which has collected data on number of hired workers in agriculture for each year during 1961–2006. Wage data for some of the period also exist, but since 1997, Statistics Canada has not conducted a wage survey for agriculture.

Census data are used to estimate the number of weeks worked per hired worker for census years and then intercensal years for this ratio are generated using linear interpolation. These ratios are then applied to employment data from the Labour Force Survey to get estimates of total weeks of hired work for the whole period.

The weekly wage from the Census of Agriculture provides the basis for calculating the hired labour wage. The weekly wage for intercensal years is then calculated using linear interpolation.[30]

Intermediate inputs: within-sector use

CROPS.   Within-sector use data for relevant crops from the supply disposition balance sheets – see Eqn 3.3, variable D_t – are used to estimate quantities of feed and seed that are both produced and used within the agricultural sector. Commodities that are used for feed within the sector are barley, corn, flaxseed (where crushed on farm), oats, rye, wheat, tame hay and fodder. Canola and

soybeans are not included as feed because these must be converted into meal, which is done by processors outside of the agricultural sector; meal is then purchased from these processors, usually as a compound feed where the meal has been mixed with other ingredients.

Within-sector use is valued at the price that would have been received had these quantities been sold outside the sector. Thus, the price includes the derived unit subsidy referred to in the section on policies and programme payments.

LIVESTOCK.   Calves moving into fed cattle production are a significant intermediate input in the livestock sector. The category 'calves' includes both young calves and stocker/feeder animals, which are not separated in the database due to incomplete data. As a result, an average price is used for this category, which likely understates the value of feeder cattle.

Eggs used for hatching (included as an output) are used within the sector to produce live chicks (also included as an output). Chicks from hatching production are used within the sector for broiler production, laying hen replacements and turkey production. Prices for within-sector use of eggs for hatching and chicks are the same as those used to value output.

Data for milk fed to livestock are only available up until 1977, after which the data are imputed. The imputed data are based on the number of calves on farms. There is no separate feed price for milk, so the same price is used as that for fluid milk. This is an upper bound on the value of within-sector milk use, because some or all of the milk fed to livestock would be that in excess of production quota and therefore cannot be sold as fluid milk or to processors (as 'industrial' milk).

Purchased inputs, custom work and incomplete data

Inputs purchased from outside of the agricultural sector include commercial feed, commercial seed, electricity, fertilizers/lime, fuel for machinery and vehicles, heating fuel, pesticides and telephone services. Data used to measure these inputs are current dollar farm expenditures, net of tax rebates, along with individual deflators from Statistics Canada's Farm Input Price Index. Implicit quantities of these inputs are measured by deflating the current dollar series with the relevant deflator.

The expense data collected to impute custom work receipts on the output side of the account are used here to obtain an implicit quantity of custom work services input. The expense data are reported for the whole 1961–2006 period and therefore no imputation is needed to measure the custom work input, which is estimated as an implicit constant dollar implicit quantity by deflating expenses by the custom work deflator, which is a component of the Farm Input Price Index.

There are some farm operating expense items reported by Statistics Canada for the whole period 1961–2006 but for which there are no corresponding price indexes. These items are: artificial insemination expenses/veterinary fees, irrigation and a category that covers the cost of twine, wire and containers. These expenses are deflated by a price index constructed using all of the other purchased inputs referred to above to give an estimate of the constant dollar implicit quantity of this group of inputs.

As with gross output, there are many inputs, both within-sector use and purchased, that are not included in the AAFC production account because of the lack of detailed data needed to measure them. The value of these 'not included' inputs are estimated to be between 1% and 3% of total cost, on the basis of the dollar value of miscellaneous components of current dollar farm expenditures.

3.4 Estimated Total Factor Productivity Growth with the AAFC Production Account

An important aspect of any production account is that it can generate a wide variety of quantity and price indexes that can then be used for many kinds of economic analysis. The most aggregate of these indexes – constant dollar implicit quantities of total gross output and total input – are needed to compute the total factor productivity index.

Although estimation of total factor productivity growth is the primary reason for constructing the account, it can be argued that measurement of economic growth (i.e. growth in total gross output) and the ingredients that lead to it are equally important. There is therefore some discussion of the relative importance of total factor productivity growth and total input growth in accounting for growth in total gross output.[31]

One way to get a feel for the components that make up total gross output and total input is to compute sub-aggregates over groupings of individual commodities and inputs. Sub-aggregate data for two output groupings and eight input groupings are presented here. The input sub-aggregates are discussed here primarily in terms of the relative importance of each in explaining total input growth.

The AAFC production account total output and total input aggregate data can be 'spliced' with data from Lok (1961) and Danielson (1975), using some simple re-scaling. These spliced data provide the basis for an estimate of a 1926–2006 time series of total factor productivity, total output and total input indexes. The trends in these series and the possible implications of them are discussed below.

3.4.1 Total gross output, total input and the total factor productivity index

The same procedure is used to derive output and input quantities for groupings (sub-aggregates) and for total output and total input. Data for individual output and input prices or price indexes (with unit payments or rebates included) are used, along data for implicit or explicit quantities (depending on the commodity or input type) to generate Fisher Ideal chain-linked price indexes for groupings such as 'fruit' or 'intermediate inputs'.[32] The year 1961 is chosen as the 'index reference period', i.e. the year that is the starting point for the chain of year-over-year price changes underlying the value of the price index in any given year. Implicit constant dollar quantity data for each sub-aggregate are then estimated by deflating the total current dollar value of the group (including the value of unit payments).

Aggregation is assumed to be consistent, since the Fisher Ideal index has been shown to have this feature. This means that price indexes for higher-level aggregates can be constructed using price indexes for sub-aggregates rather than by calculating the index over all of the commodities in the aggregate. The 'crops' price index can therefore be estimated using price indexes for the sub-aggregates over the groups 'fruit', 'vegetables', 'miscellaneous horticulture', 'grains and oilseeds' and 'other crops'. Aggregation is done in three stages in this manner to ultimately generate price indexes and constant dollar implicit quantities for total gross output and total input.

To construct the total input price index, it is necessary first to derive the implicit wage to operator and unpaid family labour using estimated net farm income and the estimated hours of labour input from operators and unpaid family members. Then, this

wage is used along with data for hired labour to generate a price index for labour, which becomes one of the three price indexes (the others being for capital and intermediate input prices) that are used in the third-stage price index calculation for total input. The estimated implicit wage is presented below as part of the input data price index series.

The total factor productivity index is estimated simply by taking the ratio of the constant dollar implicit quantity of total gross output to the constant dollar implicit quantity of total input. Estimates of all three components are presented in Table 3.2 along with the price indexes for total gross output and total input. (The price indexes are not discussed here but are included for completeness.) The AAFC production account is completely summarized by the data in Table 3.2. Data for the two quantity aggregates (scaled to 1961 = 1) and the estimated total factor productivity index are illustrated in Fig. 3.4.

Figure 3.4 shows that growth in total factor productivity was quite constant over the 1961–2006 period, averaging 1.6% annually.[33] There does not seem to have been any persistent deviation from this trend rate of growth, although there have definitely been fluctuations around trend over the whole period.

Gross output increased by an average of 2.3 % annually, whereas total input grew by an average of 0.7% annually. This indicates that nearly 70% of the average growth in Canadian agricultural gross output between 1961 and 2006 was accounted for by total factor productivity growth, the remainder being the result of input growth.

3.4.2 Gross output and input sub-aggregates

The output sub-aggregates presented here are for the groupings 'crops', 'livestock' and 'other output'. Together these comprise the whole of gross output. Constant dollar implicit quantities for these sub-aggregates are given in Table 3.3 and the corresponding

Table 3.2. Implicit quantities, chain-linked Fisher Ideal price indexes for total gross output and total input, and the total factor productivity index for Canadian agriculture, 1961–2006.

Year	Implicit quantities (Million 1961 Canadian dollars)		Chain-linked Fisher Ideal price indexes (1961=1)		Total factor productivity index
	Gross output	Total input	Gross output	Total input	(1961 = 1)
1961	3,967	3,967	1.00	1.00	1.00
1962	4,677	4,060	1.02	1.17	1.15
1963	5,153	4,137	1.02	1.27	1.25
1964	4,891	4,197	1.03	1.20	1.17
1965	5,093	4,243	1.08	1.30	1.20
1966	5,711	4,336	1.14	1.50	1.32
1967	5,186	4,414	1.12	1.32	1.17
1968	5,318	4,372	1.09	1.33	1.22
1969	5,544	4,344	1.13	1.45	1.28
1970	5,600	4,329	1.14	1.48	1.29
1971	5,976	4,368	1.14	1.55	1.37
1972	5,874	4,432	1.41	1.87	1.33
1973	6,059	4,483	2.15	2.90	1.35
1974	5,598	4,437	2.24	2.83	1.26
1975	6,179	4,461	2.20	3.05	1.38
1976	6,401	4,621	2.06	2.85	1.39
1977	6,389	4,663	2.09	2.86	1.37
1978	6,762	4,801	2.47	3.49	1.41
1979	6,640	4,965	2.87	3.84	1.34
1980	6,908	4,947	3.14	4.39	1.40
1981	7,440	4,936	3.07	4.63	1.51
1982	7,663	4,991	3.00	4.60	1.54
1983	7,416	5,023	3.15	4.65	1.48
1984	7,370	5,051	3.22	4.70	1.46
1985	7,758	5,072	2.95	4.52	1.53
1986	8,247	5,074	2.92	4.74	1.63
1987	8,188	5,128	3.00	4.79	1.60
1988	7,708	5,038	3.22	4.92	1.53
1989	8,314	5,018	3.12	5.18	1.66
1990	8,904	5,017	2.94	5.23	1.77
1991	9,094	5,051	2.91	5.25	1.80
1992	9,174	5,038	3.07	5.58	1.82
1993	9,383	5,102	3.22	5.92	1.84
1994	9,801	5,242	3.37	6.31	1.87
1995	10,079	5,238	3.59	6.92	1.92
1996	10,604	5,198	3.50	7.15	2.04
1997	10,428	5,277	3.52	6.95	1.98
1998	11,104	5,400	3.29	6.77	2.06
1999	11,509	5,377	3.24	6.94	2.14
2000	11,456	5,583	3.49	7.16	2.05
2001	11,031	5,561	3.76	7.46	1.98
2002	10,982	5,434	3.80	7.68	2.02
2003	11,541	5,456	3.45	7.30	2.12
2004	12,240	5,359	3.33	7.61	2.28
2005	12,490	5,448	3.41	7.82	2.29
2006	12,086	5,544	3.64	7.93	2.18
Average annual growth[a]	2.3%	0.7%	3.2%	4.7%	1.6%

[a]This is b, where b is the estimated coefficient from the OLS equation $\ln(Z) = a + b\,t$ and where Z is the variable for which growth is being measured and t = 1961,..,2006.

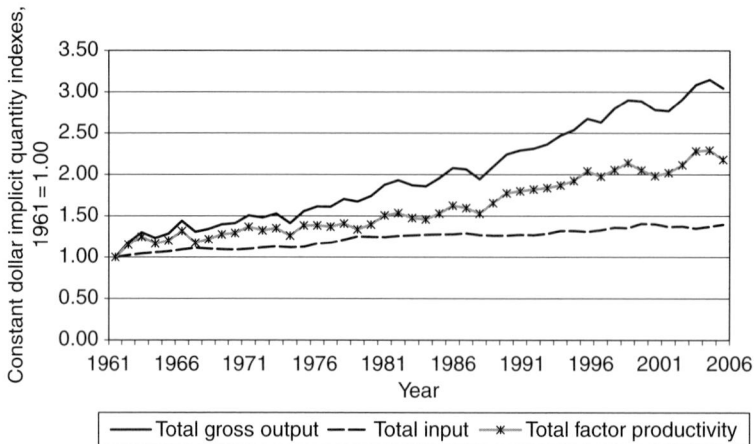

Fig. 3.4. Total gross output, total input and total factor productivity in Canadian agriculture. Source: Authors' estimates.

Fisher Ideal chain-linked price indexes are given in Table 3.4; the quantity series are also presented in Fig. 3.5. The figure shows that average annual growth in gross output of crops (2.4%) was somewhat higher than that of livestock (2.0%). Crop output was much more variable that livestock output, however, with deviations from trend often lasting for three years or more. Livestock output was less variable but deviations from trend were much more persistent, lasting 10 years or more. 'Other output' grew at a higher rate than either of the two main groups, although this category made only a small contribution to total gross output over the whole period.

The input sub-aggregate data presented in Table 3.5 and Table 3.6 are for 'building, farm machinery and vehicles services', 'livestock services', 'land services', 'inputs purchased from outside the sector' and 'within-sector use' and give constant dollar implicit quantities and price indexes respectively for the groupings. 'Raw' data for operator and unpaid family labour and hired labour are also given in these tables.

Although price indexes are generally presented here only to ensure that all of the information needed to construct total output and input is given at each level of aggregation, the implicit price to operator and unpaid family labour (see Table 3.6) needs

to be addressed. Average growth in the implicit wage between 1961 and 2006 was 5.5% annually, about 0.7 percentage points lower than growth in the hired labour wage. The implicit wage has been highly variable, which is not surprising given that it is a residual. During the 1997–2006 period, the implicit wage averaged C$9.71/h.

The quantity data from Table 3.5 are illustrated in Figs 3.6–3.8.

As Fig. 3.6 shows, the average contribution of capital services to total input growth between 1961 and 2006 was quite modest. The largest component, which is the combined capital services from buildings, farm machinery and vehicles, increased by only 0.4% annually; the large increase in services from this group of items from the early 1970s to the early 1980s was offset by a decline in services between the early 1980s and the mid-1990s. As Fig. 3.6 shows, livestock services were almost static (an increase of only 0.2% annually) as were land services, which decreased by 0.1% per year on average over this period.

As noted already with respect to the measurement of operator and unpaid family labour, this input decreased by an average 2.5% per year from 1961–2006 (Fig. 3.7). The quantity of hired labour grew, however, at an annual average of 1.6%. The overall effect of these two opposite trends was an

Table 3.3. Implicit quantities of gross output for Canadian agriculture, 1961–2006.

Year	Crops	Livestock	Other
	(Million 1961 Canadian dollars)		
1961	1610	2258	99
1962	2346	2250	99
1963	2683	2397	100
1964	2359	2454	98
1965	2575	2443	98
1966	3079	2571	97
1967	2543	2546	96
1968	2711	2531	98
1969	2900	2587	101
1970	2675	2786	103
1971	3130	2784	111
1972	2823	2917	114
1973	2980	2943	118
1974	2645	2839	113
1975	3109	2934	115
1976	3357	2902	121
1977	3494	2783	104
1978	3793	2883	89
1979	3469	3045	88
1980	3591	3179	97
1981	4127	3200	97
1982	4303	3256	99
1983	4091	3206	99
1984	3967	3256	112
1985	4333	3300	121
1986	4925	3288	133
1987	4700	3375	142
1988	3825	3574	150
1989	4491	3604	155
1990	5152	3638	178
1991	5048	3830	187
1992	4721	4070	198
1993	5280	3889	202
1994	5342	4161	221
1995	5383	4340	245
1996	5730	4489	274
1997	5345	4576	311
1998	5736	4869	312
1999	6269	4847	313
2000	5926	5014	310
2001	5046	5210	307
2002	4481	5545	310
2003	5745	5094	310
2004	6271	5327	301
2005	6512	5337	325
2006	6392	5091	324
Average annual growth[a]	2.4%	2.0%	3.2%

[a]This is b, where b is the estimated coefficient from the OLS equation $\ln(Z) = a + b\,t$ and where Z is the variable for which growth is being measured and $t = 1961,..,2006$.

Table 3.4. Chain-linked Fisher Ideal price indexes for gross output, Canadian agriculture, 1961–2006 (Base year=1961).

Year	Crops	Livestock	Other
1961	1.00	1.00	1.00
1962	0.97	1.06	1.03
1963	0.99	1.04	1.05
1964	1.02	1.02	1.08
1965	1.07	1.08	1.11
1966	1.07	1.20	1.18
1967	1.01	1.22	1.24
1968	0.92	1.26	1.29
1969	0.90	1.37	1.32
1970	0.94	1.34	1.35
1971	0.91	1.37	1.39
1972	1.27	1.56	1.45
1973	2.32	2.01	1.60
1974	2.40	2.11	1.84
1975	2.28	2.14	2.07
1976	2.00	2.13	2.24
1977	1.99	2.21	2.42
1978	2.31	2.68	2.68
1979	2.61	3.20	3.09
1980	3.09	3.24	3.45
1981	2.82	3.39	3.76
1982	2.62	3.47	3.97
1983	2.97	3.36	4.08
1984	2.97	3.53	4.21
1985	2.50	3.50	4.21
1986	2.34	3.64	4.12
1987	2.39	3.77	4.07
1988	2.82	3.74	4.14
1989	2.53	3.86	4.26
1990	2.20	3.88	4.40
1991	2.16	3.85	4.45
1992	2.39	3.91	4.57
1993	2.49	4.12	4.94
1994	2.77	4.11	5.18
1995	3.33	3.92	5.31
1996	3.11	3.97	5.48
1997	3.02	4.10	5.78
1998	2.77	3.87	5.87
1999	2.50	4.07	6.01
2000	2.59	4.53	6.29
2001	2.94	4.73	6.46
2002	3.28	4.50	6.71
2003	2.98	4.03	6.95
2004	2.65	4.14	6.93
2005	2.62	4.35	7.10
2006	2.98	4.44	7.13
Average annual growth[a]	2.7%	3.6%	4.9%

[a]This is b, where b is the estimated coefficient from the OLS equation ln(Z) = a + b t and where Z is the variable for which growth is being measured and t = 1961,..,2006.

average decline of 1.4% per year in the overall combined quantity of labour. This is reflective of the much smaller contribution to overall labour input accounted for by hired labour.

The constant dollar implicit quantity of purchased inputs exhibited the strongest growth of any component of aggregate input, increasing by 2.8% per year, on average, between 1961 and 2006 (Fig. 3.8). Within-sector use increased by about 1.9% annually. It is clear that intermediate input (combined purchased and within-sector use) was the main contributor to total input growth, because it not only increased more than either capital or labour but also its share of total cost increased substantially (from about 40% to about 55% over the period).

3.4.3 Spliced total factor productivity indexes from three studies

As noted earlier, is it possible to construct continuous annual indexes of gross output, total input and total factor productivity for Canadian agriculture for 1926–2006 by splicing together data from the Lok, Danielson and AAFC production accounts.

Although the procedures used to construct the three accounts are not exactly the same, the approaches have enough in common to justify the data splice. The main difference between the accounts is the degree to which measured output is a good estimate of gross output. Lok's data are equivalent to gross output for those commodities that have no within-sector use. Danielson's output data are equivalent to gross output in all cases but those where there is within-sector feed use.

On the input side, the three accounts also have almost the same coverage. Although measurement of operator and unpaid family labour input is done using different approaches, all of the studies show this input declining over time. Lok excludes within-sector use and Danielson excludes within-sector feed use; this will tend to show up in a higher rate of productivity growth than would have been measured had these inputs

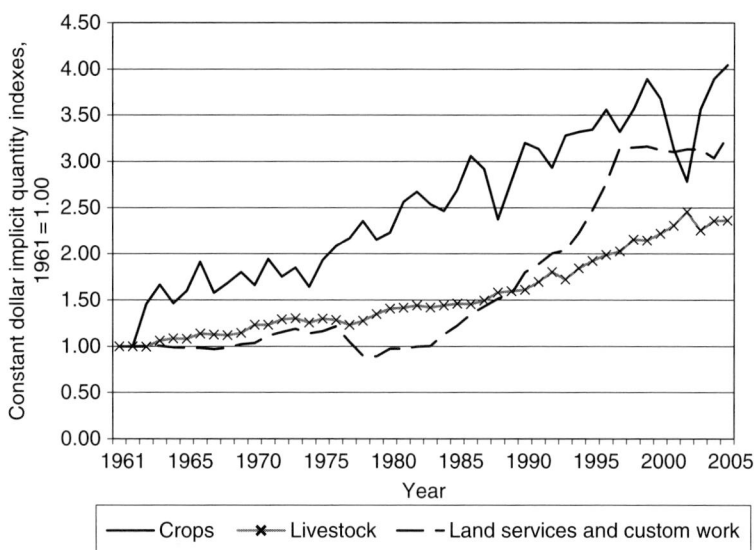

Fig. 3.5. Gross output of crops, livestock and other services in Canadian agriculture. Source: Authors' estimates.

Table 3.5. Implicit input quantities, Canadian agriculture, 1961–2006 (million 1961 Canadian dollars except where noted).

	Capital services			Labour		Intermediate inputs	
Year	Building, farm machinery and vehicle services	Livestock services	Land services	Operator and unpaid family labour (million hours)	Hired labour (million weeks)	Inputs purchased from outside sector	With-sector use
1961	642	46	179	1644	4.809	797	869
1962	644	46	180	1603	4.671	818	987
1963	661	48	180	1561	4.325	877	1062
1964	689	49	181	1518	4.111	943	1094
1965	728	49	181	1475	4.311	1005	1090
1966	774	48	180	1430	3.977	1069	1175
1967	811	47	179	1382	3.972	1140	1225
1968	825	46	178	1334	3.925	1139	1215
1969	828	47	177	1285	3.761	1146	1238
1970	807	49	176	1236	3.832	1158	1296
1971	807	50	176	1186	3.842	1207	1371
1972	838	52	175	1161	3.627	1282	1416
1973	906	53	175	1136	3.417	1295	1496
1974	982	53	174	1110	3.421	1260	1461
1975	1076	55	174	1083	3.687	1267	1442
1976	1173	55	174	1057	4.633	1316	1493
1977	1242	53	174	1028	4.539	1371	1473
1978	1307	51	173	1006	4.073	1531	1513
1979	1388	50	173	977	4.255	1634	1608
1980	1421	50	173	932	4.536	1623	1613
1981	1448	50	172	929	4.125	1622	1604
1982	1428	50	172	909	4.305	1699	1630

Continued

Table 3.5. Continued.

Year	Capital services			Labour		Intermediate inputs	
	Building, farm machinery and vehicle services	Livestock services	Land services	Operator and unpaid family labour (million hours)	Hired labour (million weeks)	Inputs purchased from outside sector	With-sector use
1983	1395	49	172	891	4.937	1736	1647
1984	1367	48	171	868	5.226	1795	1652
1985	1311	46	171	841	5.716	1854	1658
1986	1254	45	170	831	5.900	1855	1727
1987	1199	44	170	808	6.030	1883	1890
1988	1146	44	170	784	5.520	1884	1913
1989	1097	45	170	767	5.553	1939	1901
1990	1046	46	169	757	5.327	1974	1967
1991	1001	46	169	742	6.029	1969	2034
1992	959	48	169	736	5.945	2063	1978
1993	950	48	169	727	6.320	2114	2024
1994	933	48	170	698	6.371	2268	2211
1995	918	51	170	679	6.360	2313	2234
1996	903	52	171	667	6.524	2294	2215
1997	917	52	171	657	6.134	2466	2201
1998	933	52	171	641	6.932	2626	2159
1999	924	51	170	617	6.945	2596	2224
2000	919	51	170	604	7.101	2892	2307
2001	904	53	170	584	6.670	3007	2275
2002	901	54	170	577	6.663	2870	2210
2003	899	54	170	566	6.633	2942	2195
2004	898	56	170	555	6.575	2772	2256
2005	896	57	170	541	7.111	2810	2345
2006	886	57	171	533	7.650	2910	2372
Average annual growth[a]	0.4%	0.2%	−0.1%	−2.4%	1.6%	2.8%	1.9%

[a]This is b, where b is the estimated coefficient from the OLS equation $\ln(Z) = a + b\,t$ and where Z is the variable for which growth is being measured and $t = 1961,..,2006$.

been included. The impact of these exclusions on their estimates, however, depends on the level of growth in this input over the periods covered by their accounts.

The indexes in all three studies are chain linked. On the basis of Lok's description of his chain-linked indexes (Lok, 1961, pp. 42–43), it seems that these were chain-linked Paasche indexes. Both Danielson's aggregates and those constructed with the AAFC production account are Fisher Ideal chain-linked indexes.

Output, input and total factor productivity index data from Lok's production account are taken from Tables 38, Table 27 and Table 6 of Lok (1961). These data are presented here in Table 3.7 exactly as shown in Lok's tables (i.e. with 1926 = 100). Danielson's data for output, input and total factor productivity index data are all from his Table II (Danielson, 1975, p. 185) and are presented here in Table 3.7 exactly as shown in Table II (i.e. with 1946 = 1).

Splicing is done with the most recent data only. Therefore, data from Lok's account for 1946–1957 are not included in the spliced data (although they are entered in Table 3.7 as part of the original series) because Danielson's

Table 3.6. Input prices, Canadian agriculture, 1961–2006 (chain-linked Fisher Ideal indexes with 1961 = 1.00 except where noted).

Year	Capital services			Labour		Intermediate inputs	
	Building, farm machinery and vehicle services	Livestock services	Land services	Operator and unpaid family labour ($/h)	Hired labour ($/week)	Inputs purchased from outside the sector	With-sector use
1961	1.00	1.00	1.00	0.75	42.33	1.00	1.00
1962	1.03	1.08	1.04	1.10	44.25	1.03	1.07
1963	1.05	1.02	1.12	1.35	46.18	1.04	1.04
1964	1.07	0.99	1.24	1.09	48.11	1.04	1.07
1965	1.09	1.03	1.40	1.31	50.03	1.03	1.13
1966	1.12	1.12	1.57	1.77	53.52	1.07	1.20
1967	1.16	1.21	1.79	1.14	57.01	1.10	1.17
1968	1.20	1.21	1.99	1.12	60.50	1.12	1.14
1969	1.26	1.32	2.02	1.51	63.98	1.10	1.11
1970	1.29	1.36	2.04	1.58	67.47	1.10	1.11
1971	1.31	1.32	2.07	1.73	82.77	1.12	1.15
1972	1.38	1.60	2.27	2.62	98.07	1.14	1.28
1973	1.45	1.76	2.77	5.45	113.37	1.45	1.79
1974	1.63	2.27	3.66	3.52	128.67	1.86	2.39
1975	1.82	1.70	4.67	3.69	143.97	2.08	2.49
1976	1.95	1.79	5.60	2.55	154.40	2.13	2.40
1977	2.10	1.89	6.50	2.35	164.82	2.20	2.27
1978	2.35	2.36	7.81	4.40	175.24	2.30	2.33
1979	2.65	3.80	9.80	4.18	185.66	2.59	2.74
1980	2.95	4.63	12.53	4.60	196.08	3.02	3.14
1981	3.21	4.48	14.20	3.78	212.64	3.54	3.36
1982	3.38	4.28	14.35	3.68	229.19	3.58	3.09
1983	3.56	4.28	13.81	3.47	245.74	3.65	3.14
1984	3.67	4.37	13.24	2.65	262.29	3.81	3.48
1985	3.77	4.36	12.33	1.83	278.84	3.76	3.36
1986	3.93	4.40	11.49	4.45	299.66	3.57	2.84
1987	3.95	5.06	10.90	5.30	320.47	3.44	2.67
1988	4.02	5.03	11.01	4.50	341.28	3.65	3.03
1989	4.14	5.25	12.20	4.41	362.10	3.74	3.41
1990	4.22	5.32	13.02	5.43	382.91	3.65	3.06
1991	4.20	5.20	13.06	5.99	391.09	3.62	2.91
1992	4.35	5.00	12.77	7.87	399.26	3.60	3.02
1993	4.58	5.74	12.93	9.10	407.44	3.70	3.15
1994	4.87	5.98	13.65	10.35	415.61	3.91	3.24
1995	5.07	5.50	14.87	12.82	423.79	4.23	3.40
1996	5.23	4.93	15.96	11.18	432.67	4.72	3.76
1997	5.42	5.54	17.82	9.03	441.54	4.61	3.81
1998	5.60	5.78	19.07	8.31	450.42	4.29	3.74
1999	5.78	6.21	20.04	9.45	459.30	4.29	3.64
2000	5.77	7.01	20.97	11.59	468.17	4.15	3.66
2001	5.93	7.21	21.83	12.04	483.12	4.35	3.82
2002	6.08	6.67	23.37	10.56	498.06	4.71	4.06
2003	5.80	5.16	25.00	7.36	513.01	4.77	3.86
2004	5.87	4.56	26.68	8.70	527.95	5.07	3.71

Continued

Table 3.6. Continued.

Year	Capital services			Labour		Intermediate inputs	
	Building, farm machinery and vehicle services	Livestock services	Land services	Operator and unpaid family labour ($/h)	Hired labour ($/week)	Inputs purchased from outside the sector	With-sector use
2005	5.79	4.90	28.53	10.00	542.89	5.04	3.77
2006	5.69	5.71	30.62	10.09	521.02	4.99	3.99
Average annual growth[a]	4.8%	4.7%	7.5%	5.5%	6.2%	4.1%	3.3%

[a]This is b, where b is the estimated coefficient from the OLS equation $\ln(Z) = a + b\,t$ and where Z is the variable for which growth is being measured and t = 1961,..,2006.

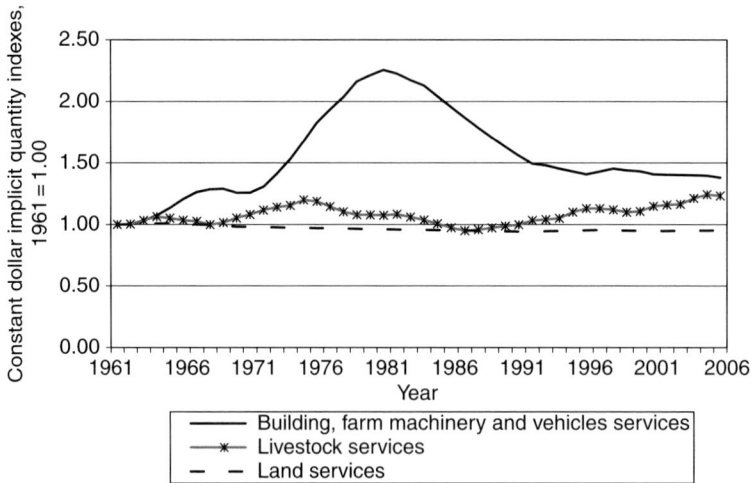

Fig. 3.6. Capital services inputs in Canadian agriculture. Source: Authors' estimates.

data are used for those years. Similarly data for 1961–1970 from Danielson's account are not included in the spliced data (although they are also entered in Table 3.7 as part of the original series) because data from the AAFC production account are used for those years. The Lok data are first scaled to 1946=1 and then spliced to the Danielson data, creating a series of Lok/Danielson data for 1926–1961 that is scaled to 1946=1. Then this series is scaled to 1961=1 and spliced to the AAFC production account series, with all data in the spliced 1926–2006 series scaled to 1961=1. Finally, the full spliced series is scaled to 1926=1; these data are given in Table 3.7.

The average annual growth rates for output, input and total factor productivity for the whole 1926–2006 period are 2.0%, 0.8% and 1.2%, respectively. These series are illustrated in Fig. 3.9. As Fig. 3.9 shows, all of the three indexes have grown quite consistently, with deviations around trend (a trend line is omitted) but no apparent systematic increase or decrease over time. Indeed, total factor productivity growth seems to have been remarkably constant over the whole 81 years between 1926 and 2006.

When data from all three production accounts are examined together, there seems to be no evidence of a slowdown (or for that matter

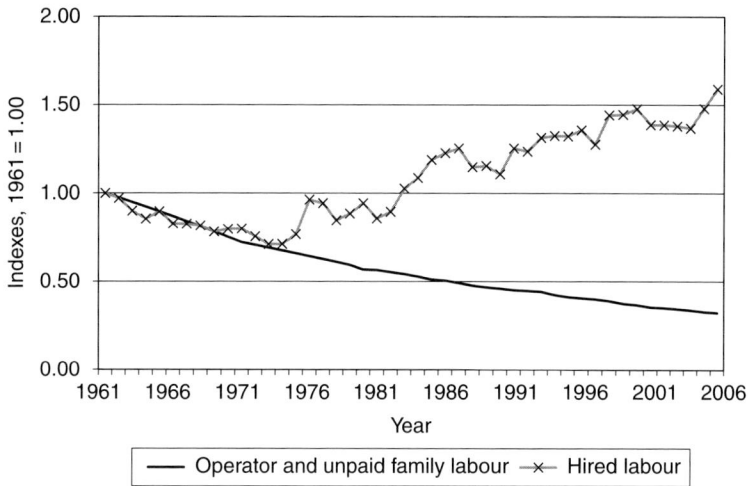

Fig. 3.7. Labour inputs in Canadian agriculture. Source: Authors' estimates.

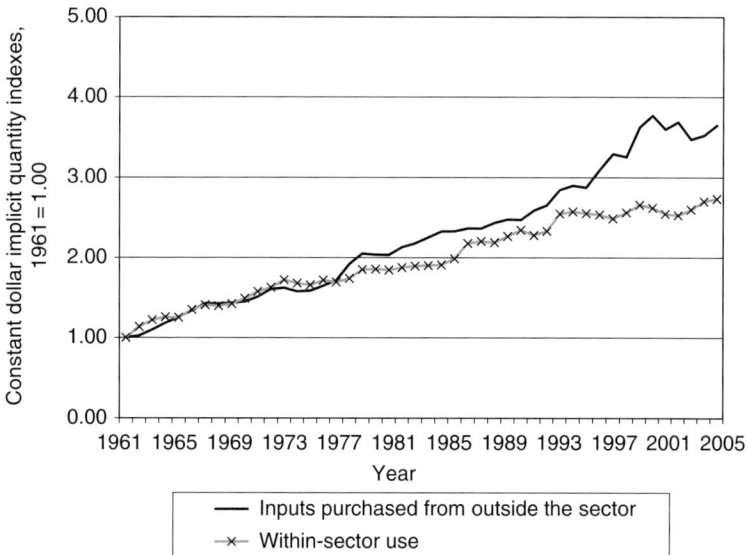

Fig. 3.8. Intermediate inputs in Canadian agriculture. Constant dollar implicit quantity indexes, 1961 = 1.00. Source: Authors' estimates.

an acceleration) of total factor productivity growth in Canadian agriculture. It is not difficult to find shorter time series where total factor productivity growth is persistently above or below trend. Examples include the Great Depression, the period between the Second World War and the Korean War, the period running from the oil crisis

(and commodity boom) to the late 1980s as well as during events specific to Canada such as periodic droughts, etc. The spliced data demonstrate the importance of looking instead at all, rather than some, of the data available for assessing long-run trends in growth. This reinforces the view expressed in the introduction to this chapter.

Table 3.7. Total output, total input and total factor productivity indexes from three studies spliced together, Canadian agriculture, 1926–2006.

Year	Total output index from				Total input index from				Total factor productivity (TFP) index from			
	Lok 1926=100	Danielson 1946=1	AAFC production account 1961=1	Spliced total output index 1926=1	Lok 1926=100	Danielson 1946=1	AAFC production account 1961=1	Spliced total input index 1926=1	Lok 1926=100	Danielson 1946=1	AAFC production account 1961=1	Spliced TFP index 1926=1
1926	100			1.00	100			1.00	100			1.00
1927	108			1.08	103			1.03	105			1.05
1928	113			1.13	105			1.05	108			1.08
1929	91			0.91	107			1.07	85			0.85
1930	96			0.96	106			1.06	90			0.90
1931	92			0.92	105			1.05	88			0.88
1932	99			0.99	103			1.03	96			0.96
1933	90			0.90	103			1.03	87			0.87
1934	92			0.92	104			1.04	89			0.89
1935	93			0.93	105			1.05	89			0.89
1936	88			0.88	104			1.04	85			0.85
1937	86			0.86	105			1.05	82			0.82
1938	103			1.03	106			1.06	97			0.97
1939	130			1.30	109			1.09	119			1.19
1940	127			1.27	107			1.07	119			1.19
1941	116			1.16	103			1.03	113			1.13
1942	158			1.58	103			1.03	154			1.54
1943	126			1.26	107			1.07	118			1.18
1944	147			1.47	108			1.08	136			1.36
1945	122			1.22	110			1.10	110			1.10
1946	131	1.00		1.31	121	1.00		1.21	108	1.00		1.08
1947	135	0.94		1.24	119	0.97		1.17	114	0.98		1.06
1948	144	1.02		1.33	116	0.95		1.14	124	1.08		1.17
1949	139	0.90		1.18	113	0.98		1.18	123	0.92		1.00
1950	133	1.02		1.33	107	0.92		1.11	124	1.11		1.20
1951	163	1.12		1.47	111	0.93		1.12	147	1.21		1.31
1952	179	1.29		1.69	109	0.93		1.12	164	1.39		1.50
1953	175	1.33		1.74	109	0.99		1.19	161	1.35		1.46

Year												
1954	149	1.08		1.41	111	1.10		1.33	135	0.98		1.07
1955	168	1.27		1.66	110	1.02		1.23	153	1.25		1.35
1956	185	1.42		1.86	109	1.06		1.28	170	1.34		1.45
1957	167	1.23		1.62	105	1.14		1.38	159	1.08		1.17
1958		1.24		1.63		1.07		1.29		1.16		1.26
1959		1.24		1.63		1.04		1.25		1.20		1.30
1960		1.31		1.71		1.02		1.23		1.29		1.40
1961		1.09	1.00	1.43		1.07	1.00	1.29		1.02	1.00	1.11
1962		1.32	1.18	1.69		0.94	1.02	1.32		1.40	1.15	1.28
1963		1.47	1.30	1.86		0.96	1.04	1.35		1.54	1.25	1.38
1964		1.37	1.23	1.76		1.00	1.06	1.37		1.37	1.17	1.29
1965		1.46	1.28	1.84		0.99	1.07	1.38		1.48	1.20	1.33
1966		1.68	1.44	2.06		0.99	1.09	1.41		1.69	1.32	1.46
1967		1.54	1.31	1.87		1.06	1.11	1.44		1.45	1.17	1.30
1968		1.70	1.34	1.92		1.06	1.10	1.42		1.61	1.22	1.35
1969		1.85	1.40	2.00		1.11	1.10	1.42		1.66	1.28	1.41
1970		1.86	1.41	2.02		1.15	1.09	1.41		1.61	1.29	1.43
1971			1.51	2.15			1.10	1.42			1.37	1.51
1972			1.48	2.12			1.12	1.44			1.33	1.47
1973			1.53	2.18			1.13	1.46			1.35	1.50
1974			1.41	2.02			1.12	1.45			1.26	1.40
1975			1.56	2.23			1.12	1.45			1.38	1.53
1976			1.61	2.31			1.17	1.51			1.39	1.53
1977			1.61	2.30			1.18	1.52			1.37	1.52
1978			1.70	2.44			1.21	1.56			1.41	1.56
1979			1.67	2.39			1.25	1.62			1.34	1.48
1980			1.74	2.49			1.25	1.61			1.40	1.55
1981			1.88	2.68			1.24	1.61			1.51	1.67
1982			1.93	2.76			1.26	1.63			1.54	1.70
1983			1.87	2.67			1.27	1.64			1.48	1.63
1984			1.86	2.66			1.27	1.65			1.46	1.62
1985			1.96	2.80			1.28	1.65			1.53	1.69
1986			2.08	2.97			1.28	1.65			1.63	1.80
1987			2.06	2.95			1.29	1.67			1.60	1.77
1988			1.94	2.78			1.27	1.64			1.53	1.69
1989			2.10	3.00			1.27	1.64			1.66	1.83

Continued

Table 3.7. Continued.

Year	Total output index from				Total input index from				Total factor productivity (TFP) index from			
	Lok 1926=100	Danielson 1946=1	AAFC production account 1961=1	Spliced total output index 1926=1	Lok 1926=100	Danielson 1946=1	AAFC production account 1961=1	Spliced total input index 1926=1	Lok 1926=100	Danielson 1946=1	AAFC production account 1961=1	Spliced TFP index 1926=1
1990			2.24	3.21			1.26	1.63			1.77	1.96
1991			2.29	3.28			1.27	1.65			1.80	1.99
1992			2.31	3.31			1.27	1.64			1.82	2.02
1993			2.37	3.38			1.29	1.66			1.84	2.04
1994			2.47	3.53			1.32	1.71			1.87	2.07
1995			2.54	3.63			1.32	1.71			1.92	2.13
1996			2.67	3.82			1.31	1.69			2.04	2.26
1997			2.63	3.76			1.33	1.72			1.98	2.19
1998			2.80	4.00			1.36	1.76			2.06	2.28
1999			2.90	4.15			1.36	1.75			2.14	2.37
2000			2.89	4.13			1.41	1.82			2.05	2.27
2001			2.78	3.97			1.40	1.81			1.98	2.20
2002			2.77	3.96			1.37	1.77			2.02	2.24
2003			2.91	4.16			1.38	1.78			2.12	2.34
2004			3.09	4.41			1.35	1.75			2.28	2.53
2005			3.15	4.50			1.37	1.78			2.29	2.54
2006			3.05	4.35			1.40	1.81			2.18	2.41
Average annual growth[a]	1.9%	1.9%	2.3%	2.0%	0.4%	0.9%	0.7%	0.8%	1.5%	1.0%	1.6%	1.2%

[a]This is b, where b is the estimated coefficient from the OLS equation $\ln(Z) = a + b\,t$ and where Z is the variable for which growth is being measured and $t = 1961,\ldots,2006$.

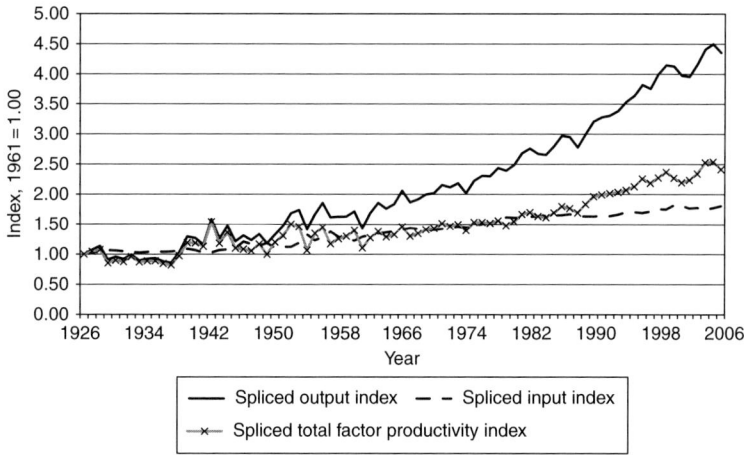

Fig. 3.9. Gross output, total input and total factor productivity in Canadian agriculture, 1926–2006. Indexes, 1961 = 1.00. The indexes are formed by splicing results from three studies: Lok's (1961) estimates for 1926 to 1957, Danielson's (1975) estimates for 1946 to 1970 and the present study's estimates from 1961 to 2006. See Table 3.7.

3.5 Conclusion

The better part of this chapter has been devoted to a description of the methodology and data used to construct the AAFC production account for Canadian agriculture. This reflects the importance of measurement in the development of the account; it is generally the case that few of the data series can be taken directly from electronic databases or from print publications without some sort of additional conversion, weighting, imputation, extrapolation, etc. Often, a variety of measurement issues must be resolved in order to quantify or value outputs and inputs. Some of these issues include:

- The derivation of intercensal data for key inputs drawn from the Census of Agriculture (e.g. hired labour, land) and validations needed to determine the reliability of derived data through comparison with other, sometimes partial, data series.
- Imputation and extrapolation of data where only partial information exists and use of proportions or other function to relate similar data series that can be used to obtain estimates that span

the whole period (e.g. output of custom harvesting services and output of land services).

- Discontinuities in data that arise because of termination of surveys and/or confidentiality restrictions (e.g. the termination of average farm price data in 1984 means that more diverse sources of price data must be linked together).
- Quantifying gross output using the supply–demand identity where production data are not available (e.g. for several types of livestock).
- Estimation using indirect information (e.g. estimating operator and unpaid family labour input using census data on off-farm work).

Each of these measurement issues can be resolved using a combination of theory and pragmatism. These challenges sometimes point to improvements needed in surveys and in census questions. An example of this would be a more explicit question on the Census of Agriculture regarding on farm work by operators and unpaid family members.

The results obtained with the AAFC production account data show that total factor productivity growth in Canadian agriculture was positive and quite consistent between 1961 and 2006. The average annual rate of growth was 1.6%. Total input grew an average of 0.7% annually during 1961–2006 and the average annual increase in total gross output over the same period was 2.3%. These results indicate that growth in gross Canadian agricultural output between 1961 and 2006 was achieved by combination of total factor productivity and total input growth. Of these two components, total factor productivity growth made the largest contribution.

When output, input and total factor productivity indexes from Lok (1961), Danielson (1975) and the AAFC production account are spliced together, the result is a 1926–2006 time series of data that gives a picture of total factor productivity growth over the long run. The average annual growth rate in total factor productivity over the 81 years is 1.2%. Total output and input grew at an average annual rate of 2.0% and 0.8% over the same period. These series, which are illustrated in Fig. 3.9, have grown quite consistently, with deviations around trend but no apparent systematic increase or decrease over time. The spliced data suggest that productivity growth must be viewed as a long-run phenomenon and that year to year or even decade to decade changes in growth rates can be misleading if they are viewed in isolation and interpreted as fundamental shifts in long-run trends.

Notes

[1] There have been several related studies that measure productivity growth with frameworks that are: (i) for specific regions of Canada (e.g. the prairies), such as Stewart, Veeman and Unterschultz (2009); (ii) for specific components of agriculture within specific regions, such as Rahumal and Veeman (1988); or (iii) for a limited set of outputs without accounting for the rest of output from the sector, such as Fantino and Veeman (1997). The results from these studies are not discussed here because each study incorporates only part of Canadian agriculture and/or only part of Canada. For much the same reason, studies that have looked at productivity growth based on value-added measures, such as Mackenzie (1962) are also excluded from the discussion because they net out all intermediate input (not just within-sector use) on the input side of the account and include only marketings on the output side of the account.

[2] In particular, 'care [was] … taken to net out intra-sectoral transfers of commodities and services' (Danielson, 1975, p. 23).

[3] CANSIM is an electronic database of socioeconomic statistics maintained by Statistics Canada.

[4] The data Furniss used were probably taken from Dominion Bureau of Statistics (1949) and subsequent issues. This publication gave index numbers of production 'after adjustments'. The adjustments effectively removed from grain production figures the amount subsequently used as feed in livestock production. In this regard, these adjustments would appear to have actually measured production as something less than cash receipts, as sales of grain from crop producers to livestock producers were netted out. The commodity coverage was very broad, however, so although not a measure of gross output, the index did provide a very complete accounting of most of Canada's agricultural output. The series was ended in 1985.

[5] For simplicity, imports are not included in the discussion around Eqn 3.3. For a complete supply–demand balancesheet, these should be included as a separate item. To incorporate imports, just interpret marketings as the sum of all market demand (by processors, consumers, exporters and farm households) less imports, i.e. marketings are equal to total market demand less imports.

[6] For most crops, crop year t begins on August 1 of the same calendar year and ends on July 31 in the following year ($t + 1$). Harvest takes place after 1 August and well before 31 December, so production in crop year t is also production in calendar year t.

[7] Note that with this method, it is possible to have within-sector feed for any crop year that exceeds production in that year. The difference will be made up by stocks held over from the previous year or from imports. Calculations with data in the disposition tables show corn is the only commodity where within-sector feed use exceeded production in more than one or two years between 1961 and 2006. This occurred primarily in the 1960s and early 1970s; since then within-sector use has typically been lower than production.

[8] Data from the farm income accounts are used to estimate home consumption. See Statistics Canada (Catalogue No. 21-525-XIE, 'Understanding Measurements of Farm Income', 2000) for a description of those accounts. Coverage includes cattle, hogs, sheep, dairy products, eggs and poultry meat.

[9] Carcass weight is the average warm (or cold) dressed carcass weight, which reports the average weight of the usable portion of a slaughtered animal (the implicit unit is carcass weight per head) while carcass yield is the proportion of the animal which becomes a usable carcass.

[10] This method was incorporated through consultations with AAFC market specialists.

[11] The yield figure for both sheep and lamb is approximately 50%.

[12] Quantities of milk fed to livestock are estimated from 1978 forward because of termination of the data series by Statistics Canada.

[13] The cost of land services, whether explicit (as rent paid) or implicit (as opportunity cost) is estimated using a cost of capital formula (see Capital Services section later in the chapter).

[14] The ratio of receipts to expenses increased to a peak of nearly 4/1 in 1997. These data suggest that work done outside of the sector using agricultural equipment and other farm resources has been an important part of custom work. The upward trend in revenue relative to expenditures between 1981 and 1997 suggests that the approach used here to impute revenue with expense data may give a slight overestimate if that trend started, for example, in the 1970s.

[15] Hog prices are available for various carcass grades over the time period, but in the production account, the price series used is for 'index 100' hogs.

[16] See Schmitz *et al.* (2002, pp. 197–207) for some further information about the historical context and linkages between the policies and programmes discussed here.

[17] See OECD (2001), which has a substantial bibliography and is a manual devoted to measuring capital alone.

[18] These are referred to as asset groupings because they comprise many different types of assets of many different vintages.

[19] Although disaggregated data are available from Statistics Canada for this period, they are very noisy. Within Statistics Canada, investment data are reconciled for the whole economy and then adjustments and allocations are made to sectoral investment data. The more disaggregated the sectoral data, the more likely it is that this procedure will bring about large year-to-year variations, which is the case for the investment data for Canadian agriculture at the disaggregated asset level.

[20] Cahill (2011) provides a complete treatment.

[21] The model underlying these derivations is similar to the one outlined in Ball *et al.* (2008, pp. 1258–1259) but is based on the assumption of geometric, rather than hyperbolic decay; it is also based on the premise that farmers are constrained by the capital that is made available to them (and which they rent from) the farm household.

[22] Note that there have been three other periods in the past where a large accumulation in farm machinery and vehicles was followed by a large diminution: (i) accumulation 1927–1929 then diminution 1930–1933 (the Great Depression); (ii) accumulation 1945–1953 followed by diminution from 1954–1961; and (iii) accumulation from 1961–1968 then diminution 1969–1971.

[23] These prices, even if converted to a live weight basis, are not equivalent to prices for market animals, which have different characteristics in both weight and size. The 'value per head' price reflects the fact that breeding livestock are more valuable than market animals because of their role in generating future production. It is important, therefore, to distinguish between the method used to estimate the value of livestock gross output and that used to estimate the value of livestock services.

[24] In this approach there is no decay element in measuring livestock services. An animal is assumed to provide the same level of services until it is replaced.

[25] Of the studies cited, only Danielson explicitly made an adjustment to capture the effect that off-farm work may have had on the operator labour input to agriculture. Danielson notes that the estimated farm family labour force was 'deflated by 20 percent to correct for off-farm employment by family members' (Danielson, 1975, p.10).

[26] Here, 313 is the maximum number of days worked in a year, irrespective of whether these were days working only on the farm or whether they are a total of combined off-farm and on-farm work. Alston *et al.* (2010, p. 41) use a similar set of assumptions to estimate hours of work for US operators with agriculture census data. Their estimate is 2800 h/year for a full-time operator. This is somewhat lower than the equivalent estimate with the approach used here: hours per full-time operator (no off-farm employment) is $10 \times 313 = 3130$.

[27] If the farm is not a partnership, operators other than the primary operator may be paid a wage/salary (as a farm expense) or they may receive some proportion of net farm income (i.e. an implicit wage) as unpaid family labour. If, however, the farm is a sole proprietorship or a partnership, neither the primary operator's nor any partner's (other operator's) labour will be included in farm expenses, since wage expenses of this type cannot be deducted from revenue when arriving at taxable income for the farm operation. If the farm is a corporation, wages and salaries paid to other operators could be included as an expense, or one or more of the additional

operators could be residual claimants, receiving what is equivalent to a 'dividend'. The relative importance of operators other than the primary operator, both in terms of the overall labour input (i.e. the allocation between hired versus unpaid labour) and in terms of cost, remains ambiguous, even though the number of these additional operators is known. The accuracy of the approach used here therefore depends both on the number and size of farms that are partnerships and on the labour input from operators other than the primary operators. According to information in Statistics Canada (2007, Table 2.3), 12% of Canadian farms were partnerships in 1986 and this proportion increased to 28% of farms in 1991. It is unclear whether this increase is due to the change in questions posed in the Census questionnaire or because there was an actual shift towards partnerships as an operating arrangement. It is also unclear what the implications are for the overall allocation of labour input across unpaid and hired labour categories from this apparent shift in operating arrangements.

[28] The assumption of 5 days per week seems reasonable. The number of hours per day was chosen to match the ratio of unpaid family workers to operators in 1961.

[29] Note that, from 1986 on, the Census of Agriculture has collected data on the number of hired workers that are family members. Between 1986 and 2006, an average of 43% of all wage expenditure was paid to family workers. So unpaid family work underestimates the total labour contribution to agriculture made by family members over this period.

[30] Linear interpolation ensures that the wage data match census expenditure data when multiplied by weeks of work from the census. The wage data referred to above (i.e. the series up to 1997) do not match census expenditures when multipled by weeks worked. For this reason, interpolated data are preferred; another is that it is possible to derive estimates of wages for hired labour for 2001 and 2006 whereas the 'observed' wage data do not exist for these years.

[31] At first glance, this seems to be a bit circular because total factor productivity growth is equal to the difference between gross output and input growth. This description skips an important step, however, namely the analysis needed to explain input and total factor productivity growth. Input growth can be explained in terms of changes in derived demand, which in turn is determined by changes in input prices, fixed inputs and output prices (if derived from a profit function). Total productivity growth has typically been attributed to investments in public R&D (or R&D spillovers), investment in public infrastructure and technological change. When viewed in this light, the relationship is not at all circular because the explanators of gross output growth are all exogenous.

[32] The 'INDEX' command in the econometrics computer program *SHAZAM* is used to generate all of the price indexes reported here. The Fisher Ideal index is well documented: Ball *et al.* (1997, p. 1046) provide an explanation of the index in the context of the production account for US agriculture. The advantages of chain linked (as opposed to fixed base) indexes are outlined in Schreyer (2004); the term 'index reference period' is used by him to define the starting point for a chain linked index (Schreyer, 2004, p. 3). The reference period (year) can be chosen arbitrarily from within the range of years covered by the data.

[33] All growth rates from the AAFC production account are based on the equation $ln(Z) = a + b\,t$, $t = 1961,...,2006$ where Z is the variable for which growth is being measured and the coefficients a and b are estimated using OLS. Annual growth is then estimated as b_{est}, where b_{est} is the OLS estimate of b.

References

Alston, J., Anderson, M., James, J. and Pardey, P. (2010) Persistence Pays: U.S. Agricultural Productivity Growth and the Benefits from Public R&D Spending. Springer, New York, NY.

Andersen, M., Alston, J. and Pardey, P. (2009) Capital service flows: Concepts and comparisons of alternative measures in U.S. agriculture. Staff Paper P09-8, Department of Applied Economics, University of Minnesota, St. Paul, MN.

Baldwin, J. and Harchaoui, T. (eds) (2002) *Productivity Growth in Canada – 2002*. Statistics Canada Catalogue no. 15-204-XPE, Ottawa, Canada.

Ball, V.E., Bureau, J.-C., Nehring, R. and Somwaru, A. (1997) Agricultural productivity revisited. *American Journal of Agricultural Economics* 79, 1045–1063.

Ball, V.E., Lindamood, W., Nehring, R. and Mesonada, C. (2008) Capital as a factor of production in OECD agriculture: Measurement and data. *Applied Economics* 40, 1253–1277.

Bollman, R. (1982) Part-time farming in Canada: Issues and non-issues. *GeoJournal* 6, 313–322.

Bowlby, G. (2005) Use of the Canadian labour force survey for collecting additional labour-related information. Paper presented at the International Seminar on the Use of National Labour Force Surveys. International Labour Office, Geneva, October 24–26, 2005.

Brinkman, G. and Prentice, B. (1983) Multifactor productivity in Canadian agriculture: An analysis of methodology and performance. Paper prepared under contract for Agriculture Canada, Regional Development Branch, Development Policy Directorate.

Cahill, S. (2011) Estimating capital stock and capital services input for the Canadian agricultural production account, 1961–2006. Mimeo, Research and Analysis Directorate, Strategic Policy Branch, Agriculture and Agri-Food Canada.

Danielson, R. (1975) *Three Studies in Canadian Agriculture*. Unpublished M.A. Thesis, Department of Economics, University of British Columbia, Canada.

Dominion Bureau of Statistics (1949) Index of Farm Production: Index Numbers of the Physical Volume of Agricultural Production, Canada, 1935–48. Catalogue 21-203, August.

Fantino, A. and Veeman, T. (1997) The choice of index numbers in measuring agricultural productivity: a Canadian empirical case study. In: Rose, R., Tanner, C. and Bellamy, M.A. (eds) *Issues in Agricultural Competitiveness: Markets and Policies*. Dartmouth Publishing, Aldershot, UK, pp. 222–231.

Furniss, I. (1964) Productivity of Canadian agriculture, 1935–1960: A quarter of a century of change. *Canadian Journal of Agricultural Economics* 12, 41–53.

Furniss, I. (1970) Agricultural productivity in Canada: Two decades of gains. *Canadian Farm Economics* 5, 16–27.

Griliches, Z. and Jorgenson, D. (1962) Sources of measured productivity change: Capital input. *American Economic Review* 56, 50–61.

Gu, W., Kaci, M., Maynard, J.-P. and Sillamaa, M.-A. (2002) The changing composition of the Canadian workforce and its impact on productivity growth. In: Baldwin, J. and Harchaoui, T. (eds) *Productivity Growth in Canada – 2002*. Statistics Canada Catalogue no. 15-204-XPE, Ottawa, pp. 67–93.

Harper, M., Berndt, E. and Wood, D. (1989) Rates of return and capital aggregation using alternative rental prices. In: Jorgenson, D. and Landau, R. (eds) *Technology and Capital Formation*, MIT Press, Cambridge, MA.

Islam, T. (1982) *Input Substitution and Productivity Change in Canadian Agriculture*. Unpublished PhD. Thesis, Department of Economics, University of Alberta, Alberta, Canada.

Lok, S. (1961) An enquiry into the relationships between changes in overall productivity and real net return per farm, and between changes in total output and real gross return, Canadian agriculture, 1926–1957. Technical Publication No. 61/13. Department of Agriculture, Ottawa, Canada.

Mackenzie, W. (1962) The impact of technological change on the efficiency of production in Canadian agriculture. Canadian *Journal of Agricultural Economics* 10, 41–53.

OECD (2001) *Measuring Capital, OECD Manual, Measurement of Capital Stocks, Consumption of Fixed Capital and Capital Services*. Organisation for Economic Co-operation and Development, Paris, France.

Patry, A. (2007) Economic depreciation and retirement of Canadian assets: A comprehensive empirical study. Statistics Canada Catalogue no 15-549-XIE, Ottawa, Canada.

Rahumal, A. and Veeman, T. (1988) Productivity growth in the Prairie grain sector and its major soil zones, 1960s to 1980s. Canadian Journal of Agricultural Economics 36, 857–870.

Schmitz, A., Furtan, H. and Baylis, K. (2002) *Agricultural Policy, Agribusiness and Rent-Seeking Behavior*. University of Toronto Press, Toronto, Canada.

Schreyer, P. (2004) Chain index number formulae in the national accounts. Paper presented at the 8th OECD-NBS Workshop on National Accounts, 6–10 December, Organisation for Economic Co-operation and Development, Paris, France.

Statistics Canada (2007) *Selected Historical Data from the Census of Agriculture*. Statistics Canada Catalogue no 95-632-XWE, Ottawa, Canada.

Stewart, B., Veeman, T. and Unterschultz, J. (2009) Crops and livestock productivity growth in the Prairies: The impacts of technical change and scale. *Canadian Journal of Agricultural Economics* 57, 379–394.

4 Measuring Productivity of the Australian Broadacre and Dairy Industries: Concepts, Methodology and Data*

Shiji Zhao, Yu Sheng and Emily M. Gray

Australian Bureau of Agricultural and Resource Economics and Sciences (ABARES), Canberra

4.1 Introduction

Since the mid-1990s, the Australian Bureau of Agricultural and Resource Economics and Sciences (ABARES) has maintained a programme to compile and publish estimates of aggregate productivity growth for the agricultural broadacre (non-irrigated grains, beef and sheep) and dairy industries in Australia (Knopke *et al.*, 2000; Knopke *et al.*, 1995; Nossal *et al.*, 2008; Nossal *et al.*, 2009; Zhao *et al.*, 2008). These statistics are released at the national and industry levels and for the three agroecological zones (the northern, southern and western zones) as defined by the Australian Grains Research and Development Corporation (GRDC). Productivity growth estimates are widely used by government agencies, industry bodies and the wider research community for the purposes of informing decision makers about agricultural policies and analysing economic issues in the agriculture sector (Mullen and Crean, 2007).

In addition to providing this statistical information, ABARES has also conducted a series of economic analyses on the basis of productivity estimates calculated at the

farm level (Alexander and Kokic, 2005; Kokic *et al.*, 2006; Zhao *et al.*, 2009). These analyses provide additional insights into the productivity performance of the broadacre and dairy industries for individual farms and at aggregate levels.

The studies mentioned above focused on the statistical properties or economic interpretations of the productivity estimates, with only minimal description of the underlying concepts, theories and methodology. The absence of documentation to describe the theories and methodology has made it hard for users to understand and interpret the productivity statistics. The lack of background documentation has also made it difficult for ABARES to communicate with professionals in the field of productivity analysis, and to engage with them to improve the methods underlying ABARES measurement of productivity growth.

This chapter has two purposes:

1. To enhance the ability of non-technical users to understand, interpret and use ABARES productivity estimates, by providing a description of the underlying concepts, theories and methodology; and

---

* The authors owe great thanks to Alistair Davidson, Peter Gooday, Peter Martin and an anonymous referee for their valuable comments and contribution. Any remaining errors are the responsibility of the authors.

2. To facilitate engagement with professionals in the field of productivity analysis to improve the methods underlying ABARES productivity estimates. Accordingly, this chapter includes more technical details than would be required by an audience unfamiliar with the measurement of productivity.

ABARES productivity estimates are based on an index method similar to that used by official statistical agencies, such as the Australian Bureau of Statistics and the US Bureau of Labor Statistics. The index method belongs to a family of so-called non-parametric methods (Hulten, 2000, provides a good overview of the differences between parametric and non-parametric methodologies). The underlying theoretical issues and statistical properties of productivity indexes are widely discussed and debated in the literature. It is beyond the scope of this chapter to explore all of those issues. Instead, it focuses on specific technical decisions that ABARES has made in developing its productivity index methodology and, in doing so, explains the theoretical underpinnings that have implications for the interpretation of its statistics.

Section 4.2 provides an overview of the total factor productivity (TFP) statistics that ABARES releases. Section 3 introduces the concept of productivity. Section 4 describes the methods used by ABARES to estimate agricultural productivity. Section 5 outlines ABARES data sources. Section 6 describes how outputs and inputs are measured and calculated. The final section adds some concluding remarks.

4.2 Agricultural Productivity in Australia: An Overview

ABARES maintains a programme to measure, analyse and report changes in the productivity of Australia's broadacre agricultural sector, including the specialized cropping, mixed crop–livestock, beef and sheep industries, and the dairy industry. These industries are defined in Box 4.1. The data from ABARES farm surveys are available from

---

Box 4.1. Definitions of broadacre and dairy industries.

Broadacre and dairy industries are defined by the Australian and New Zealand Standard Industrial Classification (ANZSIC; Australian Bureau of Statistics, 2006, cat no. 1292.0). The industries are:

- Crops industry (ANSCIC06 Classes 0146 and 0149) – farms engaged mainly in growing cereal grains, coarse grains, oilseeds, rice and/or pulses.
- Mixed livestock–crops industry (ANZSIC06 Class 0145) – farms engaged mainly in running sheep and/or beef cattle, and growing cereal grains, coarse grains, oilseeds and/or pulses.
- Beef industry (ANZSIC06 Class 0142) – farms engaged mainly in running beef cattle.
- Sheep industry (ANZSIC06 Class 0141) – farms engaged mainly in running sheep.
- Sheep–beef industry (ANZSIC06 Class 0144) – farms engaged mainly in running both sheep and beef cattle.
- Dairy industry (ANZSIC06 Class 0160) – farms engaged mainly in farming dairy cattle.

Using these definitions, farms with crop production accounting for the majority of agricultural output will be included in the 'cropping' industry for productivity analysis. It follows that, in measuring relative inputs and outputs (to estimate TFP), the 'cropping' industry output includes not only crop outputs, but all other agricultural outputs produced by farms classified within the industry. This clarification is particularly relevant as many Australian farms produce multiple products.

1977–1978 for broadacre industries and from 1978–1979 for the dairy industry.

Agriculture, which is included in ANZSIC Division A, also includes nursery and floriculture production, mushroom and vegetable growing, fruit and tree nut growing, other crop growing, such as sugar and cotton, and deer, horse and other livestock farming. In recent years, the broadacre sector has accounted for about 50–60% of the gross value of total agricultural production.

Rice growers are also included in the scope of the broadacre sector. In Australia, dry-land farming predominates in the broadacre sector, whereas rice farming, mainly located in the Murray–Darling Basin, typically accounts for a very small proportion. For example, around 61,000 tonnes of rice was produced in 2008–2009, which was about 0.08% of the total output value of the broadacre sector in that year (Australian Bureau of Statistics, 2010).

The ABARES productivity work programme consists of two components. The first involves compilation of aggregate TFP statistics for the broadacre industries and the dairy industry. The second entails economic analyses to identify determinants of TFP using farm-level TFP statistics.

Recent estimates of aggregate inputs, outputs and TFP for Australia's broadacre sector are shown in Fig. 4.1, and the corresponding data are in presented in Table 4.1. These statistics suggest that the broadacre sector increased its TFP by 1.4% per year between 1977–1978 and 2008–2009. The growth of TFP was accompanied by farms expanding output, while simultaneously reducing their inputs. Output growth in the broadacre sector averaged 0.5% per year over the period, whereas the decline in input use averaged 0.9% per year. According to the Productivity Commission (2005), agricultural productivity growth has typically exceeded productivity growth for

many other market sector industries in the economy, although productivity growth in recent years has shown a slowing trend (Sheng et al., 2010).

TFP growth in the specialized cropping, mixed crop–livestock, beef and sheep industries is shown in Fig. 4.2. Within the broadacre sector, the cropping industry has outperformed beef, sheep and mixed crop–livestock industries in TFP growth. Cropping specialists achieved annual productivity growth of 1.9% per year over the past three decades. These farms expanded output only marginally, and productivity gains were mostly achieved through more efficient use of inputs (Nossal et al., 2009). The mixed crop–livestock industry improved productivity by an average of 1.4% per year, and beef specialists achieved a similar average performance as the mixed crop–livestock industry over the past three decades. Productivity growth in the beef and mixed crop-livestock industries coincided with high output growth and marginal growth in inputs. The sheep industry lagged behind the broadacre sector in terms of long-term productivity growth. Between 1977–1978 and 2008–2009, the sheep industry experienced a decline in both output and input use.

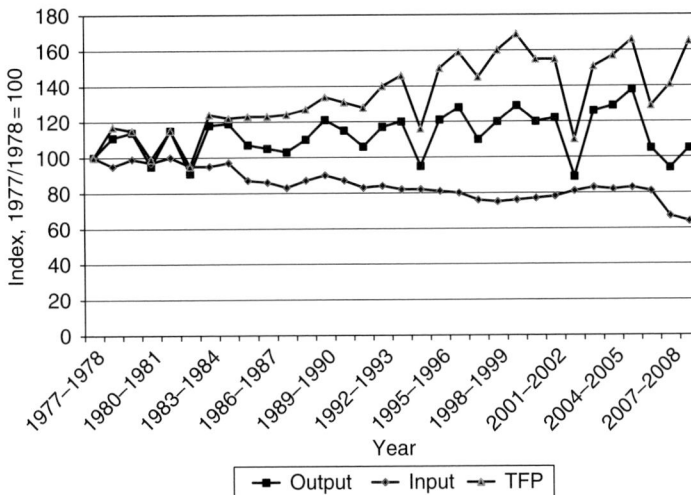

Fig. 4.1. Australian broadacre agriculture: indexes of inputs, outputs and TFP. Source: Gray et al. (2011).

Table 4.1. Input, output and TFP indexes for Australian broadacre agriculture.

Year	Input	Output	TFP
	Index, 1977–1978=100		
1977–1978	100	100	100
1978–1979	95	111	117
1979–1980	99	114	115
1980–1981	97	95	99
1981–1982	100	115	115
1982–1983	95	91	95
1983–1984	95	118	124
1984–1985	97	119	122
1985–1986	87	107	123
1986–1987	86	105	123
1987–1988	83	103	124
1988–1989	87	110	127
1989–1990	90	121	134
1990–1991	87	115	131
1991–1992	83	106	128
1992–1993	84	117	140
1993–1994	82	120	146
1994–1995	82	95	116
1995–1996	81	121	150
1996–1997	80	128	159
1997–1998	76	110	145
1998–1999	75	120	160
1999–2000	76	129	169
2000–2001	77	120	155
2001–2002	78	122	155
2002–2003	81	89	110
2003–2004	83	126	151
2004–2005	82	129	157
2005–2006	83	138	166
2006–2007	81	105	129
2007–2008	67	94	141
2008–2009	64	105	165
Annual growth rate			
1977–1978 to 2008–2009	−0.91	0.46	1.37
1977–1978 to 1993–1994	−1.26	0.67	1.92
1993–1994 to 2008–2009	−0.70	−0.41	0.31

Source: Authors' own calculation.

Productivity growth in the dairy industry averaged 0.7% per year between 1979–1980 and 2008–2009 (Fig. 4.3). As both input and output growth were high, productivity gains can be attributed to the ability of farms to increase output faster than inputs. Recent dairy industry productivity growth can be partially attributed to some farms exiting the industry following deregulation and the remaining farms increasing in size and intensity of operations.

The productivity growth statistics presented in Figs 4.1–4.3 are variable and volatile. Fluctuations in broadacre TFP estimates are common, largely reflecting the short-term impacts of seasonal conditions. As inputs used in agricultural production are often relatively stable in the short run, TFP fluctuations typically correlate to movements in output. From Fig. 4.1, it is apparent that while input use declined steadily in broadacre agriculture, output growth was more variable, occasionally returning to its level in 1977–1978 (an index level of 100 or lower). These fluctuations were often sizeable, with the volatility stemming mainly from external conditions, such as climate and factors beyond the control of farmers (for example, the droughts in 1982–1983, 1994–1995, 2002–2003 and 2006–2007), economic conditions and statistical noise resulting from measurement error. Consequently, long-term TFP trends are considered to be more reliable indicators of productivity growth.

Interestingly, Australian agricultural productivity seems to have run independently from the business cycles of the industries analysed. Climate conditions, such as droughts, seem to be more likely to affect the pattern of TFP series than other macroeconomic factors associated with business cycles, such as investment and employment rate. This behaviour contrasts with the strongly pro-cyclical productivity observed in other sectors of the national economy (Basu and Fernald, 2001).

Some of the features mentioned above are determined by the method or the data used in the estimation of productivity. They can be better understood once the underlying concepts and methodology are explained.

4.3 The Concept of Productivity

Productivity is broadly defined as a ratio of a measure of total output to a measure of a single or multiple inputs used in the production

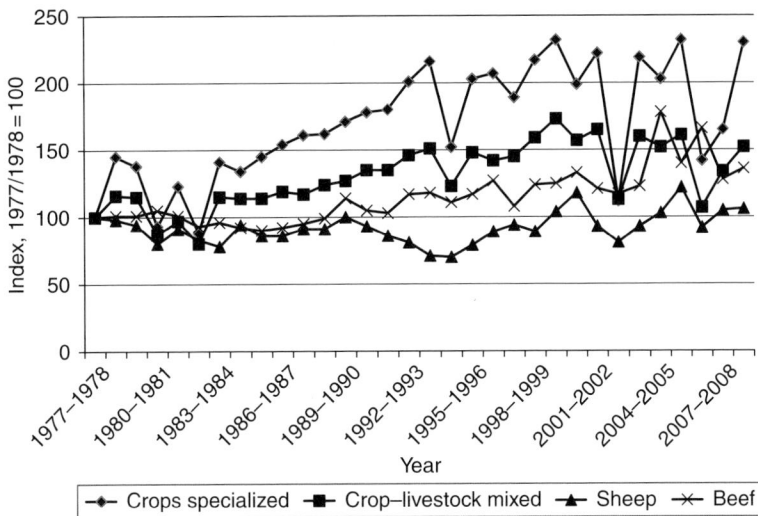

Fig. 4.2. TFP Index for Australian broadacre agriculture by sector. Source: Gray *et al.* (2011).

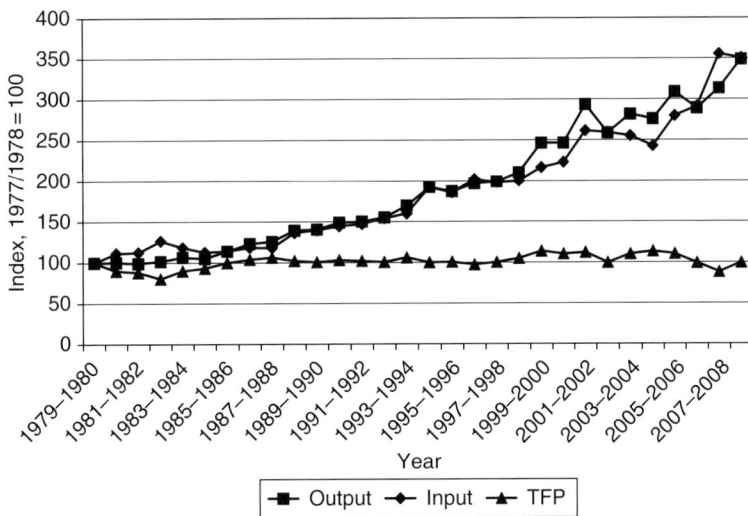

Fig. 4.3. Input, output and TFP indexes for the Australian dairy industry. Source: ABARES estimation.

process. In theory, it is a physical measure related to the *quantities* (some are adjusted for *qualities*) of outputs and inputs, independent of changes in the prices of outputs and inputs. In reality, the measures of total inputs and outputs are indirectly affected by changes in prices, which are used as weights to aggregate different components. Details on weight construction are discussed in section 4.6 on measurement of outputs and inputs. To estimate agricultural productivity statistics, ABARES includes *labour, capital, land* and *purchased materials and services* as inputs.

In Fig. 4.4, Q_0 is a technical relationship between output and input. In economics, this relationship is referred to as the production function. This function suggests

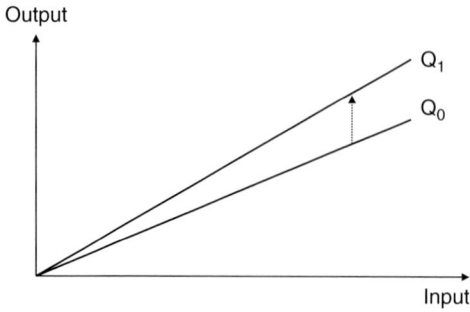

Output

Q_1

Q_0

Input

Fig. 4.4. An improvement in productivity.

Box 4.2. Scenarios for changes in productivity.

Productivity will **increase** if:

- Output increases, while input decreases or remains unchanged;
- Both output and input increase, but the former increases at a faster rate; or
- Both output and input decrease, but the former decreases at a slower rate.

Productivity will **decrease** if:

- Output decreases, while input increases or remains unchanged;
- Both output and input decrease, but the former increases at a faster rate; or
- Both output and input increase, but the former decreases at a slower rate.

Productivity will **remain unchanged** if:

- Both output and input remain unchanged; or
- Output and input increase or decrease at the same rate.

---

that output will follow when input is increased or decreased and that the two variables are positively related. A change in productivity is caused by a shift in the production function, and in Fig. 4.4 it is represented by a shift in the profile from Q_0 to Q_1.

When the technical relationship between output and input in Fig. 4.4 is presented over a period of time, the change in the ratio captures productivity *growth*. In effect, productivity growth measures the change in output that cannot be accounted for by changes in input (that is, the gap between Q_1 and Q_0 in Fig. 4.4). Several scenarios are described in Box 4.2, in which productivity increases, decreases or remains unchanged, depending on how input and output quantities change.

While there is a general consensus about the broad notion of productivity, disagreements remain on the preferred measure(s) of productivity – that is, total factor productivity versus partial productivity – how outputs and inputs are measured and aggregated and, in particular, the interpretation of productivity estimates. This section addresses the concept of productivity and the interpretation of productivity estimates. The choice of an appropriate productivity measure and the issues related to the measurement and aggregation of inputs and outputs are addressed in Sections 4.4–4.6.

What then, are the causes of a change in productivity – a shift in the profile of the input–output relationship? The most common interpretation of productivity is that it

is a measure of technology change. When productivity is expressed in terms of a change over time, or *growth*, for individual producers or at aggregate levels, such as an industry or economy, it is interpreted as a measure of *technological progress*. Similarly, one can also compare the productivity estimates of individual economic units – for example, farms, industries, regions or economies – at a particular point in time, to infer differences in the *level* of technology. In addition to measuring technology, productivity estimates have other applications. According to the Organisation for Economic Co-operation and Development (OECD, 2001), TFP can be used to measure efficiency, real cost savings and living standards, or to benchmark production processes. Discussion of those applications is beyond the scope of this chapter.

The literature on productivity makes a distinction between *embodied* and *disembodied* technologies (Chavas, 2001; OECD, 2001). Conceptually, the former refers to technologies embodied in new products, such as advances in the design and quality of new vintages of capital goods and intermediate inputs, whereas the latter is about new blueprints, findings from scientific research, and new organizational or management techniques. In practice, when it comes to specific technologies, making the distinction is not always straightforward.

In theory, TFP measures disembodied technology. Although both embodied and disembodied technologies alter the ability of producers to convert resources into outputs desired by the economy (Grilliches, 1987), their impacts on the *measure* of productivity are different. Although disembodied technologies will increase TFP in so far as their impacts are not captured by any one input, changes in embodied technologies should be reflected in the measure(s) of input(s). Further details will be discussed in Section 4. This is because producers are often required to pay for access to embodied technologies in order to produce better products or improve the quality of their capital, land or other inputs. In a competitive market, producers are expected to pay for the full benefits of using such technologies. Therefore, if the benefits and costs to producers are fully captured in the measures of outputs and inputs, then embodied technologies will have no impact on the productivity estimates.

If individual producers are not required to pay for the full benefits of using disembodied technologies, however, such as findings from scientific research, the remainder will be captured by the disembodied technology and, hence, the impact is expected to be captured by TFP measures. For example, although there is an increasing trend toward the use of professional consultants, particularly agronomists in cropping industries, the cost to the farmer typically does not nearly reflect the opportunity cost of their advice relating to disembodied technologies, such as best management practices. In the economic literature, the benefits from the disembodied technology are sometimes figuratively compared with 'free gifts' or 'manna from heaven'. In such literature, R&D-induced innovations are often used as examples of disembodied technology

(Alston and Pardey, 1996; Hayami and Ruttan, 1970; Mullen, 2007).

In practice, productivity is also influenced by a range of factors that are not directly related to technology and the effects of these factors will be picked up by productivity estimates. These factors include increasing returns to scale (Diewert and Fox, 2008), economic cycles and capacity utilization (Dennison, 1972; Jorgenson and Griliches, 1972; Jorgenson and Griliches, 1967), accumulation of human capital (Bureau of Labor Statistics, 1993; Reilly *et al.*, 2005), trade liberalization (Parham, 2004), changes in the technical efficiency within firms (Coelli *et al.*, 1998; O'Donnell, 2008), efficiency gains resulting from resource allocation between firms and industries, and measurement errors. In the estimation of TFP based on index methods, it is assumed that firms are operating on the production frontier, meaning that all firms use their resources efficiently to produce the maximum outputs. This assumption is implied in standard economic theory, and it means that firms do not systematically suffer 'technical inefficiency'. However, the literature includes some empirical studies that have relaxed this assumption and assumed that firms are technically inefficient. Interested readers may refer to Coelli *et al.* (1998) for more information about those studies.

Given its dependence on a wide range of natural resources, productivity in the broadacre agricultural sector is influenced by a set of environmental factors. For example, Alexander and Kokic (2005), Kokic *et al.* (2006) and Zhao *et al.* (2009) found that broadacre productivity is significantly influenced by the amount of moisture retained in the soil and the various natural resource management practices adopted by farmers. Those findings suggest that the productivity of this sector is particularly sensitive to climate variability and the condition of natural resources, such as land quality. The influence of natural resources on productivity is not unique to the broadacre agricultural sector. For example, Kompas *et al.* (2008) found that fish stocks were an important determinant of the productivity of a fishery and

Topp *et al.* (2008) suggested that the quality of mineral had influenced the productivity of Australia's mining industry.

4.4 Methods for Estimating Productivity

While the concept of productivity is reasonably straightforward, choosing an appropriate method to measure it is not. Two types of productivity measures are commonly used in the literature – TFP measures, which are also referred to as multifactor productivity measures, and partial factor productivity (PFP) measures. TFP is defined as a ratio of total outputs and total inputs and PFP measures as a ratio between total outputs and a single input. Comparing the two types of productivity measures, PFP measures, such as labour productivity and yield, are of limited use for summarizing the overall productivity performance of the sector. This is partly because PFP measures can result in a misleading assessment of an industry's productivity performance if the effects of technological and efficiency changes, input substitution, and technological improvements embodied in other inputs are incorrectly attributed to improvements in a particular input. ABARES measures the productivity of the broadacre and dairy industries using a TFP

measure. The following section covers the methods ABARES uses to estimate TFP.

4.4.1 Productivity estimation process: a graphical representation

Figure 4.5 illustrates the process used by ABARES to estimate the productivity of the broadacre industries. The process for the dairy industry is similar, differing only in the mix of products included in the measure of total outputs: that is, predominantly milk and dairy cattle and, to a lesser extent, the composition of inputs. Thus, the discussion of the concept, theory and methodology is equally applicable to both the broadacre and dairy industries. The differences in the outputs and inputs between broadacre and dairy industries are described in section 4.6.

Generally speaking, the estimation of TFP by ABARES involves three steps. Figure 4.5 is an approximation of this estimation process.

In the first step, specific items are aggregated to form broad types of outputs (crops and livestock) and inputs (labour, capital, land, and materials and services.) Following the OECD (2001), 'capital' in Fig. 4.5 is defined as the 'flow' of services provided by the stock of capital, including machinery, improvements to infrastructure, buildings for production purposes and the value of livestock herds. The definition of *materials and services* also follows the OECD manual and

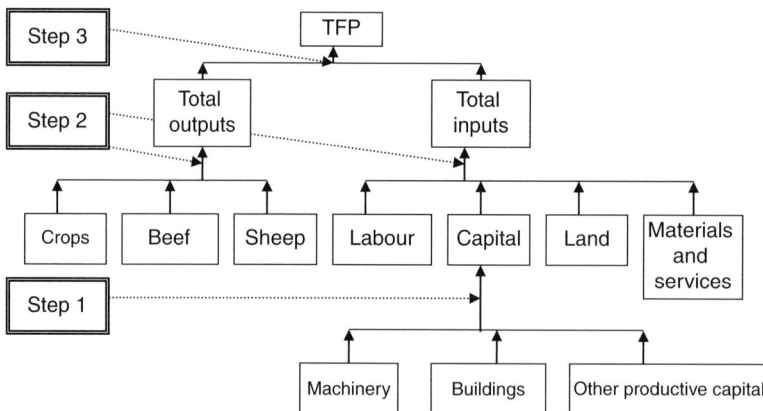

Fig. 4.5. Process of estimating TFP in Australian broadacre agriculture.

could include fuel, electricity, fertilizer, seed, fodder, local government taxes and fees, and the real costs of financial or other services. On the output side, crops could include wheat, barley, canola, hay and silage.

It is evident from the previous examples that the specific items covered by the broad types of outputs and inputs are sometimes quite heterogeneous. Therefore, it is not appropriate to aggregate such items simply by adding up quantities of different types of goods and services. The aggregation needs to be done using an index formula, as discussed in the Method of estimation section below.

The second step involves aggregation of the broad types of outputs (crops, beef and sheep) and inputs (labour, capital, land, and materials and services) to form *total outputs* and *total inputs*, respectively. Again, this is achieved using a specific index formula. In the third step, a ratio of total outputs to total inputs is calculated to form the TFP estimate.

Productivity can be also calculated for individual farms, for a specific industry or at regional and national levels. Figure 4.5 is a representation of the estimation process for individual farms. The estimation of TFP at aggregate levels involves similar processes. For example, to estimate TFP at the national level, it is necessary to construct national estimates for the broad types of outputs (crops, beef and sheep) and inputs (labour, capital, land, and materials and services) and then of total outputs and total inputs. This is achieved by aggregating specific outputs and inputs across farms to the national level and then calculating TFP by taking a ratio of total outputs to total inputs. The same principle is applied to TFP estimation at other levels of aggregation. As for the estimation of TFP at the national, regional and industry levels, outputs and inputs are aggregated using sample structure weights (see the section on Data Sources for more details).

4.4.2 Method of estimation

This section covers three issues. First, there is an outline of various types of productivity estimates. This is followed by a brief description of the methods that may be applied to the estimation of TFP. Because ABARES uses an index method, this chapter will focus on the index methodology – index formulae and an outline of their characteristics, including the 'axiomatic' and economic properties of the relevant index formulae. Finally, practical issues in the application of the Fisher index will also be discussed.

Characteristics of various types of productivity estimates

Although the methodology used by ABARES to estimate TFP is broadly similar to those used by other agencies, it does have several unique characteristics (OECD, 2001). They are discussed in detail below.

In the literature, two types of TFP measures are frequently reported and used in economic analysis. The first type is based on a measure of gross output (Eqn 4.1) and the second uses value added as total outputs (Eqn 4.2) (see equations 4.1 & 4.2 at bottom of page):

where value-added is measured by subtracting *materials and services* from gross output.[2]

Both *TFP(g)* and *TPF(v)* are valid measures of TFP, but they nearly always result in different estimates even if they are applied to the same set of data. Domar (1961) demonstrated that when using the same dataset, the gross output-based TFP measure is less than the value-added-based measure by a factor equal to the ratio of the industry value added to its current value of gross output.

$$TFP(g) = \frac{Quantity\,(index)\,of\,gross\,output}{Quantity\,(index)\,of\,labour,\,capital,\,land,\,materials\,\&\,services} \quad (4.1)$$

$$TFP(v) = \frac{Quantity\,(index)\,of\,value\,added}{Quantity\,(index)\,of\,labour,\,capital\,and\,land} \quad (4.2)[1]$$

TFP(v) provides a more reliable indication of an industry's productivity performance when only the relative contributions of capital and labour are of interest. However, *TFP(v)* is not, in general, considered to be an accurate measure of technology change, as it attributes all measured technology to capital and labour. Conversely, *TPF(g)* captures disembodied technological change because it fully accounts for the contribution of intermediate inputs. However, *TFP(g)* also reflects economies of scale, changes in technical efficiency, capacity utilization, impact of environmental and policy variables, and measurement errors (see Section 4.3).[3] TFP, in the remainder of this chapter, refers to productivity measures calculated as a gross-output measure.

Estimation of total factor productivity: index method

In a summary of the literature on the methodology of estimating productivity, Griliches (1996) classified the empirical work into two main types: parametric and non-parametric methods. The former involves econometric modelling of a production function and often uses regression techniques to estimate the relationships between total outputs and major types of inputs, such as labour, capital, land, and materials and services. The residual of these regressions can be used as a measure of total factor productivity. The more fully the production technology is characterized in the regression equation – such as by including economies of scale and capacity utilization – the smaller the residual and the smaller the increase in output explained by disembodied technological progress and other uncharacterized factors.

Non-parametric methods for productivity estimation include data envelope analysis (DEA), and Malmquist (Caves *et al.*, 1982; Diewert, 1992) and conventional index methods (Diewert, 1992). Conventional index methods are preferred by most official statistical agencies for the compilation of TFP estimates (Australian Bureau of Statistics, 2007; Bureau of Labor Statistics, 1983; OECD, 2001) and are used by ABARES in its estimation of agricultural TFP statistics.

This chapter will focus only on the conventional index method. Interested readers may refer to Hulten (2000) for a discussion of the strengths and weaknesses of this and other methods.

The ABARES estimates of TFP growth are derived from a formula called the Fisher index. This formula is applied to construct indexes of gross output and total inputs (see Fig. 4.1). In various analytical projects, ABARES has also estimated and compared *levels* of TFP between farms or regions. The estimation of productivity levels involves incorporating an additional step into the original Fisher index formula. This section provides an outline of that index method.

The Fisher index is composed of two other types of indexes. The first is called the Laspeyres index (Q_{0t}^{L}; Eqn 4.3):

$$Q_{0t}^{L} = \frac{\sum_{i=1}^{N} p_{i0}q_{it}}{\sum_{i=1}^{N} p_{i0}q_{i0}} = \sum_{i=1}^{N} W_{i0}\frac{q_{it}}{q_{i0}} \quad (4.3)$$

where $W_{i0} = \frac{p_{i0}\,q_{i0}}{\sum_{i=1}^{N} p_{i0}\,q_{i0}}$ is the share of ith item in the total value of outputs or inputs in the *base* period (denoted by *0*).

The second is called the Paasche index (Q_{0t}^{P}; Eqn 4.4):

$$Q_{0t}^{P} = \frac{\sum_{i=1}^{N} p_{it}q_{it}}{\sum_{i=1}^{N} p_{it}q_{i0}} = \left\{ \sum_{i=1}^{N} W_{it}\left(\frac{q_{i0}}{q_{it}}\right) \right\}^{-1} \quad (4.4)$$

here $W_{it} = \frac{p_{it}q_{it}}{\sum_{i=1}^{N} p_{it}q_{it}}$ is the share of ith item in the total value of outputs or inputs in the *current* period (denoted by *t*).

In Eqns 4.3 and 4.4, p_{i0} and p_{it} represent the prices of ith output or input items in the base and current periods, and q_{i0} and q_{it} are the quantity of ith item in the two periods.

Laspeyres and Paasche are called fixed weight indexes because, in the construction of index numbers between the base and current periods, the former uses the prevailing value shares in the base period (W_{i0}) as weights, whereas the latter uses those in the current period (W_{it}). Laspeyres (Paasche) quantity index is known to be biased downward (upward) because, when relative prices of individual inputs or outputs in the index change, it tends to place too much (too little) weight on the inputs for which relative prices have fallen and too little (too much) emphasis on items for which relative prices have risen. It can be inferred from the literature on price indexes that the Laspeyres and Paasche indexes represent the upper and lower limits and the 'true' index should be somewhere between them (Pollak, 1989). The divergence between the Laspeyres and Paasche indexes is called the 'Laspeyres–Paasche spread'.

The Fisher index $\left(Q_{0t}^{F}\right)$ is the geometric mean of Laspeyres and Paasche indexes, defined as shown in Eqn 4.5:

$$Q_{0t}^{F} = \sqrt{Q_{0t}^{L}Q_{0t}^{P}} \qquad (4.5)$$

In the construction of TFP estimates, ABARES has applied the Fisher index to aggregate specific items (Step 1 in Fig. 4.5) and broad types of outputs and inputs (Step 2) to create measures of gross output and total inputs.

The Fisher quantity index has a number of desirable properties. In the literature, index experts and economists have used two approaches to select and assess the quality of index formulae. The first is called the 'axiomatic test' approach, first developed by Fisher (1922). In that approach, index formulae are tested against a set of desirable properties, or axioms. Index formulae that pass more axioms, and especially the most important axioms, are considered to be superior to or 'ideal' compared with others. Diewert (1992) compared Fisher and other index formulae and found that the Fisher quantity index satisfies 21 reasonable tests – significantly more than any other index – and based on that finding, recommended using the Fisher index for the construction of productivity indexes.

An alternative method to assess the quality of indexes is the 'economic' approach first developed by Diewert (1976), which also proved that the Fisher index is 'exact' for a quadratic cost function. Diewert (1992) showed that the Fisher index provides a close second-order approximation for any arbitrary production function and, under some reasonable assumptions, is an 'exact' representation of a homogeneous quadratic

production function (OECD, 2001). In the literature, indexes characterized with this property are deemed 'superlative' (Diewert, 1976). These findings are important because they establish a connection between the Fisher index and the production function and, thereby, some key economic assumptions and behaviours – such as, cost minimization, constant returns to scale and perfect competition – can be examined using the index approach.

The Fisher index is not the only 'superlative' index. The Tornqvist index (Eqn 4.6) shares many desirable characteristics with the Fisher index. Diewert (1976) demonstrated that this index is 'exact' for a translog cost function. The OECD (2001) recommended the Tornqvist index as a preferred formula for calculating productivity indexes because the translog functional form of the production function is more popular than the quadratic form in economic analysis.

The Tornqvist index formula is shown in Eqn 4.6:

$$Q_{0t}^{T} = \prod_{i=1}^{N}\left(\frac{q_{it}}{q_{i0}}\right)^{\frac{1}{2}(w_{io}+w_{it})} \qquad (4.6)$$

It is difficult to choose between the Fisher and Tornqvist index formulae, and either may be preferred for different but equally valid reasons. The empirical literature has demonstrated that the two formulae lead to very similar index numbers (Zheng, 2005). ABARES has followed Diewert (1992) and uses the Fisher index formula in its construction of productivity indexes because of an important operational consideration – the Fisher index can easily handle values of zero which are often found in the farm survey data. It is quite clear from Eqn 4.6 that the Tornqvist index cannot operate on zeros.

Application of the Fisher index: two practical issues

Equation 4.5 was designed to calculate a Fisher index between two observations. In reality, however, productivity indexes are usually calculated over more than two periods and sometimes they are applied to cross-sectional data that contain more than three observations. In both circumstances, additional considerations are required. Two important issues are considered in this section. The first relates to the choice between 'direct' and 'chained' indexes, which is relevant to Eqn 4.5 when it is applied to time series data. The second is about how to make the Fisher index 'transitive' when it is applied to cross-sectional or panel data (so that a direct comparison of two farms produces the same TFP estimate as an indirect comparison through a third farm).

CHAINING OF FISHER INDEXES. The main difference between direct and chained indexes is that, while the former simply calculate the difference between the fixed base period *0* and period *t*, a chained index calculates and weights the changes ($\frac{q_{it}}{q_{i,t-1}}$) within the intervening period. In a chained index, the current period is compared to the previous period for all observations, rather than comparing each period to a fixed base period. For example, when a Fisher index is calculated between periods *0* and *2*, the direct index is calculated as shown in Eqn 4.7:

$$Q_{02}^{F} = \sqrt{Q_{02}^{L} Q_{02}^{P}} \qquad (4.7)$$

and the chained index is calculated as as shown in Eqn 4.8:

$$Q_{02}^{F} = \sqrt{Q_{01}^{L} Q_{01}^{P}} \times \sqrt{Q_{12}^{L} Q_{12}^{P}} \qquad (4.8)$$

where Q_{02}^{L} is a Laspeyres index between periods *0* and *2*, Q_{02}^{L} between periods *0* and *1* and Q_{12}^{L} between periods *1* and *2*. Q_{02}^{P}, Q_{01}^{P} and Q_{12}^{P} are the Paasche counterparts.

Direct and chained indexes measure the same thing, but are likely to provide different values for the change between periods *0* and *2*. Equations 4.7 and 4.8 relate to an index that is chained together every period. Alternatively, the indexes can be linked together on a regular but less frequent basis, or they can be linked on an irregular basis.

Chained indexes are preferred by analysts for several reasons. First, assuming that the purpose of the TFP index is to measure the most recent changes in technology, chained indexes allow a closer match between technologies in consecutive time periods than is obtainable with direct indexes, where periods might be considerably further apart. It is clear from Eqn 4.7 that as the current period (*t*) moves farther away from the base period (*0*) the weights used to calculate Q_{0t}^{L} become increasingly less relevant to the current situation. Even though the weights for the current period (*t*) are used in the calculation of the index number for period *t* (through the Paasche index), the geometric mean of the Laspeyres and Paasche indexes is more likely to reflect the 'average' of the change in technology between periods *0* and *t*, instead of that between periods *t−1* and *t*.

Second, the Laspeyres–Paasche spread is likely to be smaller between adjacent periods (Diewert, 1978) than the spread between periods a long way apart, and as a result the chained Fisher index has a better chance of representing the 'true' measure of the underlying technology.[4]

However, there are also disadvantages associated with chained indexes (Szulc, 1983). For example, when the movements are non-monotonic (quantities 'bounce'), it can be shown that even when there are identical prices and quantities in the base and current periods, a chained index might take a value different from 1, whereas an unchained fixed-weight index is trivially and identically equal to 1 (OECD, 2001). In other words, chained indexes might drift away from the 'true' value.

ABARES originally chose to use a direct Fisher index in its compilation of TFP and has maintained that practice to date. Because the Fisher index uses weights for both the base and current periods, it is unlikely to suffer significant biases (as discussed in Section 4.4) but, as the interval between the base and current period increases, it is unable to benefit from the advantages that chaining could offer. This is an area where a change might lead to an improvement to the index.

MULTILATERAL FISHER INDEX. The Fisher index in Eqn 4.7 is a *bilateral* index, which makes it only suitable for estimating productivity growth using time series data. In order to apply the Fisher index to cross-sectional or panel data,[5] the index must be converted to a *multilateral* index formula. A multilateral index is required if one intends to analyse productivity *levels* for individual farms. ABARES has estimated and analysed multilateral productivity indexes in a number of research projects (Alexander and Kokic, 2005; Kokic *et al.*, 2006; Zhao *et al.*, 2009).

When the bilateral Fisher index is applied to cross-sectional or panel data, it might lead to inconsistent results because it does not pass what is called the *transitivity* or *circularity* test (Diewert, 1988). Assuming there are three farms, called A, B and C, if Eqn 4.5 is used to estimate TFP, the index numbers for the three farms will depend on the choice of base farm and the sequence with which Eqn 4.5 is applied to the other farms. In this example, there are potentially six sets of TFP estimates that can be generated by Eqn 4.5. More importantly, if the three farms use very different technologies, the relative index numbers will be significantly different, and in some circumstances the ranking of their TFP estimates can be reversed. This is undesirable and confusing to those who use the estimates. Interestingly, this problem does not arise when the bilateral index is applied to time series data, because these types of data have a natural starting, or base, period and a unique sequence to follow.

A transitive index that is internally consistent across different farms is needed in order to avoid potential confusion that might occur as a result of inconsistent index numbers being derived when the TFP of a farm is compared directly or through a third farm. Although there are several types of multilateral indexes (Kravis *et al* 1982), ABARES uses the EKS index because it is relatively simple and transparent.

The EKS index was first proposed by Gini (1931) and named after Elteto and Koves

(1964) and Szulc (1964). It is an extension of the Fisher bilateral index.[6] If there are N farms, an EKS index between farms A and B can be expressed as shown in Eqn 4.9:

$$Q_{AB}^{EKS} = \left(\prod_{r=1}^{N} Q_{AC}^{F} Q_{CB}^{F}\right)^{1/N} \tag{4.9}$$

where Q_{AC}^{F} is a Fisher bilateral quantity index using farm (A) as the base farm, and Q_{CB}^{F} is a Fisher quantity index using farm (C) as the base farm.

In the example mentioned above about the three farms A, B and C, an index between farms A and B is calculated based on Eqn 4.10 (see equation 4.10 at bottom of page):

The EKS index has many desirable properties (Diewert, 1988). As explained above, it is *transitive* and *base-farm invariant*, that is, independent of the choice of base farm. It is not, however, consistent in aggregation, and one should not attempt to derive higher-level indexes by aggregating EKS indexes (Waschka *et al.*, 2003).

In the estimation of TFP for individual farms, Eqn 4.10 is applied to generate estimates of gross output and total inputs before productivity estimates are calculated (that is, Step 3 in Fig. 4.5). This is the main difference between the method used to estimate productivity growth at aggregate levels and that used for individual farms.

4.5   Data Sources

The ABARES productivity estimates are derived from data collected annually through the Australian Agricultural and Grazing Industries Survey (AAGIS) of broadacre industries and the Australian Dairy Industry Survey (ADIS). The industries covered by these two surveys accounted for around three-quarters (72%) of the gross value of Australia's agricultural production in 2008–2009; 62% and 10% for each industry, respectively.

$$Q_{AB}^{EKS} = \left(Q_{AC}^{F} \times Q_{CB}^{F} \times Q_{AA}^{F} \times Q_{AB}^{F} \times Q_{AB}^{F} \times Q_{BB}^{F}\right)^{1/3} = \left(Q_{AB}^{F} \times \left(Q_{AC}^{F} \times Q_{CB}^{F}\right) \times Q_{AB}^{F}\right)^{1/3} \tag{4.10}$$

Using the survey data, TFP can be estimated for the broadacre and dairy industries. These industries constitute a subset of the agriculture sector that belongs to Division A of the Australian and New Zealand Standard Industrial Classification (ANZSIC06) (Australian Bureau of Statistics, 2006, cat no. 1292.0). A short description of these industries was provided in Box 4.1.

ABARES has conducted surveys of selected Australian agricultural industries since the 1940s. Data are collected through face-to-face interviews and information is drawn from a variety of sources, including farm accounts and farm operators' knowledge. AAGIS is an on-going annual survey, and data from this survey have been used to estimate TFP for broadacre industries from 1977–1978 onwards. ADIS is an annual survey that focuses on the dairy industry. Most data used to derive TFP estimates – such as the quantity and value of *farmers' inputs and outputs* – are sourced from these two surveys. More details about AAGIS and ADIS can be found in ABARES (2009).

AAGIS and ADIS collect information from farming establishments that make a significant contribution to the total value of agricultural output. Each year, ABARES selects samples of about 1500 to 1600 cropping and livestock farms and 300 dairy farms to represent the whole broadacre and dairy industries. Data from AAGIS and ADIS are used to undertake economic research and to assess the economic performance of the broadacre and dairy industries. The way the surveys are designed has two implications for the estimation of productivity.

First, a stratified random survey design is used to select sample farms. The target population is stratified so that the chance of a particular farm being selected depends on its location, the industry it belongs to, and the size of its operation. The target farm population is stratified by region, industry and size. Consequently, when the samples are aggregated to form estimates for an industry, farms have to be given individual weights. For instance, large farms have smaller weights and smaller farms have larger weights, to reflect both the sampling strategy of selecting proportionally more

samples from among the large farms and the fact that there are fewer large farms in the industries. (Larger farms are given higher probabilities of being selected partly because they are considered more heterogeneous than smaller farms.) This means that when productivity at the industry, regional and national levels is estimated, it is necessary to use the sample structure weights to aggregate the data for specific farms to the relevant level before TFP is calculated (Step 3 in Fig. 4.5).

Second, surveyed farms are rotated to ensure that the AAGIS and ADIS samples are representative. This results in an (unbalanced) panel dataset that has a unique value for productivity estimation and analysis at an aggregate level. The sample in each year retains a high proportion of the farms surveyed in previous years, in order to accurately measure change, while meeting the requirement to introduce new sample farms. This accounts for changes in the target population, as well as reducing the burden on survey respondents and replaces non-responding farms. The majority of farms are retained in the surveys for between three and four years, with larger and more unique farms typically retained for substantially longer periods. The form of the data also makes it necessary to impose transitivity – that is, to apply the EKS process – before productivity estimates are compared between farms included in the surveys at different points in time. See Appendix 4.2 for the sample size of the AAGIS and ADIS surveys.

4.6 Measurement of Outputs and Inputs

This section focuses on the measurement of gross output and total inputs. It starts with a discussion of several issues important for the measurement of productivity. This is followed by a description of how specific outputs and inputs are measured in the calculation of ABARES productivity estimates. Finally, methods to calculate and interpret average productivity growth are introduced.

4.6.1 Measurement issues

When Fisher index formulae (Eqns 4.3– 4.5) are applied to the calculation of gross output and total inputs, two types of data are required. The first type is used to estimate the quantity of outputs and inputs (q) and the second type is used in the construction of weights based on either price (p) or value share (w). This subsection provides a brief description of some important principles underlying the measurement of output and input quantities and the construction of weights.

Concepts of 'flows' and 'stocks'

To understand the measures of outputs and inputs properly and to interpret the variables correctly, it is important to distinguish between two statistical concepts: one is called 'flows' and the other 'stocks'. According to the Australian Bureau of Statistics (2000), economic flows reflect the creation, transformation, exchange, transfer or extinction of economic value. Productivity, and indeed all the variables included in the measures of output, is based on the concept of flows.

Stocks refer to economic positions, such as holdings of assets and liabilities at a given point in time. Stocks are usually recorded at the beginning and/or end of each accounting period. For example, typical measures of stocks are farmers' assets and holdings of land, recorded at the beginning or the end of the financial year. Stocks are connected with flows in the sense that changes in stocks result from the accumulation of transactions and other flows over the relevant accounting period.

The distinction between the concepts of flows and stocks is particularly important for the measurement of inputs. Although the measurement of most input variables in ABARES productivity calculations is based on the concept of flows, flows of land and capital inputs cannot be measured directly because they are not observable or measurable. Instead, the measurement of land and capital is based on the concept of stocks, and these measures serve as a close approximation of their flow counterparts. It is important to note that this is a standard practice adopted by official statistical agencies and the literature on productivity, when flows cannot be measured directly.

Measurement of quantity

Productivity is about the relationship between quantities of outputs and inputs. The way these variables are measured therefore directly affects the quality of productivity estimates. On the basis of the following accounting, two approaches can be used to estimate the quantities of a *single* output and input (Eqn 4.11):

$$q = \frac{v}{p} \tag{4.11}$$

where q is the quantity measure in physical units, v is total value, and p is the price per unit of q.

If q can be measured consistently and the required data are available, the preferred approach would be to use the data directly for the measure of quantity. In circumstances where quantity cannot be observed or measured directly, a measure of quantity (or quantity index) can be obtained by using the right-hand side of Eqn 4.11, which is the ratio of total value to price.

The situation can be complex if the calculation involves multiple outputs and inputs, particularly when the outputs or inputs consist of heterogeneous products, or products of different qualities. In those circumstances, it is often not feasible to measure quantity on the basis of physical units, and it will be necessary to use index formulae (such as the Fisher index) to construct quantity indexes. For the Fisher index, weights are required to aggregate specific outputs and inputs.

Construction and interpretation of weights

In most index formulae, a set of specific weights is used explicitly in the calculation of productivity.[7] Laspeyres and Paasche indexes (Eqns 4.3 and 4.4) can be calculated in two different ways: one using corresponding prices (p_{io} or p_{it}) and quantities (q_{io} or q_{it}), and the other on the basis of the share of values (W_{io} and W_{it}). The value shares always add up to 1.[8] The weights can be easily constructed if there are data on the values of all the commodities measured by TFP.

Three points are worth making. First, Laspeyres and Paasche indexes do not require all the data for quantity, value and price. Eqn 4.11 implies that these indexes can be calculated so long as there are data on any two of the three variables.

Second, although prices or values are used in the construction of weights, the resulting Fisher index is still a measure of changes in quantity, rather than value, because the weights are used solely for the purpose of aggregating items of different measurement units included in the measures of gross output and total inputs.

Finally, the weights for labour, capital, land, and materials and services can be interpreted as the elasticities of the inputs[9] if the output and input markets are perfectly competitive and production exhibits constant returns to scale.

Adjustment for differences in quality

It often happens that each output or input includes multiple products of different qualities. For example, farmers might produce wheat of different specifications (such as protein content) that are valued differently by consumers or they might employ labour with different skills. A good productivity measure should reflect differences in the quality of outputs and inputs. The ABARES methodology has the capacity to capture quality differences, if the necessary data on specific outputs or inputs are available.

Two approaches are available for quality adjustment, and each requires detailed information on quantities, quality attributes or prices of specific items. The first approach involves 'direct' adjustment of q in Eqn 4.11. This can be achieved through two methods. With the first method, it is necessary to quantify the quality attributes and estimate their impact on the measure of output or input. For example, the quality attributes of labour might include education and work experience. An econometric technique called the hedonic method can be used to estimate the

relative importance of such attributes. Then a quality-adjusted measure for labour inputs can be calculated using the coefficients of the regression (as weights). Alternatively, if data are available for wages of labour with different qualities (for example, education and work experience), direct adjustment may also be performed simply by using the information directly as weights in the aggregation process (for more details refer to Bureau of Labor Statistics, 1983; Reilly et al., 2005). Essentially, through a quality-adjustment process, outputs or inputs of different qualities are converted (from their original units) into the same quality unit.

The second approach involves the construction of a 'quality-adjusted' price index (which substitutes for p in Eqn 4.11). This requires data on both prices and quantities. For example, as shown in Eqn 4.12, the following Fisher price index for an output or input that consists of products of different qualities may be considered:

$$P_{0t}^F = \sqrt{P_{0t}^L P_{0t}^P} \qquad (4.12)$$

where P_{0t}^L and P_{0t}^P are Laspeyres and Paasche price indexes, defined as (Eqns 4.13 and 4.14):

$$P_{0t}^L = \frac{\sum_{i=1}^{N} P_{it} q_{i0}}{\sum_{i=1}^{N} P_{i0} q_{i0}} = \sum_{i=1}^{N} W_{i0} \frac{p_{it}}{p_{i0}} \qquad (4.13)$$

$$P_{0t}^P = \frac{\sum_{i=1}^{N} P_{it} q_{it}}{\sum_{i=1}^{N} P_{i0} q_{it}} = \left\{ \sum_{i=1}^{N} W_{it} \left(\frac{P_{i0}}{P_{it}} \right) \right\}^{-1}$$

$$(4.14)$$

and

$$W_{i0} = \frac{P_{i0} q_{i0}}{\sum_{i=1}^{N} P_{i0} q_{i0}}$$

is the share of ith product in the total value of a particular output or input observed in the base period (denoted by 0) and

$$W_{it} = \frac{P_{it} q_{it}}{\sum_{i=1}^{N} P_{it} q_{it}}$$

is the corresponding share in the current period (denoted by t).

To the extent that prices are a good indicator of quality, products of higher quality will be given greater weights, and as a result they will have a greater influence on the resulting price index. If the price index is applied to the right-hand side of Eqn 4.11 to measure the quantity of output or input, then differences in quality will be captured by the quantity index. To illustrate, consider a hypothetical situation in which a grower decides to produce one more unit of a high-quality product and reduce the production of a low-quality product by the same amount. If both prices remain unchanged, the price index will be the same, but the total value will increase. It becomes clear that when the right-hand side of Eqn 4.11 is applied to the calculation, the quantity of the item will also increase. In this way, any changes in the quality of products are captured by productivity measures through the indexes of prices, outputs and inputs.

4.6.2 Output and input measures

As explained in section 4.4, TFP is the ratio of two Fisher quantity indexes of gross output and total input. The main building blocks of the two indexes are a series of Fisher indexes that represent broad categories of outputs (for example, crops, livestock, wool and milk) and inputs (labour, capital, land, and materials and services). The specific items used in the calculation of these output and input categories are described in this section.

ABARES productivity statistics relate to the broadacre and dairy industries. Most farms in broadacre industries produce both cropping and livestock products, and as a result their outputs and inputs can be defined in a similar way. Dairy farms, however, mainly produce milk, using some inputs unique to that industry. The main focus in this section is on the broadacre industries, with only brief coverage of the outputs and inputs specific to the dairy industry.

Output measures

In ABARES productivity measures, outputs include all goods and services produced on farm, using the resources available to farmers. For the broadacre industry, gross output includes four broad categories: crops, livestock, wool and 'other outputs'. These four categories are composed of 19 specific outputs. Detailed descriptions of these outputs are presented in Table A4.1.2 of Appendix 4.1.

CROPS.   The category of *crops* includes wheat, barley, oats, grain sorghum, oilseeds and 'other crops', which refers to miscellaneous cropping products. When calculating the output of specific crops, two methods are used for measurement. In the first method, *quantity harvested* (based on a physical unit, such as tonnes) is used to measure outputs, and this measurement is applied to most crops. Quantity harvested is a representation of total output, most of which is sold on the market but some of which is retained on farm for consumption or future production. But this method is not suitable for the 'other crops', which cover a diverse range of products. For these outputs, a quantity index is derived using the total value of 'sales' divided by a price index (i.e. using the right-hand side of Eqn 4.11).[10]

To apply the Fisher index (Eqns 4.3–4.5) to the calculation of the quantity of crops, prices or values of individual crops are required in order to construct weights. In the farm survey, two data items are available for this purpose. One is 'gross receipts from sale' and the other is 'net receipts from sale', where gross receipts equal net receipts plus 'market expenses paid'. In the ABARES TFP calculation, 'net receipts from sale' is used in the calculation of weights.[11]

LIVESTOCK.   Livestock consists of four types of output: beef cattle, sheep, lambs and 'other livestock'. Each is represented by three items: sales of livestock, net natural growth of on-farm livestock (because of births and deaths) and net transfers out.

The outputs of beef, sheep, lambs and 'other livestock' are measured using the Fisher index formula. In this method, quantity indexes for the three types of livestock are derived by dividing the sum of the values of sales, net natural growth and net transfers out by relevant price indexes. The total values are also used in the construction of weights in the aggregation of the three types of livestock to form a quantity index for livestock. The price indexes are derived from the ABARES database (e.g. see Table 22 in ABARES, 2008).

WOOL.   Wool is a specific output of the sheep industry. The quantity of wool is represented by the amount of wool shorn or produced, measured in kilograms. The value of wool is equal to the quantity multiplied by a price index. The price index is constructed using the net value of receipts from the sale of wool (i.e. gross receipts from the sale of wool adjusted for government support and levies) divided by the quantity of wool sold through brokers and private marketing mechanisms.

OTHER FARM INCOME.   'Other farm income' covers the remaining income received by farmers. Most items in this category are returns to farmers using other economic resources, such as for hiring of plant and equipment or for extraction of natural resources (such as royalties on gravel quarried). Government assistance (such as drought assistance and other production support) has been excluded from the 'other farm income' category since 2000 to avoid biasing TFP estimates.

Values of farm income are available from ABARES farm surveys, and they can be used to construct weights. The 'quantity' is equal to the value of income divided by a price index.

It is worth expanding on the treatment of government assistance (including government compensation and subsidies) in calculating 'other farm income'. In theory, government subsidies should be excluded from the calculation of farmers' output, whereas compensation payments should be included (OECD, 2001). However, in practice, survey data limitations have, until recently, prevented ABARES from handling this issue appropriately.

In the 1980s and 1990s, the ABARES farm survey programme did not collect data on government assistance separately. Rather, it was included as part of 'other farm output income not elsewhere included'. Therefore, government subsidies could not be excluded

from productivity estimates. Although this approach is potentially problematic, in practice it is unlikely to have significantly influenced measured productivity for those years. For example, between 1977–1978 and 2000–2001, 'other farm income' (which also includes, for example, income from off-farm contracts and receipts from lease of plant) as a proportion of total farm income typically averaged around 4.5% and was no more than 6%. Since 2000–2001, data on government subsidies have been collected separately, thus allowing them to be removed from the ABARES TFP estimates altogether.

SPECIFIC OUTPUTS FOR THE DAIRY INDUSTRY.   There are four major types of output from dairy farms: milk, dairy cattle gain, dairy cattle sold and other outputs (such as insurance recoveries and fees from agisting livestock owned by third parties on the farm). Although milk is the main output of dairy farms, changes in the value of dairy cattle (through trading or raising cattle bred on farm) should also be included in the output of dairy farms because dairy cattle are also marketable assets. Some dairy farms run a livestock breeding, raising and trading enterprise as a minor part of their business. The incomes from these enterprises are included in the TFP calculation and classified under the category of 'livestock'.

Among the specific outputs for the dairy industry, only milk is measured in physical units (litres). Others are based on quantity indexes that use output values as weights. For instance, the value of 'cattle sold' is defined as the net value of dairy cattle sold, which is equal to the gross receipts from sales adjusted for marketing costs.

The quantity indexes for the four major outputs of the broadacre sector – crops, livestock, wool and other farm income – and their corresponding value shares are presented in Tables 4.2 and 4.3. These numbers were derived on the basis of the method discussed in this section. During the past three decades, the structure of the broadacre sector has changed substantially. Between 1977–1978 and 2008–2009, the share of crops increased from 30% to 47%. In contrast, livestock and

Table 4.2. Quantity indexes for the four major outputs in the broadacre agricultural industry.

Year	Crops	Livestock	Wool	Other farm income
		Index, 1977–1978 = 100		
1977–1978	100	100	100	100
1978–1979	111	96	96	87
1979–1980	114	99	108	74
1980–1981	95	88	94	83
1981–1982	115	91	93	123
1982–1983	91	81	100	101
1983–1984	118	89	103	141
1984–1985	119	96	113	115
1985–1986	107	85	108	136
1986–1987	105	86	110	126
1987–1988	103	88	105	122
1988–1989	110	95	112	111
1989–1990	121	103	129	145
1990–1991	115	92	125	139
1991–1992	106	84	103	164
1992–1993	117	89	108	173
1993–1994	120	87	99	186
1994–1995	95	86	88	151
1995–1996	121	89	91	162
1996–1997	128	93	87	186
1997–1998	110	80	83	188
1998–1999	120	84	87	174
1999–2000	129	91	83	166
2000–2001	120	92	86	166
2001–2002	122	87	79	174
2002–2003	89	84	75	177
2003–2004	126	90	74	175
2004–2005	129	120	67	210
2005–2006	138	104	62	237
2006–2007	105	99	56	233
2007–2008	94	71	49	166
2008–2009	105	72	45	198

Source: Authors' own calculation.

wool production has been declining in both quantity and relative share. The quantity of 'other farm income' increased significantly over the period, but its share in total output value remained low (about 5%).

Input measures

Total inputs are supposed to include all the economic resources used in the production of the outputs defined in *output measures*. In the literature on agricultural productivity, inputs are usually grouped into four types: land, capital, labour, and materials and services. ABARES follows this convention for the classification of inputs. Inputs consist of 27 items for broadacre farms and three additional items for dairy farms. Details of the inputs are presented in Appendix 4.1, Table A4.1.1.

LABOUR. A standard practice for measuring the input of labour is to use the actual time spent on farm production. This approach is recommended by OECD (2001). As mentioned earlier, some official statistical

Table 4.3. Value shares of the major outputs in the broadacre agricultural industry.

Year	Crops	Livestock	Wool	Other farm income
		Output value share		
1977–1978	30.3	41.7	24.5	3.5
1978–1979	30.6	48.4	18.4	2.6
1979–1980	29.4	49.7	18.8	2.1
1980–1981	34.4	44.9	18.1	2.6
1981–1982	45.7	35.4	15.7	3.2
1982–1983	43.0	35.3	18.6	3.1
1983–1984	37.5	40.1	18.5	4.0
1984–1985	33.0	43.0	20.9	3.1
1985–1986	30.0	41.2	24.6	4.2
1986–1987	25.5	42.8	27.7	4.0
1987–1988	26.6	37.6	32.2	3.6
1988–1989	26.8	37.9	32.1	3.2
1989–1990	27.1	37.1	31.8	4.0
1990–1991	27.0	38.8	30.0	4.1
1991–1992	35.1	38.9	20.7	5.2
1992–1993	35.5	40.9	18.5	5.0
1993–1994	35.3	42.9	16.5	5.3
1994–1995	34.2	39.6	21.4	4.8
1995–1996	45.2	34.6	15.6	4.7
1996–1997	44.9	33.7	16.2	5.2
1997–1998	41.8	34.1	18.2	5.9
1998–1999	42.0	38.5	14.3	5.1
1999–2000	42.0	40.8	12.6	4.6
2000–2001	35.9	44.6	15.0	4.6
2001–2002	40.0	43.1	12.6	4.4
2002–2003	41.0	38.7	15.9	4.5
2003–2004	39.0	44.5	12.3	4.2
2004–2005	29.2	56.3	10.0	4.5
2005–2006	36.2	50.1	8.2	5.5
2006–2007	37.4	48.9	8.2	5.5
2007–2008	44.7	41.5	9.2	4.7
2008–2009	47.1	40.6	6.9	5.3

Source: Authors' own calculation.

agencies have gone a step further to produce 'quality-adjusted' labour inputs (Bureau of Labor Statistics, 1983; Reilly *et al.*, 2005), using the hedonic method. ABARES uses actual time spent on farm production without further adjustment because data required for the hedonic adjustment are unavailable.

The labour input consists of three items: hired labour, owner–operator and family labour, and shearing costs. *Weeks worked* is used to measure the quantity of labour for hired workers, the owner–operator and family members. In the construction of a Fisher index for labour inputs, the price of hired labour is derived from survey data and is equal to the total cost of hired labour divided by the quantity used. The price of owner–operator and family labour is derived using a combination of the price of hired labour (for the owner–operator) and a relevant Federal Pastoral Award rate (for family members). For shearing, quantity data (weeks worked) are unavailable. As a result, the quantity of labour used for shearing is derived by dividing the total cost of shearing by a labour cost index.

CAPITAL.   For productivity analysis, capital inputs should be measured as *capital services*, the flow of productive services from the cumulative stock of past investments. As capital services constitute the actual input in the production process (OECD, 2001), they should be measured in physical units or with a quantity index. For example, the flow of services from a tractor may be measured by the transportation of goods from one place to another and the accomplishment of other tasks (such as tillage and sowing) over a particular period.

In reality, flows of capital services are not always observable or measurable. As a result, capital services are often approximated by assuming that *service flows are proportional to the 'productive stock'*, which is the sum of capital assets of different vintages after adjusting for 'retirement' (the withdrawal of assets from service)[12] and 'decay' (the loss in productive capacity as capital goods age),[13] and converting quantities to a standard 'efficiency' unit (see Jorgenson and Griliches, 1967; OECD, 2001).

Productive stock is a measure of the *physical capability* of capital stock used in production. It is not feasible, however, to construct a measure of the productive capital stock for individual farms using AAGIS data. Estimating productive capital stock is a very data-demanding exercise that typically requires a long series of historical data on investment in specific types and vintages of capital goods over the entire service life of an asset (which can be 20 years for certain machinery and 60 years or more for buildings). This is much longer than the typical period farms are retained in AAGIS, which is generally no more than five years. Therefore, it is not possible for ABARES to collect the farm-level data on capital expenditures necessary to construct a measure of productive capital stock.

In practice, when such data are unavailable, statisticians can use *wealth capital stock* as a substitute for the *productive capital stock*. Wealth capital stock, which is sometimes referred to as net capital stock in the literature, is the market value (a value counterpart) of the productive capital stock. This measure adjusts for economic depreciation or the loss in value of an asset as the capital stock ages, resulting in shorter service life (that is, retirement), and loss of productive capacity (that is, decay). In this sense, the wealth capital stock is considered to be a close substitute for the productive capital stock.[14]

In the case of broadacre agriculture, the ABARES estimate of capital service flows (or capital inputs) comprises eight types of capital inputs:

- Building and other farm improvements;
- Plant and machinery (both leased and self-owned);
- Beef cattle;
- Sheep;
- Other livestock;
- Changes in beef cattle numbers;
- Changes in the sheep flock;
- Changes in other livestock.

Estimates of capital service flows for buildings and other farm improvements, and plant and machinery are derived by assuming they are proportional to the *wealth capital stock*. These data are collected through AAGIS, where respondents are asked to estimate the market values of capital items. The present values of *building and other farm improvements* and *plant and machinery* are deflated by their respective prices paid indexes and then used directly as capital inputs.

The numbers of beef cattle, sheep and other livestock are measured in physical units, namely the average of opening and closing numbers. Changes in this variable reflect operating gains and net transfers-in during the financial year. These items are included on the input side because, in Australia, most broadacre farms are mixed enterprises which produce both crops and livestock, and livestock itself is an important input to the production process.

The input 'livestock' changes when farmers purchase additional beef or sheep for breeding or restocking. Unfortunately, there are no data that can be used to measure these variables in physical units, and they have to be represented by quantity indexes. The quantity indexes are derived by deflating the present value of livestock purchases by a price index for livestock products.

To aggregate specific capital items, the Fisher index requires information on the prices (or values) of *capital services* to construct weights. For capital items that are hired or leased, market rental prices are an appropriate price to use. However, in Australia, many farmers own their capital items and so their use of capital does not involve market transactions. Therefore, the 'prices' of capital services are unobservable. In this circumstance, economists have recommended a 'user costs' approach, which yields an 'implicit' value of the services that the owners of the capital 'pay' themselves. In other words, it is the amount of rent that would have been charged in order to cover the cost of using an asset. User costs can be calculated using Eqn 4.15 (OECD, 2001):

$$\mu_t = q_t(r_t + d_t) - (q_t - q_{t-1}) \qquad (4.15)$$

where μ_t is the cost of using the asset at period t, q_t is the market price of a new asset, r_t is the market rate of interest and d_t is the rate of depreciation.

The first term of Eqn 4.15 has two components. The first $(q_t r_t)$ measures the interest payment if a loan was taken out to acquire the asset. It also represents the opportunity cost of not employing capital elsewhere if the acquisition of the asset was financed from equity capital.[15] The second component $(q_t d_t)$ is the cost of depreciation, or the loss in the value of the asset due to ageing. The second term $(q_t - q_{t-1})$ measures capital gains or losses (revaluation of existing assets) that might happen owing to changes in the market value of assets, general inflation or obsolescence (see OECD, 2001, pp. 62–68).

Just as data constraints make it necessary to approximate *productive capital stock* by *wealth capital stock*, the same constraints mean it is not always feasible to collect historical data on the value of capital items to derive user costs using Eqn 4.15. The data limitation is overcome by using a simplified version of the model in Eqn 4.16:

$$\hat{\mu}_t = \hat{q}_t(r_t + d_t) \qquad (4.16)$$

This equation departs from Eqn 4.15 in several ways. First, $\hat{q}_t$ is an estimate of the total market value of all the vintages of capital goods purchased. Second, for *building and other farm improvements* and *plant and machinery*, Eqn 4.16 does not explicitly adjust for capital gains or losses. It is reasonable, however, to expect that the adjustment should have been done implicitly when survey respondents provided estimates of $\hat{q}_t$. Thus, an advantage of this approach is that the resulting value of $\hat{\mu}_t$ may have implicitly adjusted for the value of repairs, maintenance and new features or functions added to capital goods. This is hard to achieve using Eqn 4.15 because data required for the adjustment are difficult to collect.

LAND.   Quantity of land is measured as an average of the areas operated at the beginning and end of each financial year. This is a measure of stocks and used as an approximation for the quantity of services 'delivered' by the land. There is no independent information about the market value of land that may be used for the purpose of TFP estimation. This variable is represented by the 'opportunity cost' of the land as a capital asset, which is equal to the total value of land (estimated by farmers) minus the value of land attachments in the relevant period. Specifically, the value of land attachments includes:

- Buildings and other farm improvements;
- Plant and machinery, both leased and self-owned;
- Operator's house(s);
- Outputs of crops, beef cattle, sheep and other livestock.

The opportunity cost is equal to the capital asset value multiplied by a real interest rate.[16] This variable provides an estimate of the value of services provided by land as an input to production.

MATERIALS AND SERVICES.   This category covers a large number of diverse inputs that can broadly be divided into the two groups of materials and services. Materials include seven items: fertilizer, fuel, crop chemicals, livestock materials, seed, fodder and other materials. Services include nine items: contract services, rates and taxes, administrative services, repair services, expenses on motor vehicles, insurance, stores and rations, vet services and other services.[17]

For several reasons, it is impractical to measure the inputs of materials and services in terms of physical units. First, if using physical units, it is crucially important to make sure the quantities are measured on the basis of identical, or at least comparable, qualities in terms of the physical attributes of inputs (for example, nutrient contents of fertilizers or pesticide toxicity). However, collecting information on the qualities of materials and services is very costly and extremely difficult. Second, the input of services is often hard to quantify. Insurance, contract services and administrative services are good examples.

Because of these practical constraints, the input of materials and services is measured using quantity indexes (the right-hand side of Eqn 4.11). This method requires the value of expenditure on these inputs and appropriate indexes of prices paid. For example, fertilizer is one of the most important materials used in cropping. The input quantity of fertilizer is the farm's total expenditure on fertilizer divided by a price paid index for fertilizer. The data on total expenditure are also used to construct weights in the calculation of the quantity index.

The data on expenditures are available from the AAGIS and ADIS surveys. Before 2001–2002, the prices paid indexes were derived from ABARES *Prices Received and Paid (PRP) by Farmer Survey*. The PRP survey has been phased out, and currently the price indexes used in the productivity estimation are constructed with data from Australian Bureau of Statistics and ABARES databases.

INPUTS SPECIFIC TO DAIRYING.   Similar to the measurement of outputs, there are several inputs specific to the dairy industry. These include the existing stock and purchases of dairy cattle and costs associated with dairy supply. Both existing stock and purchases of dairy cattle are classified as capital inputs. The quantity of existing dairy cattle stock is measured as the average of the number of

dairy cattle at the beginning and end of the financial year. The changes in this variable reflect operating gains and net transfers-in during the financial year. The value of the existing dairy cattle stock is equal to the average number of dairy cattle multiplied by a price paid index for dairy cattle. The quantity of the purchased dairy cattle is calculated as the value of purchased dairy cattle deflated by a prices paid index for dairy cattle. The value of purchases is used to construct the weight for this item. 'Dairy supply' (such as purchased feed) is classified as a part of materials and services. The value of dairy supply is obtained from farm accounts, and it is collected in ADIS. The quantity is equal to the value deflated by a prices paid index.

The quantity indexes for *land, capital, labour* and *materials and services* used for broadacre agriculture and their relative value shares are presented in Tables 4.4 and 4.5. Between 1977–1978 and 2008–2009, the use of land, labour and capital decreased, whereas the use of materials and services increased. Although the share of the land input in total cost has fluctuated significantly, it has typically exceeded 50% and increased from 32.9% in 1977–1978 to 79.8% in 2008–2009. This in part reflects continued increases in land prices.

4.6.3   Calculation of average productivity growth

So far, this chapter has focused on the methodology used to derive estimates of productivity for individual farms, regions and industries. Estimates of productivity growth over time – for a farm or at industry, regional or national levels (such as the estimate presented in Fig. 4.1) – are particularly interesting because they are considered good indicators of technological progress and the estimates are very useful in the analysis of the performance of public R&D investment.

In productivity analysis, it is often of interest to compare productivity growth rates between different farms, industries, or regions, and to examine productivity growth over different time periods. In both circumstances, it

Table 4.4. Quantity indexes for land, capital, labour and materials and services for broadacre agriculture.

Year	Land	Capital	Labour	Materials and services
		Index, 1977–1978 = 100		
1977–1978	100	100	100	100
1978–1979	89	89	99	112
1979–1980	85	98	103	125
1980–1981	74	117	99	117
1981–1982	82	122	98	124
1982–1983	76	123	93	114
1983–1984	82	110	94	115
1984–1985	86	112	95	118
1985–1986	78	99	84	100
1986–1987	77	97	82	97
1987–1988	73	90	81	102
1988–1989	77	95	82	113
1989–1990	78	99	92	120
1990–1991	78	97	87	101
1991–1992	77	84	79	104
1992–1993	77	85	74	110
1993–1994	77	80	70	110
1994–1995	79	78	67	109
1995–1996	75	77	71	113
1996–1997	73	78	70	113
1997–1998	72	71	61	106
1998–1999	71	71	60	104
1999–2000	71	72	61	108
2000–2001	72	71	63	107
2001–2002	75	64	62	118
2002–2003	71	73	65	124
2003–2004	76	75	67	124
2004–2005	70	90	55	126
2005–2006	70	82	63	137
2006–2007	66	88	58	140
2007–2008	61	58	51	108
2008–2009	59	54	50	105

Source: Authors' own calculation.

is necessary to calculate *average* productivity growth over multiple periods. For example, several ABARES reports (Nossal *et al.*, 2008; Nossal *et al.*, 2009; Zhao *et al.*, 2008) analysed the performance of the broadacre sector and dairy industry on the basis of the average growth rates of TFP for these industries.

Generally speaking, two methods can be used to calculate average productivity growth. In the first method, average productivity growth over *N+1* periods is calculated on the basis of Eqn 4.17:

$$Average\ TFP\ Growth = \frac{\sum_{t=1}^{N} \ln(\frac{TFP_t}{TFP_{t-1}})}{N} \qquad (4.17)$$

where TFP_t is TFP estimates for period *t*.

In the second method, average productivity growth can be derived econometrically with the following regression equation (Eqn 4.18):

$$\ln(TFP) = \alpha + \beta t \qquad (4.18)$$

Table 4.5. Cost shares of land, capital, labour and materials and services for broadacre agriculture.

Year	Land	Capital	Labour	Materials and services
				Input cost share
1977–1978	32.9	26.6	20.2	20.2
1978–1979	36.4	26.9	17.1	19.6
1979–1980	35.2	28.9	15.6	20.2
1980–1981	40.8	25.7	14.4	19.1
1981–1982	57.3	18.6	10.1	13.9
1982–1983	55.1	19.5	10.5	14.9
1983–1984	60.4	16.8	9.4	13.4
1984–1985	70.1	13.8	6.7	9.4
1985–1986	67.5	14.9	7.4	10.1
1986–1987	55.4	21.4	9.8	13.4
1987–1988	55.4	19.0	10.3	15.3
1988–1989	67.6	13.8	7.4	11.2
1989–1990	70.8	11.8	7.2	10.2
1990–1991	72.8	10.7	7.5	9.1
1991–1992	74.5	9.7	6.9	8.9
1992–1993	71.0	11.2	7.3	10.5
1993–1994	64.3	13.6	8.8	13.3
1994–1995	67.7	12.1	7.9	12.2
1995–1996	65.6	12.0	9.0	13.4
1996–1997	62.6	12.7	9.9	14.8
1997–1998	65.5	12.2	8.7	13.6
1998–1999	60.9	13.9	10.2	15.1
1999–2000	56.0	14.9	11.4	17.8
2000–2001	50.7	17.0	12.4	19.9
2001–2002	45.7	18.9	12.5	22.9
2002–2003	33.8	28.4	13.0	24.8
2003–2004	48.1	19.0	12.0	20.9
2004–2005	48.4	22.8	9.1	19.7
2005–2006	51.4	18.5	9.8	20.3
2006–2007	63.1	15.2	6.7	15.0
2007–2008	64.0	12.6	8.0	15.4
2008–2009	79.8	7.7	4.3	8.2
Average	57.8	16.9	10.0	15.2

Source: Authors' own calculation.

where ln(*TFP*) is the logarithm of TFP estimates and t is a linear time trend.[18] Equation 4.18 is estimated using Ordinary Least Squares regression and the coefficient (β) is the value of average TFP growth rate.[19]

Both methods are valid for calculating average productivity growth and may produce similar results. ABARES uses the second method. Table 4.6 shows average growth rates of TFP, outputs and inputs for the broadacre sector between 1977–1978 and 2008–2009, which can be used to estimate

the contribution of various factors of production to TFP growth.

Before leaving the discussion of calculating average productivity growth, a warning is worth noting. Productivity estimates for the Australian broadacre industries tend to fluctuate significantly from year to year (Fig. 4.1), partly because of the influence of variable climate conditions. Because of this volatility, analysts should be careful in determining the periods over which average TFP growth is calculated and compared,

Table 4.6. Growth of TFP, inputs and outputs for broadcare agriculture: 1977–1978 to 2008–2009.

	Average annual growth rate	Average value or cost share
	% per year	%
TFP	1.37	
Total outputs	0.46	100.0
Crops	0.71	36.0
Livestock	0.24	41.4
Wool	−0.72	18.3
Other outputs	1.10	4.3
Total inputs	−0.91	100.0
Land	−0.88	57.8
Capital	−1.74	16.9
Building and structure	−4.86	2.1
Plant and machinery	−2.58	6.1
Livestock capital	−0.53	8.7
Labour	−2.19	10.0
Purchased material	1.02	5.4
Feed/ fodder	3.52	1.2
Fuel	−0.16	1.3
Fertilizer	0.81	2.5
Services	−0.36	9.8

Source: Authors' own calculation.

because the values of the TFP estimates at the beginning and end of the period selected can be influential in the outcome. It is important to make sure that the periods under comparison make economic and statistical sense and, in particular, that they do not start from, or end at, observations that are far from the long-term productivity trend.

4.7   Concluding Remarks

For many years, ABARES has maintained a programme to report and analyse productivity estimates for the Australian broadcare sector and the dairy industry. Productivity is broadly defined as a ratio of output to input and, according to economic theory, productivity

growth represents a shift in the production function. TFP is a particular measure of productivity, usually expressed as a ratio of total outputs to total inputs. Growth in TFP is often used as an indicator of the progress of disembodied technology, because it measures the change in total outputs that is not accounted for by the change in total inputs. Among economists, it is widely accepted that innovations arising from research and development are the most important determinants of the trend of technological progress and that they are the main force underlying long-term productivity growth. In practice, scale economies and efficiency improvements arising from, for example, better use of existing capacity and institutional and organizational changes that remove constraints on production will also increase productivity.

Productivity growth in Australian broadcare agriculture tends to be volatile, particularly in the short term, because it is significantly influenced by the country's highly variable climate conditions.

In measuring the productivity of the broadcare sector and the dairy industry, ABARES uses a methodology that is recommended by OECD (2001) and is comparable with methods used by international official statistical agencies. This means ABARES productivity statistics can be interpreted in a similar way to those produced by the Australian Bureau of Statistics, as well as by the OECD and other international statistical agencies.

In common with official statistical agencies, ABARES has used an index method to estimate TFP and partial factor productivity. The estimation procedure for TFP is a three-step process. In the case of the broadcare sector, step one involves the aggregation of specific items to form four types of outputs (crops, livestock, wool and other farm income) and four types of inputs (labour, capital, land and materials and services). Step two involves aggregating the outputs to create an index of 'total outputs' and the inputs to create an index of 'total inputs'. Finally, in the third step, TFP is calculated by taking the ratio of the index of total outputs to total inputs.

The Fisher index formula is used in the process of aggregating both outputs and inputs. The Fisher index is considered to outperform many other index formulae by passing significantly more axiomatic tests. It has also proven to be 'exact' for the homogeneous quadratic production function, which is considered a good representation of a wide range of possible production functional forms. The Fisher index also has the advantage, for Australian estimates, of being able to satisfactorily handle zeros in data. When the Fisher index is applied to the time series data to estimate productivity growth, ABARES uses a direct, rather than a chained, index. When the index formula is applied to the cross-sectional or panel data to calculate productivity levels, the Fisher index is converted into a multilateral index based on the EKS formula.

Most of the data used in the compilation of productivity statistics are sourced from ABARES farm surveys: AAGIS for the broadacre sector and ADIS for the dairy industry. In the calculation of the Fisher productivity index, any additional data required for the construction of price deflators and weights are obtained from the ABARES database, and Australian Bureau of Statistics and Reserve Bank of Australia (RBA) publications.

In ABARES TFP measures, outputs include all the crops (wheat, barley, oats, grain sorghum, oilseeds and other crops),

livestock (beef, sheep, lamb and other livestock), wool, milk (for dairy farms) and other income generated by farmers in their business activities (such as hiring of plant and equipment and entitlements). Inputs include labour (both family and hired), capital (including buildings, machinery and livestock), land, and materials and services.

ABARES has applied conventional index methods to measure both outputs and inputs. For example, specific items of outputs and inputs are measured in either physical units or by quantity indexes and, for most output and input items, variables measuring economic 'flows' are used in the productivity calculation and, for the rest, 'stock' variables are used as approximations. However, the measurement of capital inputs is based on a non-conventional approach. In ABARES methodology, the measurement of 'buildings and other farm improvements' and 'plant and machinery' is based on the concept of 'wealth capital stock' rather than that of 'productive capital stock'. The latter has been used by most statistical agencies to measure capital input. In contrast, 'wealth capital stock' is the market value of the 'productive capital stock'. Nevertheless, in a competitive market, the two variables are considered to be close substitutes for the purpose of productivity estimation.

Notes

[1] When official statistical agencies estimate total factor productivity, land is often not treated separately from capital and therefore Equation 4.2 is sometimes expressed as: $TFP(v) = \dfrac{Quantity\ (index)\ of\ value\ added}{Quantity\ (index)\ of\ labor\ and\ capital}$. In many empirical studies on the productivity of non-agricultural industries, land is also not treated as a distinct input.

[2] In the literature, materials and services are known as *intermediate inputs*, which include inputs of materials, energy and services and, therefore, Eqn 4.1 is sometimes called the KLEMS model because the numerator covers capital (K), labour (L), energy (E), materials (M) and services (S).

[3] Note that the gross-output-based measure of TFP has a drawback: it is not consistent in aggregation if nominal values are directly used as weights. To overcome this problem, a set of so-called Domar weights (Domar, 1961) that add up to more than 1 should be used to aggregate TFP estimates. This treatment is necessary for official statistical agencies to account for the trade within industries when they report TFP estimates at various aggregation levels. This is not considered to be a major problem for ABARES estimation of total factor productivity as trade within the broadacre sector represents an insignificant proportion of total output.

[4] OECD (2001) suggested that fixed-weight indexes, such as the Laspeyres and Paasche formulae, have a greater danger of suffering biases than flexible weight indexes, such as Fisher and Tornqvist, when chaining is not applied. But if indexes are chained, the choice of index formulae becomes inconsequential.

[5] The samples used in the estimation of the TFP for broadacre and dairy industries take the form of unbalanced panel data. Section 4.5 provides more information on this topic.

[6] The original EKS system uses the bilateral Fisher index, but the works of Caves, Christensen and Diewert (1982) and Rao and Banerjee (1984) recognized that other bilateral indexes could also be used in conjunction with the EKS techniques.

[7] In circumstances where no data are available for constructing weights, index formulae without explicitly defined weights may also be used in the construction of indexes. They are called 'micro-indexes' in the literature and are used by official statistical agencies in the compilation of the consumer price index (Balk, 1996; Diewert, 1986).

[8] This is an externally imposed condition on the estimates of TFP, which implies that the production technology is characterized by constant returns to scale.

[9] In economics, input elasticity is a measure of output response to a change in inputs.

[10] The price indexes are derived from the average prices of the crops sold by the farmers, and they are partly determined by the quality of the products. Hence, when these indexes are used as deflators, the differences in the quality of outputs are implicitly adjusted for in the resulting quantity indexes.

[11] Using 'net receipts from sale' in the calculation implicitly assumes that farmers have retained a similar proportion of total output for their own consumption and future production.

[12] In practice, statisticians need to assume a retirement function and apply it to adjust for the withdrawal of assets from service as they are scrapped or discarded.

[13] An age-efficiency function is required to adjust for the decay that results from the effects of wear and tear. Official statistical agencies have typically used hyperbolic, geometric or linear declining functions for this adjustment.

[14] In reality, wealth capital stock may deviate from productive capital stock for at least four reasons. First, the market value of capital stock may be more strongly affected by obsolescence (Triplett, 1989). If, for example, new vintages of capital use superior technologies, then the market prices of older vintages tend to decline faster than the loss of their productive capacity. Second, if capital goods are designed for certain specific applications and, consequently, opportunities for resale are reduced, then prices of second-hand capital goods may be suppressed and under-valued. This can be true of certain types of capital goods used in agriculture, such as harvesters and farm improvements. Third, according to the OECD (2001), unless a geometric function is applied to the age-efficiency function, the age-price function will be intractable. In this circumstance, changes in the value of capital (owing to economic depreciation) and the productive capacity (owing to physical decay) will not coincide. Finally, owners nearly always need to repair, maintain or add new features or functions to capital goods, which will increase both the productive capacity and economic value. Unless the renewed or additional productive capacities are accurately valued by the market, the measures of wealth capital stock and productive capital stock will differ.

[15] In the literature, r_t has been called 'internal rate of return' or 'net rate of return'.

[16] Quality of land can vary significantly between farms. If information on the price of land was available, an adjustment could be made to account for the impact of land quality on productivity. It could be argued, however, that inclusion of output values from the farms partially achieves this quality adjustment.

[17] 'Other materials' and 'other services' contain a range of items that are diverse in nature and cannot be meaningfully classified into one category.

[18] It is also possible to derive average TFP growth by regressing TFP (without taking the logarithm) on a time trend. However, the result from this approach involves a different assumption about the pattern of TFP growth over time. Although Eqn 4.18 assumes that TFP *growth rate* over the whole period $\left(\dfrac{TFP_t - TFP_{t-1}}{TFP_{t-1}} \right)$ remains constant, this method assumes that the changes in TFP level ($TFP_t - TFP_{t-1}$) are constant.

[19] The estimates from Eqns 4.17 and 4.18 are average productivity growth in their original units. They can be transformed into growths in percentage term simply by multiplying by 100.

References

ABARES (2008) *Australian Commodity Statistics 2008*, Canberra, Australia.

ABARES (2009) *Survey Methods and Definitions,* Canberra. Available at http://www.abare.gov.au/publications_html/economy/economy_09/survey_methods.pdf

Alexander, F. and Kokic, P. (2005) Productivity in the Australian grains industry, *ABARES eReport No. 05.3.*

Alston, J.M. and Pardey, P.G. (1996) *Making Science Pay: The Economics of Agricultural R&D Policy,* American Enterprise Institute, Washington, DC.

Australian Bureau of Statistics (2000) *Australian System of National Accounts: Concepts, Sources and Methods,* ABS Catalogue no. 5216.0, Canberra, Australia.

Australian Bureau of Statistics (2006) Australian and New Zealand Standard Industrial Classification (ANZIC), 2006 Revision. ABS Catalogue no. 1292.0, Canberra, Australia.

Australian Bureau of Statistics (2007) Experimental estimates of industry multifactor productivity. Information Paper, ABS Cat. 5260.0.55.001, Canberra, Australia.

Australian Bureau of Statistics (2010) Experimental estimates of the gross value of irrigated agricultural production, 2000–2001 – 2008–2009. ABS Catalogue no. 4610.0.55.008, Canberra, Australia.

Balk, B.M. (1996) A comparison of ten methods for multilateral international price and volume comparison. *Journal of Official Statistics,* 12, 199–222.

Basu, S. and Fernald, J. (2001) Why is productivity procyclical? Why do we care? In: Hulten, C., Dean, E. and Harper, M. (eds) *New Development in Productivity Analysis.* University of Chicago Press, Chicago, IL, pp.225–302.

Bureau of Labor Statistics (1983) Trends in multifactor productivity, 1948–1981, Bulletin 2178, US Department of Labor, Washington DC.

Bureau of Labor Statistics (1993) *Labor Composition and U.S. Productivity Growth, 1948–1990,* Bulletin 2426, US Department of Labor, Washington DC.

Caves, D., Christensen, L. and Diewert, W. (1982) The economic theory of index numbers and the measurement of input, output, and productivity, *Econometrica* 50, 6, 1393–1414.

Chavas, J. (2001) Structural change in agricultural production: Economics, technology and policy. In: Gardner, B. and Rausser, G. (eds) *Handbook in Agricultural Economics,* Elsevier, Amsterdam, the Netherlands, pp. 263–285.

Coelli, T., Rao, D.S.P., O'Donnell, C. and Battese, G. (1998) *An Introduction to Efficiency and Productivity Analysis,* 2nd edn, Springer, New York, NY.

Dennison, E. (1972), Some major issues in productivity analysis: An examination of the estimates by Jorgenson and Griliches. *Survey of Current Business* 49, 1–27.

Diewert, E. (1976) Exact and superlative index numbers. *Journal of Econometrics* 4, 115–145.

Diewert, W.E. (1986) Microeconomic approaches to the theory of international comparisons. Technical Working Paper No. 53, National Bureau of Economic Research, Cambridge, MA. Abridged version: Test approaches to international comparison. In: Eichorn, W. (ed.) (1988) *Measurement in Economics,* Physica–Berlag, Heidelberg and in: Diewert, W.E. and Nakamura, A.O. (eds) (1993) *Essays in Index Number Theory,* Volume 1, North-Holland, Amsterdam.

Diewert, E. (1988) Microeconomic approaches to the theory of international comparisons. In: Eichorn, W (ed.) *Measurement in Economics,* Physica-Berlag, Heidelberg, Germany.

Diewert, E. (1978) Superlative index numbers and consistency in aggregation. *Econometrica* 4, 883–900.

Diewert, E. (1992) Fisher ideal output, input, and productivity indexes revisited. *Journal of Productivity Analysis* 3, 211–248.

Diewert, E. and Fox, K. (2008) On the estimation of returns to scale, technical progress and monopolistic markups. *Journal of Econometrics* 148, 174–192.

Domar, E. (1961) On the measurement of technological change. *Economic Journal* 71, 709–729.

Elteto, O. and Koves, P. (1964) On a problem of index number of computation relating to international comparison. *Statsztikai Szemle* 42, 507–518.

Fisher, I. (1922) *The Making of Index Numbers,* Houghton-Mifflin, Boston, MA.

Gini, C. (1931) On the circular test of index numbers. *International Review of Statistics* 9, 3–25.

Griliches, Z. (1996) The discovery of the residual: A historical note. *Journal of Economic Literature* 24, 1324–1330.

Grilliches, Z. (1987) Productivity: measurement problems, In: Eatwell, J., Milgate, M. and Newman, P. (eds) *A Dictionary of Economics,* Cambridge Press, Cambridge, MA.

Hayami, Y. and Ruttan, V.W. (1970) Factor prices and technical change in agricultural development: The United States and Japan, 1880–1960. *Journal of Political Economy* 78, 1115–1141.

Hulten, C. (2000) Total factor productivity: A short biography. In: Hulten, C., Dean, E. and Harper, M. (eds) *New Developments in Productivity Analysis,* University of Chicago Press, Chicago, IL, pp.1–54.

Jorgenson, D. and Griliches, Z. (1972) Issues in growth accounting: A reply to Edward F. Denison. *Survey of Current Business* 52, 65–94.

Jorgenson, D. and Griliches, Z. (1967) The explanation of productivity change, *Review of Economic Studies* 34, 3, 249–283.

Knopke, P., Strappazzon, L. and Mullen, J. (1995) Productivity growth: Total factor productivity on Australian broadacre farms. *Australian Commodities* 2, 4, 486–497.

Knopke, P., O'Donnell, V. and Shepherd, A. (2000) Productivity growth in the Australian grains industry, ABARES Research Report, Canberra, Australia.

Kokic, P., Davidson, A. and Rodriguez, V.B. (2006) Australia's grain industry – factors influencing productivity growth. ABARES Research Report No. 06.22, Canberra, Australia.

Kompas, T., Che, N. and Gooday, P. (2008) Analysis of productivity and the impacts of swordfish depletion in the eastern tuna and billfish fishery. ABARE Report to the Fisheries Resources Research Fund, Canberra, Australia.

Kravis, I., Heston, A. and Ross, S. (1982) *International Comparisons of Real Product and Purchasing Power.* Johns Hopkins University Press, Baltimore, MD.

Mullen, J. (2007) Productivity growth and the returns from public investment in R&D in Australian broadacre agriculture. Presidential Address at the 51st Conference of the Australian Agricultural and Resource Economics Society, Queenstown, New Zealand, 14–16 February.

Mullen, J. and Crean, J. (2007) Productivity growth in Australian agriculture: Trends, sources, performance. Report Prepared for the Australian Farm Institute, RIRDC, GRDC and MLA.

Nossal, K., Sheng, Y. and Zhao, S. (2008) Productivity in the beef cattle and slaughter lamb industries. ABARE Research Report 08.13, Canberra, Australia.

Nossal, K., Zhao, S., Sheng, Y. and Gunasekera, D. (2009) Productivity movements in Australian agriculture. *Australian Commodities* 16, 206–216.

O'Donnell, C. (2008) An aggregate quantity–price framework for measuring and decomposing productivity and profitability change. Working Paper WP07/2008. Centre for Efficiency and Productivity Analysis, University of Queensland, Brisbane, Australia.

OECD (2001) *Measuring Productivity: Measurement of Aggregate and Industry-level Productivity Growth.* Organisation for Economic Co-operation and Development, Paris, Frances.

Parham, D. (2004) Sources of Australia's productivity revival. *Economic Record* 80, 239–257.

Pollak, R. (1989) *The Theory of the Cost-of-living Index.* Oxford University Press, New York, NY.

Productivity Commission (2005) Trends in Australian agriculture. Research Paper, Commonwealth of Australia, Canberra, Australia.

Rao, D.S.P. and Baneerjee, K.S. (1984) A multilateral system of index numbers based on factorial approach, *Statiche Hefte*, 27, 297–312.

Reilly, R., Milne, W. and Zhao, S. (2005) Quality-adjusted labour inputs. Research Paper, ABS Catalogue no. 1351.0.55.010, Australian Bureau of Statistics, Canberra, Australia.

Sheng, Y., Mullen, J. and Zhao, S. (2010) Has growth in productivity in Australian broadacre agriculture slowed? 2010 Conference (54th), 10–12 February 2010, Adelaide, Australia 59266, Australian Agricultural and Resource Economics Society.

Szulc, B. (1964) Indices for multiregional comparisons. *Statistical Review* 3, 239–254.

Szulc, B. (1983) Linking price index numbers. In: Diewert, E. and Montmarquette, C. (eds) *Price Level Measurement*, Statistics Canada, Ottawa.

Topp, V., Soames, L., Parham, D. and Bloch, H. (2008) Productivity in the mining industry: Measurement and interpretation. Staff Working Paper, Productivity Commission, Canberra, Australia.

Triplett, J. (1989) Price and technological change in a capital good: a survey of research on computer. In: Jorgenson, D.W. and Landau, R. (eds) *Technology and Capital Formation,* MIT Press, Cambridge, MA.

Waschka, A., Milne, W., Khoo, J., Quirey, T. and Zhao, S. (2003) Comparing living costs in Australian capital cities. ABS presentation to 32nd Conference of Economists, 29 September–1 October, Canberra, Australia.

Zhao, S., Nossal, K., Kokic, P. and Elliston, L. (2008) Productivity growth: Australian broadacre and dairy industries. *Australian Commodities* 15, 236–242.

Zhao, S., Sheng, Y. and Kee, H. (2009) Further exploring determinants of the total factor productivity in Australia's broadacre grain farms. ABARE paper prepared for the Australian Conference of Economists, Adelaide, Australia.

Zheng, S. (2005) Estimating industry-level multifactor productivity for the market sector industries in Australia: Methods and experimental results. *ABS Research Paper 1351.0.55.004,* Canberra, Australia.

Appendix 4.1 Measures of Outputs and Inputs

Table A4.1.1. Inputs for the estimation of TFP in broadacre and dairy industries.

Input type	Specific inputs	Data items
Land	Land	Average area operated in the financial year
Capital	Buildings and other farm improvements	Farm buildings excluding operator's house, water supply structure and fencing (real opportunity cost)
		Structural improvements including water supply infrastructure, fencing and yards (real opportunity cost)
		Real depreciation cost of buildings and other farm improvements
	Plant and machinery	Plant and equipment owned by the operator (real opportunity cost)
		Plant and equipment leased by the operator (real opportunity cost)
		Real depreciation of plant and equipment
		Real depreciation of structural improvements
	Beef cattle	Real value of beef cattle
	Other livestock	Real value of dairy and other cattle
	Beef cattle purchased	Real costs of purchasing beef cattle
	Sheep purchased	Real costs of purchasing sheep and lambs
	Other livestock purchased	Real cost of purchasing other livestock
Labour	Hired labour	Weeks worked on farm by hired permanent and casual workers
	Owner operator and family labour	Total weeks worked on farm by owner operator and family members
	Services by shearers	Real costs of shearers
Materials	Seed	Real costs of seed
	Fodder	Real costs of fodder and purchase of non tree and vine crops
	Crop chemicals	Real costs of crop and pasture chemicals, such as pesticides and herbicides
	Fuel	Real costs of fuel, oil and grease
	Livestock materials	Real costs of livestock materials such as dips and drenches
	Other materials	Materials used to pack fruits
		Materials for packing crops other than tree and vine crops
		Tree and vine replacements
		Purchase of water for livestock
		Wool packs
		Livestock agisted off-farm (excluding lot feeding costs off-farm)
		Materials not included elsewhere (for example, vermin control and protective clothing)
Services	Contract services	Real cost of plant hire and non-capital development contracts such as mustering, harvesting
	Rates and taxes	Real costs of rates paid including drainage and water etc.
	Administrative services	Real costs of accountancy, banking and legal services
		Real costs of electricity
	Repair services	Real cost of repair for building and structure (for example, fences etc.)
		Real costs of repair of motor vehicles, plant and equipment
	Motor vehicle	Real costs for vehicle registration and third party insurance etc.

Continued

Table A4.1.1. Continued.

Input type	Specific inputs	Data items
	Other services (1)	Real costs of advisory services Real costs of artificial insemination, herd testing and stud fees Real costs of veterinary services
	Other services (2)	Real costs of crutching Real costs of stores and rations provided to workers Real costs of tree and vine crops packed off-farm for sale Real costs of packing crops other than tree and vine crops Real costs of (work related) travelling and entertaining expenses Real costs of services not included elsewhere
	Insurance	Real (net) costs of insurance for crops, livestock, buildings, improvements, motor vehicles and workers' compensation
Inputs specific to dairy farms	Dairy cattle	Dairy cattle (real opportunity cost), as capital
	Real value of gain on purchased dairy cattle	Dairy cattle operating gain (the change in the value of the herd, in real terms)
	Fertilizer	Real costs of fertilizer (for pastoral), as materials
	Dairy supplies	Real costs of dairy supplies, as services

Source: Authors' own definition.

Table A4.1.2. Outputs for the estimation of TFP in broadacre and dairy industries.

Output type	Specific inputs	Data items
Crops	Wheat	Real receipts from the sale of wheat (net of marketing expenses)
	Barley	Real receipts from the sale of barley (net of marketing expenses)
	Oats	Real receipts from the sale of oats (net of marketing expenses)
	Grain sorghum	Real receipts from the sale of grain sorghum
	Oilseeds	Real receipts from the sale of oilseeds
	Other crops	Real receipts from the sale of other crops
Livestock	Beef	Real receipts from the sale of beef cattle Operating gain (changes in the value of the herd, in real terms)
	Sheep	Real receipts from the sale of sheep Operating gain (changes in the value of the flock, in real terms) Transfer out (net), in real terms
	Lamb	Real receipts from the sale of prime lambs
	Other livestock	Real receipts from the sale of other livestock
Wool	Wool	Real receipts from the sale of wool through broker Real receipts from the private sale of wool
Other outputs	Off-farm contracts	Real receipts of income from off-farm contracts
	Hire of plant	Real receipts from the hire of plant
	Private use of farm equipment	Value of private use of farm equipment (telephone and vehicles, in real terms)

Continued

Table A4.1.2. Continued.

Output type	Specific inputs	Data items
	Rebates and refunds	Real rebates and refunds received on purchases of farm inputs
	Royalties	Real royalties received
	Government assistance	Real farm income received in the form of government payments
	Other farm receipts not included elsewhere	Real other farm receipts not included elsewhere
Outputs specific to dairy farms	Beef	Real receipts from the sale of beef cattle
		Beef operating gain (the change in the value of the herd, in real terms)
	Sheep	Real receipts from the sale of sheep and lambs
	Dairy cattle gain	Dairy cattle operating gain (the change in the value of the herd, in real terms)
	Dairy cattle sold	Real receipts from sale of dairy cattle sold
	Other livestock	Real receipts from the sale of other livestock
	Other outputs	Real income from off-farm contracts
		Real other farm receipts not included elsewhere
		Real receipts from hire of plant
		Value of private use of farm equipment (telephone, vehicles, in real terms)
		Real income received in the form of insurance recoveries
		Real livestock compensation received
		Real income from agistment of livestock on-farm
		Real receipts from wool sold through broker
		Real receipts from wool sold privately

Source: Authors' own definition.

Appendix 4.2 Sample Size of Broadacre and Dairy Industries

Table A4.2.1. Sample points from Australian agricultural and grazing industries survey (AAGIS): 1977–1978 to 2006–2007.

Year	Cropping specialist	Cropping–livestock mixed	Sheep specialist	Beef specialist	Sheep–beef mixed	All farms
1978	224	376	221	288	266	1,375
1979	155	255	169	189	149	917
1980	145	251	148	193	155	892
1981	150	244	175	202	152	923
1982	138	217	172	249	145	921
1983	125	215	180	259	145	924
1984	129	171	171	224	83	778
1985	164	150	173	203	80	770
1986	127	225	168	188	86	794
1987	175	283	225	250	114	1,047
1988	122	322	239	238	122	1,043
1989	117	303	298	259	137	1,114
1990	99	234	301	266	141	1,041
1991	221	423	428	366	216	1,654
1992	191	419	395	384	222	1,611
1993	264	341	291	393	236	1,525
1994	329	343	231	452	286	1,641
1995	260	306	228	410	231	1,435
1996	277	238	223	386	179	1,303
1997	354	330	210	328	165	1,387
1998	301	288	205	348	180	1,322
1999	262	269	213	316	164	1,224
2000	260	266	198	304	158	1,186
2001	301	246	180	334	160	1,221
2002	260	237	160	291	143	1,091
2003	343	332	179	354	134	1,342
2004	280	274	188	343	141	1,226
2005	334	324	218	443	189	1,508
2006	336	288	226	425	183	1,458
2007	234	345	243	453	181	1,456
Total	6,677	8,515	6,656	9,338	4,943	36,129

Table A4.2.2. Sample points from the Australian dairy industry survey (ADIS): 1978–1979 to 2006–2007.

Year	Dairy	Percentage[a]
1979	301	3.27
1980	325	3.53
1981	323	3.51
1982	331	3.60
1983	331	3.60
1984	287	3.12
1985	293	3.18
1986	281	3.05
1987	293	3.18
1988	292	3.17
1989	293	3.18
1990	295	3.21
1991	297	3.23
1992	302	3.28
1993	307	3.34
1994	402	4.37
1995	332	3.61
1996	339	3.68
1997	334	3.63
1998	310	3.37
1999	309	3.36
2000	308	3.35
2001	307	3.34
2002	310	3.37
2003	307	3.34
2004	219	2.38
2005	290	3.15
2006	296	3.22
2007	291	3.16
2008	298	3.24
Total	9,203	100

[a]Percentage of total population of dairy farms.

5 Is Agricultural Productivity Growth Slowing in Western Europe?

Sun Ling Wang, David Schimmelpfennig and Keith O. Fuglie
Economic Research Service, US Department of Agriculture, Washington, DC

5.1 Introduction

Agricultural production in Western Europe[1] has been nearly stagnant for the past 25 years. According to the Food and Agriculture Organization (FAO), agricultural output in Western Europe in 2009 was only about 4% higher than it was in 1984. World Bank data report agricultural gross domestic product (GDP) of the region grew by 26% (in constant US dollars) over the same period, but this was almost certainly a terms-of-trade or exchange rate effect, rather than a change in real production. The stagnation in real output could be due to rising costs of production (falling productivity) and/or fewer resources being employed in production. The slow growth of agriculture in Western Europe has meant that its share of global agricultural output has been falling steadily, from about 20% in the 1960s to less than 10% by the late 2000s, enough to significantly affect global agricultural markets and trade flows.

In a recent, wide-ranging global assessment of agricultural production and productivity trends, Alston, Babcock and Pardey (2010) concluded that 'agricultural productivity growth has slowed, especially in the world's richest countries,' and they point to stagnation in agricultural research and development (R&D) spending as the most

likely cause. Given that agricultural resources in these countries are also not growing (and in some cases shrinking), a declining rate of productivity growth implies slower growth in output. Their conclusion drew on data from the USA, the prairie provinces of Canada, Australia and the UK. But apart from the UK, they were not able to look in detail at Western Europe. This chapter employs the dataset Eldon Ball and his colleagues developed to measure agricultural total factor productivity (TFP) in 11 EU countries between 1973 and 2002 (Ball et al., 2010). We apply statistical tests to the individual country TFP series to see whether any of them experienced a significant slowing of TFP growth over this period. In addition, we construct aggregate TFP indexes for the 11 countries as a whole, as well as for the Mediterranean and northern continental sub-regions and examine productivity trends in these regions. We also look at partial productivity indicators (land and labour productivity, labour–capital and land–labour ratios) and examine trends in agricultural R&D spending. We would expect to see stagnant R&D spending to precede stagnant TFP growth by one to two decades, if this were indeed the case.

Although the evidence cited by Alston et al. (2010) does point to an agricultural productivity slowdown in some developed countries, previous research on this issue is not unanimous. In their previous analysis with the EU dataset, Ball *et al.* (2001, 2010) found evidence of productivity convergence: countries with initial low levels of productivity grew more rapidly, tending to catch up to productivity leaders over time. Schimmelpfennig and Thirtle (1999, 2000) partially explained EU country differences in productivity growth in terms of technology spillovers from public research systems. Thirtle *et al.* (2004) documented a significant decline in UK agricultural TFP growth since the 1980s and attributed this to a precipitous drop in public support for productivity-oriented agricultural research. Using a different estimate of agricultural TFP growth for the USA, Wang (2010) does not find evidence of a slowdown, in contrast to Alston *et al.* (2010). Zhao, Sheng and Gray's TFP estimates for Australia (Chapter 4, this volume) do show stagnation since 2000, but Cahill and Rich (Chapter 3, this volume) find no significant divergence from long-run TFP growth trends for Canada through 2006. Fuglie (2008, 2010) does not find support for a global slowdown in productivity. Given the stagnation of output growth in Western European agriculture and the mixed evidence on productivity trends for the region, a re-examination of the data for this region seems worthwhile. Before presenting formal statistical tests for TFP trends, we describe some salient features of agriculture in the EU, where about 98% of the agricultural production in Western Europe occurs.

5.2   Key Features of EU Agriculture

When six countries in Western Europe first established a common market for goods and services in 1957, they created a Common Agricultural Policy (CAP) to harmonize agricultural policies among member states. The principal features of CAP included price stabilization and support for important agricultural commodities, free internal trade, tariff protection against external competition, and income support for farmers. CAP was initially successful in raising agricultural production, and eventually supply controls such as production quotas, land retirement schemes and export subsidies were introduced to reduce production of the surpluses that accumulated for some commodities. As the common market evolved into the European Union and more countries joined, CAP was extended to new members. Several of the new members had relatively large agricultural sectors, especially in terms of the number of farms, and their inclusion in CAP put considerable budgetary pressure on EU resources. The high cost of CAP has provided a stimulus to reform, and its policies have gradually shifted from commodity price supports to direct payments to farmers and a requirement for recipients to comply with environmental, animal welfare and food quality standards. Most of these reform measures were introduced after 1992 (Economic Research Service, a).

The structure of agriculture is highly diverse across EU member states. Compared to the USA, the EU has only about one-third as much agricultural land but about three times as many farms (Normile and Letmaa, 2005).[2] More than 50% of EU farms, however, were less than 5 hectares as of 1997. The EU has a higher proportion of its agricultural land in crops (about 60%) than the USA (44%), with the remainder in permanent pastures. The preponderance of many small farms in the EU has meant that more labour has remained active in agriculture. Although the USA had about 3.3 million people working in agriculture in the 1990s, Western Europe had more than 11 million. As a share of total employment, EU agricultural labour was about 5% in the mid-1990s, but this ranged from less than 3% in Belgium and the UK to more than 30% in Greece when it joined the EU in 1981.

The integration of highly industrialized and significantly rural countries in Western Europe into the EU added to the strains on CAP but also accelerated the processes of structural change. When the Maastricht Treaty came into force in 1993, establishing the EU, it created a common labour market among member states. Workers in any member state could move and seek employment in other member states. This facilitated the exit of labour from agriculture in regions where labour productivity in agriculture was low. Between 1973 and 2002, the number of workers economically active in agriculture in Western European countries declined from 18.3 million to 7.4 million (FAO). Another consequence of the Maastricht Treaty was the movement toward monetary union. In 1994, the European Exchange Rate Mechanism was introduced to fixed exchange rates among participating countries and, in 1999, the Euro replaced several national currencies. The UK and Sweden are exceptions, remaining outside the fixed exchange rate regime and keeping their own currencies. (Denmark has also kept its own currency but is part of the European Exchange Rate Mechanism.)

Besides differences in farm size and structure, agricultural production in the EU is similarly diverse. Northern continental Europe (Germany, France, Denmark, the Netherlands and Belgium) has a relatively high share of agricultural land in annual crops (64%), including 34% of land in cereal grains. The Mediterranean region (Greece, Italy, Spain and Portugal) has 16% of land in permanent crops (vineyards, fruit and olive orchards, etc.) and a lower proportion in annuals (47%) and cereals (21%). The Mediterranean region also has a much higher share of its cropland under irrigation (27%) compared with other Western Europe countries (<10%). About a third of agricultural land in both the Mediterranean and northern continental states is in permanent pasture, whereas more than 55% of the land in the Britain and Scandinavia is in pasture (FAO).

Over time, the EU has moved toward more coordinated regulatory policies, which have implications for the adoption of agricultural practices and technologies. The EU recognizes intellectual property protection over biological life forms, including genetically modified (GM) crops, but has been more restrictive than the USA or Canada in allowing GM crops to be grown commercially or imported from abroad. The EU also restricts the use of animal antibiotics and hormones in feed additives. Finally, the EU has introduced an EU-wide organic food label (enforced by national bodies) to facility regional trade in organic foods. A large and growing share of the EU's food production is certified as organic. In 2003, more than 5 million hectares of agricultural land (about 4% of the total) was certified as organic area (Eurostat, 2006) compared with less than 1 million hectares in the USA (Economic Research Service, b).

In agricultural innovation, EU countries maintain separate agricultural research systems. The organisation of these systems and the roles played by the public and private sectors in generating and disseminating new technologies for agriculture vary widely. Germany maintains a federal system, with both the federal and state governments funding and operating separate (and joint) research institutes (von Braun and Qaim, 2000). The Netherlands merged its government and university agricultural research institutions into one centralized unit, the Wageningen University and Research Centre, with substantial financial support from government, producer groups and other private sources (Roseboom and Rutten, 1999). In the 1980s, the UK redefined government's role in supporting agricultural research and reduced funding for public agricultural research while privatizing some public research centres (Thirtle, Piesse and Smith, 1999; Piesse and Thirtle, 2010). As in other Organisation for Economic and Co-operative Development (OECD) countries, many public agricultural research systems in the EU have been undergoing reform. Alston, Pardey and Smith (1999) identified a number of general features of these changes, namely: (i) reduced rate of growth in government funding of agricultural research with more emphasis on basic science; (ii) broadening of the public agricultural research agenda beyond productivity concerns to include environmental, food quality and social issues; (iii) expanded use of producer levies to fund commodity research; and (iv) greater use of competitive mechanisms instead of institutional block grants in allocating government funds to research.

According to the OECD, these Western European members of the EU collectively spent about Intl\$4.1 billion per year in agricultural research during 2006–2008, with the public sector accounting for about 80% of the total (Table 5.1). Agricultural R&D spending was concentrated in large countries, with Germany and France accounting for more than **Intl**\$1 billion each. As a share of agricultural GDP, research spending in Western Europe averaged 2.0%, but ranged widely, from >3% in Germany and Ireland to <1% in Italy, Greece and Sweden. During the past 20 years, from 1986–1990 to 2006–2008, real agricultural R&D expenditures rose in some countries and fell in others, but for EU countries as a whole R&D spending increased by about one-third and R&D as a share of sector GDP doubled. As Alston, Pardey and Smith (1999) noted, however, a growing share of agricultural research expenditures in the public sector were being allocated across a wider agenda. We are unable to say from available information what the trends have been for spending on productivity-oriented agricultural research.

5.3   Data and Methods

5.3.1   Data

To assess agricultural productivity trends in Western Europe, we draw on a data set developed by Ball *et al.* (2010) for comparing TFP among 11 EU countries and the USA from 1973 to 2002. Provided are estimates of the growth and relative levels of output, inputs and TFP. Agricultural output is defined as gross production leaving the farm; inputs include labour, capital, land and intermediate inputs.

We focus on the 11 European countries in the data set and examine productivity trends in individual countries, as well as for three regions defined as Mediterranean countries (Spain, Italy and Greece), northern continental countries (Germany, France, Denmark, the Netherlands and Spain), and others (the UK, Ireland and Sweden). The regional demarcation is based on geographic proximity, as well as on similarities in the structure of agriculture within the regions.

Table 5.1. Agricultural R&D spending in Western European countries.

Country	1986–1990			2006–2008			Changes between 1986/90 and 2006/08	
	Total agricultural R&D (Million 2005 PPP\$/year)	Public share of R&D (%)	R&D/GDP (%)	Total agricultural R&D (Million 2005 PPP\$/year)	Public share of R&D (%)	R&D/GDP (%)	Total R&D	R&D/ GDP
Austria	47	91	0.67	73	98	1.45	Up	Up
Denmark	94	91	1.75	65	99	2.97	Down	Up
Finland	91	96	1.30	138	100	2.85	Up	Up
France	725	74	1.31	1111	67	2.71	Up	Up
Germany	760	95	2.31	938	89	3.81	Up	Up
Greece	71	97	0.40	70	98	0.69	Down	Up
Ireland	61	94	1.12	75	98	3.22	Up	Up
Italy	281	99	0.56	251	99	0.72	Down	Up
Netherlands	271	67	1.73	253	74	2.04	Down	Up
Portugal	67	100	0.41	84	98	1.45	Up	Up
Spain	228	87	0.55	752	83	2.13	Up	Up
Sweden	51	23	0.61	23	34	0.45	Down	Down
UK	276	52	1.16	269	54	1.83	Down	Up
All	3023	82	1.04	4103	80	2.04	Up	Up

Source: Public and private agricultural R&D spending is from OECD. Agricultural GDP in constant 2005 PPP\$ is from the World Bank (PPP, purchasing power parity).

Mediterranean countries have a larger share of their population employed in agriculture, devote a significant share of land to permanent crops, and have considerable areas under irrigation. Northern continental countries have a greater proportion of agricultural land in annual crops and a lower share of labour and GDP in the agricultural sector. Agriculture in countries in the 'other' group also makes a relatively small contribution to employment and GDP, but in addition these countries devote most of their agricultural land to pasture rather than to crop production.

To construct their multilateral TFP indexes, Ball *et al.* (2010) followed the methods developed by Caves, Christensen and Diewert (1982). This approach yields transitive, multilateral comparisons among countries and over time. They first constructed relative output and input prices expressed in US dollars. Assuming competitive conditions, production technology is represented by price or unit cost function that is dual to a linearly homogeneous production function. Productivity growth between two points of time for a given country is calculated as the negative of the rate of growth of the output price less the rate of growth in input prices. More details concerning construction of the data can be found in Ball *et al.* (2010).

5.3.2　Productivity measurement

In addition to using the TFP index of Ball *et al.* (2010), we also construct indexes of land and labour productivity to characterize the evolution of productivity growth among Western European countries and regions. Examining both total and partial productivity growth provides different insights into the nature and direction of productivity and structural change. This study defines labour productivity as the ratio of the implicit quantity of aggregate output divided by the implicit quantity of labour input. The output and labour quantities data are from Ball *et al.* (2010). For land productivity, we divide the implicit quantity of aggregate output by agricultural land area (square km) from FAO.

For all of the three productivity estimates, we not only measure the indexes at the country level but also construct an aggregate Western Europe index as well as two regional productivity indexes, one for the Mediterranean (including Spain, Greece and Italy) and one for northern continental Europe (Germany, France, Denmark, the Netherlands and Belgium). The regional aggregates are derived by taking the weighted average of country-specific output, input and TFP indexes, where the weights are the country's share of total revenue (or cost) for the region. To examine long-run productivity trends, we compare growth rates over three periods of time: 1973–1982, 1983–1992 and 1993–2002 as well as over the whole period.

5.3.3　Statistical tests of a productivity growth slowdown

To examine changes in the long-run rates of productivity growth, we use two tests of statistical significance: the sample-mean difference test and the trend coefficient test. We apply these tests to the TFP indexes.

1. For the sample mean difference test we use Cochran and Box's method (undertaken in SAS 9.2 using Cochran's t-test). We apply two alternative break points to divide the whole time period into two sub-periods and test for differences of average growth rates between each pre- and post-period. If the statistics are significant, we can reject the null hypothesis of no difference (equal mean) in average growth rates between sub-periods. If the difference is significant and positive (post-period minus pre-period), it indicates there is faster TFP growth in the second period, otherwise, there is a productivity slowdown.

2. For the trend coefficient test we first fit two ordinary least squares (OLS) regression models shown as the following two equations:

$$LnTFP = \alpha_0 + \beta_0 t + \beta_1 tD \qquad (5.1)$$

$$LnTFP = \alpha_0 + \beta_0 t + \alpha_1 D + \beta_1 tD \qquad (5.2)$$

where $LnTFP$ is the logarithm of TFP, t is a time trend and D is a dummy variable. $D = 1$ when the observation is located in the post-break period. α_0, α_1, β_0, β_1 are parameters to be estimated. The difference between the two equations is that Eqn 5.1 allows for a trend break, whereas Eqn 5.2 allows for structural change in both intercept and trend. Taking a first derivative with respect to t we get the average TFP growth rate τ_0:

$$\frac{d \ln TFP}{dt} = \tau_0 \qquad (5.3)$$

If β_1 is significant, we can reject the hypothesis of no change in TFP growth between sub-periods. If β_1 is significant and positive, it indicates that productivity grew faster in the post-period. On the other hand, if β_1 is significant and negative, it indicates productivity grew more slowly in the post-period. The intercept dummy in Eqn 5.2 captures short-term shocks, such as sudden drops or temporary increases that might be caused by weather conditions or other transitory factors. This allows us to examine the trend dummy for deterministic TFP growth changes.

5.4 Productivity Growth and Structural Change in Western European Agriculture

5.4.1 Changes in labour productivity, land productivity and land–labour ratios

Table 5.2 presents average labour productivity growth in sub-periods and during the entire 1973–2002 period for these Western European countries and regions. For the 11 countries as a whole, labour productivity grew at an average annual rate of 4.14%, with country averages ranging between 6.16% (Spain) and 2.37% (the Netherlands). Northern continental countries had lower productivity growth than the Mediterranean countries. Yet among the Mediterranean countries, only Spain demonstrates extraordinary growth of more than 6% compared with an approximately 3.5% growth rate for Greece and Italy. Spain's growth in labour productivity was especially high in the 1973–1982 period.

Among all figures in Table 5.2, Germany is the only country showing a period with negative labour productivity growth and that was in the 1982–1992 period. The major reason for this is the 1990 German reunification. Prior to 1990, 'Germany' included agricultural data from only the former West

Table 5.2. Average annual growth in agricultural labour productivity (%).

Countries	1973–1982	1982–1992	1992–2002	1973–2002
Western Europe	4.42	3.26	4.75	4.14
Mediterranean	4.31	3.23	6.16	4.57
Greece	4.46	2.99	3.1	3.48
Italy	1.1	1.77	7.75	3.62
Spain	8.24	5.63	4.83	6.16
Northern continental	4.54	3.12	3.34	3.64
Belgium	3.06	4.91	2.31	3.44
Denmark	7.73	3.68	4.12	5.09
France	4.17	4.49	3.13	3.92
Germany	5.02	−0.04	4.56	3.12
Netherlands	4.05	2.4	0.82	2.37
Others				
Ireland	4.38	5.31	5.06	4.94
Sweden	7.38	2.61	4.21	4.64
UK	3.49	2.57	2.75	2.92

Source: Authors' calculations.

Germany, whereas after 1990 its statistics were merged with those of East Germany. This causes a significant drop in the average labour productivity measure.

Higher labour productivity growth was often associated with capital deepening, or growth in the capital–labour ratio (Table 5.3). This accounts for much of the labour productivity growth in Denmark, France, Spain and Ireland. Yet there are exceptions, such as for Sweden, which ranks fourth in labour productivity growth but tenth in capital–labour ratio growth. This suggests that changes in partial productivity measures such as output per worker could be greatly influenced by increases in other inputs, as well as by technical advances that economize on inputs.

As for the productivity growth trends among the 11 countries, five showed lower average annual labour productivity growth in the 1992–2002 period than in either of the previous two decades (Table 5.2). These include the Netherlands, Belgium, France, Spain and Ireland. For the 11 countries as a whole, however, the last decade actually showed the most rapid growth, and no general slowdown is evident in either the Mediterranean or northern continental regions. Average labour productivity grew 4.75% annually in Western Europe during 1992–2002, rebounding by 1.5% over the previous decade. These data indicate that if there was any Western European slowdown in agricultural labour productivity, it seems to have been in the 1980s and not in the more recent decade (1992–2002).

Compared to labour productivity, land productivity has grown more slowly at 1.60% per year for Western Europe as a whole (Table 5.4). Although land productivity growth was practically equal in the Mediterranean and northern continental regions, it did vary more among countries, from 0.88% annually in the UK to more than 2.5% in Ireland and Spain (during 1973–2002). There does seem to have been a slowdown in land productivity growth for Western Europe as a whole, from around 2% a year in the first two decades to less than 1% in the last decade. This slowdown affected most countries, with the exceptions of Spain and Sweden and possibly Greece.

Figure 5.1 plots land and labour productivity trends together for the 11 countries for the 1973–2002 period using a framework popularized by Hayami and Ruttan (1985). The two indicators together can also be used to examine the relationship between factor endowments and agricultural output. By mapping each country's labour productivity (horizontal axis) against

Table 5.3. Trends in the agricultural capital–labour ratio (%).

Countries	1973–1982	1982–1992	1992–2002	1973–2002
Western Europe	5.34	2.13	3.91	3.74
Mediterranean	6.03	2.38	4.81	4.35
Greece	4.63	1.79	0.99	2.39
Italy	3.44	1.6	8.08	4.41
Spain	9.3	3.66	0.96	4.48
Northern continental	5.5	1.68	2.67	3.21
Belgium	6.55	1.67	1.35	3.08
Denmark	9.96	2.67	2.69	4.94
France	6.75	4.02	3.04	4.53
Germany	4.18	−2.57	3.25	1.53
Netherlands	7.7	1.86	0.94	3.36
Others				
Ireland	7.03	3.22	5.5	5.19
Sweden	3.06	1.92	1.3	2.06
UK	1.59	2.47	2.83	2.32

Source: Authors' calculations.

Table 5.4. Trends in agricultural land productivity.

Countries	1973–1982	1982–1992	1992–2002	1973–2002
Western Europe	2.01	1.9	0.93	1.6
Mediterranean	1.81	1.63	1.47	1.63
Greece	2.46	0.56	1.13	1.35
Italy	1.21	1.64	0.3	1.04
Spain	2.25	2.38	2.89	2.51
Northern	2.1	2.18	0.65	1.62
continental				
Belgium	0.27	3.06	−0.56	0.94
Denmark	3.39	1.62	1.26	2.05
France	1.88	1.37	0.89	1.36
Germany	2.24	3.1	0.49	1.93
Netherlands	3	2.37	0.3	1.85
Others				
Ireland	2.26	5.11	0.2	2.53
Sweden	3.94	0.02	2.46	2.08
UK	2.01	0.71	0.03	0.88

Source: Authors' calculations.

its land productivity (vertical axis) in logarithmic scale, each 45° line represents a constant land–labour ratio. Each point on Fig. 5.1 is a combination of labour productivity and land productivity. The distance from one point to the next on each arrow shows the progress in productivity. Longer arrows indicate faster progress within the period. If the productivity curve slopes less than 45°, it implies labour productivity is growing faster than land productivity and that the land–labour ratio has been increasing over time. Growth in output per worker is a combined effect of growth in output per area plus growth in area per worker.

The Netherlands, Belgium and Denmark are three countries on the frontier (the highest combinations of land and labour productivity, or the furthest toward the upper right of Fig. 5.1). Spain, with relatively lower productivity in both labour and land shows the greatest progress in both directions but especially in labour. Ireland and Sweden are the other two countries that demonstrate faster productivity growth compared with the others, but from a low productivity starting point. In fact, these three countries (Spain, Ireland and Sweden) have relatively lower levels of labour and land productivity compared with all the other Western European countries except

for the UK. All the countries except for the Netherlands show substantial movement to higher land–labour ratios, indicating increasing average farm size. Because the arrows in Fig. 5.1 all show smaller slopes than the 45° lines, it indicates that labour productivity grew faster than land productivity in all these countries.

5.4.2   Changes in total factor productivity and input use

Figure 5.2 shows average output, input and TFP growth for the 11 European countries and for these countries combined during the 1973–2002 period. Although real agricultural output increased in all countries, the use of inputs declined and TFP growth was the sole driver of output growth in most countries. The only countries that saw agricultural resources expand were the Netherlands and Germany (and in Germany's case this was due to the 1990 reunification). TFP growth rates by sub-period for countries and regions can be found in Table 5.5. Those countries located close to the frontier of labour and land productivity (Belgium, Denmark and the Netherlands; see Fig. 5.1) had slower TFP growth, especially in the most recent period,

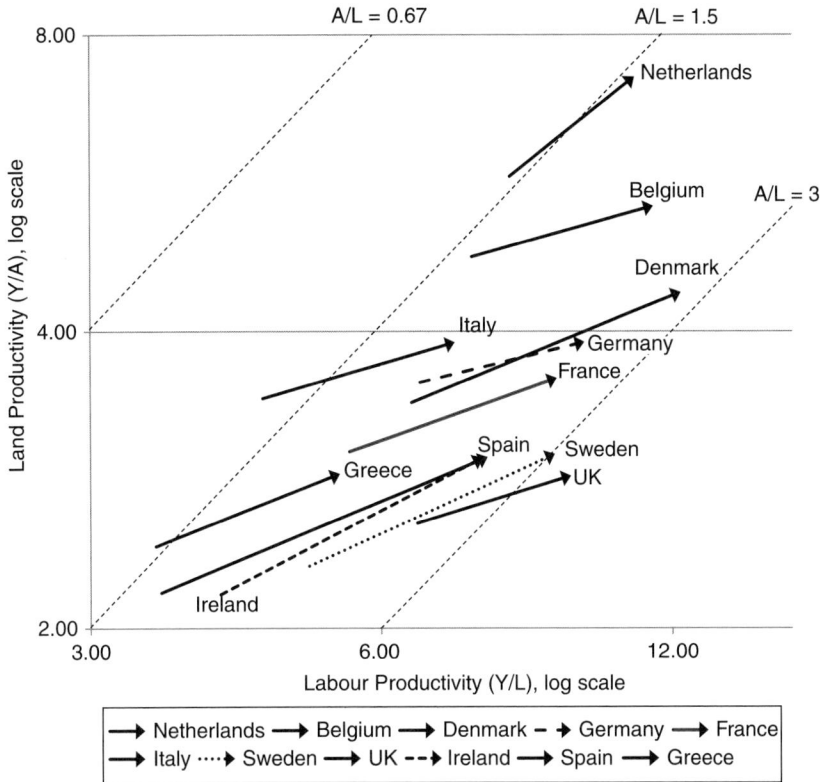

Fig. 5.1. Labour and land productivity in Western Europe, 1973–2002. Source: Output (Y) to labour (L) and output to agricultural land (A) ratios are estimated from the data described in Ball *et al.* (2001, 2010). Output is measured in constant PPP dollars. Land and labour quantities are determined by dividing total expenditures (in constant PPP dollars) for these inputs by relative price indexes for these inputs. PPP, purchasing power parity.

1992–2002. On the other hand, Spain, with its lower initial TFP level in 1973, experienced the highest TFP growth within the 1973–2002 period. Ball *et al.* (2010) referred to this phenomenon as the 'catch up' effect.

As for a possible productivity slowdown, Table 5.5 shows that among the 11 countries, six had lower average TFP growth during 1992–2002, compared with the mean for the 1973–2002 period. However, regional aggregate TFP growth rates do not show signs of decline. Agricultural TFP growth in the Mediterranean region accelerated during the three decades, whereas TFP growth in the northern continental region and for the 11 countries as a whole declined somewhat in the middle decade but then recovered in the last decade.

Changes in inputs and input cost shares reveal useful information about the sources and direction of technical change. Table 5.6 compares the input cost shares for labour, land, capital and intermediate inputs in 1973 and 2002. All countries except Greece witnessed a decline in the labour cost share, while capital cost shares increased. All countries reduced their agricultural labour and the decline in labour cost share implies that labour use fell faster than the rate by which (relative) labour wages rose. Spain's labour force fell by an average of almost 4% per year, so that by 2002 its agricultural labour input had been reduced to one-third of its 1973 level. The cost share of intermediate inputs also rose everywhere except in Germany and France, where they remained

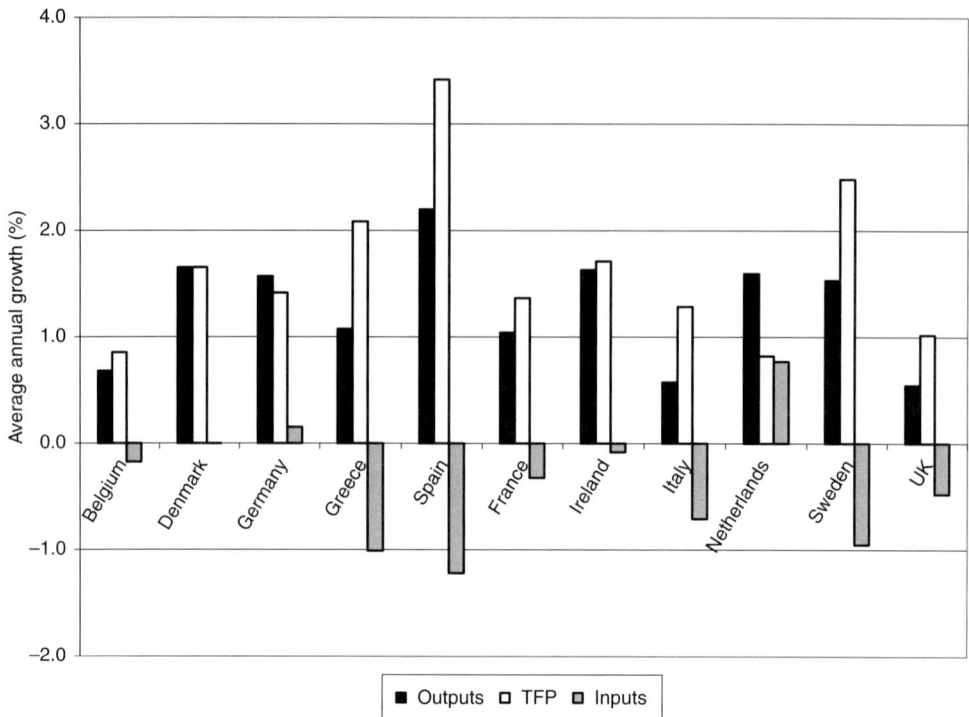

Fig. 5.2. Growth rates for agricultural outputs, TFP and inputs. Average annual growth rate during 1973–2002. *Source*: Ball *et al.* (2010).

Table 5.5. Trends in agricultural total factor productivity.

Countries	1973–1982	1982–1992	1992–2002	1973–2002
Western Europe	1.77	1.36	1.73	1.58
Mediterranean	1.60	1.75	2.87	2.16
Greece	2.78	1.88	1.30	2.08
Italy	−0.56	0.99	3.05	1.28
Spain	4.43	2.78	2.99	3.42
Northern	1.76	1.19	1.25	1.29
continental				
Belgium	0.18	1.87	1.07	0.86
Denmark	1.72	1.87	0.91	1.65
France	1.70	1.44	1.49	1.36
Germany	2.78	0.47	1.39	1.41
Netherlands	0.54	1.43	0.52	0.82
Others				
Ireland	1.76	2.60	1.25	1.71
Sweden	4.32	1.29	1.50	2.48
UK	1.96	0.58	0.68	1.02

Source: Authors' calculations.

essentially unchanged and accounted for just half of total costs. The land cost share remained fairly constant across Western Europe as a whole, although it rose slightly in some countries and declined in others. These shifts in cost structure imply that farm labour was being replaced by capital, intermediate inputs and rising TFP.

Table 5.6. Agricultural input cost shares, 1973 and 2002.

Countries	Labour		Land		Capital		Intermediate inputs	
	1973	2002	1973	2002	1973	2002	1973	2002
Western Europe	0.42	0.3	0.07	0.08	0.07	0.15	0.44	0.47
Mediterranean								
Greece	0.54	0.57	0.19	0.09	0.03	0.07	0.24	0.27
Italy	0.5	0.39	0.06	0.08	0.1	0.2	0.33	0.34
Spain	0.66	0.48	0.12	0.08	0.04	0.08	0.17	0.36
Northern continental								
Belgium	0.32	0.23	0.03	0.07	0.05	0.08	0.6	0.62
Denmark	0.35	0.18	0.01	0.02	0.05	0.18	0.58	0.63
France	0.38	0.27	0.04	0.07	0.05	0.14	0.53	0.52
Germany	0.31	0.18	0.05	0.11	0.11	0.19	0.53	0.52
Netherlands	0.48	0.21	0.03	0.07	0.04	0.16	0.45	0.55
Others								
Ireland	0.45	0.26	0.05	0.05	0.07	0.14	0.43	0.55
Sweden	0.3	0.17	0.01	0.02	0.12	0.16	0.57	0.65
UK	0.26	0.24	0.11	0.08	0.1	0.13	0.52	0.55

Source: derived from unpublished data in Ball *et al.* (2010).

Comparisons of average growth rates across Western European regions and sub-periods show that although land productivity growth slowed in the Mediterranean, northern continental Europe and Western Europe as a whole, labour productivity growth and TFP growth did not (Fig. 5.3a). But there are substantial exceptions to each of these generalities. Labour productivity growth fell in the most recent decade in Spain, Belgium, France and the Netherlands. Likewise, TFP growth fell in Greece, Belgium, Denmark and the Netherlands. Labour productivity and TFP both fell in Ireland in the 1990s. So if we look at the trends for Western Europe over this 30-year period, we do not see clear evidence of an overall productivity slowdown. Using simple averages or mean productivity growth rates might be misleading, however, and Fig. 5.3b clearly shows how much transitory factors such as the weather can cause temporary fluctuations in annual growth rates, even across the Western Europe as a whole. A statistical test is necessary if we want to address this issue more rigorously. The next section examines the productivity slowdown issue using this approach.

5.4.3 Is there a slowdown in TFP growth?

Sample means tests for national and regional agricultural TFP growth rates are reported in Table 5.7. We use 1983 and 1993 as the break points and compare mean growth rates among the periods demarcated by the break points. When using 1983 as the break point, only three countries (Belgium, Italy and the Netherlands) exhibit accelerating TFP growth in the post-1983 period, whereas the other countries all demonstrate lower growth rates. However, none of these changes in mean growth rates is statistically significant. We cannot reject the hypothesis of no differences in TFP growth rates between the 1973–1982 and post-1983 periods. When using 1993 as the break point we find that all countries show a decline in TFP growth in the post-1993 period except Italy, and Italy's acceleration in TFP growth is significant at a 10% significance level. Again, we cannot reject the hypothesis of no TFP growth slowdown for any of the other ten countries in the post-1993 period.

We then use the time trend coefficient models to see if they show a slowdown in productivity in Western Europe in recent

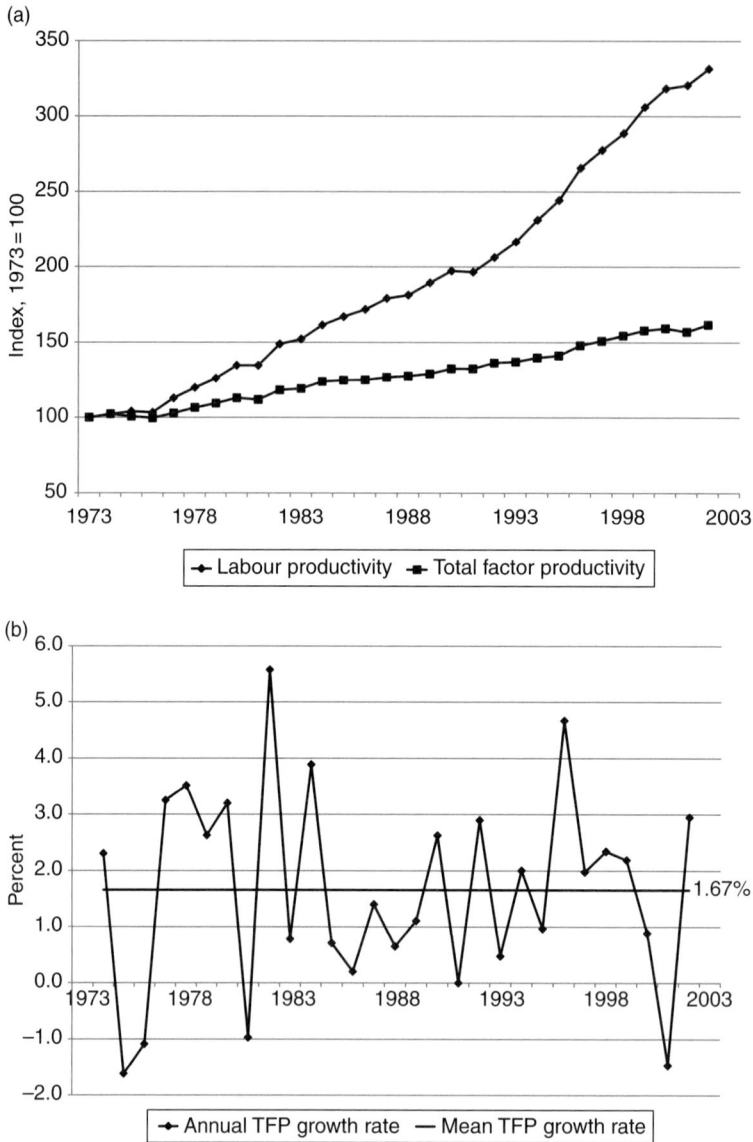

Fig. 5.3. Agricultural productivity growth in Western Europe (11 country aggregate). (a) Trend of labour productivity and TFP. (b) Rate of change in agricultural TFP. Source: Derived from data described in Ball *et al.* (2010).

years. Because *LnTFP* is a time series, we first conduct the Phillips-Perron (1988) (PP) unit root test for the series. Table 5.8 presents these results. When we apply the unit root test without trend, we cannot reject a unit root for any of the series. When we include a trend in the unit root tests, five countries (Belgium, Denmark, Germany, Greece and Ireland) and Western Europe as a whole reject the hypothesis of non-stationarity (significance levels vary and are shown in the table), whereas the others are still non-stationary. As would be expected when we take first differences of the non-stationary series, δ *LnTFP* (the *TFP* growth rate) for those countries is stationary.[3] Because half

Table 5.7. Testing for a TFP growth slowdown using the sample mean test.

Countries	Using 1983 as the break point		Using 1993 as the break point	
	Mean change	Significance	Mean change	Significance
Belgium	+	N	–	N
Denmark	–	N	–	N
Germany	–	N	–	N
Greece	–	N	–	N
Spain	–	N	–	N
France	–	N	–	N
Ireland	–	N	–	N
Italy	+	N	+	*
Netherlands	+	N	–	N
Sweden	–	N	–	N
UK	–	N	–	N
Western Europe	–	N	+	N

+ Indicates increase in the second period, – indicates decrease in the second period.
*Indicates significant at 10% level, **indicates significant at 5% level, ***indicates significant at 1% level.
N indicates not significant on the basis of the sample mean difference test undertaken using Cochran methods.

Table 5.8. Unit root tests of stationarity for the agricultural TFP series.

Countries	PP unit root test (level)			PP unit root test (first difference)	
	Without trend	With trend		Without trend	With trend
Belgium	NS	–3.82	**	S	S
Denmark	NS	–3.34	*	S	S
Germany	NS	–3.33	*	S	S
Greece	NS	–7.73	***	S	S
Spain	NS	–2.62		S	S
France	NS	–3.16		S	S
Ireland	NS	–4.17	**	S	S
Italy	NS	–2.09		S	S
Netherlands	NS	–2.34		S	S
Sweden	NS	–2.19		S	S
UK	NS	–2.55		S	S
Western Europe	NS	–14.32	***	S	S

*Indicates significant at 10% level, **indicates significant at 5% level, ***indicates significant at 1% level.
NS indicates non-stationary, S indicates stationary.

the series are stationary with a deterministic trend, we report the results using *LnTFP* as the dependent variable and include a deterministic time trend to maintain comparability.[4] Since the results for Eqns 5.1 and 5.2 are similar, we only report results that are based on Eqn 5.2, which allows for series drift through the intercept dummy.

Table 5.9 shows the time trend regression test results for a TFP slowdown. It uses three alternative break points to test for a productivity slowdown. When using 1983

as the break point, the TFP growth rate significantly increased after 1983 for Belgium, Denmark, Italy and the Netherlands. Yet Germany, Spain, Sweden and the UK exhibit productivity slowdowns in the post-1983 period. When using 1987 as the break point, only Italy and the Netherlands still exhibit significantly faster TFP growth in the post-1987 period, whereas Germany, Spain, France, Sweden and the UK show significant productivity slowdowns. When using 1993 as the break point, most of the tests

Table 5.9. Tests for agricultural TFP slowdown using the time trend regression model.

Countries	Using 1983 as the break point			Using 1987 as the break point			Using 1993 as the break point		
	TFP growth	t-statistics	F-statistics	TFP growth	t-statistics	F-statistics	TFP growth	t-statistics	F-statistics
Belgium[a]	+	**	*	+			−		
Denmark[a]	+	*	***	−			−		
Germany[a]	−	***	***	−	***	***	−		
Greece[a]	−			−			+		
Spain	−	***	***	−	***	***	−		***
France	−		**	−	**	**	−	*	*
Ireland[a]	+		*	−			−		
Italy	+	***	***	+	***	***	+	***	***
Netherlands	+	***	**	+	***	**	−		
Sweden	−	***	***	−	***	***	−	***	***
UK	−	***	***	−	***	***	−		***
Western Europe[a]	−		**	−		**	+		

[a]Belgium, Denmark, Germany, Greece, Ireland and Western Europe are stationary series.
+ Indicates increase in the post- period, − indicates decrease in the post- period; *indicates significant at 10% level, **indicates significant at 5% level, ***indicates significant at 1% level. t-Statistics is for the coefficient of the trend dummy. F-statistics is for the joint hypothesis test of the intercept dummy and trend dummy.

show lower productivity growth in the post-period but most of them are insignificant except in France and Sweden. On the other hand, Italy is the only country demonstrating significantly higher TFP growth in the post-period as it did in the previous two break periods.

Taken together, the results in Table 5.9 tell several important stories. As the arbitrarily chosen break point moves closer to the end of the series, fewer countries have significant results, so we would not expect a structural break later in the series in formal tests. Our purpose here is to comment on whether a TFP slowdown might have taken place, and using the 1983 results that have the largest number of significant results there are as many countries that increased TFP growth from this point as experienced a slowdown (four countries increased while four slowed down). Spain slowed down probably because of the phenomenal growth it experienced early in the sample period, consistent with the previously discussed 'catch-up' hypothesis. Italy, on the other hand, never very high in TFP growth, had significant positive growth over all three 'post-periods' as the 'catch-up' worked in its favour. But even the 'catch-up' is not

ubiquitous, as the Netherlands, a high TFP-level country, also had increasingly significant TFP growth over the first two sub-periods. Because the aggregate Western Europe results are not significant, these country results seem to indicate that, although some European countries are increasing their TFP growth, others are decreasing, and these trends average out for Western Europe as a whole. Local factors probably account for what is taking place in each country in the region.

5.5 Conclusions and Discussion

The declining rate of growth in agricultural output in Western Europe over the past several decades has cut the region's share of global production in half, to about 10% by 2008–2009 according to FAO. But our analysis of productivity patterns suggests this slowdown in output growth is entirely due to withdrawals of resources from agriculture, especially labour, and not to a slowdown in productivity growth. To the extent that increases in TFP reflect the rate of technical change, trends in agricultural R&D

funding are consistent with trends in TFP growth: neither shows major departures from long-term trends. However, our findings also show considerable regional and national differences in the rate and direction of productivity growth and in the organization and support for agricultural research in Western Europe.

The results show that the exit of labour from agriculture, leaving more resources for workers remaining in the sector, has been an important source of agricultural labour and TFP growth for many Western European countries. This has been especially true for less industrialized or urbanized countries with relatively large shares of labour employed in agriculture but with low output per worker. The EU's integration of product and labour markets has probably facilitated the process of structural change in these economies and enabled them to achieve or maintain relatively high rates of agricultural productivity growth. These structural changes are reflected in the changing cost structure of EU agriculture, with falling cost shares for labour and rising cost shares for capital and intermediate inputs over time for most countries, while land cost shares have remained relatively stable.

Agricultural R&D spending by Western European countries does seem to be positively correlated with their agricultural TFP, although we have not rigorously examined this relationship here. Several countries with the highest levels of agricultural TFP – the Netherlands, Denmark, France and Germany in particular – also make large investments (relative to agricultural GDP) in agricultural research, whereas several countries with the lowest TFP levels (Greece and Spain) spend considerably less on research.

The UK, Ireland and Italy, however, don't quite fit this mould; in Italy's case spending on agricultural research is low but TFP is relatively high, whereas the UK and Ireland both have relatively high R&D–GDP ratios but TFP levels below the Western European average. In the UK's case, since the 1980s there was a significant redirection of public R&D funds away from productivity-oriented research, which might explain its low level of TFP. Detailed information is lacking on how R&D funds are allocated in other European countries.

The analysis of productivity trends did not reveal any significant slowing down of TFP or labour productivity growth rates, although land productivity growth did decline after 1993, especially for northern continental states and Britain. Regarding TFP, just as many countries in the sample have had lower TFP growth rates since 1983 as those that have had higher rates. The trend regression models show that, when using either 1983 or 1987 as break points, Germany, Spain, Sweden and the UK experienced productivity slowdown, whereas Italy and the Netherlands exhibited accelerated productivity growth. These findings are not, however, statistically robust using a 1993 break date. The high year-to-year variability in agricultural TFP growth rates, primarily resulting from weather-induced fluctuations in output, introduces a serious signal-to-noise problem in constructing valid statistical tests for growth trends. Future work could explore factors linking productivity and research in each country's circumstances, keeping in mind time lags before research can be expected to help productivity and the potential for cross-country R&D spillovers.

Notes

¹ 'Western Europe' refers to the countries that have had market-oriented economies since World War II, thus excluding the 'transition' countries that were centrally planned before 1991. These countries make up what is known as the EU15, plus Norway and Switzerland which are not part of the EU. Our empirical analysis concentrates on 11 countries: Belgium, Denmark, France, Germany, Greece, Ireland, Italy, the Netherlands, Spain, Sweden and the UK, all members of the EU or its predecessor, the European Economic Community. According to the FAO, these 11 countries account for about 93% of the agricultural output in Western Europe and 95% of the agricultural output in the EU15.

[2] The comparative statistics in this paragraph refer to the USA and the EU15 in the mid-1990s unless otherwise noted.

[3] We do not pursue first difference versions of Eqns 6.1 and 6.2 any further because the tests become similar to sample mean tests looking at average growth rates as was already done in Table 5.7. We tried regressing the growth rate (first difference of *LnTFP*) on a deterministic time trend, implying a non-linear relationship between *LnTFP* and *t* but these results are all insignificant.

[4] It is a well-known time-series phenomenon in spurious regressions that the sign and significance of a deterministic trend are inflated when the underlying series has a stochastic trend (Phillips, 1998), but we are reporting and discussing the trend break variable that should have no such bias.

References

Alston, J., Pardey, P. and Smith, V. (1999) A synthesis. In: Alston, J., Pardey, P. and Smith, V. (eds) *Paying for Agricultural Productivity*. Johns Hopkins University Press, Baltimore, MD, pp. 276–282.

Alston, J. Babcock, B. and Pardey, P. (2010) Shifting patterns of global agricultural productivity: Synthesis and conclusion. In: Alston, J., Babcock, B. and Pardey, P. (eds) *The Shifting Patterns of Agricultural Production and Productivity Worldwide*. The Midwest Agribusiness Trade Research and Information Center, Iowa State University, Ames, IA, pp. 449–482.

Alston, J., Andersen, M., James, J. and Pardey, P. (2010) *Persistence Pays: U.S. Agricultural Productivity Growth and the Benefits from Public R&D Spending*. Springer, New York, NY.

Ball, V.E., Bureau, J., Butault, J. and Nehring, R. (2001) Levels of farm sector productivity: An international comparison. *Journal of Productivity Analysis* 15, 5–29.

Ball, V.E., Butault, J., Mesonada, C. and Mora, R. (2010) Productivity and international competitiveness of agriculture in the European Union and the United States. *Agricultural Economics* 41, 611–627.

Caves, D., Christensen, L. and Diewert, W. (1982) Multilateral comparison of output and productivity using superlative index numbers. *Economic Journal* 92, 73–86.

Economic Research Service (a). *European Union: Common Agricultural Policy Briefing Room*. U.S. Department of Agriculture, Washington, DC. Available at: http://www.ers.usda.gov/Briefing/EuropeanUnion/policy.htm (Accessed 15 September 2011).

Economic Research Service (b). *Organic Production Data Sets*. U.S. Department of Agriculture, Washington, DC. Available at: http://www.ers.usda.gov/Data/Organic/ (Accessed 20 September 2011).

Eurostat (2006) *Food: From Farm to Fork Statistics, Statistical Pocketbook*. European Commission, Luxembourg.

FAO. FAOSTAT Database, Food and Agriculture Organization of the United Nations, Rome, Italy. Available at: http://faostat.fao.org/ (Accessed October 2010).

Fuglie, K. (2008) Is a slowdown in agricultural productivity growth contributing to the rise in commodity prices? *Agricultural Economics* 39, supplement, 431–441.

Fuglie, K. (2010) Total factor productivity in the global agricultural economy: Evidence from FAO data. In: Alston, J., Babcock, B. and Pardey, P. (eds) *The Shifting Patterns of Agricultural Production and Productivity Worldwide*. Midwest Agribusiness Trade and Research Information Center, Iowa State University, Ames, IA, pp. 63–95.

Hayami, Y. and Ruttan, V.W. (1985) *Agricultural Development: An International Perspective*. Johns Hopkins University Press, Baltimore, MD.

Jorgenson, D.W. and Nishimizu, M. (1978) U.S. and Japanese economic growth, 1952–1974: An international comparison. *Economic Journal* 83, 707–726.

Jorgenson, D.W. and Nishimizu, M. (1981) International differences in levels of technology: A comparison between U.S. and Japanese industries. In: *International Roundtable Congress Proceedings*. Institute of Statistical Mathematics, Tokyo, Japan.

Normile, M. and Leetmaa, S. (coordinators) (2004) *U.S–EU Food and Agriculture Comparisons*. Agriculture and Trade Report WRS-04-04, Economic Research Service, U.S. Department of Agriculture, Washington, DC.

OECD. Science, Technology and R&D Statistics. Organisation for Economic Co-operation and Development, Paris, France. Available at: http://www.oecd-ilibrary.org/content/datacollection/strd-data-en (Accessed 12 September 2011).

Phillips, P. (1998) New tools for understanding spurious regressions. *Econometrica* 66, 1299–1325.

Phillips, P. and Perron, P. (1988) Testing for a unit root in time series regression. *Biometrika* 75, 335–346.

Piesse, J. and Thirtle, C. (2010) Agricultural R&D, technology and productivity. *Philosophical Transactions of the Royal Society B* 365, 3035–3047.

Roseboom, J. and Rutten, H. (1999) Financing agricultural R&D in the Netherlands: The changing role of government. In: Alston, J., Pardey, P. and Smith, V. (eds) *Paying for Agricultural Productivity*. Johns Hopkins University Press, Baltimore, MD, pp. 215–246.

Schimmelpfennig, D. and Thirtle, C. (1999) The internationalization of agricultural technology: Patents, R&D spillovers and their effects on productivity in the European Union and United States. *Contemporary Economic Policy* 17, 457–468.

Schimmelpfennig, D. and Thirtle, C. (2000) Significance of international spillovers from public agricultural research. In: Fuglie, K. and Schimmelpfennig, D. (eds) *Public–Private Collaboration in Agricultural Research: New Institutional Arrangements and Economic Implications*. Iowa State University Press, Ames, IA.

Thirtle, C., Piesse, J. and Smith, V. (1999) Agricultural R&D policy in the United Kingdom. In: Alston, J., Pardey, P. and Smith, V. (eds) *Paying for Agricultural Productivity*. Johns Hopkins University Press, Baltimore, MD, pp. 172–214.

Thirtle, C., Lin, L., Holding, J. and Jenkins, L. (2004) Explaining the decline in UK agricultural productivity growth. *Journal of Agricultural Economics* 55, 343–366.

Von Braun, J. and Qaim, M. (2000) Research and technology in German agriculture. In: Tangermann, S. (ed.) *Agriculture in Germany*, DLG-Verlag, Frankfurt am Main, Germany, pp. 255–282.

Wang, S.L. (2010) Is U.S. agricultural productivity growth slowing? *Amber Waves* 8, 3.

World Bank. World Development Indicators Databank. Washington, DC. Available at: http://data.worldbank.org/data-catalog/world-development-indicators (Accessed 12 September 2011).

6 Agricultural Productivity Paths in Central and Eastern European Countries and the Former Soviet Union: The Role of Reforms, Initial Conditions and Induced Technological Change

Johann Swinnen, Kristine Van Herck and Liesbet Vranken
*Centre for Institutions and Economic Performance (LICOS),
KU Leuven, Belgium*

6.1 Introduction

Economic and institutional reforms have dramatically affected agricultural organization, output and production efficiency in Central and Eastern Europe and the former Soviet republics. Not only did farm output fall dramatically in the transition countries of Europe and the former Soviet Union, some studies find that efficiency decreased as well during the transition. In a review of the evidence, Rozelle and Swinnen (2004) conclude that productivity started increasing early on during the transition in Central Europe and parts of the Balkan and the Baltic States, but continued to decline much longer in other parts of the former Soviet Union. Initial declines in productivity were associated with initial disruptions resulting from land reforms and farm restructuring in Eastern Europe (Macours and Swinnen, 2000a) or with poor incentives and soft budget constraints in some of the countries of the former Soviet Union (Sedik *et al.*, 1999; Lerman *et al.*, 2004) and with disorganization in the supply chains (Gow and Swinnen, 1998).

However, there are several problems in comparing efficiency studies and drawing implications from them. First, a limitation is that those studies that include more countries and a longer time horizon use aggregate data (Mathijs and Swinnen, 2001), whereas studies using farm-level data are restricted to one country and short time periods, often only one year. Second, cross-country comparisons and conclusions are complicated by differences in data samples. Third, linking efficiency changes to specific reforms is difficult, but important in understanding which factors have been crucial in constraining or stimulating efficiency growth.

To get a comprehensive picture of productivity developments and to accommodate important data constraints, we analyse three sets of productivity indicators: labour productivity (output per unit of labour use) as it relates to wages; yields (output per unit of land) as it relates to land rents; and aggregate total factor productivity (TFP) as it relates to the rate of cost reduction.

This chapter has six parts. The next section provides a conceptual framework for the evolution of productivity and efficiency measures and links this evolution to the issue of factor abundance, taking into account specific transition characteristics. The third part illustrates strong differences in the evolution of agricultural productivity measures since the start of the reforms. The fourth section describes initial factor endowments, progress of reforms and credit market imperfections, as well as the overall level of economic development. The fifth section groups countries according to their initial factor endowments and progress in reforms and for each group discusses how a combination of different factors affected changes in agricultural productivity.

Much of the discussion will refer to and compare productivity trends in four periods: the pre-reform period (before 1990), early transition (years 1–5 after reforms were initiated – roughly the first half of the 1990s), mid transition (years 6–10; the second half of the 1990s), and the late transition (years 11–15; roughly the first half of the 2000s and the recent period after 2005). In Central Europe and Balkans, the first year of the reforms is assumed to be 1989, whereas in the former Soviet Union countries (the Baltics, the European Commonwealth of Independent States, Transcaucasia and Central Asia) the start is assumed to be in 1990.[1]

6.2 Resource Endowments, Reforms, Technical Change and Productivity in Transition: a Conceptual Framework

There were (and still are) major differences in resource endowments and the nature of technology in agriculture among the transition countries. It is well known that resource endowments (factor proportions) can play an important role in agricultural productivity growth. Most famously, Hayami and Ruttan (1970) showed how technological adaptation can occur through a sequence of induced innovations in technology biased toward saving the limiting factors.[2] Hence, in countries where land is relatively scarce and labour is abundant, innovations in technology will be biased towards using land,

the scarce factor, more efficiently – such as through biological innovations (e.g. seed improvements). In contrast, in countries where labour is relatively scarce and land is abundant, technological innovations will be induced that contribute to a more efficient use of labour, the limiting factor. For example, this can involve mechanical innovations, which substitute labour by land and/or capital, leading to a decrease in the labour–land ratio.

It is also well known that resource endowments (or initial technology) also played an important role in the institutional reforms and productivity changes (Macours and Swinnen, 2002). However, the very nature of the transition process, and the inherent policy and institutional reforms which interact with pre-reform distortions, makes the impact of resource endowments on productivity growth considerably more complex than the original Hayami–Ruttan model (or its more recent extensions) would predict in transition economies.

First, under the Communist system, government regulations caused distortions in many output and factor prices and in company and household allocation decisions. Hence, the removal of those distortions in the transition process resulted in important readjustments in factor allocations which primarily reflected pre-reform distortions (Liefert *et al.*, 2003). For example, under the central planning system, labour was used inefficiently in most sectors of the economy, but especially in the case of the agricultural sector (Brada, 1989; Bofinger, 1993; Jackman, 1994). Thus, reforms which allow a more efficient labour allocation would induce an outflow of agricultural labour from agriculture to other sectors, and a decline in the labour–land ratio. The extent of this would be strongly affected by the pre-reform distortions, rather than by induced technical change.

Second, actual factor adjustments depend on several factors. One is the actual implementation of the reform policies. In many transition countries, reforms were implemented with delays, or not, which affected the reallocation of production factors to more efficient uses. Hence, many of the predicted factor adjustments were conditional on sufficient progress in the reform process.

Similarly, in order to induce technical innovations to use land and labour more productively, access to credit for investment in agricultural machinery and working capital to buy such things as fertilizer and chemicals is crucial. In the early transition years, external credit was scarce and expensive, which resulted in problems accessing physical inputs (Arnade and Gopinath, 1998; Petrick, 2004). Consequently, the technical innovations one would expect, based on the initial resource endowments, might not be observed. Again, progress in reforms was essential either to make domestic finance available to farmers and the food industry or to attract international capital through foreign direct investment in the food industry and agribusiness with positive vertical spillovers on farms. The latter process, which was conditional on the progress of reforms, has been a very important source of access to inputs, technology, credit and markets for transition-country farmers, resulting in substantial productivity growth (Gow and Swinnen, 1998; Dries and Swinnen, 2004).

Third, restructuring is a timely and costly process, and factor reallocation will be determined by opportunity costs in the transition process itself, which could be affected by government policies and the level of development. For example, mobility out of agriculture is determined by the probability of finding employment in other sectors and by the social benefits an individual would receive upon leaving agriculture. In richer transition countries, labour that was laid off (also from large farms) received sufficient unemployment payments or pensions to become effectively unemployed or to retire. In poorer countries, such social payments were an insufficient income source for households, and the agricultural sector played a buffer role ('labour sink') during transition, absorbing and providing an income-generating activity to those laid off in other sectors (Seeth et al., 1998).

Fourth, an important element of 'induced innovation' – more in line with the broader

Hayami and Ruttan (1985) framework – that strongly affected productivity change during transition was the endogenous adjustment of the structure of farms. The relative efficiency of farm organizations, and thus incentives for farm restructuring, is significantly affected by initial resource endowments because they affect the costs and benefits of shifting from corporate farms to family farms. If labour–land ratios are high, i.e. if agricultural production processes are relatively labour intensive, the benefits of shifting from corporate to family farms are greater, whereas the costs of shifting are lower (Swinnen and Rozelle, 2006). As a result, land productivity gains that stem from the shift to household farming are particularly observed. On the other hand, if the labour–land ratio is low, i.e. if agricultural production is relatively land intensive, the benefits of shifting to family farms are lower so that large-scale corporate farming remains. In that case, labour productivity gains that stem from laying off corporate farm workers will be observed.

Again, it should be emphasized that these adjustments and induced farm restructuring processes were conditional on land reforms and the implementation of farm privatization policies. In many countries, these effects were delayed (or did not occur) because such reforms were delayed or were never implemented at all.

Finally, an important element of 'induced innovation' during transition was the endogenous choice of land reforms. That effect is beyond the focus of this study. See Rozelle and Swinnen (2004) for a discussion and analysis of the subject.

In summary, resource endowments played an important role in affecting productivity changes in the transition process. They affected the incentives for endogenous farm restructuring and adjustments in factor proportions. These adjustments and the mechanism of productivity growth and technical change were, however, also strongly affected by the level of a country's development and its progress and implementation of various institutional and policy reforms. The choice and implementation of reforms were often affected by the level of development and resource endowments, making the set of interactions in determining the pace and mechanism of productivity growth complex.

(Georgia, Armenia and Azerbaijan); and Central Asia (Kazakhstan, Kyrgyzstan, Tajikistan, Turkmenistan and Uzbekistan).

6.3 Evolution of Agricultural Output and Productivity

This section summarizes and updates earlier studies and empirical data on the evolution of partial productivity measures, such as labour productivity and yields, and total factor productivity. For more details on the data source and calculation methods, see Macours and Swinnen (2000a and 2000b), Swinnen and Vranken (2010) and Swinnen *et al.* (2010a, 2010b).

To keep the discussion more tractable, we organize countries into six regional groups. These groups not only differ in terms of the countries' geographical locations, but also according their initial resource endowments, the choice of and progress in the land reform process, and overall economic development. The groups are: Central Europe (the Czech Republic, Hungary, Poland and Slovakia); the Baltic States (Estonia, Lithuania and Latvia); the Balkans (Albania, Bulgaria, Slovenia and Romania); the European Commonwealth of Independent States or CIS (Russia, Ukraine and Belarus); Transcaucasia

6.3.1 Changes in agricultural output

Figure 6.1 shows the evolution of agricultural output in the past two decades. The pattern is similar in all countries. In general, there is a U effect: an initial decline in agricultural output followed by a recovery later on.

In the first years of transition, agricultural output decreased in all countries by at least 20%. In all countries, the transition from a centrally planned economy to a market-orientated economy coincided with subsidy cuts and price liberalization, which in general caused input prices to increase and output prices to decrease. Purchased inputs were no longer affordable at the new relative prices, and the decrease in input use caused a decrease in agricultural output.

In the Baltic States and the European CIS, agricultural output decreased to about 50–60% of the pre-reform output. In Central Europe and Central Asia, output declined by 25–30%. Output stabilized and started to recover in the mid-1990s in Central Europe and later in the other regions.

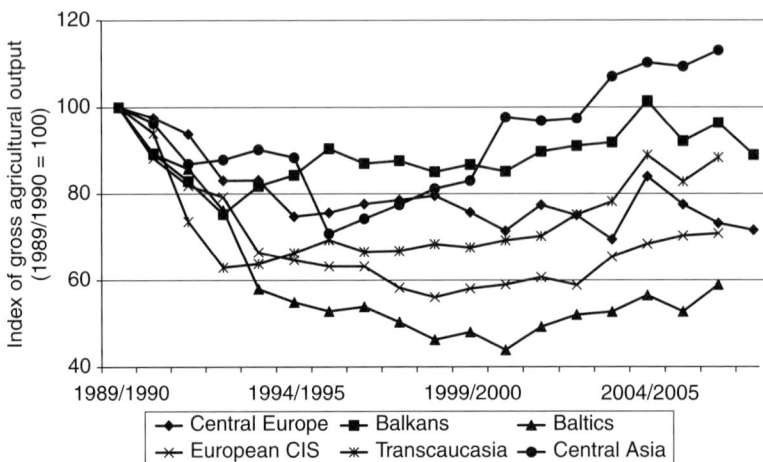

Fig. 6.1. Evolution of agricultural output since the start of reforms. Agricultural output is the FAO's index of gross agricultural output set to 100 in the base year. Reforms started in 1990 in the FSU republics and in 1989 in Central Europe and the Balkan countries. Source: FAO.

Currently, agricultural output is close to the pre-reform output level in most countries.

6.3.2 Changes in agricultural productivity

Agricultural labour productivity

Figure 6.2 illustrates the evolution of agricultural labour productivity, measured as output per farm worker. Despite a decrease in agricultural output in total, output per worker in Central Europe increased strongly during the past two decades. This increase was driven by the dramatic decrease in agricultural employment in the first years of transition from centrally planned to more market-oriented economies. As output stabilized at the end of the 1990s and agricultural employment continued to decline, the increase in labour productivity continued.

That pattern was not followed by other countries. In the Balkan countries, the agricultural sector acted as a social buffer and absorbed rural labour in the first years of transition (Swinnen *et al.*, 2005). Agricultural labour productivity decreased initially, as much labour was absorbed into agriculture. In the late 1990s, labour began to flow out of agriculture and, in combination with increased investments in the farming and agri-food industries, this resulted in a gradual but steady improvement in agricultural labour productivity.

Farther east, labour productivity fell sharply in the first decade of transition. On average, it decreased by 33% in the European CIS and by 30% in Central Asia in the first five years of transition. The strong decline was the result of two effects. First, agricultural output declined strongly in both regions and second, the outflow of agricultural labour was limited and in some regions agricultural employment even increased. From the mid-1990s, however, the decline in agricultural labour productivity started to level off and since the beginning of the 2000s it has recovered slowly.

Agricultural land/ animal productivity

Figure 6.3 illustrates the time path of land and animal productivity, or yield, which evolved in conjunction with agricultural labour productivity. Everywhere in the region, average yields fell during the first years of transition and recovered later. However, the depth and length of the fall

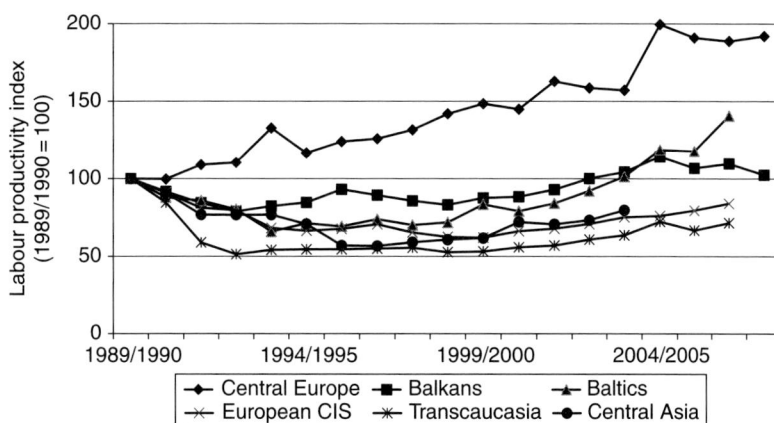

Fig. 6.2. Evolution of agricultural labour productivity since the start of reforms. Labour productivity is measured as gross agricultural output per farm worker, indexed to 100 at the start of the reforms. Reforms started in 1990 in the former Soviet Union republics and in 1989 in Central Europe and the Balkan countries. Sources: Gross agricultural output is from FAO. Agricultural labour is compiled from a variety of sources including the International Labour Organization, Asian Development Bank and national statistics (see Macours and Swinnen, 2002, for full details on sources and methods).

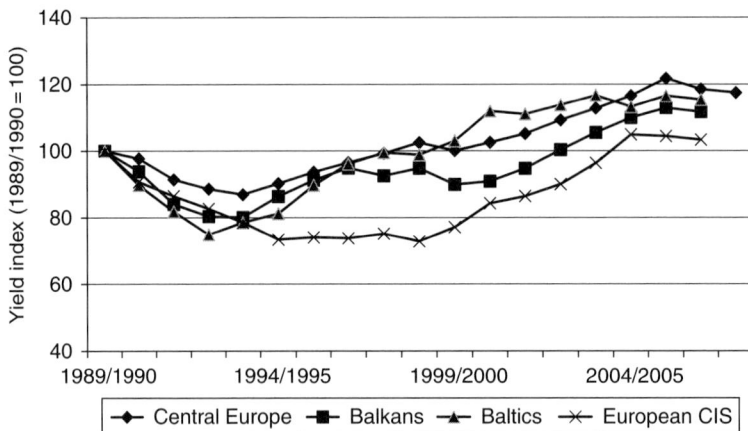

Fig. 6.3. Evolution of agricultural yield since the start of reforms. The agricultural yield index is calculated as the average yield index of grains, sugarbeets and milk, where crop yield is output per hectare and milk yield is output per cow. Given the sensitivity of grain and sugarbeet yields to weather conditions, these are calculated as three-year moving averages. The Balkans include Albania, Bulgaria and Romania. Source: Derived from FAO data.

differed strongly among countries. Average yields recovered considerably from the mid-1990s onwards in countries such as Hungary, nations with relatively more large-scale farming and investments in the food industry. In contrast, productivity recovered more slowly in countries such as Romania, which has a large number of small-scale family farms with difficult access to inputs. Yields declined the most in the European CIS and Central Asia in the years after transition and only started to increase in the beginning of the 2000s. The recovery of yields in the European CIS and Central Asia was so slow that they only recently reached their pre-reform levels.

Total factor productivity

Of course, partial productivity measures might exhibit very different patterns than would be found using measures of total factor productivity (TFP), the most comprehensive measure of productivity. Unfortunately, only a few studies have measured total factor productivity, and consequently only limited comparisons can be made between countries and over time (Macours and Swinnen, 2000b; Swinnen and Vranken, 2010; Lerman *et al.*, 2004). The available evidence on TFP (see

Fig. 6.4) is, however, roughly consistent with the evidence from the partial productivity indicators.

In Central Europe, TFP[3] grew slightly in the first years of transition – 0.4% annually between 1989 and 1992 – and significantly afterwards – by 2.2% annually between 1992 and 1995 and by 4.4% annually between 1995 and 1998. Studies find a slowdown of TFP growth in the period 1998–2001, probably as a result of substantial investments in agricultural machinery and capital inputs in that period.

In the Balkan countries, TFP fluctuates much more. From 1989 to 1992, TFP decreased by 4.1% per year. Later there was a strong recovery when TFP increased by 7.5% per year in the period 1992–1995, but it fell again in the late 1990s when bad macro-economic policies resulted in TFP declines of 1.3% annually from 1995 to 1998. After 1998, when a series of important reforms were implemented in the region, there was a strong recovery in productivity: from 1998 to 2001, TFP grew on average by 2.3% per year.

In the former Soviet Union, TFP declined by approximately 5% per year in the first years after transition. After 1994, TFP remained relatively stable at approximately

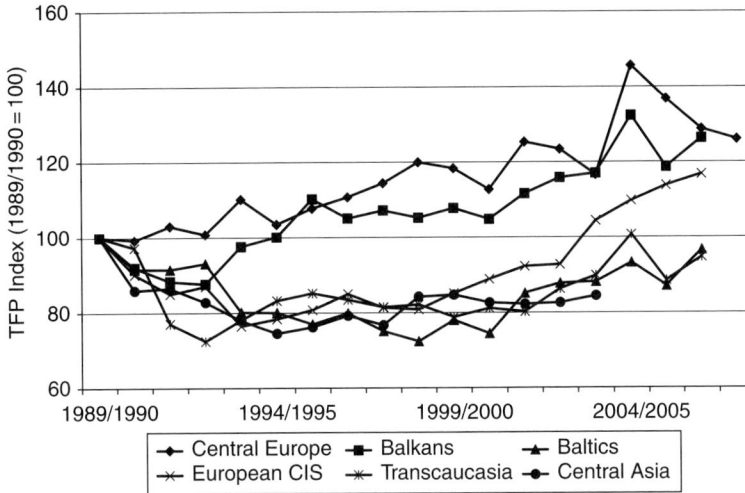

Fig. 6.4. Changes in agricultural total factor productivity since the start of reforms. Reforms started in 1990 in the former Soviet Union republics and in 1989 in Central Europe and the Balkan countries. Source: Agricultural TFP is derived using the methodology described in Swinnen and Vranken (2010) and extended through to 2006/2007 or 2007/2008.

80% of the pre-reform level. Only in the beginning of 2000s did TFP start to increase in all former Soviet Union regions, except for Central Asia. The European CIS have experienced an especially strong growth in TFP, and in 2003 TFP exceeded the pre-reform level. Growth in TFP continued and, in 2007, TFP in the European CIS was 15% higher than the pre-reform level.

6.4 Stylized Facts about Initial Resource Endowments and the Reform Pathway

6.4.1 Initial resource endowments

There are huge differences among these transition countries when it comes to their initial resource endowments at the start of the reform, and this plays an important role in explaining the development of farm structures and the evolution of agricultural productivity. Such differences in resource endowment (or technology) reveal themselves in differences in relative factor ratios. For example, the pre-reform ratio of farm workers per hectare of arable land in Russia and Kazakhstan is many

times lower than in Albania, Azerbaijan or Romania (Table 6.1). If labour–land ratios are high, i.e. if agricultural production processes are relatively labour intensive, then the benefits of shifting from corporate to family farms are greater, whereas the costs are likely to be lower for several reasons.

First, the returns to incentives that boost effort are larger in labour-intensive agricultural systems. Because of the sequential and biological nature and spatial dimensions, effort in agricultural production is difficult to measure, and corporate farms in particular cope with labour supervision problems. Because farm individualization boosts labour effort, the productivity gains of breaking up the large-scale agricultural production units into individual farms are greater when pre-reform labour–land ratios are higher.

Second, scale economies vary by commodity, and diseconomies of scale in production are typically characterized by high labour intensity (Mathijs and Swinnen, 1998). For instance, a less labour-intensive sector such as grain production tends to have more economies of scale because it is more suitable for mechanization than more labour-intensive sectors such as dairy or vegetable production. The losses in scale

Table 6.1. Agricultural labour–land ratios in transition countries at the start of reforms.

Labour-extensive	Farm workers per hectare of arable land in 1989–1991
Kazakhstan	0.03
Russian Federation	0.06
Latvia	0.12
Estonia	0.13
Hungary	0.14
Ukraine	0.14
Slovakia	0.15
Lithuania	0.16
Belarus	0.18
Czech Republic	0.19
Labour-Intensive	
Bulgaria	0.20
Moldova	0.27
Poland	0.31
Romania	0.33
Kyrgyzstan	0.42
Georgia	0.50
Armenia	0.53
Slovenia	0.56
Azerbaijan	0.61
Uzbekistan	0.65
Tajikistan	0.82
Albania	1.33

Source: Macours and Swinnen (2002).

economies and disruption costs are therefore lower in labour-intensive systems. On the other hand, in land- and capital-intensive systems, households often lack the financial means and inputs to farm more efficiently, so they are less inclined to start farming on their own. Therefore, breaking up the large-scale agricultural production units results in less productivity loss or might even result in productivity gains. And farm individualization will increase in labour-intensive systems, whereas the opposite holds true for land- and capital-intensive systems, where the share of corporate farms will be higher.

6.4.2 The reform pathway

Countries have chosen different land reform processes. Some chose restitution of agri-cultural land to former owners, whereas others chose to distribute land in kind (in actual plots) as shares of former cooperative farms, or at first as shares and later in kind. Although most Central European countries and the Baltic States chose to restitute land, most former Soviet Union republics chose to distribute land among rural inhabitants and particularly among former state and collective farm workers. The reform choice had important implications for the development of farm structures and for agricultural productivity.

With restitution of land to former owners, such as in the Czech and Slovak Republics, Bulgaria, the Baltic States and large parts of Romania and Hungary, a significant share of the land was (potentially) allocated to individuals who were not (or no longer) active in agriculture. They may have been retired or living in urban areas and less likely to use land than rural households still active in agriculture. With only limited information about the sales price and the expected increase in land prices upon accession to the European Union, most of these new land owners were unwilling to sell their newly acquired assets and chose instead to rent it out. Because identifying potential tenants involves search and negotiation costs, the easiest way for the new land owners to employ their land was to rent it back to the corporate farms, the historical users of the land (Mathijs and Swinnen, 1998). The corporate management was closely involved in the land reform process, and its search and negotiation costs to contract with the new land owners were significantly lower than costs faced by newly emerging structures (particularly family farms and *de novo* companies). As a result, restitution was more likely than other land reform procedures to contribute to a consolidation of the large-scale farming structures (collective and state farms in the past, now corporate farms) through the land rental market.

Distribution of land was done by allocating physical plots (such as in Albania) or in shares (such as in Russia and Kazakhstan) or first in shares and later in physical plots (as in Transcaucasia). These methods carry important implications for the development of the farm structure.

The distribution of land in specific plots (boundaries) created stronger property rights for the new owners, whereas the distribution of land shares often implied uncertain property rights and high transaction costs.[4] The stronger rights with distribution in plots caused stronger growth of family farms as it was easier for these new owners to access their land. This was particularly true for family farming in Albania in the early 1990s and for Azerbaijan in the second half of the 1990s. Within a few years of the start of the land reform, around 90% of all agricultural land in both countries had shifted to family farms.

In contrast, where land shares were distributed (for example, in Russia, Kazakhstan and pre-2000 Ukraine) the result was much weaker property rights. As Uzun (2000, p. 8) observes: 'land share owners do not know where their land shares are located; managers of agricultural enterprises have an opportunity to use the land owned by citizens freely and without controls; and workers, still, after nine years of reforms, do not clearly understand their choices.' Weak property rights constrained the shift of land use to family farms and contributed to the consolidation of large-scale corporate farming organizations.

Not only did the choice of the reform process differ, so did the speed with which countries actually pursued, or delayed, reforms. This is nicely illustrated by the World Bank's policy indexes developed to assess the status of agricultural reform (Csaki and Kray, 2005). A score of one means no reform, a situation comparable with a centrally planned economy. The maximum score a country can reach is ten, which means market reforms have been completed and the nation is a free-market economy. Table 6.2 shows the evolution of market reform in the period 1997–2005 and how, in 2005, the average of the land reform and market reform indexes ranged from less than three in Belarus to nine in some Central European countries, such as Slovenia, Latvia, Hungary and Estonia.

Table 6.2. Status of agricultural reforms in transition countries.

	1997	2001	2005
	Average of land and market reform indexes		
Central Europe			
Czech Republic	8.5	9	9
Hungary	9	9	9
Poland	8.5	8	8.5
Slovakia	7	8	8.5
Balkans			
Albania	8	8	8.5
Bulgaria	6.5	8.5	8.5
Romania	7	7.5	8.5
Slovenia	8.5	9	9
Baltics			
Estonia	8	9	9
Latvia	8	9	9
Lithuania	7.5	8	8
Transcaucasia			
Armenia	7.5	7	8
Azerbaijan	6	5.5	6
Georgia	7	8	7
European CIS			
Belarus	2	2	2.5
Russia	6	6.5	6
Ukraine	6	5.5	5.5
Central Asia			
Kazakhstan	6	8	7.5
Kyrgyzstan	6	7	8
Tajikistan	3	6	5
Uzbekistan	2.5	4	5

Source: Csaki and Kray (2005).

The development of farm structure, which in turn affects agricultural productivity changes, also depends on a country's overall economic development. In order to continue farming, a producer needs to generate a threshold income, and this will equal the opportunity cost of his labour, which may either be off-farm wages or retirement or unemployment benefits. Thus, a producer stays active in agriculture only if his human capital can generate an income greater than a threshold income that could be earned outside of agriculture. The latter increases when a country is more advanced in the transition process, i.e. when the opportunity cost of labour increases because off-farm labour opportunities increase as well as retirement or unemployment benefits.

Table 6.3. Evolution of GDP per capita in transition countries.

	Year after the start of the reform			
	Year 1	Year 5	Year 10	Year 15
	(constant 2000 US$ per capita)			
Central Europe				
Czech Republic	NA	4710	5322	6285
Hungary	4422	3638	4452	5626
Poland	NA	3041	4251	5045
Slovakia	3828	2844	3749	4467
Balkans				
Albania	1093	734	1115	1450
Bulgaria	1859	1481	1456	1972
Romania	2013	1558	1616	2165
Slovenia	NA	7392	9480	11264
Baltics				
Estonia	3891	2806	4106	6213
Latvia	3901	2356	3302	5047
Lithuania	4337	2462	3263	4873
Transcaucasia				
Armenia	795	423	620	1127
Azerbaijan	1251	559	655	1183
Georgia	1493	438	648	973
European CIS				
Belarus	1410	1024	1273	1871
Russian Federation	2602	1686	1775	2444
Ukraine	1389	759	636	962
Central Asia				
Kazakhstan	1612	1095	1229	1978
Kyrgyzstan	465	243	279	321
Tajikistan	485	196	159	234
Turkmenistan	965	569	645	1297
Uzbekistan	685	514	558	684

NA, not available. *Source*: World Bank.

Less productive producers will exit agriculture, thereby raising the average productivity of country's agriculture. Table 6.3 illustrates how GDP per capita, which can be seen as a proxy for a country's overall development, differs among transition countries.

6.5 Causes of Agricultural Productivity Changes[5]

6.5.1 Advanced reformers with labour-extensive agriculture

Central Europe

At the start of the reforms, three Central European countries – the Czech Republic, Hungary and Slovakia[6] – were characterized by labour-extensive agricultural production systems, relatively high levels of pre-reform economic development, good progress in land reform and farm restructuring policies, and agri-food chains that were not dependent on the former Soviet Union for their exports. According to our conceptual framework, we would expect these countries to adjust factor ratios in order to use the most limited factor, labour, more efficiently. The model predicts that adjustments in labour–land ratios would drive a change in labour productivity.

Data confirms it. Despite strong decreases in aggregate output, agricultural labour productivity – output per worker – increased strongly immediately after the start of the reforms. The dramatic reduction in the use of labour drove the rise of agricultural labour productivity in Central Europe. Official employment data show an average reduction in labour use of 45% during the first five years of transition. The strongest reductions occurred in Hungary (57%) and the Czech Republic (46%). Utilized agricultural area remained stable, so that land productivity (gross agricultural output per hectare of arable land) decreased, but the labour–land ratio decreased tremendously immediately after the start of the transition period. Because of the high level of economic development, which is probably accompanied by off-farm employment opportunities, labour can flow out of the agricultural sector. Strong land property rights were established, and soft budget constraints (i.e. credit made available on easy terms) were introduced for the large-scale farms due to the progress in land reform and farm restructuring. Consequently, factor ratios were able to adjust. In Central Europe, a tremendous decrease in the labour–land ratio drove the increase in labour productivity. The land-intensive character of the agricultural sector in Central Europe gave larger production units – typically corporate farms that succeeded the communist collective and state farms – an advantage over smaller family-operated farms so that the former large-scale farms still continued to cultivate a large share of the agricultural area.

Figure 6.5 shows that it was mainly the strong increase in agricultural labour productivity that drove TFP changes in Central Europe. TFP there started increasing from the start of the reforms and continued to increase at rapid rates during the first decade. Average TFP growth was around 2.5% per year in the 1992–1995 period and around 4% per year during 1995–1998. What is remarkable is that in Central Europe, after robust growth through 1992–1998, TFP growth then slowed down and even became negative in the Czech Republic over 1998–2001. TFP growth was stronger again during 2001–2004, but became negative during 2004–2007. This fluctuating pattern in the past decade was probably due to alternating good and bad crop years. However, with an average annual growth rate of 1.7% over the past decade, the overall trend in TFP growth remains positive.

The Baltic States

Like Central Europe, the Baltic States were characterized by a labour-extensive agricultural sector and relatively high levels of pre-reform economic development. Immediately after the start of the transition process, the Baltic States made good progress in implementing the land reform and farm restructuring process. Land rights were restituted, resulting in strong property rights.

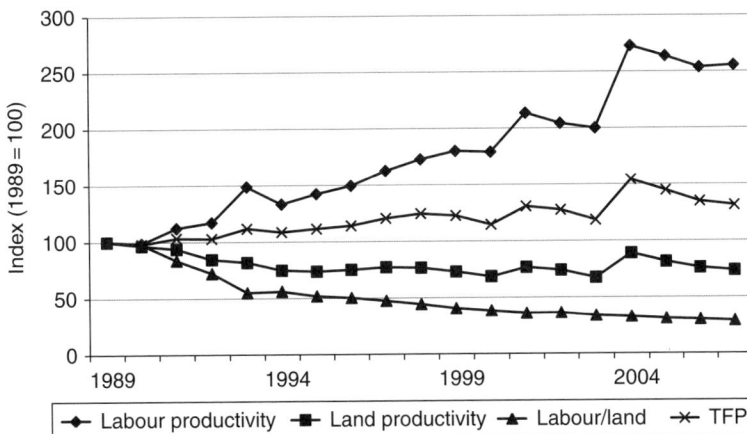

Fig. 6.5. Evolution of agricultural productivity measures in Central Europe. Central Europe includes Czech Republic, Hungary and Slovakia. Sources: See sources to Figs 6.1–6.4.

At the start of the reforms, the collective and state farms were considerably less efficient than those in Central Europe. Agricultural labour was used inefficiently on these large-scale farms, and many gains could be reached by organizing agricultural labour differently. As a result, the large-scale collective and state farms were broken up into smaller production units – typically family farms – that face fewer moral hazard problems and provide good labour incentives. In contrast to Central Europe, the agri-food chain in the Baltic States was much more dependent on the former Soviet Union for both input supplies and export of agricultural products. Consequently, they were heavily affected by the USSR's disruption. According to our conceptual model, we expect that changes occur to use the relatively scarce factor, labour, more efficiently. This can be done by substituting labour for land so that the labour–land ratio decreases.

As expected, we observe a strong decrease in the labour–land ratio in the Baltic States. Immediately after the start of the reforms, labour flowed out of agriculture. The agriculture structures in the countries changed completely. At the beginning of the reforms, the large-scale collective and state farms were cultivating more than 90% of the agricultural area, but by the mid-2000s that amount fell to around 10% in Latvia and Lithuania and around 45% in Estonia. This disruption of farm structures and institutional environment, which was strongly organized towards large-scale collective and state farm structures together with the break-up of the USSR, resulted in a very strong decrease in agricultural output at the start of the reforms.

Figure 6.6 shows that the decrease in the labour–land ratio could not compensate for the decrease in agricultural output so that agricultural labour productivity, as well as land productivity (gross agricultural output per hectare) and the TFP index, decreased at the start of the reforms. By the mid-1990s, however, the Baltic States started to reap the benefits of improved agricultural labour management on family farms and labour productivity began to recover. A decade after the start of the reforms, output started to recover, probably not only in response to improved farm labour management but also because of an adjustment of the sectors surrounding agriculture and the fact that, by the late 1990s to early 2000s, institutions had had time to adjust to the new farm structures in place. As a consequence, land productivity and TFP measures started to increase.

6.5.2   Advanced reformers with labour-intensive agriculture

The Balkans

The Balkan countries[7] were characterized by very labour-intensive agriculture at the start of the reforms. Resource endowment

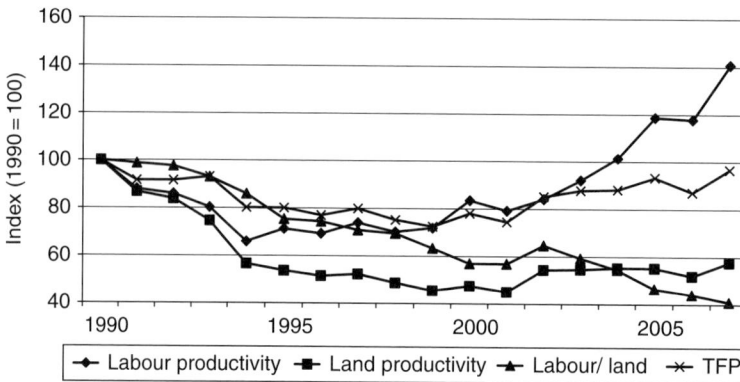

Fig. 6.6. Evolution of agricultural productivity measures in the Baltics. The Baltics include Estonia, Latvia and Lithuania. Sources: See sources to Figs 6.1–6.4.

affects the costs and benefits of shifting from corporate farms to family farms, and if the labour–land ratio is high, such as in the Balkans, the benefits from better labour governance are greater, whereas the losses in scale economies of shifting to smaller farms are lower. These productivity incentives resulted in a strong shift to small-scale farming. Similar to Central Europe, the collectivization of agriculture and the introduction of central planning in the Balkans occurred after World War II, in contrast to the former Soviet Union where it had begun in the 1920s. Consequently, rural households in Central Europe and the Balkan countries had much more experience with private farming than their counterparts in most of the former Soviet Union. This difference affected not only the emergence and dynamics of the new private farms but also the preferences for land reforms: as in Central Europe, land in the Balkans was restituted to the former owners. On the basis of our conceptual framework, we expect gains in productivity to come from using the most limited factor, land, more efficiently.

These expectations are reflected in the data: overall, labour outflow from agriculture was limited in the early transition period because of poor overall development and limited job opportunities outside of agriculture. In these countries, agriculture served as a social buffer in times when overall unemployment was high and social benefits were low. The restitution of land to former owners constrained young farmers' access to land because land was given to older people who took up farming to supplement their small pensions. Official employment data show that in the first decade of transition there was almost no outflow of labour from agricultural employment. For example, in 2000 agricultural employment in Romania had actually increased by 17% compared with its pre-reform level. Hence, in the first decade of transition, gains in productivity stemmed mainly from improved property rights and the incentives that came along with the shift to individual farming. Only in the 2000s did the labour–land ratio start to adjust as farmers moved out of agriculture and gains in productivity accelerated.

Figure 6.7 shows that TFP decreased slightly in the first years of transition in Balkan agriculture. However, by the mid-1990s TFP began to rise again as agricultural labour productivity and land productivity (gross agricultural output/land) improved. By the early 2000s, TFP growth accelerated as employment in agriculture started to decline. This resulted in a strong increase in labour productivity and a decrease in the

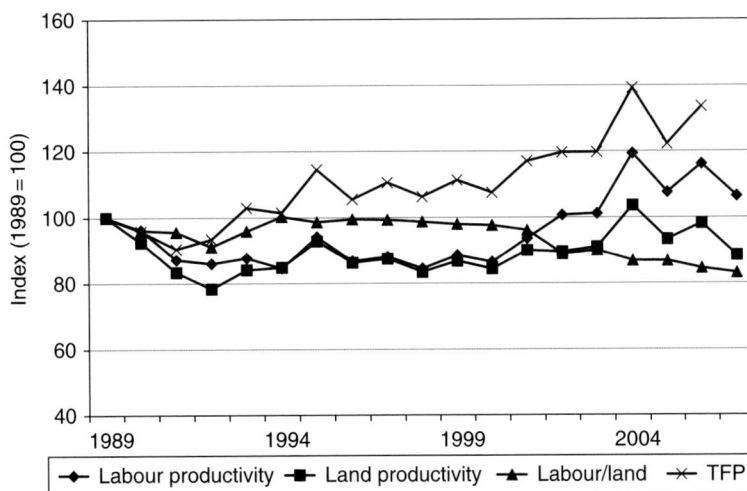

Fig. 6.7. Evolution of agricultural productivity measures in the Balkans. The Balkans include Albania, Bulgaria and Romania. Sources: See sources to Figs 6.1–6.4.

labour–land ratio, which has driven recent TFP growth in the Balkans.

6.5.3 Slow reformers with labour-intensive agriculture

Transcaucasia

Like the Balkans, the agricultural sector in Transcaucasia is characterized by a labour-intensive agricultural sector and a relatively low level of economic development. Similarly, we expect gains in productivity to come from a more efficient use of the scarce factor, land.

The gains from a shift to small-scale individual farming are conditional on land policy and, unlike in the Balkans, the Transcaucasian countries made only slow progress in the implementation of land reform in the first years after transition. This constrained the shift towards individual farming. In the beginning of the 1990s, land rights were transferred as paper shares or certificates, without any direct link between the individual and a specific plot of land. Owners had little incentive to put in effort or undertake investments because property rights on specific plots were not clearly defined. This is reflected in a decline in land productivity in the first years after transition. At the end of the 1990s and the beginning of the 2000s, land reforms distributed physical plots instead of shares. This resulted in improved property rights and incentives. The consequent shift towards small-scale individual farming was the driver behind efficiency gains in the beginning of the 2000s.

Given the level of economic development, the agricultural sector served as a social buffer in the first years after transition and mainly attracted unemployed, often unmotivated individuals and old people that took up farming to supplement their pensions. Since Transcaucasia had an even lower level of overall economic development than the Balkans, the inflow of agricultural labour was even more pronounced: by 2000, agricultural employment in Transcaucasia increased by 27% compared to its pre-reform level. This contributed to a decrease in agricultural labour productivity and an increase in the labour–land ratio in the first decade of transition. Recently, agricultural employment has started to decline slowly so that labour productivity and the land–labour ratio have begun to rise.

Figure 6.8 illustrates the evolution in agricultural TFP and the partial productivity measures in Transcaucasia. In the first decade after transition, the combination of slow land reform and the strong increase in agricultural employment resulted in a decline in all partial productivity measures (labour productivity, output/land and land/labour).

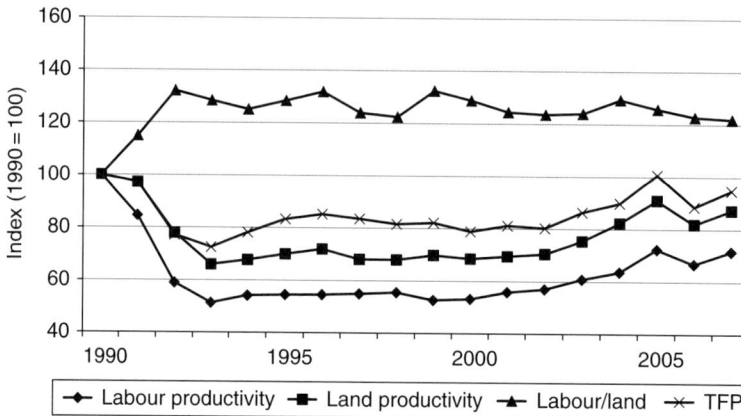

Fig. 6.8. Evolution of agricultural productivity measures in Transcaucasia. Transcaucasia includes Armenia, Azerbaijan and Georgia. Sources: See sources to Figs 6.1–6.4.

The beginning of the 2000s, when land reform distributed physical plots instead of shares, brought a shift towards small-scale individual farming. This resulted in a significant increase in agricultural output and land productivity (output/land). Growth in labour productivity was the main driver behind the recent growth in TFP.

6.5.4 Slow reformers with labour-extensive agriculture

The European CIS

At the start of the reforms, the European CIS were characterized by a very labour-extensive agricultural production system. According to our conceptual model, we expect changes to occur so that the relatively scarce factor, labour, is used more efficiently. Productivity gains can come from large farms shedding labour, resulting in a decrease in the labour–land ratio.

Gains in productivity from shedding labour, however, are conditional on farm policy. Slow progress in implementing the land reform and farm restructuring process in the European CIS resulted in an initial decline in productivity. The European CIS opted to implement a truncated reform process, which resulted in poor incentives and soft budget constraints. For example, in

the early transition period, Russia liberalized its output prices but retained some input support and, in Belarus, agricultural support remained intact until the end of the 1990s (Rozelle and Swinnen, 2004). As a result of this incomplete reform process, the outflow of agricultural labour was limited, and there was a strong decline in agricultural output and productivity.

Figure 6.9 illustrates these evolutions. In the first years after transition the labour–land ratio remained constant. The path of agricultural labour productivity for the European CIS mirrors that of the region's output, falling between 20% and 50% between 1990 and 1999. Land productivity also fell sharply. This resulted in a decline in TFP in the first decade of transition.

After the 1998 Russian financial crisis, farm restructuring processes were implemented. The introduction of hard budget constraints improved incentives and large farms started shedding agricultural labour, leading to increases in the partial productivity measures. In addition, uncertainty about property rights was reduced as land policies were further liberalized and limited land transactions became possible as, for example, in 2002 in Russia (Rozelle and Swinnen, 2004). As a consequence, TFP started to increase at the end of the 1990s.

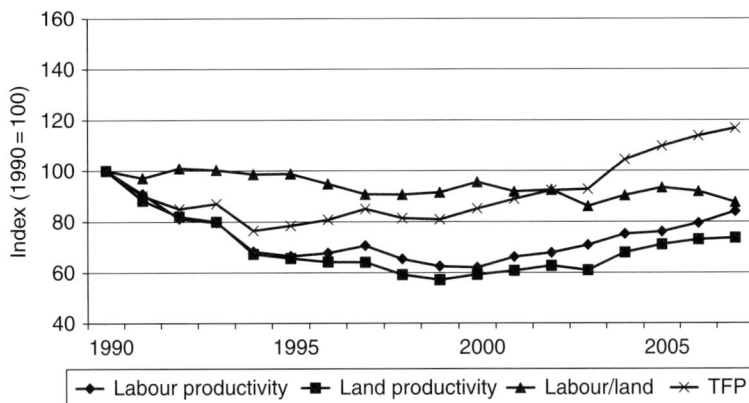

Fig. 6.9. Evolution of agricultural productivity measures in the European CIS. The European CIS include Belarus, the Russian Federation and Ukraine Sources: See sources to Figs 6.1–6.4.

6.6 Conclusions

There have been dramatic changes in productivity during the past 15 years in transition countries. In general, there is a U-shaped effect: an initial decline in productivity followed by a recovery later on. Virtually all countries witnessed an initial decline in productivity, and virtually all countries are currently experiencing increases in productivity. In several transition countries, productivity growth during the past five years has been quite spectacular.

The depth and length of the initial decline differed enormously among countries, however. Our analysis suggests that productivity changes were related to the extent of the pre-reform distortions and how countries implemented reforms. In the most advanced reformers (mostly in Central Europe), the decline was relatively mild and recovery started fairly early on; productivity growth, particularly labour productivity, has been strong there for more than a decade. In the Baltic States, recovery started relatively quickly, reflecting the fast pace of reforms, but the initial fall in productivity was much deeper than in Central Europe, possibly reflecting the larger pre-reform distortions in these former Soviet Union countries. In many of the other former Soviet Union countries, the decline of productivity was much more dramatic (around 25% in several countries) and lasted for much of the 1990s. It was only after the 1998 Russian financial crisis that a recovery started and, although hesitant at first, productivity increases have been strong, certainly for crop yields.

It is well known that initial resource endowments (factor proportions) play an important role in agricultural productivity growth. Technological adaptation can occur through a sequence of induced innovations in technology towards saving the limited factors. Policies interact with these forces. For the transition countries examined in this study, regions that implemented the necessary reforms early on, such as the Central European countries, started to witness technological adaptation and productivity growth earlier. It is clear, however, that the level of development also made a difference.

The institutional and human capital hurdles to create a market economy were higher in the countries further east, so it is no surprise that the transition productivity declines were deeper and longer in those regions.

Productivity growth in regions with labour-extensive production systems, such as in Central Europe, the Baltic States and the European CIS, came from shedding agricultural labour. The extent and speed with which the restructuring process occurred is, however, strongly correlated with the level of economic development and reform progress. In Central Europe and the Baltic States, there was a relatively high level of economic development and a rapid reform process. This resulted in a rapid increase in productivity in the years after transition, although productivity growth in the Baltic States was delayed compared with Central Europe, possibly reflecting larger pre-reform distortions as these countries were part of the former Soviet Union. In the European CIS, the government implemented a truncated reform process that slowed farm restructuring and land reform. This resulted in a large decline in productivity because agricultural labour did not leave the sector. Only after the reform process accelerated at the end of the 1990s did large farms start to shed agricultural labour and productivity measures begin to improve.

Productivity growth in labour-intensive regions such as the Balkans and Transcaucasia did not come from laying off workers because in many cases there was not a significant outflow of labour from agriculture, at least not in the first decade of transition. Initially it came from land reforms that provided better incentives for labour governance in these labour-intensive agricultural systems.[8] Once this one-off productivity shock from land reform (which typically needs to involve the distribution of physical plots to rural households) is 'consumed' it is crucial to focus on other sources of productivity growth.

Recent productivity growth has come from a combination of improved options for employment outside of agriculture (either in other sectors or through migration), improvements in social payments (pensions and unemployment benefits) allowing subsistence farmers to leave agriculture, and improved access to factor and output markets. The latter has come about through general reforms that have stimulated investments in the food industry (including the inflow of foreign investment), brought new technology and capital into the agri-food chains, and fostered a more stable general investment climate leading to off-farm job creation.

Further impetus for productivity improvement followed the 1998 Russian financial crisis. The devaluation of the rouble strengthened competitiveness. Also, the subsequent rise in international energy prices improved Russia's fiscal situation. Those factors led to a higher demand, more timely payments and higher farm profits, all of which improved farmers' financial situation and thereby their ability to purchase inputs and invest in new technology. There have also been fewer disruptions in the market exchanges between farms and upstream and downstream companies. All of these factors have contributed to increased productivity.

Notes

[1] Central Europe includes the Czech Republic, Hungary, Poland and Slovakia. The Balkans include Albania, Bulgaria, Romania and Slovenia. Besides Slovenia, other states of former Yugoslavia are not included in the analysis. The former Soviet Union includes the Baltic States, the European Commonwealth of Independent States (CIS), Transcaucasia and Central Asia. The Baltic States refer to Estonia, Latvia and Lithuania. The European CIS refer to Russia, Ukraine, Belarus. Transcaucasia includes Armenia, Azerbaijan and Georgia. Central Asia refers to Uzbekistan, Turkmenistan, Kazakhstan, Kyrgyz Republic and Tajikistan.

[2] See, for example, Koppel (1995) for a critical assessment and more recent interpretations of the Hayami–Ruttan induced innovation model.

[3] Agricultural TFP is derived using the methodology described in Swinnen and Vranken (2010) and extended through 2006/2007 or 2007/2008. Swinnen and Vranken (2010) use a growth accounting approach in which TFP growth is the difference between the growth in output (measured by GAO) and the aggregate growth in land, labour, fertilizer, machinery and livestock capital inputs. Input weights are the production elasticities estimated from a Cobb–Douglas production function from Cungu and Swinnen (2003). Data are obtained from FAO except labour data, which come from a variety of sources.

[4] Individuals usually had to declare their intention to start up their own farms in order to take physical possession of their land. The barriers to exit were severe as leaving the farm was often discouraged by farm managers and local officials. In several countries, the share distribution system was accompanied by continued soft budget constraints for the large farms (e.g. in Ukraine, Russia and Kazakhstan), further reducing incentives for restructuring farms.

[5] The story is mixed in Central Asia with Kazakhstan following a pattern similar to the European CIS and Kyrgyzstan, Tajikistan, Turkmenistan and Uzbekistan that follow a similar pattern to the Transcaucasia countries. In addition, because of missing capital data, TFP measures could not be calculated for Tajikistan, Turkmenistan and Uzbekistan. The evolution of the productivity measures of Central Asian countries will therefore not be discussed.

[6] In the section, Poland is not included because its pre-transition structure differed markedly from other Central European countries. Unlike those countries, Poland's agriculture never underwent significant collectivization and most land was still used by individual holdings during the pre-transition era.

[7] Slovenia is not included in this analysis because, similar to the situation in Poland, agricultural land in the countries of the former Yugoslav Republic was mainly used by individual holdings during the pre-transition era.

[8] The Balkan countries implemented land reforms rapidly after transition, whereas land reforms in Transcaucasia were implemented only at the end of the 1990s.

References

Arnade, C. and Gopinath, M. (1998) Capital adjustment in U.S. agriculture and food processing: A cross-sectoral model. *Journal of Agricultural and Resource Economics* 23, 85–98.

Bofinger, P. (1993) The output decline in Central and Eastern Europe: A classical explanation. *CEPR Discussion Paper,* London, UK.

Brada, J. (1989) Technical progress and factor utilization in Eastern European economic growth. *Economica* 56, 433–448.

Csaki, C. and Kray, H. (2005) The agrarian economies of Central-Eastern Europe and the CIS: An update on status and progress in 2004. ECSSD Working Paper No. 40, June, World Bank, Washington, DC.

Cungu, A. and Swinnen, J. (2003) Transition and total factor productivity in agriculture, 1992–1999. Working Paper 2003/2, Research Group on Food Policy, Transition and Development, Katholieke Universiteit Leuven, Belgium.

Dries, L. and Swinnen, J. (2004) Foreign direct investment, vertical integration and local suppliers: Evidence from the Polish dairy sector. *World Development* 32, 1525–1544.

FAO. FAOSTAT Database. Food and Agricultural Organization, Rome. Available at: http://faostat.fao.org

Gow, H. and Swinnen, J. (1998) Up- and downstream restructuring, foreign direct investment, and hold-up problems in agricultural transition. *European Review of Agricultural Economics* 25, 3, 331–350.

Hayami, Y. and Ruttan V.W. (1970) Factor prices and technical change in agricultural development: The United States and Japan, 1880–1960. *Journal of Political Economy* 78, 5, 1115–1141.

Hayami, Y. and Ruttan V.W. (1985) *Agricultural Development: An International Perspective*. Johns Hopkins University Press, Baltimore, MD.

Jackman, R. (1994) Economic policy and employment in the transition economies of Central and Eastern Europe: What have we learned? *International Labour Review* 133, 327–345.

Koppel, B. (1995) *Induced Innovation Theory and International Agricultural Development: A Reassessment*. Johns Hopkins University Press, Baltimore, MD.

Lerman, Z., Csaki, C. and Feder, G. (2004) *Agriculture in Transition: Land Policies and Evolving Farm Structures in Post-Soviet Countries*. Lexington Books, Lanham, MD.

Liefert, W., Gardner, B. and Serova, E. (2003) Allocative efficiency in Russian agriculture: The case of fertilizer and grain. *American Journal of Agricultural Economics* 85, 1228–1233.

Macours, K. and Swinnen, J. (2000a) Impact of initial conditions and reform policies on agricultural performance in Central and Eastern Europe, the former Soviet Union, and East Asia. *American Journal of Agricultural Economics* 82, 5, 1149–1155.

Macours, K. and Swinnen, J. (2000b) Causes of output decline during transition: The case of Central and Eastern European agriculture. *Journal of Comparative Economics* 28, 172–206.

Macours, K. and Swinnen, J. (2002) Patterns of agrarian transition. *Economic Development and Cultural Change* 50, 365–395.

Mathijs, E. and Swinnen, J. (1998) The economics of agricultural decollectivization in East Central Europe and the former Soviet Union. *Economic Development and Cultural Change* 47, 1–26.

Mathijs, E. and Swinnen, J. (2001) Production efficiency and organization during transition: An empirical analysis of East German agriculture. *Review of Economics and Statistics* 83, 100–107.

Petrick, M. (2004) Farm investment, credit rationing, and governmentally promoted credit access in Poland: A cross-sectional analysis. *Food Policy* 29, 275–294.

Rozelle, S. and Swinnen, J. (2004) Success and failure of reforms: Insights from transition agriculture. *Journal of Economic Literature* 42, 2, 405–456.

Sedik, D., Trueblood, M. and Arnade, C. (1999) Corporate farm performance in Russia, 1991–1995: An efficiency analysis. *Journal of Comparative Economics* 27, 514–533.

Seeth, H., Chachnov, S. Surinov, A. and von Braun, J. (1998) Russian poverty: Muddling through economic transition with garden plots. *World Development* 26, 1611–1623.

Swinnen, J. and Rozelle, S. (2006) *From Marx and Mao to the Market: The Economics and Politics of Agricultural Transition*. Oxford University Press, Oxford, UK.

Swinnen, J. and Vranken, L. (2010) Reforms and agricultural productivity in Central and Eastern Europe and the former Soviet Republics: 1989–2005. *Journal of Productivity Analysis* 33, 241–258.

Swinnen, J., Dries, L. and Macours, K. (2005) Transition and agricultural labour. *Agricultural Economics* 32, 15–34.

Swinnen, J., Van Herck, K. and Vranken, L. (2010a) Shifting patterns of agricultural production and productivity in the former Soviet Union and Central and Eastern Europe. In: Alston, J., Babcock, B. and Pardey, P. (eds) *The Shifting Patterns of Agricultural Production and Productivity Worldwide*. CARD-MATRIC, Ames, IA, pp. 279–313.

Swinnen, J., Van Herck, K. and Vranken, L. (2010b) Agricultural productivity in transition economies. *Choices* 24, 4.

World Bank. World Development Indicators Database. Washington, DC. Available at: http://publications.worldbank.org/WDI

7 Total Factor Productivity in Brazilian Agriculture

José Garcia Gasques,[1] Eliana Teles Bastos,[1] Constanza Valdes[2] and Miriam Rumenos Piedade Bacchi[3]

[1]Ministry of Agriculture, Brazil; [2]Economic Research Service, US Department of Agriculture, Washington, DC; [3]Centre for Applied Economics (CEPEA), University of São Paulo, Brazil

7.1 Introduction

The objective of this study is to estimate total factor productivity (TFP) indexes for Brazilian agriculture during the 1970–2006 period on the basis of the Agricultural Censuses of 1970, 1975, 1980, 1985, 1995–1996 and 2006. The release of the Brazilian Institute for Geography and Statistics (IBGE) information in the 2006 Agricultural Census has made it possible to update, improve and extend earlier estimates of agricultural TFP growth that covered the period of 1970–1995 (Gasques and Conceição, 2001). With the latest Census information we are able to examine the behaviour of Brazilian agriculture in greater detail and over a longer period of time. The database provided by the Census allows for a greater coverage of the products included in the productivity estimate and provides more complete information regarding agricultural inputs. Furthermore, it allows us to obtain productivity estimates not only for the country as a whole, but also for each state in the federation. The methodology is growth accounting: indicators of aggregate output and input are created on the basis of all products and factors of production covered by the Census.

The difference in the growth between output and input is defined as growth in TFP. TFP reflects improvements in productivity efficiency brought about by technical change, economies of scale, quality improvement in the inputs and other factors that raise the amount of output produced from a given bundle of inputs.

Besides TFP, the study looks at structural changes taking place in Brazilian agriculture, namely the composition of different products. This is represented by an index of structural change. We also examine whether Brazilian agriculture is moving toward specialization or diversification in its product mix.

Before we describe our analysis of productivity and structural change, some long-term trends in Brazilian agricultural are shown in Table 7.1. The data are drawn from agricultural Censuses beginning in 1920 through to the most recent conducted in 2006. The first thing to note is that the number of farms/lands increases strongly until 1980, revealing the expansion and occupation process that took place up to that time. Since the 1980 Census, the number of farms has remained relatively stable and was at 5.2 million in 2006. Average farm size declined until 1970 and since then has averaged 60–70 ha/farm.

Table 7.1. Long-term changes in Brazilian agriculture from the agricultural censuses, 1920–2006.

Description	Census year									
	1920	1940	1950	1960	1970	1975	1980	1985	1995	2006
Number of farms	648,153	1,904,589	2,064,642	3,337,769	4,924,019	4,993,252	5,159,851	5,801,809	4,859,865	5,175,636
Average farm size (ha)	270	104	112	75	60	65	71	65	73	64
Total area in farms (ha)	175,104,675	197,720,247	232,211,106	249,862,142	294,145,466	323,896,082	364,854,421	374,924,929	353,611,239	333,680,037
Crop land in farms (ha)	6,642,057	18,835,430	19,095,057	28,712,209	33,983,796	40,001,358	49,104,263	52,147,708	41,794,455	60,592,576
% of land in crops	4	10	8	11	12	12	13	14	12	18
Pasture land in farms (ha)	–	88,141,733	107,633,043	122,335,386	154,138,529	165,652,250	174,599,641	179,188,431	177,700,472	160,042,062
% of land in pastures	–	45	46	49	52	51	48	48	50	48
Forest area in farms (ha)	48,916,653	49,085,464	55,999,081	57,945,105	57,881,182	70,721,929	88,167,703	88,983,599	94,293,598	100,040,933
% of land in forests	28	25	24	23	20	22	24	24	27	30
Total persons employed in agriculture	6,312,323	10,159,545	10,996,834	15,633,985	17,582,089	20,345,692	21,163,735	23,394,919	17,930,890	16,568,205
Average employment per farm	9.7	5.3	5.3	4.7	3.6	4.1	4.1	4.0	3.7	3.2
Total number of tractors in use	1,706	3,380	8,372	61,345	165,870	323,113	545,205	665,280	799,742	820,718
Average crop land per tractor (ha)	3,893	5,573	2,281	468	205	124	90	78	52	74
Livestock										
Total number of cattle (bovines)	34,271,324	34,392,419	46,891,208	56,041,307	78,562,250	101,673,753	118,085,872	128,041,757	153,058,275	176,147,501
Number of bovines per ha of pasture	–	2.6	2.3	2.2	2.0	1.6	1.5	1.4	1.2	0.9
Total number of pigs	16,168,549	16,839,192	22,970,814	25,579,851	31,523,640	35,151,668	32,628,723	30,481,278	27,811,244	31,189,351
Total number of poultry (1000)	–	59,274	77,830	132,275	213,623	286,810	413,180	436,809	718,538	1,143,456
Cow milk production (1000 litres)	–	1,829,755	2,750,892	3,698,260	6,303,111	8,513,783	11,596,276	12,846,432	17,931,249	20,567,869
Egg production (1000 dozen)	–	112,557	184,300	268,376	556,410	878,337	1,248,083	1,376,732	1,885,415	2,781,619
Wool production (tonnes)	–	4,464	13,453	22,015	33,617	31,519	30,072	23,877	13,724	10,210

Crops										
Coffee										
Production (tonnes)	788,488	1,201,186	1,952,774	4,069,493	1,140,510	2,502,219	2,117,351	3,700,004	2,838,165	2,421,478
Harvest area (ha)	2,215,658	—	2,465,450	4,030,614	1,635,666	2,266,372	2,449,225	2,636,704	1,812,250	1,687,479
Yield (kg/ha)	356	—	792	1,010	697	1,104	864	1,403	1,566	1,435
Cocoa										
Production (tonnes)	66,883	108,076	146,728	169,050	204,478	301,821	352,998	422,737	242,104	199,172
Harvest area (ha)	197,129	—	303,347	398,958	419,965	457,962	474,837	691,026	679,778	515,828
Yield (kg/ha)	339	—	484	424	487	659	743	612	356	386
Orange										
Production (tonnes)		1,273,972	875,490	1,347,134	3,081,997	4,584,517	7,844,649	11,841,691	15,628,487	12,175,593
Harvest area (ha)		—	57,135	103,009	207,457	252,098	456,458	632,525	946,886	596,668
Yield (kg/ha)		—	15,323	13,078	14,856	18,185	17,186	18,721	16,505	20,406
Grape										
Production (tonnes)		114,411	196,651	358,529	509,361	546,026	426,598	728,423	653,275	828,892
Harvest area (ha)		—	34,657	50,419	60,856	52,869	46,878	58,657	56,370	63,290
Yield (kg/ha)		—	5,674	7,111	8,370	10,328	9,100	12,418	11,589	13,097
Paddy rice										
Production (tonnes)	831,495	1,196,500	2,784,989	3,762,212	5,271,272	7,548,930	8,086,747	8,986,289	8,047,895	9,687,838
Harvest area (ha)	532,384	—	2,163,653	2,950,043	4,312,134	5,662,875	5,712,072	5,173,330	2,968,126	2,415,582
Yield (kg/ha)	1,562	—	1,287	1,275	1,222	1,333	1,416	1,737	2,711	4,011
Beans										
Production (tonnes)	725,069	681,147	1,240,075	1,419,602	1,518,846	1,598,252	1,732,044	2,235,810	2,063,723	3,088,082
Harvest area (ha)	672,912	—	2,363,631	3,566,218	4,081,950	3,895,498	4,361,467	5,928,033	4,069,615	4,205,619
Yield (kg/ha)	1,078	—	525	398	372	410	397	377	507	734
Corn (maize)										
Production (tonnes)	4,999,697	5,359,863	6,660,680	8,374,406	12,770,216	14,343,556	15,722,581	17,774,404	25,511,889	41,427,610
Harvest area (ha)	2,451,382	—	5,311,799	7,791,314	10,670,188	10,741,210	10,338,592	12,040,441	10,448,537	11,598,576
Yield (kg/ha)	2,040	—	1,254	1,075	1,197	1,335	1,521	1,476	2,442	3,572
Wheat										
Production (tonnes)	87,180	96,885	364,108	503,715	1,905,961	1,562,819	2,411,724	3,824,286	1,433,116	2,233,255
Harvest area (ha)	136,069	—	515,661		2,057,898	2,301,145	2,638,320	2,518,086	842,730	1,298,317
Yield (kg/ha)	641	—	706		926	679	914	1,519	1,701	1,720

Continued

Table 7.1. Continued.

Description	Census year									
	1920	1940	1950	1960	1970	1975	1980	1985	1995	2006
Soybean										
Production (tonnes)		1,928	45,023	216,033	1,884,227	8,721,274	12,757,962	16,730,087	21,563,768	46,195,843
Harvest area (ha)					2,185,832	5,656,928	7,783,706	9,434,686	9,240,301	17,882,969
Yield (kg/ha)					862	1,542	1,639	1,773	2,334	2,583
Sugar cane										
Production (tonnes)	13,985,999	17,920,711	22,920,101	39,857,707	67,759,180	79,959,024	139,584,521	229,882,037	259,806,703	407,466,569
Harvest area (ha)	414,578		853,270	1,165,572	1,695,258	1,860,401	2,603,292	3,798,117	4,184,599	5,679,833
Yield (kg/ha)	33,736		26,861	34,196	39,970	42,979	53,618	60,525	62,086	71,739
Whole cotton										
Production (tonnes)	332,338	1,168,130	769,528	956,249	1,261,704	935,979	1,170,597	2,178,455	814,188	2,491,586
Harvest area (ha)	378,599		2,037,413	2,180,800	1,485,280	1,014,005	1,044,457	2,048,772	610,704	858,882
Yield (kg/ha)	878		378	438	849	923	1,121	1,063	1,333	2,901

Source: IBGE – Agricultural Census, 2006.

Pastures account for about half of all agricultural land, although there has been a steady rise in the share of land devoted to crop production. By 2006, cropland accounted for about 18% of all agricultural land, whereas pastures accounted for another 48%. The remainder was in natural or planted woodlands (30%) or for other uses. Labour per farm declined from 9.7 people per farm in 1920 to 3.6 in 1970 and to 3.2 people in 2006. This trend is a result of technology innovations in the production systems used, introduction of new products and changes in Brazilian labour policies. The rapid rise in the total number of tractors also reflects the introduction of technology innovations. Crop area per tractor fell from 3893 ha in 1920 to 205 in 1970 and 73 in 2006.

Table 7.1 also shows the production increase obtained in livestock and crops. By observing the relationship between lands for pasture per total number of cattle, we observe that stocking rates have increased over the Census years. The average number of cattle per 100 ha of pasture increased from 39 in 1940 to 51 in 1970 and to 108 in 2006. This relationship reveals improvements in land capacity and pasture and livestock management, and means that land can be freed for other purposes. The data regarding crop production also show improvement in productivity levels of crops. For example, the coffee yield multiplied by six between 1920 and 2006; grape production almost tripled between 1920 and 2006; corn production almost doubled between 1970 and 2006; wheat production tripled; soybeans tripled between 1970 and 2006; and sugarcane more than doubled between 1950 and 2006. Albuquerque and Silva (2008) detail this in a study about tropical crops, showing that yield improvements were largely due to new technologies made available through investments in agricultural research, especially in the decades since the 1970s, which are the focus of this study.

7.2 Methodology

Total factor productivity is understood to be an increase in output that cannot be explained by an increase in input, but by productivity gains. TFP measures the relationship between total output and total input (Fuglie *et al.*, 2007). Details on the concepts involved and how to calculate the index can be found in Jorgenson (1996), Christensen (1975) and E. Alves (unpublished observations).

Equation 7.1 defines the Tornqvist index used to obtain TFP. This index is a discrete approximation of the Divisia index (Chambers, 1998), and is therefore ideal for the analysis of economic variables, considering that these are presented in a discrete format, not continuous, as would be the case of the Divisia index (see equation 7.1 at bottom of page).

In Eqn 7.1, variables Y_i and X_j are the quantities of output and input, respectively. S_i and C_j are the shares of product i in total production value and of input j in the total cost of input, respectively. The left side of the equation defines the TFP variation between two subsequent periods of time.

The first variable is the logarithm of the mean of the quantities during two subsequent periods of time, weighed out by the average share of each product in the total production. The second variable is the logarithm of the mean of the quantities of input during two subsequent periods of time, weighed out by the average share of each input in the total cost. Thus, to create the Tornqvist index, we must have prices and quantities for all output and input.

The relationship between the total factor productivity *(TFP)* during period t and the total factor productivity in the period before that *(TFP(t-1))* is found by calculating the exponentiation of Eqn 7.1. Having done so, in order to obtain the TFP for each year, we assume a base-year as being 100 and link the indexes for the subsequent years. This process is called the chain-link index, and you can read more

$$\ln\left(TFP_t \Big/ TFP_{(t-1)}\right) = \frac{1}{2}\sum_{(i=1)}^{n}\left(S_{it}+S_{i(t-1)}\right)\ln\left(\frac{Y_{it}}{Y_{i(t-1)}}\right) - \frac{1}{2}\sum_{(j=1)}^{m}\left(C_{jt}+C_{j(t-1)}\right)\ln\left(\frac{X_{jt}}{X_{j(t-1)}}\right) \tag{7.1}$$

about how to obtain it in Thirtle and Bottomley (1992) and Hoffmann (1980, p. 325).

Another indicator used to analyse transformation in agriculture is the structural change index (Ramos, 1991). It is obtained from a dissimilarity measure on the basis of the cosine, as expressed in Eqn 7.2. This representation measures the angle θ, made between two vectors representing time periods.

$$\cos\theta = \frac{\sum\limits_{i=1}^{n}(S_{it} \cdot S_{i(t-1)})}{\sqrt{\sum\limits_{i=1}^{n}(S_{it})^2 \cdot \sum\limits_{i=1}^{n}(S_{i(t-1)})^2}} \qquad (7.2)$$

Here S_{it} and $S_{i(t-1)}$ are shares of product i in the total production value for subsequent periods. These shares act as structural parameters in calculating the indicator proposed. The measurement of the angle, measured in degrees referring to structural changes, is located between 0 (zero) and 1 (maximum), $0 \leq \cos\theta \leq 1$. This indicator should be interpreted in the following manner (Ramos, 1991): the closer to 0, the greater the structural changes between two periods; the closer to 1, the lesser the changes.

Another indicator used in this study to assess transformations in agriculture is the diversification index. This index is also based on the shares of each product in total gross production, and is defined by Eqn 7.3 (Hoffmann *et al.*, 1984). It is equal to 1 when there is only one single activity (culture or livestock), and it increases with diversification. This definition encompasses Hoffmann's comment (personal communication) made during a seminar at the Brazilian Ministry of Agriculture, in March, 2010.

$$D = \frac{1}{\sum S_{it}^2} \qquad (7.3)$$

S_{it} represents the shares of activity i in the total value of production. The greater the index, the greater the degree of diversification.

7.3 Data Sources and Definition of the Variables

The data used to produce the indicators for this study are almost all provided by the Instituto Brasileiro de Geografia e Estatística (IBGE). Because the chapter seeks to update a previous study based on the Agricultural Census data from 1970 to 1995–1996, the main sources of information are the censuses of 1970, 1975, 1980, 1985, 1995–1996 and 2006. In this study, we kept the results obtained in the previous study (Gasques and Conceição, 2001) and incorporated information from the 2006 Census, following as precisely as possible the procedures used in the previous study. The 2006 Agricultural Census encompassed new activities and also changed the measuring units of many products (IBGE). The treatment given to these issues is described throughout the chapter.

Considering that TFP is a relationship between a total output index and a total input index, we will first present the information needed to obtain the output index and then refer to the index for the input used in production.

The output index was obtained by adding the figures for livestock, crops and rural agroindustry. The IBGE considers the Brazilian livestock sector to include meat, milk, eggs and fibre from cattle, goats, bubaline (buffalo), hybrid species, rabbits, swine, poultry, beehives and silk cocoons. Crop output includes harvests from permanent and temporary crops, horticulture, floriculture and woodlands. Finally, rural agroindustry encompasses the transformations of products raised on farms, among them cassava (manioc) flour, vegetable oils and meals, butter and cheese, sausages and stuffed meat, fruit pulp and others. This study uses 367 products.

To determine the product index requires information on the amounts produced and their values. From those data, we obtain revenue shares (S_{it}), and growth rates (Y_i/Y_{it}) for constructing the Tornqvist output index. Activities for which the Census did not have information on the quantities produced but just a value, such as floriculture, were not taken into account when calculating the output index.

Livestock production in 2006 was measured differently among commodities, but in each case the outputs represent the product flow during the year of the Census.

For bovine, swine, sheep, goat, rabbit, bubaline and hybrid species, we used the quantities sold and values of sales. For other products, such as honey, cocoons, eggs, milk and others, we took the amounts produced and the value of that production. For activities that are part of crop-raising and rural agroindustry, the quantities produced and the production values were obtained directly from the Census.

As with the output index, the input index requires information about the quantities used and the cost of input, in order to calculate the shares of input in cost and also the relationship between two periods of time. The variables used are presented in the flow-like manner because they represent the quantities and costs of inputs used throughout the year.

The long list of inputs used was constructed by combining the information on input use contained in the Census with corresponding information from the tables of expenses. These inputs refer to the activities included in the main groups covered by the Census, such as livestock, crops and rural agroindustry. There are some inputs such as feed, veterinary products and others that we did not include because the Agricultural Census captured only the cost of inputs, not the quantity. This is a restriction for introducing all the inputs in the TFP index measurement.

The lands considered for the study include those used for temporary and permanent crops, those with natural and planted pastures, woodland areas and planted forests. These categories are in the group called 'use of lands in farms per type of use'. Therefore, the amount of land was estimated by adding the areas of the lands committed to the different usages referred to above.

Because the land used is considered a flow variable, the price used to obtain its value is the rental price because that best expresses the cost of using the land. The rental prices of land were not the prices published in the Census when presenting the expenses with rentals, because we noted a difference in the price per hectare in states where the quantity of rented lands was more than the total quantity of lands. We chose, therefore, to use the average rental prices for crops and pasture lands provided by the Getúlio Vargas Foundation (FGV). This procedure was maintained in the 2006 Agricultural Census. The price considered for woodlands and planted forests was equal to the average price of pasture lands because the FGV does not survey average prices of woodlands and forests. In the group of states for which the FGV does not publish rental prices, we used the rental prices observed in Brazil.

Regarding labour, we considered the total quantity of labour, which includes permanent and temporary employees, those who are managers and relatives, and those with family ties to the producer. The previous study, Gasques and Conceição (2001), obtained the cost of labour in a different manner than this chapter does. Until 2006, the IBGE Agricultural Census did not register farming costs associated with a family labour force, i.e. people with family ties to the producer who carry out agricultural duties and functions. The only available data concerned employees who were formally employed, temporarily or permanently. Gasques and Conceição (2001) attributed a cost to family labour force obtained through the average wage of non-family labour force. Following the 2006 Agriculture Census, we adopted its cost of family labour. Our impression is that this new methodology might accelerate the trend of diminishing the total allotted to remunerating the labour force in general.

For soil fertilizers, soil conditioners and pesticides, we used expense data released by the Agricultural Census. The quantities were obtained from the IBGE Brazil Statistics Yearbook and refer to the active ingredient because it better reveals the amount of such input consumed. The consumption of these inputs in each federation unit was obtained by estimating the share of each federation unit in the total value of the country's agricultural production and multiplying this share by the quantity of input consumed in the country. We therefore obtained estimates for the consumption of fertilizers, soil conditioners and agrotoxins for each unit of the federation. In 2006, we used the quantities of fertilizers provided by the Associação

Nacional para Difusão de Adubos (ANDA) database and the quantity of limestone provided by the Associação Brasileira dos Produtores de Calcário Agrícola (ABRACAL) database. The quantities of insecticides were provided by the Associação Nacional de Defesa Vegetal (ANDEF).

The information on the number of tractors used and that cost were obtained from Barros (1999) because the Census did not provide this information in an appropriate manner for calculating the input index. Barros's detailed study calculated series for the quantities and capital value of agricultural machinery. We used the series referring to tractor supply expressed in units, not in power, and the value of the tractor fleet was estimated on the basis of a 7% depreciation rate per year. For 2006, we used the number of tractors surveyed by the Census, and obtained the value by correcting the 1995 value, as estimated by Barros. That adjustment was based on the IGP-DI – general price index – domestic availability – from FGV. More details about the procedure are available from the authors.

The quantities found in the 2006 Census were used for all types of fuel: alcohol, bagasse, gas, gasoline, lumber, diesel oil and kerosene. We used the ANP prices for alcohol, diesel and gasoline because no data were published on the value of those inputs.

7.4 Results

The following presentation of findings is made in two parts. First, we present the results of TFP indexes for Brazil and federative units. Second, we present the results for structural change indicators and the diversification index. The two-part approach is merely for presentation purposes; conceptually, there is a direct relationship between both.

7.4.1 Total factor productivity

Brazil's agricultural TFP follows a rising path in the 36 years between 1970 and 2006

(Table 7.2). The TFP rose continuously between Census years, from an index value of 100 in 1970 to 224 in 2006 (124% growth). The output index went from 100 in 1970 to 343 in 2006. The input index went from 100 to 153, within the two points of comparison. Note that although agricultural output, which includes production of crops, livestock and rural agroindustry, increased 243% between 1970 and 2006, the use of input increased by only 53%. This result shows that the growth of Brazilian agriculture has been based mainly on productivity gains.

Figure 7.1 further illustrates these results, showing the product index, input index and TFP. The difference in these lines reveals that, until 1995, Brazilian agriculture production was propelled mainly by increasing the use of input. That was, indeed, a period marked by stressed growth because of the occupation of land in new regions, such as the central-west. Thanks to the increasing introduction of technology into agriculture (Graziano da Silva, 1998), it was also a time of major subsidies towards rural credit and of new growth patterns.

If we observe the results through annual growth rates, rather than through indexes, we note that the product index increased from 1970 to 2006 by an average of 3.48% a year (Table 7.3).

Between 1995 and 2006, product growth was 3.14% a year. The states of Mato Grosso and Rondônia had the highest growth rates during both periods. In Rondônia, the index increased 10.24% a year, from 1970 and 2006, and 7.15% between 1995 and 2006. In Mato Grosso, it rose more than 6% during the former time period and 8.68% between 1995 and 2006. The TFP average annual growth rates were of 2.27% for the 1970 to 2006 time period and of 2.13% from 1995 to 2006.

By observing how much of the product growth was due to productivity, we note that between 1970 and 2006 65% of agricultural and livestock product growth was due to increases in TFP, and 35% was due to increases in the amount of input. In the 1995–2006 period, 68% of product growth was due to additional productivity and 32% was due to increases in the amount of input.

Table 7.2. National, regional and state-level agricultural output, input and TFP indexes for Brazil.

Regions and states	Output index					Input index					TFP index				
	1975	1980	1985	1995	2006	1975	1980	1985	1995	2006	1975	1980	1985	1995	2006
Brazil	139	173	211	244	343	122	142	149	137	153	114	122	142	178	224
North region															
Acre	101	129	132	152	258	117	151	182	184	201	87	86	72	82	128
Amapá	151	134	121	143	153	140	98	170	155	67	108	137	71	92	228
Amazônas	103	127	131	87	63	137	169	196	152	88	75	75	67	58	72
Pará	135	207	225	226	320	143	199	236	208	238	95	104	95	109	135
Rondônia	308	547	1043	1566	3346	404	1051	1342	1715	2230	76	52	78	91	150
Roraima	108	178	226	329	328	129	150	132	192	102	84	119	171	171	320
Tocantins	–	–	100	134	147	–	–	100	88	145	–	–	100	151	101
Northeast region															
Alagoas	153	183	238	233	383	126	158	163	134	114	121	115	146	174	336
Bahia	119	132	143	141	265	121	143	167	142	147	99	92	86	99	180
Ceará	164	151	194	242	355	99	112	116	102	91	166	135	168	238	391
Maranhão	118	146	146	153	309	126	144	144	124	127	94	102	101	123	243
Paraíba	155	139	183	187	187	126	113	123	90	78	123	123	149	207	241
Pernambuco	138	164	203	187	259	107	118	118	97	84	130	139	172	193	308
Piauí	142	132	172	201	375	116	142	145	115	150	123	93	119	174	249
Rio Grande do Norte	153	153	188	244	256	110	125	121	99	83	139	122	156	247	310
Sergipe	113	142	160	182	237	113	118	133	125	109	100	121	120	145	217
Southeast region															
Espírito Santo	110	116	161	220	319	111	141	171	202	108	99	83	94	109	296
Minas Gerais	140	163	214	236	312	158	205	205	172	169	89	79	105	137	185
Rio de Janeiro	150	159	168	139	134	118	125	124	90	75	127	127	135	156	180
São Paulo	139	176	215	209	257	119	146	134	128	139	117	120	160	164	184
South region															
Paraná	203	256	313	337	477	127	133	134	119	139	160	192	234	284	343
Rio Grande do Sul	132	155	173	199	278	135	159	141	133	167	98	97	123	149	167
Santa Catarina	137	205	254	343	516	115	134	134	135	148	119	153	189	253	349
Centrewest region															
Distrito Federal	166	390	644	992	1493	128	266	308	382	512	130	146	209	260	292
Goiás	155	192	219	282	358	131	151	107	109	125	119	127	204	258	287
Mato Grosso	44	80	155	378	944	51	69	78	111	182	85	117	198	341	518
Mato Grosso do Sul	100	144	204	338	412	100	111	113	111	131	100	130	180	304	315

Source: authors estimates. (Base year 1970 = 100 except for Tocantins where base year = 1985 and Mato Grosso do Sul where base year = 1975.)

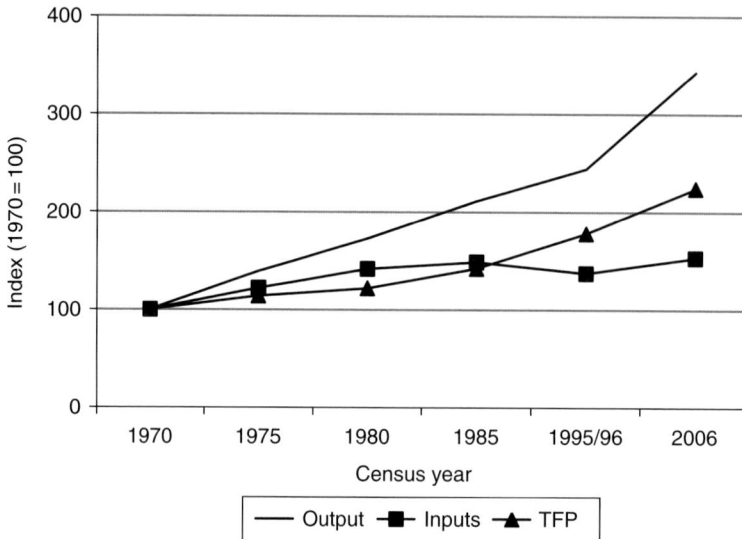

Fig. 7.1. Agricultural output, input and TFP indexes for Brazil. Source: Authors' estimates.

Productivity has therefore been the main growth stimulator for Brazilian agriculture. Referring to the data about land and labour productivity growth in Table 7.3, we note that in both the 1970–2006 period and the 1995–2006 period, the annual growth rate for labour productivity was greater than land productivity growth. This does not mean the increase in labour productivity was more decisive than the increase in land productivity in determining the TFP growth, because land productivity is also a component in labour productivity.

As some studies have shown, there has been an increase in the quality of the agricultural labour force, as measured by the average level of schooling (Del Grossi and Graziano, 2006; Balsadi, 2006; De Negri *et al.*, 2006). This is one cause of the rise in labour productivity. The process of bettering the labour force also includes improved management of rural facilities, as shown by studies carried out by the Brazilian Confederation of Agriculture and Livestock (CAN).

An increase in machine and equipment efficiency during the past years is undoubtedly another decisive aspect in increasing agricultural labour productiv-

ity. A study by Albuquerque and Silva (2008) describes the operational capacity of machines and agricultural equipment and their effects on the yield of sugarcane operations. There has also been a trend to reduce the use of less powerful tractors and expand the use of medium- and high-capacity tractors (ANFAVEA Statistic Yearbook, 1972 to 2008).

The increase in land productivity is due to the adoption of new technologies, developed through agricultural research, and to the addition of new, more productive lands, which took place during this >30-year period. A primary source of technology innovations has been EMBRAPA, the national agricultural research organization, which was established in 1973. Productivity-improving innovations have been particularly significant in rice, corn, coffee, sugarcane and livestock products. In addition to research innovations that improve crop and livestock quality and yield, many others have enhanced production processes, such as no-tillage farming, inoculating soil with nitrogen-fixing bacteria, integrated pest management, and adapting crop varieties and animal species to different environmental conditions. The increase in research spending has direct effects on productivity.

Table 7.3. Growth rates of agricultural output, input, TFP, and land and labour productivity in Brazil.

States	Output 2006/1970	Output 2006/1995	Input 2006/1970	Input 2006/1995	TFP 2006/1970	TFP 2006/1995	Land productivity 2006/1970	Land productivity 2006/1995	Labour productivity 2006/1970	Labour productivity 2006/1995
				(Average % per year over period)						
Brazil	3.483	3.138	1.189	0.991	2.267	2.126	3.316	3.158	3.528	3.409
North region										
Acre	2.669	4.931	1.958	0.783	0.697	4.115	1.315	2.609	1.606	4.862
Amapá	1.195	0.641	-1.101	-7.319	2.322	8.589	0.941	-0.069	0.879	1.423
Amazônas	-1.266	-2.906	-0.367	-4.872	-0.902	2.066	-2.201	-5.199	-1.609	-2.168
Pará	3.287	3.242	2.434	1.229	0.833	1.988	2.245	1.158	2.005	3.580
Rondônia	10.242	7.147	9.007	2.416	1.133	4.619	7.174	4.743	4.844	7.469
Roraima	3.351	-0.035	0.064	-5.581	3.285	5.874	3.876	4.485	2.905	0.706
Tocantins		0.873		4.614		-3.576		2.284		1.446
North-east region										
Alagoas	3.804	4.647	0.365	-1.449	3.426	6.186	3.637	4.583	3.677	4.377
Bahia	2.742	5.873	1.077	0.305	1.647	5.551	2.286	5.959	2.485	6.037
Ceará	3.580	3.537	-0.272	-1.047	3.863	4.633	3.884	3.129	3.425	3.679
Maranhão	3.184	6.623	0.672	0.239	2.495	6.369	2.526	5.334	3.172	7.450
Paraíba	1.756	0.000	-0.698	-1.369	2.471	1.388	2.022	0.186	2.052	-0.038
Pernambuco	2.678	3.017	-0.477	-1.246	3.170	4.317	2.703	2.749	2.962	3.236
Piauí	3.737	5.808	1.140	2.432	2.568	3.296	3.591	4.627	2.939	4.852
Rio Grande do Norte	2.647	0.426	-0.525	-1.627	3.190	2.087	2.952	0.338	3.001	1.591
Sergipe	2.431	2.467	0.248	-1.225	2.178	3.737	2.413	2.883	2.285	2.898
South-east region										
Espírito Santo	3.276	3.429	0.208	-5.537	3.062	9.492	3.300	4.052	3.219	3.770
Minas Gerais	3.209	2.580	1.463	-0.182	1.721	2.767	3.486	3.280	2.685	3.013
Rio de Janeiro	0.826	-0.330	-0.805	-1.628	1.644	1.320	1.214	0.199	1.301	0.062
São Paulo	2.654	1.875	0.925	0.780	1.713	1.086	2.752	1.962	3.103	1.861
South region										
Paraná	4.436	3.196	0.921	1.455	3.482	1.716	4.228	3.347	4.952	3.564
Rio Grande do Sul	2.884	3.100	1.432	2.052	1.432	1.026	2.984	3.207	2.903	3.444
Santa Catarina	4.666	3.787	1.095	0.805	3.532	2.958	4.620	3.998	4.926	4.487
Central-west region										
Distrito Federal	7.799	3.788	4.638	2.689	3.021	1.070	7.777	3.799	6.464	2.553
Goiás	3.606	2.185	0.620	1.223	2.968	0.950	4.015	2.661	3.800	2.590
Mato Grosso	6.436	8.679	1.685	4.631	4.672	3.869	6.702	8.101	6.647	8.661
Mato Grosso do Sul		1.819		1.498		0.317		1.851		1.932

Source: Authors' estimates.

We observed that a 1% rise in research spending by EMBRAPA raised the TFP index by 0.2% (Gasques *et al.*, 2009).

Looking at the TFP growth in the 1995–2006 period, we observe significant differences among the Brazilian states (Table 7.4). Two states in the northern region, Pará and Tocantins, presented yield growth below the TFP growth for Brazil. In the north-east, only Paraíba and Rio Grande do Norte recorded TFP growth below Brazil's average. In the south-east, only Espírito Santo and Minas Gerais recorded productivity growth above Brazil's. In the south, Rio Grande do Sul and Paraná recorded productivity growth below the Brazilian average, and in the central-west, only Mato Grosso's TFP growth was above the average for Brazil.

Table 7.4. Agricultural TFP growth rates by state, 1995/96 to 2006.

State	Abbreviation	TFP Growth (annual %)
BRAZIL	BR	2.126
Acre	AC	4.115
Amapa	AP	8.589
Amazônas	AM	2.066
Pará	PA	1.988
Rondônia	RO	4.619
Roraima	RR	5.874
Tocantins	TO	−3.576
Alagoas	AL	6.186
Bahia	BA	5.551
Ceará	CE	4.633
Maranhão	MA	6.369
Paraiba	PB	1.388
Pernambuco	PE	4.317
Piaui	PI	3.296
Rio Grande do Norte	RN	2.087
Sergipe	SE	3.737
Espírito Santo	ES	9.492
Minas Gerais	MG	2.767
Rio de Janeiro	RJ	1.320
São Paulo	SP	1.086
Paraná	PR	1.716
Rio Grande do Sul	RS	1.026
Santa Catarina	SC	2.958
Distrito Federal	DF	1.070
Goiás	GO	0.950
Mato Grosso	MT	3.869
Mato Grosso do Sul	MS	0.317

Source: Authors' estimates.

7.4.2 Structural change and diversification

This section presents the results of structural change and diversification indicators to illustrate the transformations that have taken place in Brazilian agriculture. Both indicators, as we have seen, are based on the revenue shares of the different outputs included in the Agricultural Census. They might reflect changes in the composition of input for there is a direct relationship between decisions regarding production and the use of inputs (Gasques and Conceição, 2001).

In order to make the interpretation of structural change indexes clearer, Table 7.5 provides the shares of the ten main products in terms of agricultural and livestock total production value during various years of the Agricultural Census. Note that livestock activities, on top, remain stable throughout all of the years examined. Other products start disappearing from the list, such as beans, manioc, cotton, rice and chicken eggs. Some products upgrade in position, such as sugarcane; others pass over to the list of main products. The structural change index seeks to represent this dynamic over time.

Changes in the composition of the products also led to changes in the composition of factors in agriculture (Table 7.6). Many changes may be observed in the composition of inputs, but what especially stands out is the cost of labour – in 1970, it represented 51% of the cost, while by 2006, the year in which we adopted the Census cost of family labour, the figure dropped to 16.1%. Tractors expanded considerably in their share of expenses, from 7% in 1970 to 17.8% in 2006. Electrical energy, fertilizers and soil conditioners, and diesel oil also presented expressive rises in shares of total expenses.

Figure 7.2 plots the index of structural change between different Census years for Brazil as a whole. The outer line shows a pentagon with each corner equidistant from the centre, whereas the inner line represents the degree of structural change between the Census periods. Points closer to the centre indicate greater structural change, compared with points nearer the outer pentagon period.

Table 7.5. Output shares of the ten most valuable products in Brazilian agriculture, 1995/96 and 2006.

	1995/96	Output share (%)		2006	Output share (%)
1	Beef	15.6	1	Beef	14.1
2	Sugar cane	11.4	2	Sugar cane	12.7
3	Milk	10.0	3	Soybean	11.0
4	Soybean	9.1	4	Corn (maize)	7.3
5	Corn (maize)	7.0	5	Milk	5.7
6	Poultry meat	6.3	6	Coffee beans	5.5
7	Coffee beans	5.3	7	Poultry meat	4.0
8	Pork	3.7	8	Bananas	3.3
9	Rice	3.4	9	Pork	3.0
10	Poultry eggs	2.8	10	Oranges	2.7
Share of total output		74.6	Share of total output		69.3

Source: Agricultural censuses of Brazil, 1995/96 and 2006.

Table 7.6. Agricultural input cost shares from Brazil's agricultural censuses: 1970, 1995/96 and 2006.

1970 Agricultural		1995/96 Agricultural		2006 Agricultural	
Census	Cost share (%)	Census	Cost share (%)	Census	Cost share (%)
Labour	51.0	Labour	46.5	Personnel employed	16.1
Land	33.3	Land	23.0	Land	30.7
Tractor services	7.0	Tractor services	17.1	Tractor services	17.8
Fertilizers and soil conditioners	3.7	Fertilizers and soil conditioners	6.0	Fertilizers and soil conditioners	16.3
Pesticides	1.3	Pesticides	3.0	Pesticides	9.9
Fuels	2.0	Fuels	2.7	Fuels	4.0
Electricity	0.2	Electricity	1.4	Electricity	4.6
Lumber	1.4	Lumber	0.4	Lumber	0.7
Total	100.0	Total	100.0	Total	100.0

Source: Authors' estimates from IBGE agricultural census' data.

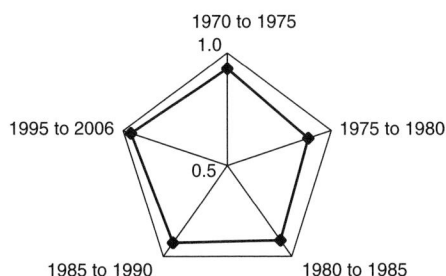

Fig. 7.2. Index of structural change in the composition of agriculture output. The structural change index ranges between 0 and 1, with 1 indicating no structural change. The movement from the outer to the inner pentagons quantifies the degree of structural change in output composition of Brazilian agriculture between Census periods. The largest structural change occurred between 1975 and 1980, and the smallest between 1995 and 2006. Source: Authors' estimates from Agricultural Census data.

The largest structural change occurred between 1975 and 1980, and the smallest between 1995 and 2006. Thus, the main changes that have occurred in the composition of the production took place prior to 1995.

Although the results of the structural change index do not show important changes for Brazil between 1995 and 2006, the results per state reveal two important transformations. The first is a reduction in the pursuit of traditional activities, such as beef, milk, cocoa, coffee, cashew nuts, manioc, corn and rice. The second is an increase in the participation, in terms of value, of new products, especially fruits such as banana, grape, mango and papaya. The enhanced importance of fruit mainly concerns the north-east, where we also observe an expressive value decrease of traditional products. In the states of Rio

Grande do Norte, Bahia and Pernambuco, the increase in shares of fruit, such as watermelon, papaya, coconut, banana, grape and mango, is particularly notable.

In states of the north, such as Rondônia and Pará, the most significant changes between 1995 and 2006 were the drop in coffee shares in Rondônia (from 16.2% of production value to 11.9%) and the strong increase in shares of bovine in the state's

production value, from 27.2% of total production value in 1995 to 48.3% in 2006. In Pará, there was also a major increase in the production value of bovine between 1995 and 2006, from 22.9% to 30.4%. Furthermore, that state recorded a major decrease in the participation of lumber in production value, from 9.6% in 1995 to 1.3% in 2006.

Figure 7.3 shows the structural change index from 1975 to 2006 for the Brazilian

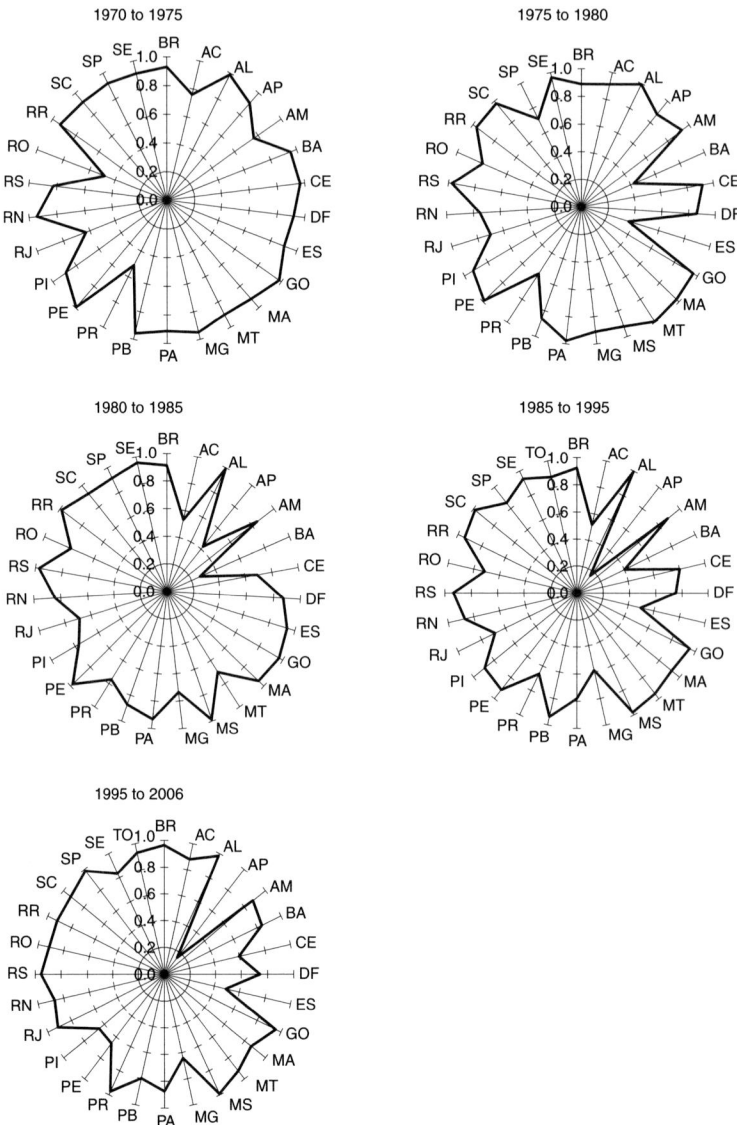

Fig. 7.3. Indexes of structural change in the output composition by state. For a description of the index measure, see Fig. 7.2 caption. See Table 7.4 for a list of state abbreviations. Source: Authors' estimates from Agricultural Census data.

states. Bear in mind that the closer to zero the index is, the greater the structural change. We may observe that between 1995 and 2006, changes in the composition of products were already relatively secured. Some states, however, were still undergoing transformations, such as Amapá and Espírito Santo. Others states, such as Paraná, Rondônia and Rio de Janeiro, went through all the transition between 1970 and 1975. Others yet, such as São Paulo, Bahia and Espírito Santo, underwent transformations between 1975 and 1980, and others from then on.

Results of the agriculture diversification index are given in Table 7.7 for Brazil and its states. According to its definition, the greater the index value, the bigger the degree of diversification. On the basis of the diversification index data, we conclude that the general trend for Brazil and most of the Brazilian states has been towards agricultural diversification, not specialization. It can be inferred by noting that the year 2006 has a greater index than 1995. Therefore, between 1995 and 2006, the trend was towards diversification, because the index rose.

A lot of information can be extracted from these outcomes about the increase of the degree of diversification. But two points are very important. First, diversification, as it is being carried out by modern Brazil, can have very positive effects on employment and income. This is because diversification is predominantly setting in by maintaining major products, while incorporating products of high added value, such as livestock and fruit. The second point is that the direction of specific policies, such as irrigation and programmes like Pronaf – National Programme for the Strengthening of Family Agriculture (Programa Nacional de Fortalecimento da Agricultura Familiar), provide a minimum of financial resources that allow new products to be introduced in agriculture.

7.5 Conclusions

The average annual growth rate for agricultural total factor productivity in Brazil between 1995 and 2006 obtained through

Table 7.7. Agricultural diversification index for Brazil and its states, 1995/96 and 2006.

Region/state	Abbreviation	1995/96	2006
BRAZIL	BR	13.62	14.69
Acre	AC	6.45	5.88
Alagoas	AL	2.53	2.98
Amazônas	AM	3.64	6.98
Amapá	AP	3.05	11.00
Bahia	BA	15.59	18.88
Ceará	CE	14.37	10.12
Distrito Federal	DF	11.78	12.07
Espírito Santo	ES	4.10	4.44
Goiás	GO	6.34	6.11
Maranhão	MA	9.83	11.47
Minas Gerais	MG	3.98	8.24
Mato Grosso do Sul	MS	2.62	3.04
Mato Grosso	MT	4.40	5.05
Pará	PA	9.65	7.55
Paraíba	PB	9.81	12.31
Pernambuco	PE	10.02	7.91
Piauí	PI	12.74	12.30
Paraná	PR	9.95	9.70
Rio de Janeiro	RJ	11.96	14.74
Rio Grande do Norte	RN	10.30	14.51
Rondônia	RO	6.85	3.62
Roraima	RR	9.12	5.17
Rio Grande do Sul	RS	11.77	11.77
Santa Catarina	SC	11.29	14.17
Sergipe	SE	10.81	11.90
São Paulo	SP	6.75	5.02
Tocantins	TO	1.71	3.85

Source: Authors' estimation from IBGE Agricultural Census.

this study is 2.13% a year. This figure is lower than those found in other studies, such as Gasques *et al.* (2009) who found average TFP growth rate for the period 1975–2008 to be 3.66% per year. One major difference between these studies is the source of output data. Whereas our results are based on Brazilian Agricultural Census data, Gasques *et al.* (2009) used annual data provided by IBGE. Output in the Agricultural Census consists of a larger number of commodities than that of other sources. However, we do not expect the difference in output growth to be large because those included in other sources represent nearly all of the value of crop and livestock

Table 7.8. Comparison of output quantities between LSPA and the agricultural census for 2006.

Products	IBGE LSPA 2006 annual survey (tonnes)	IBGE 2006 agricultural census (tonnes)	Difference (LSPA – census) (tonnes)	Difference (LSPA – census) (%)
Whole cotton	2,898,721	2,350,132	548,589	18.9
Rice (husked)	11,526,685	9,447,257	2,079,428	18.4
Coffee beans	2,573,368	2,360,756	212,612	8.6
Sugar cane	477,410,655	384,165,158	73,080,358	15.1
Cassava (manioc)	26,639,013	16,093,942	10,545,071	39.9
Corn (maize)	42,661,677	42,281,800	379,877	0.9
Soybeans	52,464,640	40,712,683	11,751,957	22.0
Wheat	2,484,848	2,257,598	227,250	9.5

Source: LSPA (IBGE) and 2006 agricultural census (IBGE).

production in the country. It is more likely that the lower output (and TFP) growth reported in this study is due to an apparent underestimation of production data provided in the 2006 Agricultural Census. There are large differences in the 2006 harvest quantities for certain products reported by the annual Levantamento Sistemático da Produção Agrícola (LSPA) published by IBGE and the Agricultural Census (Table 7.8). The most outstanding examples are whole cotton, husked rice, sugarcane, cassava (manioc) and soybeans. Because these products, especially soybeans and sugarcane, are significant components of the gross production value, if underestimation of quantities provided by the Census is confirmed, it may lead to changes in the estimated TFP growth rate. None the less, even if agricultural TFP in Brazil grew at an annual rate of 2.13%, it is still above the annual rate of 1.89% recorded for the USA in the same period, 1995–2006 (Economic Research Service, 2010).

References

Albuquerque, A. and Silva, G. (eds) (2008) Agricultura tropical – quatro décadas de inovações technological, institucionais e políticas. Embrapa, Informação Tecnológica, Brasília. Vol. I, p. 1337.

Associação Nacional para Difusão de Adubos (ANDA) National Association for the Dissemination of Fertilizers, São Paulo, Brazil. Available at: http://www.anda.org.br (Accessed 3 February 2010).

Associação Nacional de Defesa Vegetal (ANDEF) National Association for Agrochemicals, São Paulo http://www.andef.com.br/home.

Associação Brasileira dos Produtores de Calcário Agrícola (ABRACAL) Brazilian Association of Lime Producers, Porto Alegre, Brazil. http://www.calcario-rs.com.br/abracal.asp.

Associação Nacional dos Fabricantes de Veículos Automotores (ANFAVEA) *Statistics Yearbook, 1972 to 2008*. National Association of Motor Vehicle Manufacturers, São Paulo, Brazil http://www.anfavea.com.br/index.html.

Balsadi, O. (2006) O mercado de trabalho assalariado na agricultura brasileira no período 1992–2004 e suas diferenciações regionais. Tese (Doutorado), Instituto de Economia da Universidade de Campinas (IE/UNICAMP), Campinas, São Paulo, Brazil.

Barros, A. (1999) *Capital, Produtividade e Crescimento da Agricultura: O Brasil de 1970 e 1995*. PhD Thesis, Escola Superior de Agricultura Luiz de Queiroz, São Paulo University. Piracicaba, São Paulo, Brazil.

Chambers, R. (1998) *Applied Production Analysis: A Dual Approach*. Cambridge University Press, Cambridge, UK.

Christensen, L. (1975) Concepts and measurement of agricultural productivity. *American Journal of Agricultural Economics* 57, 5, 910–915.

De Negri, J., De Negri, F. and Coelho, D. (eds) (2006) *Tecnologia, Exportação e Emprego*. Institute for Applied Economic Research (IPEA), Brazil.

Del Grossi, M. and Graziano, J. (2006) *Mudanças Recentes no Mercado de Trabalho Rural*. In: Center for Strategic Studies and Management Science, Technology and Innovation (Centro de Gestão e Estudos Estratégicos) – CGEE Strategic Partnerships, n. 22, June. CGEE, Brazil.

Economic Research Service (2010) Data documentation and methods, research and productivity briefing room. U.S. Department of Agriculture, Washington, DC. http://www.ers.usda.gov (Accessed 3 February 2010).

FGV (various annual issues). Agriculture and Livestock: Average prices and indexes for rentals/leases, Land Sales, Wages and Services. Getúlio Vargas Foundation, Rio de Janeiro, Brazil. http://portal.fgv.br/en

Fuglie, K., MacDonald, J. and Ball, E. (2007) Productivity growth in U.S. agriculture. Economic Brief No. 9, Economic Research Service, U.S. Department of Agriculture, Washington, DC.

Gasques, J. and Conceição, J. (2001) Transformações estruturais da agricultura e produtividade total dos fatores. In: Gasques, J. and Conceição, J. (eds) *Transformações da Agricultura e Políticas Públicas*. Institute for Applied Economic Research (IPEA), Brazil.

Gasques, J., Bastos, E. and Bacchi, M. (2009) Produtividade e Fontes de Crescimento da Agricultura. Technical Paper Age/Mapa. Livestock and Food Supply, Ministry of Agriculture, Brazil.

Graziano da Silva, J. (1998) A *Nova Dinâmica da Agricultura Brasileira, 2nd edition*. Economic Institute, State University of Campinas, São Paulo, Brazil.

Hoffmann, R. (1980) *Estatística para Economistas*. Pioneira, São Paulo, Brazil.

Hoffmann, R., Engler, J.J. de C., Serrano, O., Thame, A.C. de M., Neves, E.M. (1984) *Administração da Empresa Agrícola, 4th edn*, Pioneira, São Paulo, Brazil.

IBGE (various annual issues) Anuário Estatístico do Brasil. Instituto Brasileiro de Geografia e Estatístic (Annual Stastistics of Brazil, Brazilian Institute of Geography and Statistics). Ministry of Planning, Brasilia, Brazil. http://www.ibge.gov.br/

IBGE (various annual issues) Levantamento Sistemático da Produção Agrícola (LSPA). Instituto Brasileiro de Geografia e Estatístic (Systematic Survey of Agricultural Production, Brazilian Institute of Geography and Statistics). Ministry of Planning, Brazil. http://www.ibge.gov.br/

IBGE (various Census years) Censo Agropecuário. Instituto Brasileiro de Geografia e Estatístic (Agricultural Census, Brazilian Institute of Geography and Statistics). Ministry of Planning, Brazil. http://www.ibge.gov.br/

Jorgenson, D. (1996) The embodiment hypothesis. *Journal of Political Economy* 74, 1, 1–17.

Ramos, R. (1991) Metodologia e cálculo de indicadores de mudanças estruturais do setor industrial. Internal Report – CPIT, number 1. Institute for Applied Economic Research (IPEA), Brazil.

Thirtle, C. and Bottomley, P. (1992) Total factor productivity in UK agriculture, 1967–90. *Journal of Agricultural/ Economics* 43, 381–400.

8 Agricultural Productivity in China: National and Regional Growth Patterns, 1993–2005

Haizhi Tong,[1] Lilyan E. Fulginiti[1] and Juan P. Sesmero[2]

[1]*University of Nebraska, Lincoln;* [2]*Purdue University, Lafayette*

8.1 Introduction

Since the economic reforms of 1978, the performance of China's agricultural sector has been impressive. According to China's Statistical Yearbook, by 2005 output from farming, forestry, the animal industry and fisheries had increased by more than four times since the reforms were initiated. China has 9% of the world's total arable land and 20% of the world's population with 70% living in rural areas.

Many studies have examined Chinese agricultural growth. Those studies point towards a rapid expansion of agricultural output and productivity during the 1980s and a slowdown during the 1990s, raising questions about the sustainability of these growth rates. Few studies cover the 2000s and most estimate productivity at the national rather than the provincial level.

The objective of this study is to examine *regional* agricultural productivity growth in China 20 years after the introduction of reforms in the sector. Because China is a country with diverse ecosystems, it is relevant to identify how productivity growth patterns differ across regions. We focus on the 1993–2005 period, which includes the year 1998 in which some of the reforms of the 1970s were due to expire, in particular

the 20-year leases on land. Most commonly, one would develop a Fisher or Tornquist productivity index from observed price and quantity data as is done by the United States Department of Agriculture (USDA) for the USA, but prices and input shares are not available. Even if they were, prices would not represent opportunity costs given the high degree of interventions in the economy, including the strong restrictions on input mobility. With only quantities used and produced available from China's Statistical Yearbook, we estimate multifactor productivity growth (TFP) with a non-parametric Malmquist index and alternatively with a stochastic and parametric production frontier. The period of analysis is crucial because during these years some of the policies and contracts implemented by the Household Responsibility System of the 1970s were due to expire. This could have increased uncertainty considerably, possibly having a stifling effect on productivity growth.

Several issues motivate the present study. First, most studies on Chinese agricultural productivity are done at the national level using FAO data. Some have focused on specific agricultural commodities. This analysis uses provincial data from an alternative source, China's Statistical Yearbook. Second, the period of analysis in this study corresponds to the 20th anniversary of the reforms implemented in the 1970s. A slight slowdown in output growth was observed during the late 1990s, raising concerns that the rapid growth of the previous 20 years had been fuelled by increases in inputs rather than by innovations. By 2003, though, the rate of output growth picked up, returning to pre-1996 levels. Third, we use two very different approaches to the measurement of productivity growth: one stochastic and parametric, the other non-parametric and non-stochastic, with the objective of identifying sensitivity to choice of technique.

We find that productivity growth in Chinese agriculture was higher in the mid-1990s than in the late 1990s, consistent with the slowdown mentioned by others. But we also find an indication of a trend reversal around 1998 with productivity picking up pace again in the 2000s.

8.2   China's Agricultural Policies

Before 1978, agriculture in China was under a collective system. The first step in China's agricultural reform was the introduction of the 'household production responsibility system' (HRS) in 1978. Under the HRS, farmland is not privately owned, but farmers have long-term user rights. They are free to allocate resources as they see fit but must deliver a quota to the government at procurement prices. The leftover output can be freely traded in the market. The objective was to align market signals to farmers' incentives to encourage them to raise output, reduce costs and adopt new technologies. Under this system, the fruit, vegetable and livestock markets have been less controlled than the grain markets. Although farmers pay taxes and local fees, the local government is responsible for extension services and for the introduction of new technologies and seed varieties.

A second step in the reform process occurred at the beginning of the 1990s, when China abandoned its food rationing system. Under the grain-rationing system, urban consumers used coupons to buy a fixed amount of grain at a low price, with more available at market prices. Because of budget pressures in 1991–1992, the government reduced the gap between controlled and market prices and finally eliminated it in 1994 as no resistance from urban consumers materialized.

An important reform called the Grain-Bag responsibility system was introduced in 1995, requiring leaders in each province to maintain an overall balance of grain supply and demand within each province and to regulate local markets. This policy advocates self-sufficiency in grain production and resulted in potentially inefficient reallocation of resources towards grain production. Additional important reforms were introduced more recently. In 1998, a second HRS wave replaced the one introduced in 1978 as land leases expired and were replaced by new ones. Starting in 2000, taxes of the farming sector were gradually eliminated. And in 2001, China became a member of the World Trade Organization (WTO).

8.3   Previous Research on Agricultural Productivity in China

There are numerous studies of agricultural productivity growth in China. They cover different periods, different aggregation levels, and use different data sets and methodologies. In discussing these studies and their findings, we focus on those that report yearly estimates of TFP growth in Chinese agriculture that overlap the years covered by this study. Table 8.1 shows a summary of these studies.

Fan (1991) used a frontier production function to separate agricultural growth into input growth, technical change, institutional reform and efficiency change. Lin (1992) employed a fixed effects model on provincial data to evaluate the effects of decollectivization (HRS), price adjustments and other factors

Table 8.1. Studies of TFP growth in Chinese agriculture.

Author	Method[a]	Studies using provincial data (estimated annual growth rate (%) in agricultural TFP)												
		1994	1995	1996	1997	1998	1999	2000	2001	2002	2003	2004	2005	Avg.
This study	SPF	4.5	4.5	4.4	4.3	4.2	4.1	4.0	3.9	3.8	3.7	3.6	3.6	4.0
	DEA-M	6.3	2.7	4.1	1.3	0.9	4.6	3.0	4.1	4.3	3.9	2.3	2.4	3.3
Lambert et al. (1998)	A-D						1993–1995							5.8
Jin et al. (2002)	A-D	-6.3	0.0											-3.2
Mead (2003)	CD	2.1	-0.5	-1.0										0.2
Dekle et al. (2006)	CD	9.1	4.2	16.0	6.9	3.2	12.5	8.9	2.0	2.0	1.0			6.6

Author	Method	Studies using national data (estimated annual growth rate (%) in agricultural TFP)												
		1994	1995	1996	1997	1998	1999	2000	2001	2002	2003	2004	2005	Avg.
Colby et al. (2000)	A-D						1995–1997							0.8
Wu et al. (2001)	DEA-M	3.9	0.6											2.3
Fan et al. (2002)	A-D	6.0	6.5	4.7	0.2									4.3
Hsu et al. (2003)	DEA-M						1993–2000							1.0
Lezin et al. (2005)	Review	1.7	1.9	-0.2	1.6									1.3
Bosworth et al. (2008)	CD						1993–2004							1.7

Author	Method	Cross-country studies (estimated annual growth rate (%) in agricultural TFP)												
		1994	1995	1996	1997	1998	1999	2000	2001	2002	2003	2004	2005	Avg.
Fuglie (2008)	CD			1990–1999		3.8							3.2	3.5
Nin Pratt et al. (2009)	DEA-M	5.9	2.8	8.1	2.5	4.9	-2.3	7.1	4.9	2000–2006 5.9	4.4			4.9
	A-D	3.6	3.4	3.7	6.0	-1.9								3.0

[a]Methods: SPF, stochastic production frontier; DEA-M, Malmquist index based on data envelopment analysis; A-D, growth accounting method with Divisia or Tornqvist–Theil index; CD, Cobb–Douglas production function estimation or growth accounting.

on productivity growth. In a follow-up paper, Lin (1993) studied the efficiency of different systems and showed that household farms outperformed cooperative farms, which gave support for institutional reform in China. Lin (1995) examined rice production and tested the induced institutional innovation theory. A study by Huang and Rozelle (1995) of 1952–1990 data found that environmental stress was an important factor in reducing TFP growth after the mid-1980s. Spitzer (1997) applied a non-parametric index number approach to decompose Chinese TFP and found that technical change was positive and efficiency change negative during the period from 1985 to 1994. Zhang and Carter (1997) constructed a Cobb–Douglas production function to separate the contribution of inputs, weather and efficiency to the growth of grain production from 1980 to 1990.

Zhang and Fan (2001) used a generalized maximum entropy approach to estimate a multi-output production technology for 25 provinces during 1979–1996. They did not, however, calculate and decompose total factor productivity growth. Jin *et al.* (2010) use a stochastic production frontier function approach to estimate the rate of change in TFP for 23 of China's main farm commodities. To do so, they rely on the National Cost of Production Data Set, finding negative rates of efficiency change outweighed by positive rates of technical change. They do not, however, report yearly productivity growth at an aggregate level.

Regarding the role of market institutions and transaction costs on productivity, Rozelle *et al.* (1997) examined market integration after the implementation of liberalized economic policies in food markets. In another study, Rozelle *et al.* (1999) used a labour migration framework to model the effects of migration and remittances on agricultural productivity growth in China. DeBrauw *et al.* (2000) examined how market liberalization influenced the behaviour of producers.

Many authors have estimated agricultural productivity growth on the basis of data from FAO in a multi-country context, including China. Coelli and Rao (2005)

used a data envelopment analysis (DEA) approach to Malmquist indexes of TFP growth for many countries on the basis of data from FAO for the period 1980–2000. They found that agricultural TFP in China grew at an average yearly rate of 1.06% during that period. Bravo-Ortega and Lederman (2004) also used FAO data to calculate agricultural TFP growth for China (among other countries) during the 1961–2000 period. They estimated a translog production function and calculated TFP as a residual. Although they found that Chinese agricultural TFP grew 1.67% per year in the period, they did not report annual figures of TFP growth after 1994. Ludena *et al.* (2007) constructed TFP indexes for Chinese agriculture on the basis of a DEA directional distance function. Using data from FAO, they calculated an average agricultural TFP growth of 3.05% per year for the period 1990–2000, consistent in sign with Bravo-Ortega and Lederman (2004). Because these studies did not report yearly TFP growth estimates they are not directly comparable to this analysis.

A number of studies have calculated and decomposed agricultural TFP growth in China within a time frame overlapping (at least partially) that considered here (see Table 8.1). Using provincial data, Lambert and Parker (1998) constructed a Divisia index for the period 1979–1995, finding an increase in TFP of 5.8% per year in the period 1993–1995. Jin *et al.* (2002) also used an accounting approach and constructed a Tornqvist index. They concluded that new technologies were the main driver of agricultural productivity growth during 1980–1995. In contrast with Lambert and Parker (1998), however, they found that TFP declined by 3.2% annually in the period 1994–1995. Mead (2003) re-examined data on Chinese agricultural productivity growth using an alternative calculation of the country's labour force. This estimate is employed in a TFP calculation on the basis of a constant-returns-to-scale Cobb–Douglas production function. The study found a strong correlation between policies and productivity growth during 1984–1999. In contrast, Dekle and Vandenbroucke (2006), calculating productivity growth in China as a residual based on a constant-returns-to-scale Cobb–Douglas approximation to the technology, found strong TFP growth in the period 1994–2003 (6.6% per year).

Using national data, Wu *et al.* (2001) constructed Malmquist indexes for 1980–1995. They found that TFP grew at an annual rate of 2.3% in 1994–1995. This is in line with Colby *et al.* (2000), Fan and Zhang (2002), Hsu *et al.* (2003), Lezin and Wei (2005), and Bosworth and Collins (2008), who found rather strong growth in agricultural TFP during different parts of the 1994–2005 period.

Colby *et al.* (2000) used a Tornqvist index to analyse the sources of output growth in grains and in four major crops in China (rice, wheat, corn and soybeans). They found that agricultural TFP grew on average at an annual rate of 0.8%. Fan and Zhang (2002) adjusted previous measures of growth in outputs and inputs and calculated a Tornqvist–Theil index of TFP at the national and provincial levels for the period 1952–1997. In particular, they found an increase in TFP during the period 1978–1997. Lezin and Wei (2005) also estimated a Cobb–Douglas production function for the province of Zhejiang and found positive TFP growth in the period 1994–1997. Hsu *et al.* (2003) calculated output-orientated Malmquist productivity indexes using a non-parametric data envelopment analysis approach covering 1984–1999. They estimated that TFP growth averaged 1% per year. Bosworth and Collins (2008) calculated productivity growth in China as a residual on the basis of a constant-returns-to-scale Cobb–Douglas approximation to the technology. They calculated average national productivity of China and India and compared their performances in the period 1978–2004. They estimated China's agricultural TFP growth at 1.7% per year in the 1993–2004 period.

In a cross-country study, Fuglie (2008) estimated an annual rate of TFP growth in Chinese agriculture of approximately 3.5% from 1990 to 2006. Nin Pratt *et al.* (2009), using a multi-country context and FAO data, calculated both a Tornqvist–Theil index and a Malmquist index of TFP growth for China. They found increases in Chinese agricultural productivity in the post-reform period up until 2003: growth averaged 5% per year when calculated with a Malmquist index and 3% with a Tornqvist–Teil index. They also found that both efficiency and technical change were important drivers of productivity growth and that returns to agricultural research and development had been high.

From the above studies, which use data from the China's Statistical Yearbook and from FAO, we learn several things. Studies using FAO data generally show higher TFP growth rates for 1970–2000. Methodologies used include econometric estimation of production functions, some of them stochastic frontiers, growth accounting TFP indexes and data envelopment analysis. Studies cover different periods that extend from the introduction of policy reforms up until the early 2000s.

Estimates indicate that agricultural productivity growth in China was higher immediately after the introduction of the HRS (from 1978 to the mid-1980s), thus making institutional reform the main contributor to TFP growth in that period. There is, however, evidence that TFP growth slowed after that period and towards the end of 1990s, a trend considered to be due to the exhaustion of the institutional effect, the introduction of the procurement price system, environmental stress or lack of agriculture investments and innovations.

8.4   Regional Productivity Growth

Productivity refers to output per unit of input and can be measured using different approaches. We care about productivity growth because it indicates an increase in output in perpetuity. The most direct way to measure productivity is by constructing indexes of outputs and inputs with costs and revenue shares as weights. Given the difficulty in obtaining these shares, two alternative methods are used in this chapter: a non-stochastic Malmquist index and a stochastic production frontier. Both methods allow for estimation of the rate of productivity growth when no reliable information on prices is available. We refer to this rate as the total factor productivity growth rate.

8.4.1 Data

Data used are from China's Statistical Yearbook for 30 'provinces' during 1993–2005. Some of the provinces are municipalities such as Beijing, Shanghai and Tianjin. Sichuan includes Chongqing, considered a provincial-level municipality since 1997. Others are not provinces but autonomous regions such as Inner Mongolia, Tibet, Xinjiang, Ningxia and Guanxi. Hainan is an island province. We use agricultural output (gross output value for the agriculture sector, including farming, forestry, the animal industry and fisheries) in constant 1993 Yuans and the corresponding quantity indexes converted using 1993 as the base year. We obtain the 1993 data from the 1994 *China Statistical Yearbook* (p. 330, Agriculture,

Table 11-4 Gross Output Value and Indices of Farming, Forestry, Animal Husbandry and Fishery), and the 1994 data from the same table in the 1995 China Statistical Yearbook. The data for 1995–2005 can be found online. The following inputs used in the analysis are also obtained from corresponding tables in the Agriculture chapter of the China Statistical Yearbook: total sown areas, agricultural machinery, labour and fertilizer. Fertilizer includes nitrogen, phosphate, potash and compound fertilizer. Table 8.2 summarizes the data at the national level. (Data by province are available from the authors.)

8.4.2 The Malmquist Index

The Malmquist index used here is the version specified by Färe *et al.* (1994). Productivity change is decomposed into efficiency change (first term) and technical change (second term; Eqn 8.1) (see equation 8.1 at bottom of page):

Table 8.2. Major agricultural output and input measures for Chinese agriculture, 1993–2005.

	Gross output value	Land sown	Fertilizer	Labour	Machinery power
	(Constant 1993 Yuan, 100 millions)	(10,000 hectares)	(10,000 metric tons)	(10,000 persons)	(10,000 kw)
1993	10,996	14,774	3,152	33,258	31,817
1994	11,931	14,824	3,318	32,690	33,802
1995	13,232	14,988	3,594	32,335	36,118
1996	14,473	15,238	3,828	32,261	38,547
1997	15,215	15,397	3,981	32,435	42,079
1998	16,119	15,571	4,086	32,627	45,208
1999	16,880	15,637	4,124	32,912	48,996
2000	17,485	15,630	4,147	32,798	52,574
2001	18,234	15,571	4,254	32,451	55,172
2002	19,136	15,464	4,339	31,991	57,930
2003	19,855	15,241	4,412	31,260	60,387
2004	21,327	15,355	4,637	30,596	64,028
2005	22,521	15,549	4,766	29,976	68,398
Annual rate of change (%)	5.97	0.43	3.46	−0.87	6.37

Source: China Statistical Yearbook, State Statistical Bureau (various annual issues).

$$M_0(x_{t+1}, y_{t+1}, x_t, y_t) = \frac{D_0^{t+1}(x_{t+1}, y_{t+1})}{D_0^t(x_t, y_t)} * \left[\frac{D_0^t(x_{t+1}, y_{t+1})}{D_0^{t+1}(x_{t+1}, y_{t+1})} * \frac{D_0^t(x_t, y_t)}{D_0^{t+1}(x_t, y_t)} \right]^{1/2} \tag{8.1}$$

where D_o is the output distance function, x and y are input and output vectors, and t indicates the time period. Data envelopment analysis, a programming approach, is used to calculate D_o as follows in Eqn 8.2 (Coelli, 2008):

$$
\begin{aligned}
&[D_0^t(x_t, y_t)]^{-1} = max_{\phi, \lambda}\phi \\
&subject\ to \quad -\phi y_{it} + Y_t\lambda \geq 0, \\
&x_{it} - X_i\lambda \geq 0, \\
&\lambda \geq 0
\end{aligned}
\qquad (8.2)
$$

where X and Y are the $(K*N)$ input and the $(M*N)$ output matrices, respectively. λ is an $N*1$ vector of constants, or intensity variables. Here $1 \leq \phi < \infty$ and $(\phi\text{-}1)$ are the proportional increase in outputs achieved by the $i\text{-}th$ region while maintaining input quantities constant. This approach constructs the production surface as an envelope of the observations in each period, defined by those representing best performance.

Technical efficiency change, also known as the 'catching up' component of the index, indicates whether a particular region is moving closer to or away from the frontier. Technical change, or the innovations component of the index, refers to a shift of the best practice frontier. Indexes <1 represent deteriorations. The Malmquist index is calculated with information from two consecutive data periods and it is very sensitive to extremes but free of specification error. We use Coelli's DEAP programming code to compute the Malmquist index and its components.

Table 8.3 shows the national average rate of productivity growth derived using the Malmquist method and breaks this down into technical change and efficiency change components. China experienced high rates of productivity growth in 1994 and 1995, followed by a decrease from 1996 to 1998. The trend reversed after 1999, with annual productivity growth rates of between 4% and 2% between 2000 and 2005. On average, TFP growth in Chinese agriculture during 1993–2005 was a robust 3.97% annually, compared with 1.73% in US agriculture during the same period.

Table 8.3. Productivity growth indexes for China's agriculture, 1993–2005; Malmquist method.

Year	Total factor productivity	Technical efficiency	Technical change	Total factor productivity[a]	Efficiency change	Technical change
	(Index, 1993=100)			(Annual growth rate in %)		
1993	100	100	100	–	–	–
1994	106	96	111	5.9	−4.1	10.4
1995	109	94	116	2.3	−2.3	4.7
1996	113	96	118	3.8	2.4	1.4
1997	114	86	135	1.2	−10.7	13.3
1998	114	87	134	0.3	1.0	−0.7
1999	119	85	143	3.8	−2.6	6.6
2000	122	85	146	2.8	0.4	2.4
2001	127	80	163	4.0	−6.3	11.0
2002	133	80	171	4.2	−0.4	4.7
2003	138	77	185	3.8	−3.5	7.6
2004	141	80	182	2.1	3.6	−1.5
2005	144	81	183	2.3	1.8	0.5
Average annual growth rate (unweighted)				3.03	−1.80	4.93
Average annual growth rate (weighted by output)				3.98		

[a]The rate of growth in total factor productivity equals the rate of efficiency change plus the rate of technical change.

Table 8.4 reports average productivity rates of growth for each province and region, also using the Malmquist method.[1] Most provinces experienced positive TFP growth, mainly due to technical change resulting from the adoption of new innovations. Jiangsu, Fuijan and Liaoning were the

Table 8.4. Agricultural productivity growth in China's provinces and regions: Malmquist method.

Regions	Provinces	Total factor productivity growth	Efficiency change	Technical change
		(Average annual growth rate in %, 1994–2005)		
East[a]		5.7	−0.8	6.8
	Beijing	5.9	0.0	5.9
	Fujian	6.9	−1.9	9.0
	Guangdong	5.5	−1.7	7.3
	Guangxi	2.1	−1.7	3.9
	Hainan	5.6	0.0	5.6
	Hebei	5.6	−0.3	5.9
	Jiangsu	7.4	−0.5	7.9
	Liaoning	6.1	0.4	5.7
	Shandong	5.4	−1.8	7.3
	Shanghai	5.8	0.0	5.8
	Tianjing	2.5	−3.2	5.9
	Zhejiang	4.9	−0.1	5.0
Central[a]		2.9	−1.9	5.4
	Anhui	2.1	−3.0	5.3
	Heilongjiang	2.8	−1.4	4.3
	Henan	5.0	−1.8	6.9
	Hubei	2.3	−3.6	6.2
	Hunan	2.1	−0.9	3.0
	Inner Mongolia	0.1	−3.4	3.6
	Jiangxi	0.5	−3.7	4.4
	Jilin	5.0	−1.5	6.6
	Shanxi	1.5	−2.2	3.8
West[a]		0.9	−2.8	4.3
	Gansu	1.6	−0.7	2.3
	Guizhou	−2.1	−5.9	4.0
	Ningxia	1.1	−2.7	3.9
	Qinghai	2.1	0.3	1.7
	Shaanxi	2.0	−3.1	5.3
	Sichuan	0.4	−3.9	4.6
	Tibet	−1.2	−1.3	0.1
	Xinjiang	2.7	−2.1	4.9
	Yunnan	0.2	−1.9	2.1

[a]Output weighted regional averages.

best performers, with average annual TFP growth rates of 6% to 7%. The rural areas around Beijing and Shanghai were also strong performers, probably due to a shift in output toward higher valued vegetable and fruit production. Provinces with productivity growth rates around 5% per year are Hebei, Hainan, Guangdong, Shandong, Jilin, Heinan and Zheijang. A second set of provinces cluster around annual growth rates of 3%. The worst performer was Tibet, which actually experienced a productivity decline.

Figure 8.1 aggregates performance in three geographical regions: East, Central and West (see Table 8.4 for a list of the provinces assigned to each region). The East (with an average annual TFP growth rate of 5.7%) outperformed the Central (2.9%) and West (0.9%) during this period. It is interesting to note that TFP growth rates in the West improved rapidly after 2000, whereas those of the East decreased slightly. By 2004–2005, TFP growth rates in all three regions had converged to about 3% per year.[2]

Among the factors that might have affected economic performance during this period are: (i) bad weather conditions in the late 1990s; (ii) government efforts to encourage diffusion and adoption of new technologies and production processes; (iii) elimination of the rationing system in years 1994 and 1995; (iv) a steady decline in procurement prices during the period; (v) introduction of the Grain-Bag Responsibility System in 1995; (vi) a new round of reforms around 1998, especially a second HRS with renegotiation of land contracts; (vii) the reinforcement of market-oriented policies and tax exemptions to the agricultural sector; and (viii) WTO membership.

8.4.3   Stochastic production frontier

As an alternative to the Malmquist index, a stochastic parametric translog production frontier is estimated econometrically, following Battese and Coelli (1992), to use in the calculation of TFP growth rates. The specification

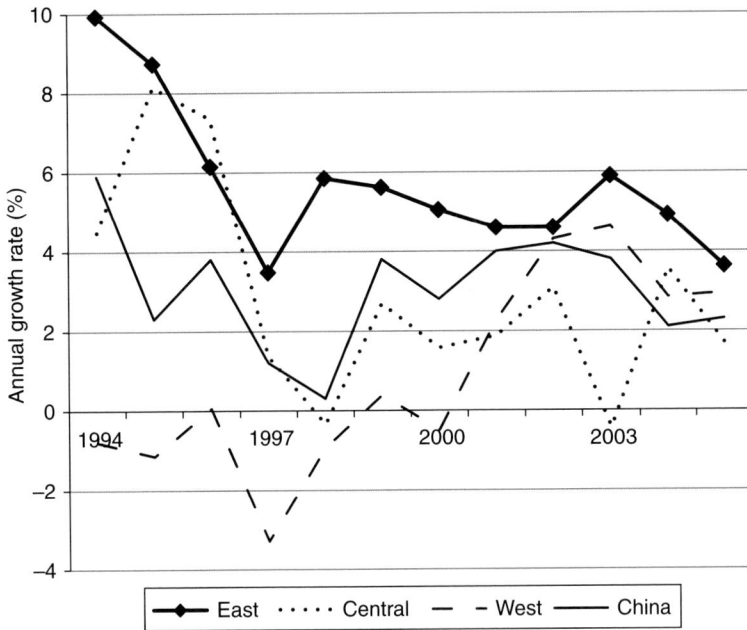

Fig. 8.1. Agricultural TFP growth rates by region 1994–2005: Malmqvist method. Source: Authors' estimates.

used is (Eqn 8.3) (see equation 8.3 at bottom of page):

Where Y_{it} is output level of the i-th province in the t-th time period, x_{it} are inputs (land, labour, fertilizer and machinery), t a time trend representing disembodied technical change, α and β are coefficients to be estimated, v are random errors assumed to be independent and identically distributed $N(0, \sigma_v^2)$, u are assumed to be one-sided errors independent and identically distributed $N(0,\sigma_u^2)$ and independent of v. $u_{it} = u_i$ $exp(-\eta(t - T))$ and accounts for technical inefficiency. η is a parameter to be estimated.

Equation 8.3 is estimated using Coelli's Frontier 4.1 econometric package with symmetry imposed (Coelli, 2007). The maximum likelihood estimates of the parameters are in Table 8.5. Yearly production elasticities evaluated at the mean of

the variables, represented in Table 8.6, indicate the estimates are not globally concave. Nevertheless, they show a decreasing production elasticity of land and labour and an increasing production elasticity for fertilizer and machinery over time.

Technical change is obtained through differentiation of Eqn 8.3 with respect to t to give Eqn 8.4:

$$\frac{\partial lnY_{it}}{\partial t} = \alpha_t + \beta_{tt}t + \beta_{tm}\,ln(x_{mit}) \qquad (8.4)$$

Technical efficiency level of region I at time t is defined as:

$$TE_{it} = exp(-u_{it}) \qquad (8.5)$$

and efficiency change is the difference in TE across years.

$$lnY_{it} = \alpha_0 + \sum_m \alpha_m\,ln(x_{mit}) + \alpha_t t + \frac{1}{2}\sum_m\sum_n \beta_{mn}ln\,(x_{mit})ln(x_{nit}) + \frac{1}{2}\beta_{tt}t^2$$

$$+ \sum_m \beta_{tm}ln(x_{mit})t + \upsilon_{it} - u_{it}, \quad i = 1,\dots 30; t = 1,\dots 9. \qquad (8.3)$$

Table 8.5. Translog stochastic frontier maximum likelihood estimates for China's agriculture.

Parameters (description)	Parameters (symbols)	Coefficient estimates	Standard error	T-ratio
Intercept	α_0	4.7258	1.3598	3.48
log(land)	α_D	−2.2786	0.5841	−3.90
log(labour)	α_L	1.7028	0.3488	4.88
log(fertilizer)	α_F	0.0667	0.3793	0.18
log(power)	α_P	0.7554	0.2997	2.52
time	α_t	0.1002	0.0214	4.69
log(land)^2	β_{DD}	0.2907	0.0976	2.98
log(labour)^2	β_{LL}	0.0214	0.0369	0.58
log(fertilizer)^2	β_{FF}	0.0644	0.0333	1.93
log(power)^2	β_{PP}	0.0030	0.0257	0.12
log(land)*log(labour)	β_{DL}	−0.1863	0.0998	−1.87
log(land)*log(fertilizer)	β_{DF}	−0.2896	0.1041	−2.78
log(land)*log(power)	β_{DP}	0.0171	0.0958	0.18
log(labour)*log(fertilizer)	β_{LF}	0.2304	0.0616	3.74
log(labour)*log(power)	β_{LP}	−0.1841	0.0565	−3.26
log(fertilizer)*log(power)	β_{FP}	0.0903	0.0493	1.83
log(land)*time	β_{Dt}	−0.0057	0.0069	−0.83
log(labour)*time	β_{Lt}	−0.0034	0.0038	−0.88
log(fertilizer)*time	β_{Ft}	0.0038	0.0041	0.92
log(power)*time	β_{Pt}	0.0015	0.0038	0.40
time^2	β_{tt}	−0.0009	0.0003	−3.45

The econometric model is estimated using annual panel data for 30 provinces during 1993–2005.

Table 8.6. Estimated production elasticities for China's agriculture.

	Land	Fertilizer	Labour	Power
1993	0.124	0.453	0.206	0.075
1994	0.110	0.464	0.202	0.084
1995	0.091	0.478	0.203	0.096
1996	0.078	0.490	0.199	0.104
1997	0.068	0.505	0.186	0.109
1998	0.061	0.517	0.174	0.112
1999	0.055	0.530	0.158	0.114
2000	0.049	0.540	0.142	0.117
2001	0.037	0.550	0.136	0.123
2002	0.025	0.560	0.129	0.129
2003	0.011	0.568	0.124	0.136
2004	0.000	0.577	0.119	0.146
2005	−0.001	0.581	0.106	0.155
Mean	0.054	0.524	0.160	0.115

The elasticities are derived from the estimated translog stochastic frontier production function evaluated at the mean values of the variables.

The growth rate of TFP change is defined as the rate of change in output not accounted for by input change (where a dot over a variable represents rate of change):

$$TFP = \dot{y} - \sum_m \varepsilon_m \dot{x}_m = Technical\ change$$
$$+ Efficiency\ change \qquad (8.6)$$

where ε_m are input production elasticities, and m = land, labour, machinery and fertilizer. National annual indexes and growth rates for TFP, technical efficiency and technical change for the years 1993 to 2005 are reported in Table 8.7. We estimate strong and positive rates of technical change and a

Table 8.7. Productivity growth indexes for China's agriculture, 1993–2005: stochastic frontier method.

Year	Total factor productivity	Technical efficiency	Technical change	Total factor productivity[a]	Efficiency change	Technical change
	(Index, 1993 = 100)			(Annual growth rate; %)		
1993	100	100	100	–	–	–
1994	105	99	106	4.5	−1.1	5.6
1995	109	98	112	4.4	−1.1	5.5
1996	114	97	118	4.4	−1.1	5.5
1997	119	96	125	4.3	−1.1	5.4
1998	124	95	131	4.2	−1.2	5.3
1999	130	93	138	4.1	−1.2	5.2
2000	135	92	146	4.0	−1.2	5.2
2001	140	91	153	3.9	−1.2	5.1
2002	146	90	161	3.8	−1.2	5.0
2003	151	89	170	3.7	−1.3	5.0
2004	157	88	178	3.6	−1.3	4.9
2005	162	87	187	3.5	−1.3	4.8
Average annual growth rate (unweighted)				4.04	−1.19	5.23
Average annual growth rate (weighted by output)				4.13	−0.89	5.02

[a]The rate of growth in total factor productivity equals the rate of efficiency change plus the rate of technical change.

negative rate of efficiency change throughout the period. This yields positive TFP growth rates throughout the period of analysis.

Table 8.8 presents the econometric estimates of TFP growth by province. Liaoning, Zhejiang, Shanghai and Fujian define the frontier throughout the period. The average TFP growth rates across provinces are less dispersed than in the Malmquist estimation. The TFP growth rates for Shanghai, Hainan, Beijing, Liaoning and Fujian are the highest, about 5–6% per year. Tibet, Ningxia, Guizhou and Qinghai show the lowest rates of growth.[3] A summary of the information in these tables is presented in Fig. 8.2, which shows the evolution of the annual TFP growth rate for the three regions, with the East (4.6%) performing better than the Central (3.7%) and West (3.5%) regions.[4] Again, the pattern of these evolutions mimics the evolution of the TFP growth rates for the whole country and does not show the variability evident in the evolution of TFP growth rates estimated non-parametrically with the Malmquist index. Although the Malmquist index is subject to extreme variability – because it relies on information from two consecutive periods only and because of its

deterministic nature – the econometric estimates might suffer from specification error from the linearity evident in Eqn 8.4.

Only a few previous studies estimated provincial agricultural productivity growth rates for a period as recent as the one studied here (see Table 8.1). For the overlapping years, our results are consistent with those of Lambert and Parker (1998), Colby et al. (2000), Wu et al. (2001), Fan and Zhang (2002), Lezin and Wei (2005), Hsu et al. (2003), Dekle and Vandenbroucke (2006), and Bosworth and Collins (2008). The differences between results in this study and those of Jin et al. (2002) and Meade (2003) are not surprising considering the many differences in terms of data, periods and sectors covered. Jin et al. (2002) used data collected by the State Price Bureau and calculated provincial and national TFP indexes based on a sampling framework. Meade (2003), on the other hand, used a time series of provincial data and obtained TFP growth rates residually from the estimation of a Cobb–Douglas production function.

Only the studies by Dekle and Vandenbroucke (2006) and Bosworth and Collins (2008) include information up to 2004. Bosworth and Collins (2008) used the same

Table 8.8. Agricultural productivity growth in China's provinces and regions: stochastic frontier method.

Regions Provinces		Total factor productivity growth	Efficiency change	Technical change
		(Annual growth rate in %)		
East[a]		4.6	−0.6	5.2
	Beijing	5.4	−1.0	6.4
	Fujian	5.1	−0.3	5.4
	Guangdong	5.0	0.0	5.1
	Guangxi	3.9	−1.0	4.9
	Hainan	5.6	−0.2	5.7
	Hebei	3.8	−1.3	5.1
	Jiangsu	4.5	−0.6	5.1
	Liaoning	5.2	−0.2	5.3
	Shandong	4.2	−0.8	5.0
	Shanghai	5.9	−0.3	6.2
	Tianjing	4.8	−1.5	6.3
	Zhejiang	4.9	−0.2	5.2
Central[a]		3.7	−1.2	5.0
	Anhui	3.6	−1.3	4.9
	Heilongjiang	3.6	−1.2	4.9
	Henan	3.4	−1.4	4.8
	Hubei	4.0	−1.1	5.1
	Hunan	3.9	−0.8	4.8
	Inner Mongolia	3.6	−1.4	5.1
	Jiangxi	4.0	−0.9	4.9
	Jilin	4.2	−1.1	5.3
	Shanxi	3.1	−2.2	5.3
West[a]		3.5	−1.3	4.8
	Gansu	3.2	−1.8	5.0
	Guizhou	2.9	−1.8	4.6
	Ningxia	3.0	−2.8	5.8
	Qinghai	2.8	−2.8	5.7
	Shaanxi	3.4	−1.7	5.1
	Sichuan	3.5	−0.9	4.4
	Tibet	3.0	−2.8	5.8
	Xinjiang	4.5	−1.0	5.4
	Yunnan	3.3	−1.4	−4.7

[a]Output weighted regional averages.

approach as Meade (2003), with data for the country as a whole obtained from China's Statistical Yearbook. Dekle and Vandenbroucke (2006) also fit a Cobb–Douglas production function but using provincial data. Consistent with our results, both studies

found evidence of positive TFP growth rates.

There are many multi-country studies that include China in their analyses. Those studies use FAO data, focus on a national aggregate and cover a period of time starting in 1961. The studies are not methodologically comparable to this one, but they are of empirical interest. We mention here two of the latest studies of this type: Nin Pratt *et al.* (2010) and Fuglie (2008). Nin Pratt *et al.* (2010) calculate Tornquist and Malmquist indexes for 59 countries, one of them China. Both indexes yield high and positive average agricultural TFP growth rates in China, but both also identify some slowdown in the 1990s. Fuglie (2008) calculated productivity growth for a large set of countries using a fixed-weight index for the 1961–2006 period. The weights were carefully calculated with expenditure data from a reduced set of countries, then applied to other countries in the same geographical area. China's weights were calculated with information from China's Statistical Bureau. For the period of interest, Fuglie (2008) estimated a TFP growth rate in Chinese agriculture of approximately 3.5% per year. The rates obtained for China in those two multi-country studies are consistent with those in this study.

8.5   Conclusions

In this study, a Malmquist index and a stochastic production frontier are estimated to examine agricultural productivity growth in China's provinces from 1993 to 2005. Three important conclusions follow.

First, China achieved high rates of agricultural productivity growth throughout the whole period. The average annual TFP growth rate is estimated at around 4%. These rates are supported by high rates of technical change, indicating that agricultural productivity growth in China was driven by technological innovations rather than increases in input use. Productivity growth rates show a slowdown during the 1990s, a rebound in the early 2000s, and a slight slowdown in 2004 and 2005.

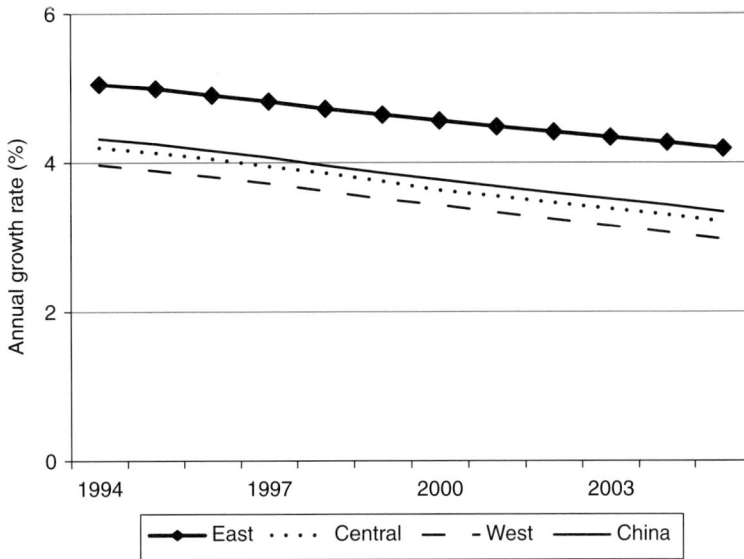

Fig. 8.2. Agricultural TFP growth rates by region 1994–2005: stochastic frontier method. Source: Authors' estimates.

Second, there have been significant regional disparities in productivity performance. Among the three regions, the East outperformed the Central and West. The evolution of productivity growth among these regions shows a different pattern across methods, however. The Malmquist index shows an important improvement in performance in the West, not much change in the Central region, and a slight deterioration through time in the East, indicating a convergence among them over time. Stochastic frontier estimates show slight deteriorations through time for all three regions.

Finally, this study shows important differences across methods in the estimation of productivity growth rates. Although the Malmquist index reveals that average annual TFP growth rates in China decreased from 6% in 1994 to 0.3% in 1998, followed by an increase to 2–4% after 1998, the econometric stochastic frontier estimates indicate a slight but continuous decline in TFP growth from 4.5% to 3.5% during the period of analysis. These differences are also notable in the evolution of regional TFP growth rates as discussed above. We suspect that the

choice of an econometric specification, such as the translog, implies a strongly maintained hypothesis on the derivatives of the level function of interest. The Malmquist results show wide variations across years, leading one to question its deterministic nature and the fact that it only uses information on two adjacent periods, possibly making it very sensitive to data errors and to any temporary changes in the data. Caution suggests that studies of productivity growth should use alternative approaches. Although the Malmquist index, because of its nature, exacerbates variability, the econometric estimates exacerbate uniformity and smoothness.

This study's findings are important because they provide a contrast with most other studies of Chinese agricultural TFP growth. By estimating rates of productivity growth for 30 provinces, using two different approaches and an alternative data source, it provides additional information that supports the results of earlier studies and extends the period of analysis. Future research should look into understanding the differences in patterns most obvious in our Malmquist estimates.

Notes

[1] Annual TFP growth indexes for province and regions from the Malmquist method are given in Appendix Table A8.1.

[2] One reviewer suggested that the Central region be divided into a North Central and a South Central region, given differences in agronomic characteristics. Doing so, we found that in later years the North Central region marginally outperformed the South Central region. On average, annual TFP growth rates are 3.3% for the North Central and 2.5% for the South Central region.

[3] Detailed information for all provinces and all years from the stochastic frontier method is provided in the Appendix Table A8.2.

[4] Separating the Central region into North Central and South Central regions shows that the North Central outperformed the South Central throughout the whole period. Average TFP growth rate in the North Central is 4.2% versus 3.7% in the South Central.

References

Battese, G. and Coelli, T. (1992) A model for technical inefficiency effects in a stochastic frontier production function for panel data. *Empirical Economics* 20, 325–332.

Bosworth, B. and Collins, S. (2008) Accounting for growth: Comparing China and India. *Journal of Economic Perspectives* 22, 1, 45–66.

Bravo-Ortega, C. and Lederman, D. (2004) Agricultural productivity and its determinants: Revisiting international experiences. *Estudios de Economia* 31, 133–163.

Coelli, T. (2007) A Guide to Frontier Version 4.1: A Computer Program for Stochastic Frontier Production and Cost Function Estimation. Centre for Efficiency and Productivity Analysis, University of Queensland, Brisbane, Australia. Available at: http://www.une.edu.au/econometrics/cepa.htm

Coelli, T. (2008) A Guide to DEAP Version 2.1: A Data Envelopment Analysis (Computer) Program. Centre for Efficiency and Productivity Analysis, University of Queensland, Brisbane, Australia. Available at: http://www.une.edu.au/econometrics/cepa.htm

Coelli, T. and Rao, P.D.S. (2005) Total factor productivity growth in agriculture: A Malmquist index analysis of 93 countries, 1980–2000. *Agricultural Economics* 32, 115–134.

Colby, H., Diao, X. and Somwaru, A. (2000) Cross-commodity analysis of China's grain sector: Sources of growth and supply response. Technical Bulletin 1884, Economic Research Service, U.S. Department of Agriculture, Washington, DC.

DeBrauw, A., Huang, J. and Rozelle, S. (2000) Responsiveness, flexibility, and market liberalization in China's agriculture. *American Journal of Agricultural Economics* 82, 1133–1139.

Dekle, R. and Vandenbroucke, G. (2006) Wither Chinese growth? A sectoral accounting approach. Working Paper, Department of Economics, University of Southern California, Los Angeles, CA.

Fan, S. (1991) Effects of technological change and institutional reform on production growth in Chinese agriculture. *American Journal of Agricultural Economics* 73, 266–275.

Fan, S. (1997) Production and productivity growth in Chinese agriculture: New measurement and evidence. *Food Policy* 22, 213–228.

Fan, S. and Zhang, X. (2002) Production and productivity growth in Chinese agriculture: New national and regional measures. *Economic Development and Cultural Change* 50, 819–838.

Färe. R., Grosskopf, S. and Lovell, C. (1994) *Production Frontiers*. Cambridge University Press, Cambridge, UK.

Fuglie, K. (2008) Is a slowdown in agricultural productivity growth contributing to the rise in commodity prices? *Agricultural Economics* 39, supplement, 431–441.

Hsu, S., Yu, M. and Chang, C. (2003) An analysis of total factor productivity growth in China's agricultural sector. AgEconSearch, Department of Applied Economics, University of Minnesota, Minneapolis, MN.

Huang, J. and Rozelle, S. (1995) Environmental stress and grain yields in China. *American Journal of Agricultural Economics* 77, 853–864.

Jin, S., Huang, J., Hu, R. and Rozelle, S. (2002) The creation and spread of technology and total factor productivity in China's agriculture. *American Journal of Agricultural Economics* 84, 916–930.

Jin, S., Hengyun Ma, H., Huang, J., Hu, R. and Rozelle, S. (2010) Productivity, efficiency and technical change: Measuring the performance of China's transforming agriculture. *Journal of Productivity Analysis* 33, 191–207.

Lambert, D. and Parker, E. (1998) Productivity in Chinese provincial agriculture. *Journal of Agricultural Economics* 49, 378–392.

Lezin, A-B. and Wei, L. (2005) Agricultural productivity growth and technology progress in developing country agriculture: Case study in China. *Journal of Zhejiang University Science*, 6A, Suppl. I, 172–176.

Lin, J.Y. (1992) Rural reforms and agricultural growth in China. *American Economic Review* 82, 34–51.

Lin, J.Y. (1993) Cooperative farming and efficiency: Theory and empirical evidence from China. In: Csaba, C. and Kislev, Y. (eds) *Agricultural Cooperatives in Transition*. Westview Press, Boulder, CO.

Lin, J.Y. (1995) Endowments, technology, and factor markets: A natural experiment of induced institutional innovation from China's rural reform. *American Journal of Agricultural Economics* 77, 231–242.

Ludena, C., Hertel, T., Preckel, P., Foster, K. and Nin, A. (2007) Productivity growth and convergence in crop, ruminant, and non-ruminant production: Measurement and forecasts. *Agricultural Economics* 37, 1–17.

Mead, R. (2003) A revisionist view of Chinese agricultural productivity? *Contemporary Economic Policy* 21, 1, 117–131.

Nin Pratt, A., Yu, B. and Fan, S. (2009) The total factor productivity in China and India: New measures and approaches. *China Agricultural Economic Review* 1, 9–22.

Rozelle, S., Park, A., Huang, J. and Jin, H. (1997) Liberalization and rural market integration in China. *American Journal of Agricultural Economics* 79, 635–642.

Rozelle, S., Taylor, J. and DeBrauw, A. (1999) Migration, remittances, and agricultural productivity in China. *American Economic Review* 89, 287–291.

Spitzer, M. (1997) Interregional comparison of agricultural productivity growth, technical progress, and efficiency change in China's agriculture: A nonparametric index approach. INTERIM REPORT IR-97-89/December. International Institute for Applied Systems Analysis (IIASA), Laxenburg, Austria.

State Statistical Bureau (1994–2000 annual issues) *China Statistical Yearbook*. China Statistical Press, Beijing. Available at www.stats.gov.cn/english/statisticaldata/yearlydata/

Wu, S., Walker, D., Devadoss, S. and Lu, Y. (2001) Productivity growth and its components in Chinese agriculture after reforms. *Review of Development Economics* 5, 375–391.

Zhang, B. and Carter, C. (1997) Reforms, weather and productivity growth in China's grain sector. *American Journal of Agricultural Economics* 79, 1266–1277.

Zhang, X. and Fan, S. (2001) Estimating crop-specific production technologies in Chinese agriculture: A generalized maximum entropy approach. *American Journal of Agricultural Economics* 83, 378–388.

Appendix 8.1. Regional and Provincial TFP Growth Rates for China's Agriculture

Table A8.1. Agricultural TFP indexes for China's provinces and regions: Malmqvist method.

Region	Province	TFP index (1993=100)											
		1994	1995	1996	1997	1998	1999	2000	2001	2002	2003	2004	2005
East		105	111	116	122	128	134	140	146	153	160	167	174
	Beijing	102	104	103	104	104	105	125	151	186	211	210	209
	Fujian	111	125	136	146	152	162	175	187	197	213	223	230
	Guangdong	103	108	116	119	124	137	145	150	170	178	188	194
	Guangxi	101	112	113	123	121	124	123	121	124	125	125	130
	Hainan	112	121	119	125	132	163	156	175	179	183	191	200
	Hebei	112	125	129	139	146	154	163	171	177	186	194	197
	Jiangsu	113	127	135	132	136	144	153	163	170	187	221	243
	Liaoning	102	113	131	134	157	162	163	174	187	197	203	212
	Shandong	122	134	142	146	159	166	173	177	179	189	192	194
	Shanghai	112	118	132	157	152	142	157	173	186	214	214	207
	Tianjing	100	99	99	104	117	114	117	126	128	137	130	137
	Zhejiang	111	116	123	127	140	145	162	168	166	176	179	181
Central		104	109	113	118	122	127	132	137	141	146	151	156
	Anhui	98	109	116	123	120	130	128	125	135	122	134	132
	Heilongjiang	109	119	129	121	107	112	112	119	123	131	137	143
	Henan	102	124	139	147	154	164	166	173	173	169	181	185
	Hubei	107	119	123	129	125	127	132	134	132	132	132	133
	Hunan	103	107	110	114	112	115	120	123	126	128	127	129
	Inner Mongolia	105	96	109	100	107	104	108	106	106	101	104	103
	Jiangxi	108	110	115	113	106	103	105	105	106	110	106	107
	Jilin	112	113	128	119	134	130	125	133	152	161	171	187
	Shanxi	103	104	115	108	117	106	115	110	123	127	130	122
West		104	108	112	117	121	125	130	134	138	143	147	152
	Gansu	99	95	97	92	104	100	105	112	113	117	122	123
	Guizhou	102	99	93	89	82	76	73	75	73	75	78	78
	Ningxia	95	94	108	107	90	91	103	109	117	122	120	119
	Qinghai	102	92	94	100	106	106	103	109	112	115	121	130
	Shaanxi	99	97	108	99	102	97	103	107	114	113	124	130
	Sichuan	97	99	99	95	90	85	85	87	94	101	103	107
	Tibet	169	126	99	103	71	147	146	156	160	170	138	142
	Xinjiang	104	102	96	99	107	113	116	119	123	128	132	139
	Yunnan	96	94	92	91	94	116	102	99	99	102	103	105
National average		107	109	114	115	117	122	126	131	137	142	145	149

Table A8.2. Agricultural TFP indexes for China's provinces and regions: stochastic frontier method.

Region	Province	TFP Index (1993 = 100)											
		1994	1995	1996	1997	1998	1999	2000	2001	2002	2003	2004	2005
East		110	121	128	133	141	149	157	164	172	182	191	198
	Beijing	106	112	119	126	133	140	148	156	164	173	182	189
	Fujian	106	112	118	124	130	137	144	152	159	167	175	181
	Guangdong	106	112	118	124	131	137	144	152	159	167	175	183
	Guangxi	104	109	114	119	123	128	133	139	144	149	154	160
	Hainan	106	113	119	126	134	141	149	157	166	175	185	195
	Hebei	104	109	113	118	123	128	132	137	142	147	152	157
	Jiangsu	105	110	116	122	127	133	140	146	152	159	166	172
	Liaoning	106	112	118	125	132	139	146	153	161	169	177	186
	Shandong	105	110	114	120	125	130	136	141	147	153	159	164
	Shanghai	107	114	121	129	136	145	153	162	172	182	192	199
	Tianjing	105	111	117	123	129	135	142	148	155	163	170	178
	Zhejiang	105	111	117	123	130	136	143	150	157	165	173	178
Central		105	113	122	124	123	127	129	131	135	135	140	142
	Anhui	104	108	113	117	122	126	130	135	140	144	149	154
	Heilongjiang	104	109	113	118	123	127	132	136	141	145	150	154
	Henan	104	108	112	116	120	125	129	133	137	141	145	151
	Hubei	104	109	114	119	124	129	134	139	145	150	156	160
	Hunan	104	109	114	119	123	128	133	139	144	149	154	163
	Inner Mongolia	104	108	113	117	122	126	131	135	140	145	149	156
	Jiangxi	104	109	114	119	124	129	134	139	144	150	155	163
	Jilin	105	110	115	120	125	131	136	142	148	153	159	165
	Shanxi	104	107	111	115	119	123	126	130	134	137	141	148
West		99	98	98	95	94	94	94	96	100	105	108	111
	Gansu	104	108	111	115	119	123	127	131	135	138	142	146
	Guizhou	104	107	111	114	118	121	124	128	131	135	138	141
	Ningxia	104	107	111	115	119	122	126	129	133	136	140	143
	Qinghai	104	107	111	114	118	121	124	128	131	135	138	141
	Shaanxi	104	108	112	116	120	124	129	133	137	142	146	150
	Sichuan	104	108	112	117	121	125	130	134	139	143	148	153
	Tibet	104	107	111	114	118	122	126	129	133	136	140	143
	Xinjiang	105	110	116	122	128	133	139	145	152	158	165	171
	Yunnan	104	108	112	116	120	124	129	133	137	141	145	149
National average		105	109	114	119	124	130	135	140	146	151	157	163

9 Structural Transformation and Agricultural Productivity in India*

Hans P. Binswanger-Mkhize[1] and Alwin d'Souza[2]

[1]*China Agricultural University, Beijing* [2]*Jawaharlal University, New Delhi*

9.1 Agriculture and the Economic Structural Transformation Process

Studies of the patterns of economic growth and transformation, carried out by Simon Kuznets, Hollis Chenery and most recently by Peter Timmer (2009), have shown important regularities in the structural composition of economic activity. Prior to economic transformation, agriculture generally accounts for the bulk of the economic output and the labour force. Because productivity in the non-agricultural sector is higher than in the agricultural sector, the share of agriculture in total GDP falls far short of its share in the labour force. As industrial growth takes off, industry becomes even more productive, and the productivity differential with agriculture increases, while the share of agriculture in GDP starts to fall even more rapidly.

Thus, the structural gap widens during periods of rapid growth as long as the share of agricultural in GDP falls much faster than the share of agricultural labour. Farm incomes visibly fall behind incomes earned in the rest of the economy. 'This lag in real earnings

from agriculture is the fundamental cause of the deep political tensions generated by the structural transformation' (Timmer, 2009, p. 6). This tension is often aggravated because productivity gains in agriculture are quickly lost to farmers via declining prices. Income convergence between the agricultural and non-agricultural sectors is therefore primarily driven by labour migration.

During structural transformation, the speed with which labour is pulled out of agriculture depends on the labour intensity of industry. With a lag, services also start to increase their share in value added and in the labour force. Structural change, by moving workers from lower to higher productive activities, accelerates economic growth. Productivity in agriculture will start increasing as technical change spreads to the agricultural sector and as labour leaves the sector and agricultural investment increases. In advanced countries, at the end point of this process, the shares of agriculture in output and employment will approximate each other, as will incomes across the sectors. Agriculture will become just like any other

* This chapter was a background paper for "India 2039: Transforming Agriculture," Centennial Group, Washington, DC, 2012. We are grateful to Centennial Group, the Syngenta Foundation, and Integrated Research and Development (IRADe) for their support and to Centennial Group for permission to publish this paper.

181

sector of the economy. Even though agriculture becomes a very small sector of the economy, in absolute terms it continues to grow throughout the transformation period and beyond.

What happens during the transition depends on labour intensity in the industrial and services sectors and on productivity growth in those sectors. If it is slow, much of the labour force remains in agriculture and convergence of incomes is delayed. The focus of policy has to be as much on urban productivity as on agricultural productivity, and especially on the absorption of labour into the urban sector and the rural non-farm sector.

During most structural economic transformation, labour productivity in agriculture, and therefore agricultural incomes, will typically fall further and further behind productivity and incomes in other sectors, opening a widening inter-sector income differential. This income inequality will often cause major political problems. The reason for the widening gap is the long time it takes before the withdrawal of labour from agriculture translates into higher agricultural productivity, wages and incomes. Only towards the end of the structural transformation do the inter-sector productivity, wage and income differences start to fall, and productivity and incomes converge across all sectors. The turning point is reached when the share of agricultural labour in the economy starts declining at a faster rate than its share of output.

Timmer (2009) uses a sample of 86 countries to measure the pace of divergence and convergence across countries from 1965 to 2000. On average, countries reach the point when labour and output shares (and sectoral productivities) start to converge only at $9133 of per-capita income (in real 2000 US dollars). The estimates of turning points are not stable, and we need to analyse the Indian data to make judgements on how soon a turning point could arise, a task done later in the chapter.[1] Timmer also shows that during the past 35 years the turning point from divergence to convergence of productivity across the sectors has been reached at later and later stages in the economic transformation of high-growth performers. This suggests that industry and services have become less able to absorb the rapidly growing labour forces of developing economies.

In Asia, political issues associated with the divergence between rural and urban areas have flared up dramatically in the past few years. China's policy makers responded strongly to the challenge by extending health insurance, free education up to year nine, and safety nets to rural areas. In addition, massive, accelerated rural infrastructure programmes were put in place, and taxation of agriculture was completely abolished. A few WTO-conforming agricultural subsidies were also introduced. Since its accession to the WTO, China has also tripled its level of expenditures for agricultural research. It is clear that China is trying to use structural policies to bridge the income gaps and thereby avoid falling into the trap of the high agricultural subsidies faced by OECD countries.

India's policy responses have also been stepped up. The 11th Five Year Plan (2007–2012) has put renewed emphasis on agriculture and rural development, with enhanced programmes for infrastructure, irrigation, research and extension, and special production programmes for food grains, horticulture and other sectors. The growth of agricultural credit has been stepped up. Self-help groups that organize and assist women with credit have been extended nationwide. The support prices have been extended to more and more crops and raised to higher levels. At the same time, irrigation, electricity and fertilizer subsidies have been maintained. Finally, the implementation of the Mahatma Ghandi National Rural Employment Guarantee Act (NREGA) has been extended nationwide. A Right to Food Act that would benefit poor people in rural and urban areas is in preparation.

Supporting agriculture and rural development during the process of structural economic transformation is not only important for political reasons. Successful agricultural and rural development has also been shown to be the essential element for successful poverty reduction during the process of transformation. This is because, while the share of agriculture in economic output is declining all along the growth path, agricultural output keeps increasing in absolute terms, although at a slower rate than economy-wide growth. In the process, agricultural growth can make a key contribution to poverty reduction, despite its declining share in output. Agricultural growth does this as follows, as Johnston and Mellor (1961) showed 50 years ago:

1. It raises agricultural profits and labour income.
2. It raises rural non-farm profits, and employment and labour income via linkage effects.
3. It leads to lower prices for (non-tradable) foods, which is especially beneficial for the poor.
4. Lower food prices raise real urban wages and accelerate urban growth.
5. A tightening of urban and rural labour markets raises unskilled wages in the wider economy.

The World Development Report 2008 summarizes the large body of literature that demonstrates how effective agricultural growth is in reducing poverty. During the previous decade to about 2005, global poverty, as measured by a two-dollar-a-day poverty line, declined by 8.7% in absolute terms. This decline was entirely attributable to the reductions in rural poverty, with agriculture as the main source of growth. At the same time, urban poverty increased. And contrary to the view that sees structural transformation of the economy away from agriculture as a major source of poverty reduction, migration has not been the main instrument for rural (or overall) poverty reduction in the recent past. This is consistent with Timmer's discussion above that during the past 50 years or so it has become more difficult to reach the point where productivities and incomes start to converge between the sectors.

While rural non-farm activities can be an engine of growth for rural development, most rural non-farm activities produce goods and services that are linked to agriculture via forward, backward and consumer-demand linkages (Hazell and Hagbladde, 1993; World Bank, 1983). The advantages of lower rural wages in terms of industrialization are frequently offset by the disadvantages of a rural location. Therefore, in most countries agriculture has been the most important driver of the rural non-farm sector over the past few decades.

In conclusion, successful poverty reduction depends not only on how much overall economic growth occurs, but also on whether or not it is based on rapid agricultural growth. Most of the world's 2.1 billion people who live on less than two dollars a day are found in rural areas and depend on agriculture for their livelihoods. However, while the number of rural poor decreased in East Asia and the Pacific (successful agricultural performers), it increased in South Asia and Africa. The next section will discuss India's performance.

9.2 The Structural Transformation in India, 1961–2009

9.2.1 Economic and agricultural growth

As shown in Table 9.1, GDP increased nearly tenfold between 1961 and 2009, accelerating to >5% from the 1980s, to 7% for the present decade and to >8% during 2006–2010. During the same period, per-capita income growth accelerated from around 3.3% to 5.8%. The agricultural sector's growth can be measured by output (an output index) or in value added that is also influenced by relative price changes. The two measures have risen by factors of 3.45 and 3.8, respectively. Except for during the 1960s, decadal growth rates by both measures were similar. It is well known that agriculture performed best during the 1980s when the Green Revolution was generalized across most of agriculture, and agriculture output (but not GDP) grew at nearly 4%. Since then, growth rates have varied between 2% and 3%.

Figure 9.1 shows the marked slowdown in agricultural growth in the early years of the century that coincided with the decline of terms of trade and some poor rainfall years. The recent recovery of agricultural growth is also visible in Fig. 9.1 and coincides with a marked improvement in agricultural prices (not shown), as well enhanced public expenditures under the

Table 9.1. Growth of the Indian economy, 1961–2009.

Indicator	Average, 1961/62 to 1963/64	Average, last three years	Growth rates in average annual per cent during period shown					
			1961–1970	1971–1980	1981–1990	1991–2000	2001–2009[a]	2006–2009[a]
GDP (constant 2000 Rupees, billion)	3,454	32,353	3.8	2.8	5.4	5.5	7.0	8.2
Per capita GDP (constant 2000 US dollars)	191	719	1.7	0.7	3.3	3.5	5.7	5.8
Gross agricultural output (constant 2000 US dollars, billion)	51	176	2.0	2.1	3.9	2.6	2.3	2.7
Agricultural GDP (constant 2000 Rupees, billion)	1,703	6,574	3.8	1.5	3.3	2.7	2.9	3.1
Terms of trade of agriculture versus non-agriculture (1990/91 = 100)					88	102.3	102.1[a]	

[a]GDP is measured through to 2010; terms of trade is measured from 1980 to 2006 only. Sources: Indian Ministry of Agriculture, except gross agricultural output which is from FAO.

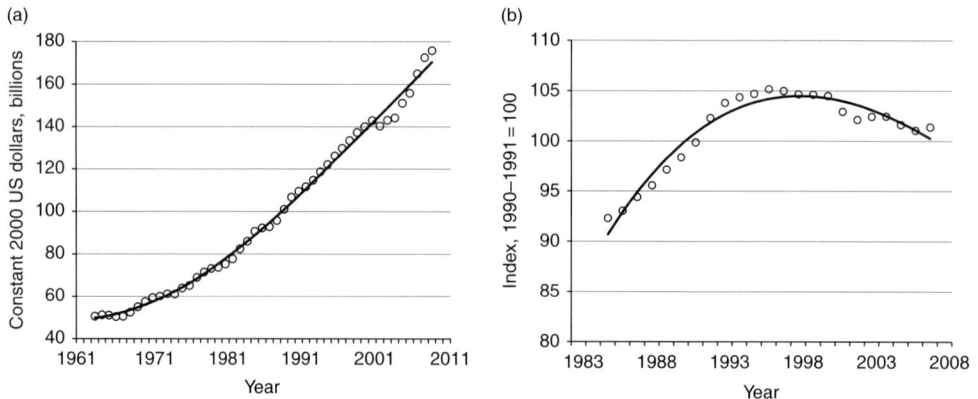

Fig. 9.1. Gross agricultural output and terms of trade for agriculture. All data are three-year averages ending in the year shown. The fitted curves are cubic. (a) Gross agricultural output. (b) Agricultural terms of trade. The terms of trade index is between agricultural and non-agricultural sectors. Source: Gross agricultural output is from FAO; agricultural terms of trade index is from the Ministry of Agriculture (2010).

11th Five Year Plan. Agricultural growth measured by GDP has grown at just over 3% during the last 5 years up to 2010.

Terms of trade in Indian agriculture are only reported from the 1980s to 2006. The slowdown of agriculture since the 1990s occurred despite the improvement in terms of trade from 88 in the 1980s to around 102 in both the 1990s and the 2000s (Fig. 9.1). The decadal averages hide the fact that, during the latter part of the 1990s, terms of trade turned against agriculture and had not yet started to recover by 2006. Since then, real food prices in India have increased, so there should be a significant recovery in terms of trade by 2010/2011 once the data are available.

During the past 5 years (2005–2010), several positive trends have emerged that might lead to an acceleration of the agricultural growth rate beyond 3%. Gross capital formation as a percentage of agricultural GDP rose from 13.46% in 2004–2005 to 20.3% in 2010–2011. Distribution of certified seeds more than doubled, from 1.13 million tons in 2004–2005 to 2.57 million tons in 2009–2010. Per-hectare consumption of fertilizers in nutrient kg increased from 112 kg in 2006–2007 to 135 kg in 2009–2010, and fertilizer subsidies changed to subsidies based on nutrients. A shift from the physical distribution of subsidized fertilizers to fertilizer coupons has been announced in the 2011–2012 budget (All data from the Economic Survey 2010–2011).

As shown in Table 9.2, the 1980s were the period of greatest total factor productivity (TFP) growth in agriculture – at a rate of 2.1%, including livestock, forestry and fisheries (as measured by Fuglie, 2008) – but that growth rate declined to <1.5% for the period 2001–2006, the latest for which there is an estimate. This is in contrast to China's TFP growth rate in agriculture of 4% in the 1990s and 3.2% in 2000–2006. As shown by Fuglie (2008), during the past 40 years, no other region or large country has ever sustained a TFP growth in agriculture of more than 3% for more than a decade.

9.2.2 The Structural transformation

The share of agriculture in Indian GDP fell from >40% in the early 1960s to around 17% by the end of the 2000s. From Fig. 9.2, it seems that the rate of decline in the agricultural share accelerated as the rate of economic growth increased. The share of industry as a whole rose from about 20% in 1960 to around 28% in 2009, whereas the share of manufacturing alone disappointingly stayed at around 15% during the entire period, again a sign of sluggish structural transformation. On the other hand, the share of services rose from <40% to around 55%.

Since the early 1960s, India's population has increased by a factor of nearly 2.5, to an average of 1181 million during 2006–2009 (Table 9.2). After peaking in the 1970s, the population growth rate declined in the 1990s to about 1.4% in the second half of that decade.

Table 9.2. Distribution of labour force and labour productivity growth across sectors.

Indicator	Average, 1961/62 to 1963/64	Average, last three years	Growth rates in average annual per cent during period shown					
			1961–1970	1971–1980	1981–1990	1991–2000	2001–2009	2006–2009
Total population (million)	472	1181	2.1	2.3	2.2	1.9	1.5	1.4
Agricultural labour force (million)	134	262	1.4	1.7	1.6	1.4	1.2	1.1
Non-agricultural labour force (million)	49	211	2.7	3.2	3.7	3.2	3.1	3.0
Agricultural labour share (%)	73	55	−0.4	−0.4	−0.7	−0.7	−0.8	−0.8
Agricultural output/ worker (constant 2000 US dollars)	380	672	0.6	0.4	2.3	1.2	1.1	1.5
Non-agricultural output per worker (constant 2000 US dollars)	752	3068	3.9	0.6	3.0	3.6	5.8	6.4

Source: FAO.

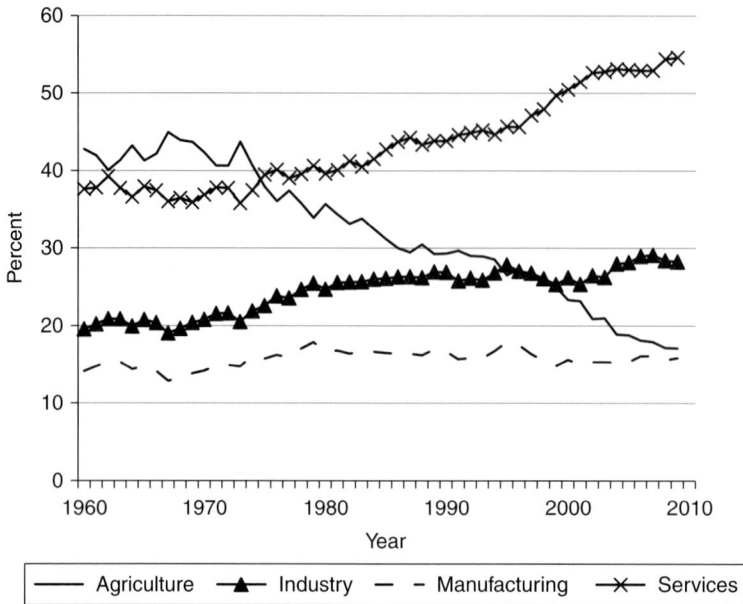

Fig. 9.2. Sector shares in the Indian economy, 1961–2010. Source: World Bank.

As a consequence of population trends, and despite the rapid change in the sectoral composition of the economy, the agricultural labour force rose from 134 million to 262 million, an increase of 95%. The annual rate of increase in the agricultural labour force has declined only very modestly since the 1970s and was still 1.1% during 2005–2009, whereas growth of the non-agricultural labour force accelerated to a little over 3%, almost three times as fast as that of the agricultural labour force. During the past 50 years, the share of labour in agriculture therefore declined from 73.2% to 55.3% in 2009, a decline of only around 24% of its former value (Fig. 9.3a). During the same period, the agriculture share of GDP in the economy declined from 41.7% in the early 1960s to 17.4%, a decline of 58% of its former value. Far from converging, the two shares are still on a diverging path that is even more striking for the two labour productivity measures.[2]

Non-agricultural labour productivity per worker rose from $752 in the early 1960s to $3068 in the late 2000s (constant 2000 US dollars), or by a factor of 4.1, with the growth rate accelerating in the early 1990s (Table 9.2; Fig. 9.3b). At the same time, agri-

cultural output per farm worker rose from $380 to $672 or by only 77%, and there was no trend of acceleration. The pace of annual productivity growth in non-agriculture accelerated from 0.6% in the 1980s to 5.8% during 2001–2009. This means the inter-sectoral productivity difference is increasing at an ever faster rate, and inter-sectoral convergence is further and further away.

It is not surprising that convergence in the Indian economy still has not started. Recall that Timmer (2009) showed convergence only starts at a per-capita income of between $1600 and $9000 (in 2000 US dollars), whereas Indian per-capita income in 2006–2009 averaged only $719. This wide gap does not provide good guidance about how soon convergence will begin to occur. We will have to examine country-specific factors to get a better idea of how far off might be the conversion point for India.

Rapid structural transformation should show up in the Indian data as a tightening of the rural labour market and an acceleration of rural–urban migration, and there are few signs that this is starting to happen. Although India's population growth has slowed down, labour force growth has

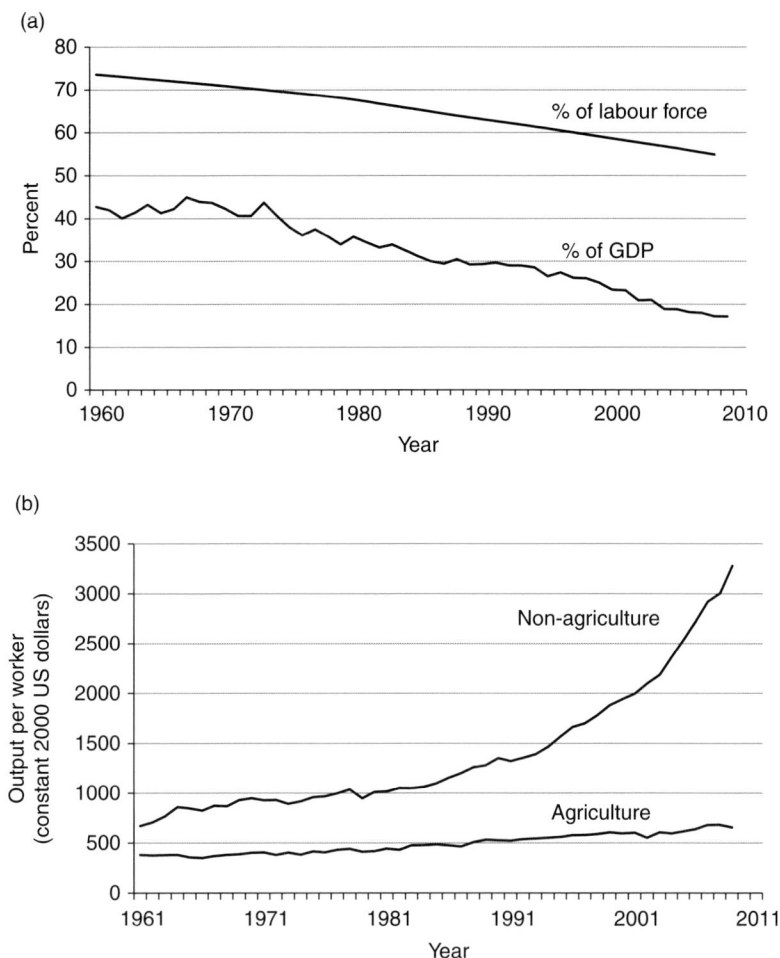

Fig. 9.3. Structural transformation in India. (a) Share of agriculture in labour force and GDP. (b) Labour productivity in agriculture and non-agriculture. Source: World Bank.

accelerated, and by the first half of the 2000s was 2.8%. Slow agricultural output and productivity growth have meant a slow-down in agricultural employment genera-tion (World Bank, 2011). The growth rate of agricultural wages declined from 1980 to the mid-2000s, but has started to increase recently. By all measures of unemployment, it has always been lower for rural than for urban areas, and higher for females than for males in both areas, with urban females fac-ing the greatest employment challenges. During the 2000s, unemployment increased slightly for rural males and more sharply for rural females. Unemployment (measured by

current daily status) among agricultural labour households was significantly higher than the overall unemployment rate. It rose from 9.5% in 1993–1994 to 15.3% in 2004–2005, twice as high as in the early 1980s. During the first decade of this century, labour force participation rates for both rural males and rural females, and for all definitions, barely changed. However, rates increased slightly for urban males and more so for females.

The data and projections suggest a con-tinuing and daunting challenge of employ-ment generation overall and especially in rural areas. Rural population and the labour force have grown rapidly and are expected to continue to grow. Hazell *et al.* (2011) cite UN population projections that suggest the rural population will peak at 900 million in 2022. They then use United Nations Development Programme labour force projections, and differences in rural and urban fertility rates and age structures to project that the rural labour force might continue to grow until 2045. It will, therefore, be a long time before the need for rural employment generation starts to slow down.

9.2.3 Rural and urban incomes and poverty and their drivers

As discussed in World Bank (2010), India's poverty rates did not start to decline until the mid-1970s, first in urban areas and then in rural areas. Using poverty rates according to the Lakdawala methodology, Datt and Ravallion (2009) show that, although poverty reduction accelerated somewhat between the early 1990s and 2004–2005, the change in the rate of decline was not statistically significant. Because of population growth, the rate of decline in the number of poor people was only 0.3%. Using the new poverty rates according to the Tendulkar committee (Planning Commission, 2009), the headcount poverty rate in India declined from 45.3% in 1993–1994 to 37.2% in 2004–2005 then to 32.2% in 2009–2010 (tentative estimates made by Ravi using grouped data cited in Ahluwahlia, 2011).

As seen in Table 9.3, the rural poverty rate declined from 50.1% in 1993–1994 to 31.8 % in 2004–2005, or by 18.3 %, whereas urban poverty declined from 41.8% to 25.7%, or by 6.1%. Therefore, in absolute

terms, the decline in rural areas was larger than in urban areas, but in relative terms the rate of poverty decline in urban areas has been slightly faster than in rural areas.

The trends in consumption poverty are consistent with the population's perceptions about whether or not their lives are getting better (World Bank, 2010). The prevalence of self-reported hunger has also declined during the past 30 years: the percentage of individuals reporting inadequate food intake in rural areas fell from 17% to 3% between 1983 and 2004–2005. The reduction in poverty rates and self-reported hunger did not, however, lead to significantly improved nutrition indicators for adults and children (what has been referred to as 'the Indian Enigma'). Moreover, the consumption of calories, protein and other nutrients declined, with higher declines in rural areas than in urban areas. A possible explanation of rising incomes, declining poverty rates and declining self-reported hunger on the one hand and declining calorie consumption on the other is that increased mechanization of agricultural ploughing and transport has reduced calorie requirements (Deaton and Drèze, 2009).

In 1983, rural and urban inequality was the same at a Gini Coefficient of 0.30 (Table 9.4). From 1983 to 2009-2010, rural inequality barely changed. However, urban consumption inequality, after being stable in the decade between 1983–1984 and 1993–1994, rose between 1993 and 2004, and then stayed constant to 2009–2010 at 0.37. So far, the increase in consumption inequality seems to have been an urban rather than a rural phenomenon.

Data on rural/urban income disparities show that the ratio of urban-to-rural per-capita income declined from 2.45 in

Table 9.3. Changes in rural and urban poverty rates.

	Rural	Urban	Difference
Period	(Percentage of population below the poverty line)		
1993–1994	50.1	31.8	18.3 = 45%[a]
2004–2005	41.8	25.7	16.1 = 48%[a]

[a]Calculated with respect to the mean percentage over the period. Source: Planning Commission (2009; Tendulkar Report).

1970–1971 to 2.3 during the 1980s and early 1990s. On the other hand, the data on consumption shown in Table 9.4 suggest that the ratio of urban consumption to rural consumption increased from 1.54 in 1983 to around 1.7 during the latter half of the 2000s. Whether rural/urban disparities have increased on average is therefore dependent on the data used and the period considered, but neither data series suggests a very sharp change in urban/rural disparities during the past 25 to 30 years.

Given the significant increases in the non-agricultural to agricultural productivity differential, it is surprising that the urban/rural per-capita income and consumption gaps have not increased sharply, and that the gap between rural and urban headcount poverty rates has not increased sharply as well.

Ravallion and Datt (1996) showed that, for the decades prior to 1991, rural growth was the most important driver of poverty reduction and reduced rural, national and even urban poverty. Urban growth reduced only urban poverty and had no impact on rural or national poverty. Datt and Ravallion (2009) updated the earlier study to 2004–2005. It showed that rural growth remains significant for reducing rural and national poverty. But since 1991, urban growth has become the major driver not only of urban poverty reduction, but of both national and rural poverty reduction as well. In the post-1991 period, the elasticities of the headcount index of national and rural poverty with respect to urban growth had increased (in absolute value) to −1.21 and −1.26, respectively, whereas those with respect to

rural growth had fallen to −0.66 to −0.90. A rising impact of urban growth on national and rural poverty is consistent with evidence over time and across countries, suggesting that a higher pace of urbanization can bring overall progress against poverty, although the incidence of urban poverty may well increase in the process (Ravallion et al., 2007).

The results above are concerned with the impact of urban and rural growth on poverty, with rural growth including agricultural and rural non-farm growth. A closer look at the impact of agricultural growth shows it has not lost its potential to reduce rural poverty. Himanshu et al. (2010) used econometric analysis to investigate the determinants of non-farm growth. Higher yields are associated with declining rural poverty, suggesting that the impact of agricultural production growth on poverty remains high. There is also a strong and negative impact of higher agricultural wage growth on rural poverty, consistent with the strong agricultural impact on rural poverty.

9.2.4   The rural non-farm sector: why the inter-sectoral productivity differential did not lead to a growing poverty differential

Despite sluggish urban absorption of labour, the most likely explanation that income, consumption and poverty differentials did not rise sharply as a consequence of the rapidly diverging non-agricultural to agricultural productivity differential is the

Table 9.4. Consumption inequality, India (1983–1984 to 2009–2010).

	Gini Coefficient of distribution of consumption				
	1983	1987–1988	1993–1994	2004–2005	2009–2010
Rural	0.3	0.3	0.28	0.3	0.28
Urban	0.3	0.35	0.34	0.37	0.37
Urban–rural ratio of mean consumption (Constant prices)[a]	1.54	1.44	1.64	1.72	1.69

[a]Original shows urban–rural ratio. Source: Ahluwahlia (2011, Table 6).

emergence in India of a dynamic rural non-farm sector that no longer seems to be driven just by agricultural linkages. Table 9.5 shows that between 1983 and 2004 the non-agricultural sector in rural areas grew significantly more rapidly than agricultural growth and even faster than economy-wide growth. This has been associated with an acceleration of non-farm employment growth. Agricultural wages grew at an average rate of 3.2% during the two decades from 1983 to 2004, but the growth rate in wages slowed down in the second decade of this period.

In the 2000s, employment growth in rural areas has come primarily from an increase in rural non-farm employment. In the 1980s, four out of ten rural jobs were in the non-farm sector. By the end of the 2000s it was six out of ten (Himanshu *et al.*, 2010). The growth in non-farm jobs has come primarily from increases in services, transport and construction. Nearly half (48%) of the average rural household's income comes from non-farm earnings (Dubey, 2008).

In 1983, close to 40% of rural non-farm jobs were in manufacturing. That share declined to just a little above 30% by the end of the 2000s. In 1983, social services, trade, transport and communications generated about 26% of non-farm jobs. Since then, social services have declined to about 18% of the jobs, whereas trade, transport and communications have grown rapidly to about 33%. In 1983, construction was by far the smallest sector, with a share of only 10%. Since then, it has grown the fastest, generating close to 19% of rural non-farm jobs. The high level of rural construction has visually transformed villages all over

India with much better village infrastructure and housing.

Within the rural non-farm sector, employment growth has largely been in the informal sector. A particularly dynamic development has been the growth in self-employment in the non-farm sector, especially by members of farming households, who diversified not only within agriculture but into the non-farm sector. Over time, employment growth in the non-farm wage sector has accelerated, whereas growth in average earnings has decreased. These two trends have cancelled each other out, and growth in total non-farm earnings has been constant for the past two decades (World Bank, 2010, p. 67).

Among the age cohorts, it is primarily 18–26 year-old males who are moving out of agriculture into non-farm jobs (Eswaran *et al.*, 2009). Education is an important determinant of access to non-farm occupations. Women are barely transitioning into the non-farm sector. The percentage of males working primarily in non-farm activities increased from 25% in 1983 to 35% in 2004–2005, but for women the increase over the same period was from 15% to 19%. In growth terms, the number of rural men working off-farm doubled between 1983 and 2004–2005; for women the increase was 73%. Individuals from scheduled castes and tribes are markedly more likely to be employed in agricultural labour than in non-farm activities, even controlling for education and land. Even a small amount of education, such as achieving literacy, improves the prospect of finding non-farm employment and with higher levels of

Table 9.5. Trends in agricultural wages and national, rural non-farm and agricultural GDP.

	Agricultural wages	Non-farm employment	National GDP	Non-farm GDP	Agriculture GDP
Period	(Average annual rate of growth, %)				
1983–2004	3.2	3.3	5.8	7.1	2.6
1983–1993	3.2	3.5	5.2	6.4	2.9
1993–2004	1.7	4.8	6.0	7.2	1.8

GDP at factor cost at 1993–1994 prices. Agriculture GDP originating in agriculture, forestry, and fishing. Non-farm GDP defined as a residual. Source: Himanshu *et al.* (2010) and Eswaran *et al.* (2009).

education the odds of employment in well-paid, regular non-farm occupations rise. Finally, those in the non-farm sector own more land on average than do agricultural labourers, except for those in casual non-farm employment, who typically own significantly less.

As discussed above, agricultural wages have risen, but the growth rate of agricultural wages between 1993 and 2004 slowed to only 1.7%. The decline in wage growth was especially sharp for females. The growth rate of average real wage rates in non-agricultural employment was negligible in the period 1999–2000 to 2004–2005 (World Bank, 2010, pp. 63–64). There is a substantial and rising premium in the casual non-farm wage over the agriculture wage. The premium rose from 25–30% in 1983 to about 45% in 2004–2005 (Himanshu et al., 2010).

In spite of the preponderance of non-farm jobs in rural employment generation, Eswaran et al. (2008) estimate the contribution of the rural non-farm sector to rural wage growth to be only about 22% of the total growth, thereby confirming the importance of agricultural productivity growth to rural wage growth. In particular, the rural non-farm sector has not contributed to wage growth among the illiterate, but only among the more educated (Eswaran et al., 2009), a topic discussed further below.

Eswaran et al. (2009) used National Sample Survey data to show that wage premiums associated with education grew over time. By the 2004–2005 National Sample Survey, these premiums had increased to 86 Rupees (Rs) for literate workers over illiterate ones, Rs 197 for those who had attended middle school, and Rs 696 for high school graduates. The authors concluded that if more middle school and high school graduates were available in 2004 they would have found employment in rural industry and services. Therefore, they concluded that the main reason why the non-farm sector has not contributed more to poverty reduction is because most of the employment it creates is for educated workers rather than illiterates and primary school graduates.

Within the rural non-farm sector, non-farm self-employment has become a major source of job growth. Because of the absence of income data for the sector, the question of how remunerative such self-employment is remains unanswered. Data from the 2007 round of the Rural Economic and Demographic Survey of the National Council of Applied Economic Research bridged the data gap for the 238 mainly agricultural villages it surveyed nationally. Between 1999 and 2007, in only 8 years, the portion of households in these villages that had self-employment in the non-farm sector rose from 10.1% to 19.7% and the net incomes they derived from these enterprises rose by 70%. By 2007, these self-employment incomes were now slightly higher than the profits of farming households (with which many of the non-farm self-employment households overlap). Within the entire village economy, the non-farm self-employment share of income rose from 7.3% to 19.6% during 1999–2007, whereas the agricultural profit share declined. At the same time, the share of non-farm wage income stayed nearly constant at around 7.5%. Clearly, non-farm self-employment has become a far more dynamic sector of the rural economy than had generally been assumed.

As discussed previously, Himanshu et al. (2010) used econometric analysis to investigate the determinants of non-farm growth. Higher yields are associated with declining rural poverty, suggesting that the impact of agricultural productivity growth on poverty reduction remains high. Higher agricultural wage growth also has a strong impact on rural poverty reduction. When state fixed effects are used, non-farm employment is positively correlated with rural poverty. This pattern is consistent with the finding of Foster and Rosenzweig (2005) that non-farm enterprises producing tradable goods (the rural factory sector) locate in settings where reservation wages are lower. If the rural factory sector seeks out low-wage areas, factory growth will be largest in areas that have not experienced local agricultural productivity growth. Thus, rural non-farm growth reduces spatial inequalities in economic opportunities and incomes. However, it is also consistent with distress-induced recourse to non-farm employment. Nevertheless, poverty decline is most rapid in regions where the non-farm sector has grown. Thus, the location of factories where wages are low has an equalizing impact on income distribution in rural areas.

So, the first major new finding is that the rural non-farm sector has compensated for slow agricultural growth nationally, and also in areas with poor agricultural potential. Second, the dynamic of production and consumption linkages with agriculture is no longer the only driver of non-farm sector growth. Rather, the non-farm sector has also expanded in regions where agriculture is in decline or where agricultural wages are low (World Bank, 2010, p. 67). Agriculture and agricultural productivity growth nevertheless remain more important drivers of rural poverty reduction than does the rural non-farm sector, probably because they alone can employ the lesser skilled labour force – in particular women unable to make the transition to the non-farm sector. The picture that emerges shows the rural non-farm sector as an important source of structural transformation in the rural economy, and therefore in India as a whole, and that agricultural growth remains a powerful force

for structural change and poverty reduction. The growing importance of the rural non-farm economy will also require a reformulation of the geographic dimension of the structural transformation. Until recently, it was equated with rural/urban differences and migration from rural to urban areas, but now it will have to include occupational changes from the farm to the rural non-farm sector.

9.2.5 Structural change in Indian agriculture

As shown in Fig. 9.4, the sectoral composition of production in agriculture has changed only slowly over time. Crops still made up 72.5% of value added in 2009, down from 79.2% in 1961, and livestock products increased commensurately from 20.9% to 27.5%. Within the crop sector, cereals declined by only 3.5% to 23.1% of agricultural value added. The disappointing performances of oilseeds and pulses are reflected in a decline in their shares of value added from 14.8% to 9.4%. All other crops (sugar, cotton, plantation crops, spices and other industrial inputs) also declined in importance, from 24.4% to 18.6%. Horticulture, on the other hand, compensated for all the

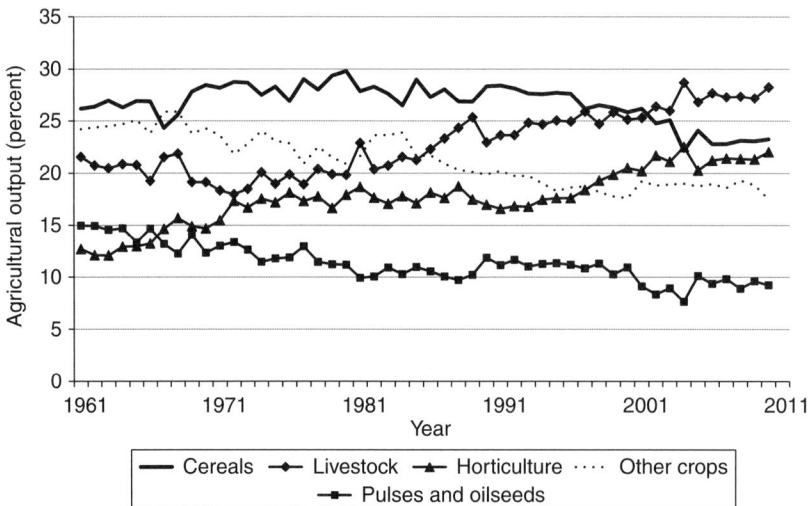

Fig. 9.4. Changing composition of agricultural output in India. Source: FAO.

other declines by sharply increasing its share from 12.4% to 21.5%, a higher absolute increase in share than even livestock products. Rising incomes have clearly shifted consumer demand significantly to horticultural and livestock products.

Horticultural production increased from 170.8 million tonnes in 2004–2005 to 214.7 million tonnes in 2008–2009, i.e. by 26% during those 5 years. As a consequence, per-capita availability of fruits and vegetables increased from 391 g/ day to 466 g/day. India is, however, unable to keep pace with the growth in demand for milk. Demand is growing by about 6 million tonnes annually, whereas supply is growing by only 3.5 million tonnes per year.

During the last half century, since 1961, fluctuations in the rate of change of output shares in agriculture were common, and it does not seem that changes in the sectoral composition of output were concentrated in specific time periods. Despite the sharp acceleration in economic growth, it is unclear whether changes in the composition of agricultural value added have accelerated between the 1990s and the past 4 years for which we have data (2006–2010).

Table 9.6 shows the rate of growth of crop output per hectare of cropland, livestock output per cattle equivalent, and TFP, all of which peaked in the 1980s with crop productivity and TFP growth declining again in the 1990s. Both India and China had no TFP growth in the 1960s and 1970s. China's TFP growth exceeded or was close to 3% in the 1980s, reached 4.2% in the 1990s and was about 3% from 2000 to 2007. This means it has had TFP growth at, close to or far greater than 3% for nearly 30 years, by far the world's longest period of such rapid growth in TFP.

The annual TFP data provided by Fuglie (2010) show that agricultural TFP in India grew at average annual rates fluctuating between 0.5% and 1.4% during the past four decades of the 20th century. It then accelerated to 1.7% over 2001–2009 and even up to 2.4% during 2006–2009, a remarkable acceleration similar to China's TFP growth rates over the past 30 years. This could be an indication that India has overcome its slump in agricultural productivity growth.

As shown in Table 9.6, output per hectare of cropland over the period increased by a factor of 3.3 from $315 to $1044, whereas livestock output per head of cattle equivalent increased by a factor of 3.5 from $43 to $148 (constant 2000 US dollars). Clearly, livestock productivity growth in the Indian economy has been slightly more impressive than crop productivity growth. Because Indian agricultural policy debates focus largely on crops, and especially on

Table 9.6. Productivity change in Indian agriculture, 1961–2009.

Indicator	Average, 1961/62 to 1963/64	Average, last three years	Growth rates in average annual per cent over period shown					
			1961– 1970	1971– 1980	1981– 1990	1991– 2000	2001– 2009	2006– 2009
Agricultural output per hectare of cropland (constant 2000 US dollars)	315	1,044	1.7	1.9	3.8	2.4	2.6	3.1
Animal output/head of cattle equivalent (Constant 2000 US dollars)	43	148	0.7	2.9	3.5	3.1	3.3	3.4
Total factor productivity growth, India			0.5	0.8	1.4	1.2	1.7	2.4
Total factor productivity growth, China			1.3	0.4	1.7	4.1	2.7	3.0

Source: Unpublished data based on Fuglie (2010).

cereals, the increase in livestock productivity is rarely appreciated.

As shown in Table 9.7, expansion of cropland ceased in the 1970s, after which it stayed practically constant at 162 million ha (and even declined slightly in the 2000s). During the 1961–2009 period, the agricultural labour force grew by 95% to 262 million, which means cropland per worker declined from 1.2 ha to 0.64 ha. Because the population growth rate is still 1.4% and rural–urban migration is relatively slow in India, the agricultural labour force keeps growing at around 1% and the decline in land per worker will undoubtedly continue for a long time. Despite the declining land–labour ratio, agricultural output per worker has grown. Labour productivity growth has been driven more by agricultural growth than by rural–urban migration. It is, therefore, not so much a consequence of structural transformation but of transformations within agriculture.

The transformation of agriculture is also seen in the use of irrigation, machines and fertilizers (Table 9.7). During the period shown, the share of cropland irrigated has more than doubled to 36.7%, but this growth has nearly stopped since 2000. Electricity consumption grew fastest during the 1980s, and slowed markedly during the early 2000s, only to resume its growth since 2006. Fertilizer use (in terms of NPK nutrients) expanded from almost none to about 140 kg/ha of cropland, with the growth rate tapering off in the 1990s and accelerating again in the 2000s. The number of tractors rose from <1 to 14 per 1000 ha of cropland, with the growth rate slowing down steadily from >12% per year in the 1960s and 1970s to near zero during 2006–2009. The recent slow growth of irrigation and of tractor use is cause for alarm.

9.3 Conclusions

During the half century since 1961, the share of labour declined from 73.2% to 55.3%, a decline of only around 25% of its former value. The agriculture share of GDP in the economy declined from 41.7% the early 1960s to 17.4% during 2006–2009, a decline of 58% of its former value. The declines were steepest from the 1970s to the 1980s and from 1991 to the 2006–2009 period. The lack of sectoral convergence is mirrored in the continued divergence of labour productivity between agriculture and non-agriculture that has accelerated as a consequence of recent rapid non-agricultural growth and slow agricultural growth. The productivity differential between labour in the non-agricultural and agricultural sectors has been widening, and now stands at a ratio of 4.2. This has been a common pattern across the world during the early parts of structural transformation between agriculture and non-agriculture. A turning point, where the productivity differential between the agricultural sector and non-agricultural sectors might start to converge, and where the agricultural labour share in the economy would decline faster than its share in GDP, is likely to be some distance away.

As consequence of rapid economic growth, poverty has continued to decline in both the rural and urban sectors, and the gap between the urban and rural poverty rates has narrowed. Real rural and urban wage rates have tended to grow at similar rates, suggesting that rural and urban labour markets are integrated. In particular, in the period from 2004–2005 to 2009–2010, rural and urban male wage rates grew at 4.5% and 4.2%, respectively. Since the early 1970s, the urban–rural per-capita income ratio has been declining slowly from 2.45 to 2.30, whereas since 1983 the corresponding ratio of consumption has been slowly increasing from 1.54 to 1.70. In the face of the widening gap in labour productivities between agriculture and non-agriculture, it is most surprising there have been no sharp increases in the disparities between rural and urban income, consumption, wages and poverty, a set of trends that requires explanation.

Our analysis suggests it is because the non-farm sector, rather than agriculture, has emerged as the dominant source of employment and income growth in the rural economy. We find a strong trend of farmers adding rural non-farm self-employment

Table 9.7. Yields and inputs in Indian agriculture, 1961–2009.

Indicator	Average, 1961/62 to 1963/64	Average, last three years	Growth rates in average annual per cent during period shown					
			1961–1970	1971–1980	1981–1990	1991–2000	2001–2009	2006–2009
Cropland (million ha)	162	168	0.3	0.2	0.1	0.2	−0.4	−0.4
Agricultural labour force (1000 persons)	134	262	1.4	1.7	1.6	1.4	1.2	1.1
Cropland per agricultural worker	1.2	0.64	−1.1	−1.5	−1.5	−1.2	−1.6	−1.5
Agricultural output per worker (constant 2000 US dollars)	380	672	0.6	0.4	2.3	1.2	1.1	1.5
Crop output per ha of cropland (Constant 2000 US dollars)	246	734	2.0	1.4	3.5	2.1	2.2	2.8
Share of cropland irrigated (%)[a]	15.6	39.0	2.1	2.2	1.9	2.7	1.2	0.8
Cropping intensity[b]	115	138	0.2	0.4	0.4	0.5	0.4	1.1
Electricity consumption (Kwh per ha)		571			12.9	7.1	1.9	5.4
Fertilizer use per hectare (kg of NPK)[a]	3.3	140	16.0	10.4	8.2	3.2	5.2	6.7
Tractors per 1000 ha of cropland[a]	0.2	15	12.7	13.2	9.4	7.4	2.6	0.0
Real credit from formal sources (Rs per ha of cropland)[a]		20,633		8.7	4.3	4.6	12.1	5.1

[a]Until 2008; [b]until 2007. *Source*: FAO and Ministry of Agriculture (2010).

income to their total incomes. A part-time farming model has emerged as the most successful approach to household income growth in the rural economy. Structural change is taking place in the rural economy but not in the form expected by the standard structural transformation model, where an increase in agricultural productivity allows for labour to be pulled into the urban economy at an accelerating rate.

The rural non-farm sector, therefore, has and will continue to provide a major source of rural income and poverty reduction for farmers and non-farmers alike. Although it continues to be driven by agriculture, it now also grows in response to opportunities arising from urban growth. Agricultural growth remains, however, an important driver of rural wages and rural poverty reduction because it alone can significantly improve the incomes of the poorest group in the Indian economy, the casual rural workers with little education and few chances to migrate or find employment in the non-farm sector.

Rural population is projected to continue to grow until at least the early 2020s if not longer. Because of the demographic dividend, it will take much longer for the rural labour force to peak. Slow agricultural output and productivity growth have meant a slowdown in agricultural employment generation. Most employment growth in the Indian economy has come from the rural non-farm sector, rather than the urban economy. Rural–urban migration remains sluggish in India, and urban labour absorption favours educated over unskilled labour. After growing steadily at around 2% for 20 years, the all-India labour force growth will start declining to about 1.6% from 2011–2012 to 2016–2017. In spite of the decline, the additional work force during the next 20 years will still be slightly more than 40 million. This is larger than labour force growth in the entire industrialized world and even that of China. The challenge of labour absorption for the urban economy, the farm and the rural non-farm sectors will remain very high.

Real agricultural output growth declined from the 1980s to around 2005. Terms of trade increased in the 1980s up to 1995 but the increase does not seem to have accelerated agricultural growth. Consistent with the decline in the agricultural growth rate, the growth rates of both labour productivity and TFP have slowed down over the 1990s and 2000s, with the TFP growth rate declining from a peak of 2.1% in the 1980s to only 1.4% between 2000–2001 and 2006–2007. This is in sharp contrast to China, which has experienced TFP growth of around 3% in the same period. The slowdown in productivity was aggravated by a slowdown in public and private investment in agriculture that has only been fully reversed in the current 11th Five Year Plan period. This means agricultural growth has still not responded to higher economic growth.

There are, however, a few positive indicators that could predict an acceleration in agricultural growth:

1. Since 2005 there has been an acceleration in agricultural capital formation;
2. The distribution of improved seeds has nearly doubled in the same period, along with higher fertilizer use per hectare; and
3. Credit volumes (per hectare) have increased rapidly.

Although these are positive signs, they have yet to translate into sustained higher growth rates.

Both productivity per animal and aggregate crop yields have grown impressively during 1961–2009, although policy attention has focused primarily on food grains. The growth in animal productivity, a little noticed factor of major importance, has been consistently under-appreciated.

Accelerating per-capita income growth in India has altered the sectoral composition within agriculture. After peaking in 1981, the share of cereals in agricultural GDP had fallen to around 23% by 2009. All other commodity groups except horticulture and livestock have also seen declining shares, with the most pronounced losses for pulses, oilseeds and 'other' crops. The major gainers have been vegetables, fruits and livestock products (including fish); these commodities now account for nearly half of value added in Indian agriculture.

Notes

[1] Timmer's estimate of a turning point is not stable across specifications and sub-samples. When the specification includes a variable for the agricultural to non-agricultural terms of trade, the turning point is estimated at US$5000; and when using only the Asian sample of countries, he finds a turning point for these countries at a per-capita income of US$1600. McMillan and Rodrick (2011) also investigate the question of the turning point. They use a sample of 38 developed and developing countries from 1990 to 2005, and regress the ratio of agricultural to non-agricultural labour productivity on the economy-wide labour productivity in purchasing power parity dollars (PPP) of 2000. They find a turning point towards convergence at $9000 PPP, which is between the 2005 economy-wide labour productivities of India ($7700 PPP) and China ($9518 PPP). They find too that Asia has been better at moving to conversion than Latin America and Africa.

[2] Timmer and Akkus (2008) show similar graphs for other Asian countries that suggest the following for the last decade of the 20th century: Sri Lanka, the Philippines and even China continued on a divergent path and the gap between output and labour shares stayed constant in Indonesia, Thailand and Nepal, whereas the gap had started to fall in Bangladesh, Malaysia and Pakistan.

References

Ahluwahlia, M. (2011) Prospects and policy challenges in the Twelfth Plan. *Economic and Political Weekly* XLVI, 21.

Datt, G.V and Ravallion, M. (2009) Has India's economic growth become more pro-poor in the wake of economic reforms? Policy Research Working Paper 5103, World Bank, Washington, DC.

Deaton, A. and Drèze, J. (2009) Food and nutrition in India: Facts and interpretations. *Economic and Political Weekly* XLIV, 7, 42–65.

Dubey, A. (2008) Consumption, income and inequality in India. Background paper prepared for India Poverty Assessment Report. World Bank, Washington, DC.

Eswaran, M., Ramaswami, B. and Wadhwa, W. (2008) How does poverty decline? Suggestive evidence from India, 1983–1999. Bread Policy Paper No 14. Available at: http://ipl.econ.duke.edu/bread/papers/policy/p014.pdf

Eswaran, M., Kotwal, A., Ramaswami, B. and Wadhwa, W. (2009) Sectoral labour flows and agricultural wages in India, 1983–2004: Has growth trickled down? *Economic and Political Weekly*, 44, 2.

FAO. FAOSTAT Agricultural Database. Food and Agriculture Organization of the United Nations, Rome.

Foster, A. and Rosenzweig, M. (2005) Agricultural development, industrialization and rural inequality. Working Paper, Department of Economics, Brown University, Providence, RI.

Fuglie, K. (2008) Is a slowdown in agricultural productivity growth contributing to the rise in commodity prices? *Agricultural Economics* 39, supplement, 431–441.

Fuglie, K. (2010) Total factor productivity in the global agricultural economy: Evidence from FAO data. In: Alston, J., Babcock, B. and Pardey, P. (eds) *The Shifting Patterns of Agricultural Production and Productivity Worldwide*. Midwest Agribusiness Trade and Research Information Center (MATRIC), Iowa State University, Ames, IA.

Hazell, P. and Hagbladde, S. (1993) Farm-nonfarm growth linkages and the welfare of the poor. In Lipton, M. and van der Gaag, J. (eds) *Including the Poor*. The World Bank, Washington, DC.

Hazell, P., Headey, D., Nin-Pratt, A. and Byerlee, D. (2011) Structural Imbalances and Farm and Nonfarm Employment Prospects in Rural South Asia. Background Report for the World Bank, Washington, DC.

Himanshu, P., Mukhopadhyay, A. and Murgai, R. (2010) Non-farm diversification and rural poverty decline: A perspective from Indian sample survey and village study data. Asia Research Centre Working Paper 44, London School of Economics, UK.

Johnston, B. and Mellor, J. (1961) The role of agriculture in economic development. *American Economic Review* 51, 566–593.

McMillan, M. and Rodrick, D. (2011) *Globalization, Structural Change, and Productivity Growth*. NBER Working Paper Series Vol. W147143, Cambridge, MA.

Ministry of Agriculture (2010) Agricultural statistics at a glance 2010. Directorate of Economics and Statistics, Department of Agriculture and Cooperation, Ministry of Agriculture, Government of India, New Delhi, India.

Planning Commission (2009) Report of the Expert Group on Methodology for Estimation of Poverty (Tendulkar Report), Government of India, New Delhi, India.

Ravallion, M. and Datt, G. (1996) How important to India's poor is the sectoral composition of economic growth? *World Bank Economic Review* 10, 1–26.

Ravallion, M., Chen, S. and Sangraula, P. (2007) New evidence on the urbanization of global poverty. *Population and Development Review* 33, 667–702.

Timmer, C.P. (2009) A world without agriculture: The structural transformation in historical perspective. American Enterprise Institute, Washington, DC.

Timmer, C.P. and Akkus, S. (2008) The structural transformation as a pathway out of poverty: Analytics, empirics, and politics. Working paper 15, Center for Global Development, Washington, DC.

World Bank (2010) *Perspectives on Poverty in India: Stylized Facts from Survey Data.* World Bank, Washington, DC.

World Bank. World Development Indicators. World Bank, Washington, DC.

10 Shifting Sources of Agricultural Growth in Indonesia: A Regional Analysis

Nicholas E. Rada and Keith O. Fuglie
Economic Research Service, US Department of Agriculture, Washington, DC

10.1 Introduction

Indonesia is a large country with diverse agriculture. One dimension of this diversity is sharp regional differences in relative resource endowments, which range from very densely populated 'inner' Indonesia, defined here as Java and Bali, to relatively land-abundant 'outer' Indonesia, including the islands of Sumatra, Kalimantan and Sulawesi. Another dimension of its agricultural diversity is its broad commodity mix. Although rice-based farming is the principal agricultural system – accounting for about half of all agricultural output – Indonesia also has extensive areas in tropical perennial or plantation crops, especially oil palm, rubber, cocoa, coconut and coffee. Recently it has emerged as a major net agricultural exporting nation, although it continues to experience a trade deficit in strategically important food grains. A majority of its agricultural exports are from tropical perennials. These two dimensions of diversity are not independent, because most of the country's rice crop is produced on the inner islands whereas perennial crops are primarily grown on the outer islands. Both geographic segments, however, do have significant resources in both annual and perennial crops, as well as in livestock production. Examining Indonesia's record

of agricultural performance might provide important lessons for other developing countries that are facing questions about the relative policy emphasis to give food crops versus non-food crops, or to land-scarce regions versus land-abundant regions, and what shape these policies might take.

Past research provides several clues on the sources and direction of Indonesia's agricultural productivity growth. Fuglie (2004) found that the 'Green Revolution'[1] period of the 1970s and 1980s saw a temporary lift in the rate of total factor productivity (TFP) growth, but this had largely petered out by the early 1990s. Evidence of renewed agricultural TFP growth in the late 1990s and early 2000s was found by Fuglie (2010b) who suggested that the growth sourced from a changing agricultural structure rather than yield growth, that is from a gradual shift in output shares to higher-valued perennial, horticultural and livestock commodities and away from food staples. Rada, Buccola and Fuglie (2011) – hereafter referred to as RBF – confirmed that the primary source of agricultural productivity growth in post-Green Revolution Indonesia had shifted away from food crops such as rice and toward export-oriented tropical perennial crops such as oil palm. They trace this growth to a gradual liberalization of the Indonesian economy, including currency devaluations, which started in the 1980s in an effort to diversify Indonesian exports. However, unlike Salmon (1991) and Evenson *et al.* (1997), who found high returns to agricultural research and development (R&D), RBF did not find strong evidence that national investments in agricultural R&D were contributing much to output growth and conjectured that international technology transfer could be a significant source of productivity-enhancing new technology for Indonesian agriculture.

In this chapter we extend the work of RBF by examining in more detail the roles of domestic and international R&D as sources of growth in Indonesian agricultural TFP. Using the same provincial-level agricultural output and input data set, we first estimate Fisher-ideal indexes for agricultural TFP by province over the 1985–2005 period. We then try to explain TFP changes in terms of policy reform, national and international agricultural research, general public spending on agriculture and water resource development, as well as local population density (to account for Boserupian productivity change). We introduce explicit measures of international R&D or technology transfer to Indonesia in terms of the share of the area planted in modern rice varieties originating from the International Rice Research Institute (IRRI) and technical change in Malaysian oil palm. As late as 1993, as much as 65% of the area planted in modern, high-yielding rice varieties in Indonesia was devoted to varieties bred at IRRI (Darwanto, 1993). And Malaysian plantation companies have made major investments in the Indonesian oil palm sector, including the transfer of improved, high-yielding oil palm seed stock (Pray and Fuglie, 2002). Furthermore, we distinguish between domestic R&D investments in plantation crops from R&D in other agricultural commodities. Fuglie and Piggott (2006) describe how the national agricultural research system in Indonesia organizes, funds and manages these research components separately: plantation crop research is the responsibility of the Indonesian Planters Association for Research and Development, whereas research on annual crops, livestock and non-commodity issues, such as natural resources and agricultural policy, is the responsibility of the Indonesian Agency for Agricultural Research and Development. This new modelling framework allows us to assess returns to each component of the national agricultural research system separately, as well as to explore interactions between domestic and international sources of new agricultural technology.

10.2 Measuring Productivity Growth

Our strategy is first to construct provincial-level agricultural TFP measures using Fisher-ideal quantity indexes for agricultural outputs and inputs (the TFP index is then simply the ratio between the input and output indexes). The Fisher index is superlative: it is an index that corresponds to a functional form capable of providing a second order approximation to an arbitrarily twice differentiable linear homogenous function (Diewert, 1976). Developed originally as a price index (Fisher, 1927), the Fisher-ideal quantity index is defined as the geometric mean of the Laspeyres and Paasche indexes (Eqn 10.1):

$$Q_F(p^0, p^1, x^0, x^1) = \\ [(p^0x^1/p^0x^0) * (p^1x^1/p^1x^0)]^{0.5} \quad (10.1)$$

Where p^r, $x^r > 0$, $r = 0,1$, are vectors of prices and quantities, respectively, at periods 0 and 1. The Laspeyres quantity index $[Q_L(p^1, p^0, x^1, x^0) = p^0x^1/p^0x^0]$ is defined similarly to the Paasche quantity index $[Q_P(p^0, p^1, x^0, x^1) = p^1x^1/p^1x^0]$, but the two measures differ in how they gauge quantity changes. The Laspeyres index allows the numerator's quantity (x^1) to vary from initial-period values (p^0x^0), whereas the Paasche index measures quantity changes by varying the denominator's quantity (x^0) from end-period values (p^1x^1). Suppose (p^0x^0) represent prices and quantities of all agricultural outputs produced in the agricultural sector. Then Eqn 10.1 provides an index measure of real output change (where the index value of output in the base period is set equal to 1.00). Similarly, if (p^0x^0) are vectors of all agricultural inputs (and their prices or service flows) used in production, then Eqn 10.1 gives an index measure of aggregate input change. We define the ratio of the two as an index of total factor productivity growth. Simply put, if output rises faster than input, TFP increases. The percentage increase in TFP between two periods is equal to the percentage change in output minus the percentage change in input between those periods. From this framework we derive annual agricultural output, input and TFP indexes for 22 Indonesian provinces between 1985 and 2005, with the value of each index set to 100 in 1985. We also aggregate provincial data into five regions representing the principal islands or island groups (Java–Bali, Sumatra, Kalimantan, Sulawesi and Eastern Indonesia). By construction, this measure of productivity compares TFP changes over time, but does not provide a geographical comparison of initial TFP levels.

10.3 Data Sources for Measuring Provincial Agricultural Productivity

Data used in our construction of provincial productivity statistics consists of 50 agricultural outputs and six inputs recorded annually over the 1985–2005 period. Although quantities of outputs and inputs are available for each province and each year, prices of outputs are national-level information (and therefore assumed to be the same for each province). Prices for inputs vary regionally and temporally, as described below. As long as internal markets are integrated, intra-country output prices should differ only by the cost of transportation. Input prices, however, could vary regionally because of large and persistent differences

in relative resource endowments. Labour-to-land cost ratios, for example, can be expected to be lower in densely populated, land-scarce regions such as Java, compared with sparsely populated regions such as Kalimantan or Sumatra.

10.3.1 Agricultural outputs

Agricultural output comprises 50 commodities, and is sub-divided into output of food crops and other annuals, plantation crops and livestock products (Table 10.1). Provincial production data are from the Ministry of Agriculture's annual Agricultural Statistics and are now also available online. National farm-level agricultural output prices are from the Food and Agriculture Organization (FAO).

10.3.2 Agricultural land

To account for differences in land quality, we differentiate among six types of agricultural land, assign quality weights to each type and then aggregate the quality-weighted areas into a single 'quality equivalent' measure of total agricultural land area. The six land classes are: (i) irrigated paddy or *sawah* (for growing wetland rice); (ii) rain-fed paddy (also *sawah* in Indonesian nomenclature); (iii) non-irrigated non-*sawah* 'upland' for other annual crops (including upland rice); (iv) land in perennial crops ('permanent cropland' in FAO's nomenclature); (v) permanent pastures or meadows; and (vi) temporary fallow. Among these land classes there is a wide variation in average or potential productivity, reflecting both their natural characteristics and man-made improvements, such as levelling, dike

Table 10.1. Principal agricultural commodities in Indonesia.

Food crops and other annual crops	Rice, corn, soybeans, peanuts, mungbean, cassava, sweet potato, potato, vegetables and fruits, fibre crops and tobacco
Plantation (perennial) crops	Cocoa, coconut, coffee, oil palm, sugarcane, rubber, tea, nuts and spices
Livestock products	Meat of cattle, buffalo, horse, swine, sheep, goats and poultry; milk and eggs

construction and tree planting. Area in each land class by province is reported annually by the Central Bureau of Statistics (a) through 2004 and extrapolated for 2005 using historical growth rates in the series.

We use the total value of agricultural yield for the entire agricultural sector as a dependent variable and each land type's share of total land as independent variables in a regression to assess each land type's relative productivity. Those estimates are then employed as 'quality weights' to adjust the land area of each type into a 'constant quality' unit. Letting rainfed paddy be the reference land quality (i.e. its quality weight is set equal to 1.00), the weights for the other land classes are: irrigated paddy = 2.00, non-irrigated upland = 0.75, perennial cropland = 0.75 and pastures = 0.20. (We also let the weight for temporary fallow equal 0.20, assuming this is used to pasture livestock.) This implies that irrigated rice land is twice as productive as rainfed paddy land (perhaps because it enables multiple cropping), which in turn is about one-third more productive than other non-irrigated cropland and about five times as productive as land in pastures and fallow. With these weights, we aggregate the total agricultural area in each province into the number of 'rainfed-paddy-equivalent' hectares.

Because no consistent series for land prices are available, we derive service flows from agricultural land as the residual payment after costs of other inputs are paid. In other words, the land rental payment is the difference between the total value of output minus costs of labour, material inputs and capital services. To obtain an estimated land rental value per hectare, we divide the residual payment by the quality-adjusted hectares of rainfed paddy equivalents.

10.3.3 Agricultural labour

Agricultural labour consists of the number of male and female workers over the age of 15 whose primary employment is in the agricultural sector. Our estimates are taken from annual national labour force or census surveys reported in the *Statistical Yearbook of Indonesia* (Central Bureau of Statistics (a)). To derive annual hours worked, we assume male agricultural labourers work on average 300 six-hour days per year and female labourers 250 six-hour days per year. Daily wages are assumed to be the same for men and women and are the simple average of the wages paid for different farming operations (planting, ploughing and weeding) reported annually for different regions by *Farm Wage Statistics in Rural Areas* (Central Bureau of Statistics (d)).

10.3.4 Agricultural capital

Crego *et al.* (1998) derive an estimate of fixed agricultural capital stock for Indonesia by applying the perpetual inventory method to annual capital investments from national accounts data. But their estimate is not reliable, for it shows sharply declining levels of capital stock at a time when the number of agricultural machines employed in agriculture rose significantly. Instead, we compute total power employed in farming by deriving total horsepower supplied by farm machinery and draft animals.

The annual stock of mechanical horsepower is calculated as the number of power machines used in agriculture multiplied by each machine's average horsepower (hp). Mean hp for different farm machinery categories are: 40 hp for four-wheel large tractors, 30 hp for four-wheel medium tractors, 25 hp for four-wheel small tractors, 12 hp for two-wheel walking tractors and 25 hp for power-threshers. The number of each type of machine in use at the provincial level is reported by the Central Bureau of Statistics (a). Machinery data at the provincial level are available annually only up to 2002, so we extrapolate estimates for 2003, 2004 and 2005 using historical growth rates in the number of machines in use. Draft animals consist of horses, buffaloes and non-dairy cattle, and each animal is assumed to provide 1 hp of draft power. The number of draft animals in stock by province is given by the Ministry of Agriculture.

The value of mechanical power services is assumed constant (one-hoss-shay) over the life of a machine. We amortize the FAO import price of a four-wheel tractor over 10 years using a 10% discount rate to derive an annual service flow per tractor, and divide that by 40hp to get an average 'rental price' per hp of machinery used in agricultural production. To obtain the annual value of animal work services, we amortize the unit price of horses and buffaloes over a three-year period, using a 10% discount rate and FAO import price data.

10.3.5 Material inputs: fertilizer and animal feed

The principal material inputs used in agriculture are fertilizers and animal feed. Fertilizer quantities (measured in metric tonnes of synthetic N (nitrogen), P_2O_5 (phosphate) and K_2O (potash) nutrients applied to agriculture) are available annually at the national level from the FAO. We do not include fertilizers supplied from animal manure. The publication on *Agricultural Indicators* from the Central Bureau of Statistics (b) includes estimates of regional fertilizer consumption for some years, and Central Bureau of Statistics (c), *Farm Cost Structure of Paddy and Secondary Food Crops*, gives fertilizer application rates per hectare for rice and secondary food crops at the provincial or regional level for most years. We apply these regional application rates to provincial-level data from the Ministry of Agriculture on crop area harvested in order to derive total fertilizer applied to food crops. Food-crop fertilizer applications in the 1980s accounted for about 80% of total fertilizer applied to agriculture. We assume the rest was applied to perennial crops and allocate it by province based on the provincial share of crop area in perennials. Farm-level fertilizer prices are reported in *Producer Price Statistics of the Agricultural Sector in Indonesia* (Central Bureau of Statistics (e)). We apply Central Java's mean farm-level fertilizer prices to provinces in Java, and Bali and North

Sumatra's mean farm-level fertilizer prices to all other provinces.

FAO reports estimates of total animal feed from rice, maize, root and tuber crops, and agricultural processing residues (oilseed meals, molasses and bran from grain milling). There is, however, no provincial-level data on animal feed quantities or expenditures. To estimate the feed input quantities at the provincial level, we use constant feed-to-product conversion ratios and then multiply these ratios by the provincial-level animal output quantity. These feed-to-meat conversion ratios reflect not only biological conversion rates but also account for pasture forage, reducing the need for purchased feed. Sheep and goats are assumed to be fed entirely on forages with no purchased feed; buffaloes are also assumed to derive a large share of their feed from forages and hay, therefore using less purchased feed than other large ruminants. The meat conversion ratios we use (tonnes of feed consumed per tonne of meat produced) are: 7.0 for cattle meat, 4.0 for buffalo meat, 6.0 for horse meat, 2.5 for poultry meat, 0.0 for sheep and goat meat, and 3.0 for pig meat. For non-meat animal products, conversion ratios are 3.0 for cows' milk and 3.0 for hens' eggs. When aggregated across the livestock sector and to the national level, total estimated feed (in grain-equivalent tonnes) is roughly similar to the total animal feed estimated by FAO. The price of a tonne of feed (measured in grain-equivalents) is assumed to be 1.5 times the price of rice (the principal grain), to account for feed manufacturing, storage and transportation costs.

10.4 Determinants of Productivity Growth

After constructing provincial TFP growth indexes, we estimate a multiple-regression model to test hypotheses regarding possible explanations for that growth. As in RBF, the model includes general public investment in agriculture and water resource development. Departing from the RBF approach, we include population density and the World

Bank's measure of the relative rate of assistance to agriculture to capture the effect of policies on agricultural input and output prices (Anderson and Martin, 2009). We then extend the RBF model by specifying a series of variables to capture domestic and international sources of new technology for agriculture. In particular, we differentiate between R&D for export-oriented plantation crops and R&D for food crops and other commodities.

10.4.1 Rate of assistance to agriculture

The rate of assistance to agriculture measures the impact of distortions to agricultural incentives resulting from policy interventions in markets. The nominal rate of assistance (NRA) is defined as the percentage by which government policies have raised (or lowered, if the NRA is less than zero) the gross return to producers above the gross returns they would have received without government intervention (Anderson and Martin, 2009). But farmers are affected not only by the prices for their own outputs, but also by changes in factor market prices and the exchange rate because of the incentives non-agricultural producers face. The economy-wide effect of distortions in agricultural incentives is measured by the relative rate of assistance (RRA) to agriculture, or the extent to which agricultural producers are assisted or taxed relative to producers of non-agricultural products (Anderson and Martin, 2009).

Specifically, RRA is the ratio of the nominal rates of assistance to agriculture and the nominal rates of assistance to non-agricultural sectors, thus accounting for agriculture's domestic, border and input price supports relative to the rest of the economy. Stated differently, the RRA is an all-economy, rather than all-agriculture, measure reflecting Indonesia's market liberalization and price policy interventions. A negative RRA implies policy discrimination against agriculture and would probably encourage movement of resources out of agriculture. A positive RRA implies greater incentive for private investment in agriculture. Annual estimates of

RRA for Indonesia from 1985 to 2004 were constructed by Fane and Warr (2009). The value for 2005 is assumed to be the same as in 2004.

Figure 10.1 plots the trend in NRAs for the agricultural and non-agricultural sectors and the RRA to Indonesian agriculture from 1985 to 2004. For most of the 1985–1999 period, economic policies in Indonesia discriminated against agriculture as both the NRA and RRA were negative. The 1997–1999 Asian financial crisis imposed significant currency devaluations and other policy reforms that led to a significant improvement in the terms of trade facing agriculture. Policy discrimination against agriculture, between 2000 and 2004, largely disappeared and the RRA became slightly positive. (See Fane and Warr, 2009, and RBF for detailed discussions of agricultural and trade policy changes in Indonesia in recent decades.)

10.4.2 Public investment in agriculture and irrigation

Agricultural input subsidies, irrigation development, farm credit, agricultural extension and marketing services are some of the areas in which the Indonesian government has made significant budgetary commitments. Until a decentralization law was passed in 2002, most of these recurring expenditures were made by the central government, either directly through its ministries or as transfer payments to provincial and district governments. These 'development' expenditures (reflecting programme expenditures but excluding civil service personnel and overhead costs) are reported in nominal Rupiah for each province by the Central Bureau of Statistics (a). They are adjusted for inflation by the World Bank's GDP implicit price deflator for Indonesia.

10.4.3 Population density

Boserup (1965) proposed that population pressure on land would induce intensification

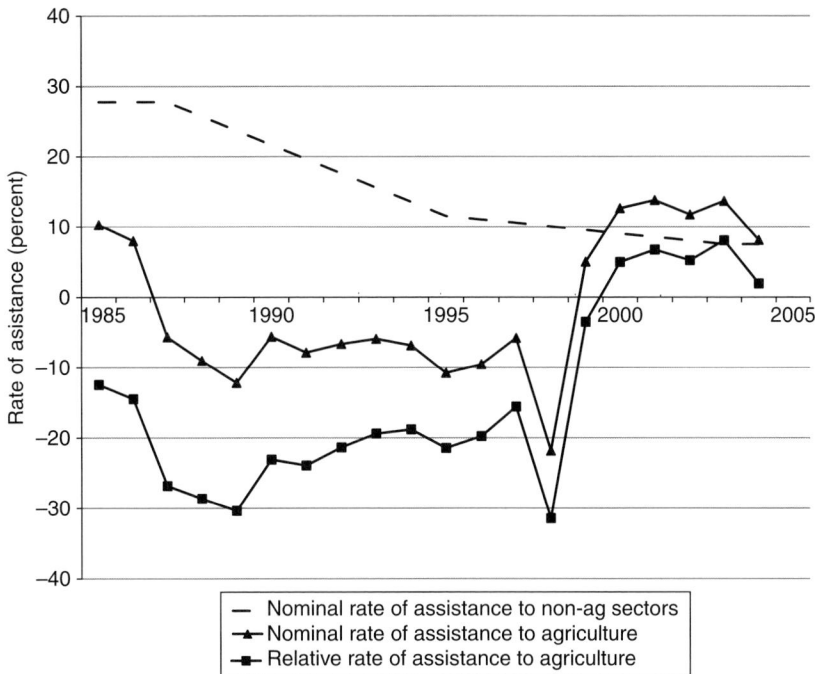

Fig. 10.1. Nominal and relative rates of assistance to agriculture for Indonesia. Source: Fane and Warr (2009).

of agricultural production systems. As previously noted, population density varies widely across the Indonesian archipelago. Java, with only 7% of the land area, contains nearly 60% of the country's population. The land frontier in Java was reached in the 1940s: average farm size is only about 0.5 ha and has been declining over time (Booth, 1988). In other regions of the country, however, agricultural land area is still expanding, both by small holders and by large plantation companies. With the rapid expansion of the area planted in oil palm and other plantation crops, agricultural land per farm worker has been rising (Fuglie, 2010a). We hypothesize that higher population pressure on land will be associated with more rapid agricultural productivity growth. To estimate annual population by province, we use decennial population census data (1980, 1990 and 2000) reported by the Central Bureau of Statistics (a) and apply a constant inter-census population growth rate to interpolate population between census years. We then divide population by land area to obtain

population densities (people per km²) for each province.

10.4.4 National agricultural research

Investments in agricultural research have been shown to be crucial for raising agricultural productivity in both developed and developing countries (Hayami and Ruttan, 1985; Evenson and Fuglie, 2010). But research investments affect productivity only with a time lag and, like physical capital investments, eventually depreciate because of technology obsolescence (Alston et al., 1998). In Indonesia, agricultural research institutes were established in the 1880s for a number of export commodities (sugarcane, coffee, tea and rubber) and later for food crops. Much of that research capacity, though, was destroyed during World War II and the subsequent war for Indonesian independence (Fuglie and Piggott, 2006). By 1966, the Central Agricultural Research

Station, which was responsible for research on food crops, employed only one PhD-level scientist.

Indonesia's modern agricultural research system can be dated to 1974, when existing agricultural research capacities were amalgamated into the Indonesian Agency for Agricultural Research and Development (IAARD). A significant expansion of the agricultural research system followed (Fuglie and Piggott, 2006). IAARD is responsible for research on annual crop and livestock commodities, as well as on natural resources, post-harvest, mechanization and policy research. Research on plantation crops is delegated to a separate institution, the Indonesian Planters Association for Research and Development (IPARD). Although IPARD is nominally under the oversight of IAARD, in practice they maintained separate funding and management arrangements. IAARD is an agency within the Ministry of Agriculture, is funded directly by the central government and its research staff is made up of civil service employees. IPARD is funded primarily by its members (state-owned and large private-estate companies) and from product sales – such as improved seed stock – and its scientists are not subject to civil service employment rules or pay scales. Although IAARD has grown much larger than IPARD, expenditures per scientist have been much higher at IPARD. Fuglie and Piggott (2006) note that in 1996 funding per scientist at IPARD centres was four times greater than that at IAARD institutes.

To account for the time lag between when research is conducted and when new technology from that research is adopted by farmers, we treat research expenditures as investments that create 'knowledge capital' stock and include the stock value in the regression model explaining TFP growth. We assume a trapezoidal lag structure to estimate knowledge capital stocks from annual research expenditures (Huffman and Evenson, 1993). We employ a one-year lag before research expenditures (RE_{ct}) begin to affect productivity, the effect rising over time and peaking between years 7 and 14, then falling until completely depreciated by year 17 through technology obsolescence.

Letting S_{ct} be knowledge capital stock of institution c in year t, we have (Eqn 10.2):

$$
\begin{aligned}
S_{ct} = {} & 0.01\ (RE_{c,t-1}) + 0.02\ (RE_{c,t-2}) \\
& + 0.03\ (RE_{c,t-3}) + 0.04\ (RE_{c,t-4}) \\
& + 0.05\ (RE_{c,t-5}) + 0.06\ (RE_{c,t-6}) \\
& + \Sigma_{j=0}^{7}\ 0.08\ (RE_{c,t-7-j}) + 0.060\ (RE_{c,t-15}) \\
& + 0.05\ (RE_{c,t-16}) + 0.04\ (RE_{c,t-17}) \quad (10.2)
\end{aligned}
$$

where the coefficients (time-lag weights) sum to 1.00 and c=IPARD and IAARD expenditures. Separate research stock variables are constructed for IPARD and IAARD to distinguish research on plantation crops from other agricultural research.[2]

Because of the 17-year time-lag structure for the research capital stock variable from Eqn 10.2, computing S_{ct} for the 1985–2005 period requires data on annual research expenditures from 1968 to 2005. Fuglie and Piggott (2006) report annual research expenditures for IAARD and IPARD from 1974 to 2003. Expenditures for 2004 and 2005 are extrapolated using the 1974–2003 average growth rate (in constant Rupiahs). To extend IAARD research expenditures back to 1968, we use the number of scientists employed at the Central Agricultural Research Station during 1968–1973 as reported in van der Eng (1996) multiplied by average expenditures per scientist at IAARD in 1974 (in constant Rupiah) from Fuglie and Piggott (2006). For IPARD research prior to 1974, we use a similar approach based on the number of scientists employed at the Sugar Research Institute (van der Eng, 1996).[3] During those early years, the Central Agricultural Research Station and the Sugar Research Institute were the largest components of what was to become IAARD and IPARD. Lacking data on employment for other IPARD institutes, we assume that research expenditures followed trends similar to the Sugar Research Institute between 1968 and 1973.

10.4.5 International technology transfer

We generally expect domestic agricultural research to be essential for developing locally suitable agricultural technology. But it is also possible that some technology can be directly transferred from abroad. This is especially true if the technology is developed for a specific environment or under conditions similar to those found locally. Indonesia in particular has been able to borrow rice varieties developed by the International Rice Research Institute (IRRI) located in nearby Los Baños, Philippines. During the Green Revolution, Indonesia directly disseminated rice varieties IR8, IR36 and other varieties developed by IRRI in Los Baños to farmers in Java and elsewhere in Indonesia. At its peak, IR36 alone was grown on >3 million ha, or 9% of Indonesia's total rice-growing area. Subsequently, Indonesia has continued to disseminate new generations of IRRI-bred varieties, as well as to develop its own locally adapted varieties (often by crossing IRRI and local parent material). As recently as 1993, 65% of Indonesia's rice area planted in modern rice varieties was sown with IRRI-bred varieties, most notably IR64, which first became available to Indonesian farmers in 1987 (Darwanto, 1993). The Mexico-based International Center for Wheat and Maize Improvement (CIMMYT) has also directly contributed improved maize varieties through its Thailand-based regional breeding programme (Morris, 2002).

Another example of international technology transfer is oil palm technology from Malaysia. In the 1970s, the Malaysian Oil Palm Board (formerly the Palm Oil Research Institute of Malaysia) made a concerted effort to adapt West African oil palm varieties to South-east Asia and it has continued to invest a portion of its palm oil export revenues in research. Although it initially restricted international transfer of its improved oil palm clones, it later eased those restrictions by allowing Malaysian companies to transfer improved genetic material to their foreign plantations (Pray and Fuglie, 2002). Malaysian plantations have invested considerable resources in Indonesia's recent oil palm expansion. As of 2009, Malaysia and Indonesia together accounted for 85% of global oil palm pro-

duction, with Indonesia's production share (46%) slightly larger than Malaysia's (FAO, 2011).

We model the impact of international research on agricultural productivity by including two variables – one for IRRI's contribution to food-crop technology and one for Malaysia's contribution to plantation-crop technology. For the former, we estimate the share of total annual crop area planted in varieties introduced directly from IRRI. Area planted in IRRI-bred rice varieties is from Darwanto (1993). For the latter, we use a five-year lag of average Malaysian oil palm yield to account for the lag time in transferring new Malaysian technology to Indonesia. Oil palm yield data are from FAO.

The multivariate regression model of provincial agricultural TFP growth contains explanatory variables that are measured at either the national or provincial level. The domestic and international R&D and the price policy (RRA) variables are measured at the national level and therefore vary only temporally. For any year, then, they have the same measure for each province. As such, these variables explain productivity changes over time but are unable to explain spatial differences among provinces. Population density and public investment in agriculture and irrigation vary by province and by year, thus are able to explain changes over time and space.

10.5 Results

10.5.1 Agricultural output, input and TFP growth

The composition and growth of agricultural output among major commodity categories is shown for Indonesia's provinces in Table 10.2. The Java–Bali region accounted for about half of the nation's gross agricultural production in 2005. With 36% of cropland under irrigation, food crops dominated agricultural production in Java–Bali. Plantation crops made up <10% of production, which compares with >30% for the rest of the country. Nationally, output from plantation crops and

Table 10.2. Agricultural output composition and growth for Indonesian provinces and regions.

Region	Province	Gross output in 2005 (billion Rupiah)	Plantation crops output share (%)	Average annual growth rate, 1985–2005 (%)		
				Food crops	Livestock products	Plantation crops
Java–Bali		209,848	8.8	1.85	3.67	−1.49
	West Java	72,365	4.9	2.16	4.94	−0.70
	Central Java	52,307	7.6	1.55	3.68	−2.60
	Yogyakarta	6,016	6.4	1.89	2.75	0.12
	East Java	72,132	13.6	1.80	3.42	−1.40
	Bali	7,027	8.3	−0.61	3.80	1.91
Sulawesi		39,739	21.3	2.65	2.27	5.98
	South Sulawesi	23,257	15.3	1.79	1.35	6.97
	SE Sulawesi	4,034	15.3	1.79	1.35	6.97
	Central Sulawesi	6,003	32.0	4.81	0.67	7.15
	North Sulawesi	6,445	31.8	4.61	4.77	0.94
Sumatra		127,943	39.3	3.38	2.98	4.86
	Aceh	9,323	27.4	2.07	1.41	0.55
	North Sumatra	32,693	42.3	3.73	1.66	1.85
	Riau	13,615	66.6	−0.30	6.18	10.77
	West Sumatra	12,629	18.9	1.87	3.85	6.35
	Jambi	8,928	52.8	1.88	3.10	8.47
	Bengkulu	4,274	33.7	4.42	−1.26	4.90
	South Sumatra	22,542	42.8	3.72	4.30	6.73
	Lampung	23,939	29.7	5.53	4.72	3.90
Kalimantan		30,609	28.9	3.69	3.82	9.09
	West Kalimantan	10,166	37.7	3.19	3.44	8.15
	Central Kalimantan	6,721	41.0	4.21	3.04	11.29
	South Kalimantan	9,030	15.6	3.58	5.35	6.22
	East Kalimantan	4,692	24.6	4.80	3.04	11.07
E. Indonesia		17,135	13.7	1.22	3.29	2.90
	Nusa Tenggara Barat	7,669	6.3	2.06	2.56	4.51
National		425,274	20.0	2.28	3.42	3.55

Provincial boundaries are defined as they existed from 1985. Data from provinces that later separated are aggregated for this analysis (i.e. West Java includes Banten; North Sulawesi includes Gorontalo; South Sumatra includes Bangko-Belitung; Riau includes Riau islands; and South Sulawesi includes West Sulawesi). Regions are the authors' groupings. East Indonesia includes Nusa Tenggara Barat, Nusa Tenggara Timor, Maluku and Papua. *Source*: Authors' estimates.

livestock grew substantially faster than output of food crops. All of the growth in plantation crops has taken place off Java and was especially high in Kalimantan (9% annual growth rate). By 2005, plantation crops made up about half of Sumatra's agri-

cultural output and a third of the total agricultural output in Sulawesi and Kalimantan.

Average annual growth rates in agricultural output, inputs and TFP by province during 1985–2005 are given in Table 10.3. The figures demonstrate the wide regional

Table 10.3. Agricultural output, input and TFP growth for Indonesian provinces and regions, 1985–2005.

Region	Province	Output	Input	TFP
		(average annual percentage change)		
Java–Bali		1.97	0.81	1.16
	West Java	2.49	0.67	1.82
	Central Java	0.80	0.79	0.01
	Yogyakarta	2.25	0.66	1.60
	East Java	1.83	1.07	0.76
	Bali	0.86	0.03	0.83
Sulawesi		3.81	1.74	2.08
	South Sulawesi	2.92	1.39	1.53
	SE Sulawesi	7.06	3.83	3.23
	Central Sulawesi	6.20	2.46	3.74
	North Sulawesi	4.02	1.60	2.42
Sumatra		4.59	2.23	2.35
	Aceh	2.72	1.61	1.11
	North Sumatra	3.50	1.55	1.95
	Riau	7.32	3.38	3.95
	West Sumatra	3.42	1.59	1.83
	Jambi	5.34	3.63	1.71
	Bengkulu	5.19	3.08	2.11
	South Sumatra	5.59	2.95	2.64
	Lampung	5.51	2.25	3.26
Kalimantan		5.46	2.14	3.32
	West Kalimantan	5.08	1.29	3.79
	Central Kalimantan	7.69	3.29	4.40
	South Kalimantan	4.30	1.97	2.33
	East Kalimantan	6.19	3.29	2.90
E. Indonesia				
	Nusa Tenggara Barat	2.77	1.65	1.12

As a result of insufficient or poor quality data, indexes cannot be estimated for provinces in East Indonesia except Nusa Tenggara Barat. *Source*: Authors' estimates.

variation in agricultural growth over this period. Output growth was highest in Kalimantan (5.5% per year on average) and lowest in Java–Bali (just less than 2% per year). All of Kalimantan's four provinces had average annual output growth of more than 4%, as did three out of four provinces in Sulawesi and five out of eight provinces in Sumatra. On the other extreme, agricultural output in Central Java and Bali grew by less than 1% per year.

There was also wide variation in resource expansion, with aggregate inputs growing by only 0.8% per year in Java–Bali and by more than 2% per year in Kalimantan and Sumatra (with several provinces there and in Sulawesi showing annual input growth of more than 3%). In Java and Bali, all of the input growth was in non-land inputs (i.e. labour, capital and materials – the land input actually fell by 11% during 1985–2005), whereas in other, more land-abundant regions of the country agricultural land continued to expand along with other inputs. Indeed, in the outer islands, agricultural land area expanded faster than agriculture labour, causing the land-to-labour ratio to increase.

Total factor productivity growth has been an important source of rising output not only for land-constrained Java–Bali, but also for regions of the country experiencing land extensification. In fact, TFP growth averaged more than 2% per year in the three outer island regions where land has expanded most rapidly (Kalimantan, Sumatra and Sulawesi) but only 1.16% per year in land-constrained Java and Bali. In each region, TFP accounted for 50–60% of total output growth and resource expansion for the other 40–50%.

10.5.2 Determinants of productivity growth

Table 10.4 presents the multiple-regression model results of Indonesia's agricultural TFP-growth determinants. The variables in the model explain about 37% of the observed variation in TFP. As expected, most of the explanatory power comes from TFP's change over time (the within R-squared is >50%), the

model explains only about 4% of the variation in TFP growth across provinces (the between R-squared). Recall that several variables in the model (RRA, domestic and international R&D) vary only temporally and therefore cannot explain spatial differences in TFP growth. The only variables that do help explain differences in TFP growth between provinces are the development expenditure and population density variables.

Table 10.4. Determinants of agricultural productivity growth in Indonesia.

Explanatory variables	Variable definition	Model				
		1	2	3	4	5
Food crop and livestock research (IAARD)	Log of R&D stock (constant 2002 Rp)	0.153** (0.02)		0.115 (0.14)	0.143** (0.03)	0.111 (0.15)
Plantation crops research (IPARD)	Log of R&D stock (constant 2002 Rp)	0.157** (0.03)		0.155** (0.03)	0.169** (0.02)	0.163** (0.03)
International technology transfer for food crops	Share of total food crop area planted to IRRI rice varieties		0.189** (0.03)	0.0951 (0.35)		0.0871 (0.4)
International technology transfer for plantation crops	Log of Malaysian oil palm yield lagged 5 years		0.00628 (0.94)		−0.0505 (0.57)	−0.0346 (0.71)
Market and trade liberalization	Relative rate of assistance (RRA) to agriculture	0.193*** (0.00)	0.215*** (0.00)	0.184** (0.01)	0.180** (0.02)	0.176** (0.02)
Public spending for agriculture and irrigation	Log of development spending (constant 2002 Rp)	0.0292*** (0.00)	0.0511*** (0.00)	0.0316*** (0.00)	0.0306*** (0.00)	0.0323*** (0.00)
Population density	Log of persons per km^2	−0.00389 (0.82)	−0.00942 (0.59)	−0.00475 (0.79)	−0.00445 (0.80)	−0.00509 (0.77)
Constant		−4.617*** (0)	−1.023 (0.32)	−4.134*** (0)	−4.043*** (0)	−3.782*** (0.01)
R-squared	Within	0.53	0.52	0.54	0.54	0.54
	Between	0.04	0.03	0.04	0.04	0.04
	Overall	0.37	0.36	0.37	0.37	0.38

Parentheses indicate p-values. ***, **, * indicate statistical significance at the 1%, 5% and 10% levels, respectively.

Consistent with RBF, agricultural market and trade liberalization (measured by the improvement in the relative rate of assistance to agriculture) had a major impact on agricultural productivity, as did public investments in agriculture and irrigation. In contrast, population density was not associated with productivity growth, contrary to Boserup's induced innovation hypothesis.

Our primary interest in this analysis, however, is the effects of R&D, both domestic and international, on agricultural productivity growth in Indonesia during this post-Green Revolution period. The impact of plantation crop research by IPARD is significant and positive, with a research elasticity of about 0.16 (i.e. a 1% increase in the stock of plantation crop research was associated with a mean 0.16% increase in agricultural TFP). Our proxy for plantation crop technology transfer from Malaysia was not significant, indicating that either our proxy measure is poor or direct transfer of Malaysian technology has not had measurable impact in Indonesia.[4] It is possible that Malaysian R&D has only indirectly affected Indonesia through local adaptive research by IPARD or by the in-house R&D of private plantations.

For agricultural research on other commodities (food crops, horticulture and livestock), the domestic IAARD research and the IRRI technology transfer variables show mixed results. Neither is significant together, but both are significant separately. The research elasticity for IAARD is a mean 0.15 when the IRRI variable is excluded from the model but falls to 0.11 (and is not statistically different from zero) when the IRRI variable is included. One explanation is that for food crops, domestic and international agricultural R&D are complementary inputs to the technology development process. The interpretation would be that the mean R&D elasticity of 0.15 represents a joint return to IAARD and international agricultural research (or at least to IRRI research relevant to Indonesia).

Given the research lag structure in Eqn 10.2 and the research elasticities in Table 10.4, we can estimate the annual internal rate of return to IAARD and IPARD research

in Indonesia over this period.[5] This is the discount rate that equates the present value of $1 in research spending in a given year with the productivity benefits that accrue over the subsequent 17 years (constant 2002 Rp). Using the mean IPARD elasticity of 0.16, we find an internal rate of return to perennial crop research of 39%. Assuming the mean R&D elasticity of 0.15 for IAARD research (which is really a joint return to IAARD and IRRI's technology contributions), we obtain an internal rate of return to research of 19%. IPARD returns to research are considerably higher than its IAARD counterpart. This might reflect a stronger institutional and incentive structure at the better-funded and more flexible IPARD than at the more bureaucratic IAARD.

10.6   Summary and Conclusions

Indonesian agriculture exhibited widely divergent patterns of regional growth in the post-Green Revolution period of 1985–2005. Java and Bali, heavily dependent on irrigated rice production and where much of the Green Revolution growth occurred in the 1970s and 1980s, experienced the slowest agricultural growth of any region since 1985. Production growth there is severely land constrained. Resource intensification (more labour, capital and materials per hectare) continued, but this by itself improved output by less than 1% per year. About 60% of output growth in Java–Bali has been due to raising TFP of agricultural resources, enlarging total gross output at an annual rate of just over 2%. Outside Java–Bali (in Kalimantan, Sumatra and Sulawesi), significant agricultural land expansion was still taking place. Productivity contributed to that growth as well. With both more rapid resource expansion and TFP growth than Java–Bali, agricultural output has grown about twice as fast in these regions compared to Java and Bali. Much of that output growth has been a response to foreign demand (especially for edible oils), enabling Indonesia to become a major agricultural net exporter even as it continued to run a significant trade deficit in food and feed grains.

An important element behind Indonesia's agricultural growth success has been its market and trade liberalization policies. These policies, implemented gradually and not necessarily with agriculture specifically in mind, improved agriculture's terms of trade. That strengthened incentives for farmers and agribusiness to invest in agriculture, expand production and raise productivity, especially in tropical perennial crops where the country has a strong international comparative advantage.

Public investment in agriculture has also been an important stimulus to growth. State spending on irrigation, general agricultural services and agricultural research is significantly associated with TFP growth in Indonesia's agriculture. Agricultural research spending on plantation crops has earned about twice the rate of return as research on food crops and livestock. One reason could be that institutional support and research incentives are stronger in the IPARD system responsible for plantation crop research than in the IAARD system responsible for research on other commodities. IPARD scientists are better funded and better paid, and are not subject to the bureaucratic rules of ministerial agencies like IAARD. Government funding for IAARD, on the other hand, remains low relative to the size of the country's agricultural sector and is heavily dependent on foreign donor assistance (Fuglie and Piggott, 2006). None the less, IAARD research seems to have generated reasonable returns by partnering with international institutes such as IRRI. Institutional reforms that improve research incentives might help IAARD further improve its performance.

Notes

[1] The Green Revolution in Indonesia was characterized by widespread adoption of new, semi-dwarf rice varieties that responded well to fertilizer and irrigation. Fertilizer use greatly increased, irrigated area expanded, and average yield of rice more than doubled from less than 2 to over 4 tonnes per hectare. There was also some penetration of improved corn (maize) and soybean varieties (Fuglie and Piggott, 2006).

[2] The research capital stock variables in our model are national-level variables. We also constructed regional research stocks for each variable as in Evenson et al. (1997) using the location of scientists to apportion total R&D spending by region and allowing for inter-regional research spillovers. But this model did not perform well. In fact, most of the research centres that make up IAARD and IPARD are centrally managed national institutions that operate regional sub-stations. Thus, we treat them here as national knowledge capital stock variables.

[3] In 1974 the Sugar Research Institute accounted for two-thirds of all IPARD scientists.

[4] We also created a research stock variable from time-series data on oil palm research expenditures in Malaysia (Gert-Jans et al., 2005), but this variable performed poorly in the model. Further examination of oil palm research spending by the Malaysia Oil Palm Board indicated that a large portion of this research is devoted to post-harvest market utilization (new product development) and the mechanization of oil palm harvesting. Neither of these research areas is of much relevance to farm-level oil palm productivity in Indonesia (low labour costs imply little demand for mechanization).

[5] See Alston et al. (1998) for the methodology deriving these rates of return. In addition to the R&D elasticity and the research lag structure, it involves estimating the economic value from a given percentage change in productivity and the annual research spending associated with a 1% change in research stock.

References

Alston, J., Norton, G. and Pardey, P. (1998) Science Under Scarcity: Principles and Practice for Agricultural Research Evaluation and Priority Setting. Cornell University Press, Ithaca, NY.

Anderson, K. and Martin, W. (2009) Introduction and summary. In: Anderson, K. and Martin, W. (eds) Distortions to Agricultural Incentives in Asia. World Bank, Washington, DC, pp. 3–82.

Booth, A (1988) *Agricultural Development in Indonesia*. Allen and Unwin, Sydney, Australia.

Boserup, E. (1965) *Conditions of Agricultural Growth: The Economics of Agrarian Change Under Population Pressure*. Aldine Publishing, New York, New York.

Central Bureau of Statistics (a). Annual Issues 1985–2005. *Statistical Yearbook of Indonesia*. Badan Pusat Statistik, Jakarta.

Central Bureau of Statistics (b). Various issues. *Agricultural Indicators*. Badan Pusat Statistik, Jakarta.

Central Bureau of Statistics (c). Various issues, *Farm Cost Structure of Paddy and Secondary Food Crops*. Badan Pusat Statistik, Jakarta.

Central Bureau of Statistics (d). Various issues. *Farm Wage Statistics in Rural Areas*. Badan Pusat Statistik, Jakarta.

Central Bureau of Statistics (e). Various issues. *Producer Price Statistics of the Agricultural Sector in Indonesia*. Badan Pusat Statistik, Jakarta.

Crego, A., Larson, D., Butzer, R. and Mundlak, Y. (1998) A New Database on Investment and Capital for Agriculture and Manufacturing. Policy Research Working Paper 2013, World Bank, November.

Darwanto, D.H. (1993) Rice Variety Improvement and Productivity Growth in Indonesia. PhD Thesis. University of Los Bānos, Philippines.

Diewert, W. (1976) Exact and superlative index numbers. *Journal of Econometrics* 4, 115–145.

Evenson, R.E. and Fuglie, K. (2010) Technology capital: The price of admission to the growth club. *Journal of Productivity Analysis* 33, 173–190.

Evenson, R.E., Abdurachman, E., Hutabarat, B. and Tubagus, A. (1997) Contribution of research on food and horticultural crops in Indonesia: An economic analysis. Ekonomi dan Keuangan Indonesia XLV, 551–578.

Fane, G. and Warr, P. (2009) Indonesia. In: Anderson, K. and Martin, W. (eds) *Distortions to Agricultural Incentives in Asia*. World Bank, Washington, DC, pp. 165–196.

FAO (2011) FAOSTAT Database. Food and Agricultural Organization, Rome. Available at: http://faostat.fao.org/

Fisher, I. (1927) *The Making of Index Numbers: A Study of Their Varieties, Tests, and Reliability*, 3rd edn, revised. Riverside Press, Cambridge, MA.

Fuglie, K. (2004) Productivity growth in Indonesian agriculture, 1961–2000. *Bulletin of Indonesian Economic Studies* 40, 209–225.

Fuglie, K. (2010a) Indonesia: From food security to market-led agricultural growth. In: Alston, J., Babcock, B. and Pardey, P. (eds) *The Shifting Patterns of Agricultural Production and Productivity Worldwide*. Midwest Agribusiness Trade and Research Information Center, Iowa State University, Ames, IA, pp. 343–381.

Fuglie, K. (2010b) Sources of growth in Indonesian agriculture. *Journal of Productivity Analysis* 33, 225–240.

Fuglie, K. and Piggott, R. (2006) Indonesia: Coping with economic and political instability. In: Alston, J., Babcock, B. and Pardey, P. (eds) *The Shifting Patterns of Agricultural Production and Productivity Worldwide*. Midwest Agribusiness Trade and Research Information Center, Iowa State University, Ames, IA, pp. 65–104.

Gert-Jans, S., Tawang, A. and Beintema, N. (2005) Malaysia. Agricultural Science and Technology Indicators (ASTI) Country Brief No. 30. November. International Food Policy Research Institute, Washington, DC.

Hayami, Y. and Ruttan, V.W. (1985) *Agricultural Development: An International Perspective*. Johns Hopkins University Press, Baltimore, MD.

Huffman, W.E. and Evenson, R.E. (1993) *Science for Agriculture*. Iowa State University Press, Ames, IA.

Ministry of Agriculture (annual issues). Agricultural Statistics, Department Pertanian, Republik Indonesia, Jakarta. Available at: http://database.deptan.go.id/bdsp/index-e.asp

Morris, M. (2002) Impacts of International Maize Breeding Research in Developing Countries. International Maize and Wheat Improvement Center (CIMMYT), Mexico, D.F.

Pray, C. and Fuglie, K. (2002) Private Investment in Agricultural Research and International Technology Transfer in Asia. Economic Research Service, U.S. Department of Agriculture, Washington, DC.

Rada, N., Buccola, S. and Fuglie, K. (2011) Government policy and agricultural productivity in Indonesia. *American Journal of Agricultural Economics* 93, 863–880.

Salmon, D. (1991) Rice productivity and the returns to rice research in Indonesia. In: Evenson, R.E. and Pray, C (eds) *Research and Productivity in Asian Agriculture*. Cornell University Press, Ithaca, NY.

Van der Eng, P. (1996) *Agricultural Growth in Indonesia: Productivity Change and Policy Impact Since 1880*. Macmillan, Basingstoke, UK.

World Bank. World Development Indicators Database. Washington, DC.

11 Total Factor Productivity in Thai Agriculture: Measurement and Determinants

Waleerat Suphannachart[1] and Peter Warr[2]

[1]*Kasetsart University, Bangkok;* [2]*Australian National University, Canberra*

11.1 Introduction

Agricultural growth is important for overall economic development and poverty reduction, particularly in developing countries, where the majority of poor people live in rural areas and depend directly or indirectly on agriculture for their livelihoods (Johnston and Mellor, 1961). Diminishing returns to conventional farm-level factor inputs, high fuel and fertilizer prices, environmental degradation, the possibility of output-reducing climate change, and the declining availability of arable land, fresh water supplies and other natural resources all point to the increased long-term importance of agricultural productivity growth.

In the case of Thailand, agriculture is an important source of export earning and rural income. Sustaining agricultural growth is thus important for maintaining export competitiveness and improving the living standard of the majority of poor people residing in rural areas and directly involved in agricultural production (Warr, 2004). Growth of total factor productivity (TFP) has been shown to contribute significantly to output growth in the Thai agricultural sector, and its contribution has been substantially greater than in the non-agricultural sectors (Tinakorn and Sussangkarn, 1996; Chandrachai *et al.*, 2004; Warr, 2009).

Recent studies have, however, indicated a slowdown in agricultural TFP growth. Thus, refocusing attention on what determines TFP in Thai agriculture is important for understanding and sustaining long-term agricultural growth and thereby maintaining its contribution to overall economic performance.

Unfortunately, there is limited empirical evidence on what determines the growth rate of TFP in Thai agriculture, with the majority of previous studies focusing on the determinants of TFP in the overall economy. Such studies include Chandrachai *et al.* (2004), Chokpaisansin (2002) and Kaipornsak (1999). Internationally, studies of this issue have generally overlooked some other potentially important determinants of productivity growth in agriculture, including international research spillovers, weather changes and epidemics. There is a strong possibility that the omission of these variables could have produced an upwards bias in the estimated effects of public research, potentially throwing into doubt the conclusion that public research has been a significant source of productivity growth. Moreover, these studies suffer from other potentially important statistical deficiencies by only investigating factors affecting TFP expressed in growth-rate terms and ignoring level or long-term information, and often imposing arbitrarily restrictive forms of lags. The latter could be responsible for further upward biases in the estimated effects of public research (Alston *et al.*, 1998b; Alston *et al.*, 2000).

This study measures TFP in Thai agriculture and examines the factors influencing it, taking account of the variables noted above that are overlooked in most previous studies, along with public infrastructure investments and extension. The analysis employs time-series data at an aggregate level, covering the period from 1970 to 2006. The scope of the study focuses on crops and livestock because these two subsectors dominate agricultural output.[1] TFP measurement and the investigation of its determinants are undertaken separately for both the crops and livestock sectors, using error correction statistical methods and without imposing predetermined lag structures. (Suppannachart and Warr, 2011, address some of these issues for the crops sector alone.)

11.2   Review of TFP Measurement and Determinants

In general, TFP measurement methods used in empirical productivity studies can be grouped into two main approaches: conventional or non-frontier methods, and frontier analysis. The first approach assumes outputs are efficiently produced on the production frontier, whereas the second allows for outputs being produced off the frontier. The frontier analysis is often applied to cross-sectional or panel data, whereas the conventional approach is mainly applied to time series macro-productivity data sets.

Both the conventional and frontier approaches can be further classified into parametric and non-parametric methods. The non-parametric method does not impose a specific functional form, whereas the parametric method imposes a functional form and employs econometric techniques in estimating a production function, a cost function or a profit function. Table 11.1 summarizes the principal methods used in measuring TFP and the corresponding data requirements.

In examining TFP determinants, TFP is generally decomposed into embodied and disembodied technical change. Embodied technical change refers to change captured in factor inputs, such as improved seeds, breeds or a new type of machinery (Alston *et al.*, 1998b). Disembodied technical change means technological change not embodied in factor inputs but arising exogenously to the firm in

Table 11.1. Summary of TFP measurement methods and data requirements.

	Conventional approach		Frontier approach	
	Non-parametric	Parametric	Non-parametric	Parametric
Principal methods	TFP index/ GA	LS/ GA	DEA	SFA
Estimation of specific functional form and statistical tests	No	Yes	No	Yes
Data used:				
Cross sectional	Yes	Yes	Yes	Yes
Time series	Yes	Yes	No	No
Panel	Yes	Yes	Yes	Yes
Basic method requires data on:[a]				
Input quantities	Yes	Yes	Yes	Yes
Output quantities	Yes	Yes	Yes	Yes
Input prices	Yes	No	No	No
Output prices	Yes	No	No	No

[a]This list applies to the production function method only. *Source*: Adapted from Coelli *et al.* (2005, p. 312); GA, growth accounting, LS, least squares, DEA, data envelopment analysis, SFA, stochastic frontier analysis.

the form of better methods and organization that improve the efficiency of factor inputs (Chen, 1997), such as more effective production methods that improve input usage.

In the context of agricultural productivity, factors found to influence TFP are public and private agricultural research, extension services, infrastructure investment, education of farmers and economic policies (Huffman and Evenson, 2005; Mundluk, 1992). Numerous studies have investigated the sources of productivity growth, but their theoretical foundations differ (e.g. Griliches, 1996; Evenson and Pray, 1991; Mahadevan, 2002; Mundlak, 1992; and Huffman and Evenson, 2005). Determining the factors that influence TFP is a matter of empirical study.

11.3 Analytical Framework

The primary concept is the production function (Eqn 11.1):

$$Q^t = h(X^t, Z^t), \qquad (11.1)$$

where Q denotes output, X denotes conventional farm-level inputs – labour, land and capital – and Z denotes unconventional inputs, such as research, extension, infrastructure, weather and disease outbreaks, all measured at time t. These unconventional inputs, especially research, extension and infrastructure, are applied beyond the farm level and are therefore normally not considered in farm-level productivity studies, which focus on the conventional inputs. Nevertheless, the unconventional inputs do affect aggregate agricultural output. It is mathematically convenient, but not essential, to assume that the function h is separable between conventional and non-conventional inputs, giving Eqn 11.2:

$$h(X^t, Z^t) = f(X^t)g(Z^t) \qquad (11.2)$$

By definition, TFP is an index of aggregate output relative to an index of aggregate conventional inputs, combined. TFP is therefore a function of the levels of non-conventional inputs. Thus (Eqn 11.3):

$$TFP^t = Q^t / f(X^t) = g(Z^t) \qquad (11.3)$$

11.4 Methodology

This section will explain three features of the empirical component of this study: the TFP measurement method, the TFP determinants model and the estimation method.

11.4.1 TFP Measurement

Although there are several approaches for measuring TFP, as shown in Table 11.1, a suitable approach depends on the objectives of the study and data availability. Because this study aims to examine sources of agricultural growth at an aggregate level, the growth accounting framework is considered the most appropriate. The competitive equilibrium conditions, which are the underlying assumptions of the growth accounting approach, are reasonable for the case of Thai agriculture. The agricultural sector is well characterized by a competitive market, in that there are a large number of farmers who maximize profit (or minimize cost) and take prices as given (Pochanukul, 1992). Compared with other industries, such as manufacturing and services, the agricultural sector is generally considered a suitable vehicle for applying the growth accounting method, and this method is widely applied in previous Thai studies, such as Tinakorn and Sussangkarn (1996), Chandrachai et al. (2004) and Poapongsakorn (2006).

Analytical framework

The method is national income-based growth accounting, in the sense that most output and input data are obtained from the national accounts. We begin with the basic production function, Equation 11.1 above, which explains the relationship between output and inputs, both conventional and non-conventional. Assuming $h(X^t, Z^t)$ to be differentiable, it is familiar that:

$$q^t = \sum_{i=1}^{I} \varepsilon_i^t x_i^t + \sum_{j=1}^{J} \eta_j^t z_j^t \qquad (11.4)$$

where q^t, x_i^t and z_j^t denote the proportional rates of change of Q^t, X_i^t and Z_i^t, respectively. Thus, $q^t = (dQ^t/dt)/Q^t$. The parameters $\varepsilon_i^t = h_{xi}X_i^t / Q^t$ and $\eta_j^t = h_{zj}Z_j^t / Q^t$ denote the elasticities of output with respect to the inputs X_i^t and Z_i^t, respectively. The growth rate of TFP is now given by:

$$TFPG^t = q^t = \sum\nolimits_{i=1}^{I} \varepsilon_i^t x_i^t + \sum\nolimits_{j=1}^{J} \eta_j^t z_j^t \quad (11.5)$$

Accordingly, a change in TFP is measured as the residual part of the movement in output left unexplained by the growth of conventional factor inputs (Solow, 1957; Jorgenson and Griliches, 1967; Jorgenson, 1995). As is well known, if the function $f(X^t)$ is linearly homogeneous (constant returns to scale in conventional inputs), and these conventional factors are paid according to their marginal value products, then the elasticity parameters ε_i are equal to the corresponding factor cost shares at time t and these factor income shares, denoted S_{Xi}^t, will sum to unity across the I conventional factor inputs.[2]

Because differentiation is applicable only to continuous variables, the growth rate terms in the above equations refer to an instantaneous rate of change. In practice, however, discrete data, especially annual data, are normally used in empirical work. Hence, the discrete annual data can be applied to approximate Eqn 11.5 by taking the average of factor shares at two consecutive periods (Oguchi, 2004):

$$TFPG^t = \ln TFP^t - \ln TFP^{t-1}$$
$$= (\ln Q^t - \ln Q^{t-1})$$
$$- \frac{1}{2}\sum\nolimits_{i=1}^{I}(S_{Xi}^t + S_{Xi}^{t-1})(\ln X_i^t - \ln X_i^{t-1}).$$
$$(11.6)$$

In this study, the conventional inputs identified are labour, capital and land. The labour and land inputs are adjusted for their quality changes, following the method developed by Tinakorn and Sussangkarn (1996), which is suitable for Thai data. For labour, the adjustment method accounts for the effect of qualitative changes in age, sex and education. The land input used in crop production is adjusted by the effect of irrigation, to account for multiple cropping.

Factor quality adjustment

LABOUR QUALITY.   to adjust for the qualitative change of labour, this study adopts the

labour quality-adjusted index computed by the Thailand Development Research Institute (TDRI). The index computation follows the method developed by Tinakorn and Sussangkarn (1996). This method modifies Denison's (1967) pioneering study on labour quality adjustment to suit the nature of Thailand's data. The rationale is that changes in age, sex and the education levels of workers over time should have an impact on output if the marginal products of different groups are not the same. For example, as long as younger workers and those with lower education earn less, the shifting of worker composition away from these groups indicates an improved contribution to output (Tinakorn and Sussangkarn, 1996). The constant growth in the number of workers may increase or decrease output because of the changing composition of workers' age, sex and education level. Adjusting for changes in labour quality therefore gives a clearer picture of the contribution of labour input to output growth, both quantitatively and qualitatively.

The adjustment method computes the age-sex-education index, or quality-adjusted labour index, on the basis of the wage differentials of workers classified according to their age, sex and education level. In the Thai government's Labour Force Survey (LFS), the male and female labour force is categorized into five age groups: 15–19, 20–29, 30–39, 40–49, and 50 and over. They are then grouped according to levels of education attainment. There are five educational classes: (i) no formal education, elementary education or lower; (ii) upper and lower secondary education; (iii) vocational education; (iv) university education or high-level technical vocational education; and (v) teacher training education.

The qualitative changes in age, sex and education are measured simultaneously because the average wage of male and female workers at different age groups is influenced by education levels and vice versa. This method is widely applied in the Thai literature (Poapongsakorn, 2006; Chandrachai *et al.*, 2004; Chockpisansin, 2002). It is summarized below. (See details in Tinakorn and Sussangkarn, 1996.)

1. Tabulate the average wage of private employees in each year by age (i), sex (j) and education levels (k) from the LFS.
2. Calculate the average wage differential index ($W_{i, j, k}$) by using the average wage of male workers, age 30–39, and using elementary education as a reference point (that is, their wage is set equal to 1).
3. Compute the share of employment ($S_{i,j,k}$) or the percentage of workers, categorized by age, sex and education levels.
4. The index is computed as the weighted average of wage differential using employment share as weights. The index formula is
$I = \sum S_{i,j,k} \cdot W_{i,j,k}$
5. This index is then used to adjust labour input.

LAND QUALITY.   With the same rationale explained above for labour quality adjustment, productivity changes arising from land quality changes can be separated out from the residual TFP. This study makes an adjustment on the stock of land in order to distinguish the effect of land quality changes over time. Access to irrigation can improve the quality of land, for example, by enabling farmers to grow rice crops during dry seasons, effectively increasing the amount of land service provided by the same physical area of land. An expansion in irrigated area is expected to increase agricultural output and productivity (Tinakorn and Sussangkarn, 1996).

To account for the impact of irrigation on land quality, it is important to consider which source of land input data is being used. The two sets of land data used in this study are stock and flow. If available data relate to the stock of land area, which does not take into account second rice – that is, rice grown during dry seasons, relying on irrigation systems – or multiple cropping, it is appropriate to adjust the land input with its access to irrigation. If available data cover land area that already includes multiple cropping, however, it is not necessary to adjust it to allow for the effect of irrigation.

Following Tinakorn and Sussangkarn (1996), the adjustment method of land input quality uses an index of irrigated area to adjust the index of cultivated area. Numbers are converted into simple indices to compare their changes with the base year. The method then adjusts the stock of land series with an irrigated area series.[3] This method is also used in Poapongsakorn (2006). The steps undertaken are as follows:

1. Tabulate the accumulated irrigation area and land area that needs to be adjusted.
2. Find the proportion of irrigated area in land area.
3. Calculate an index of proportion of irrigated area and an index of land area by using 1988 as a reference or base year.
4. The proportion of irrigated area index is used to adjust land input by multiplying it by the index of land area.

Factor cost shares

Because the growth rates of the factor inputs are so different, the calculation of TFP is sensitive to the factor cost shares (the ε_i^t parameters in Eqn 11.5 above) used. In particular, because capital inputs grow faster than other inputs, the higher the share of capital used in calculating TFP, the lower the resulting estimate of TFP growth.

The labour income share, S_L, is calculated by dividing the labour income with GDP at factor cost. This is expressed as $S_L = (W \cdot L)/GDP_{FC}$, where W denotes the imputed average wage rate, discussed below, L denotes the number of employed persons, and GDP_{FC} denotes GDP at factor cost, all in current prices. The average wage rate reported by the LFS includes only that of private or hired workers. Since it does not incorporate the wage rate of self-employed, own-account workers or unpaid family workers, which are all important sources of labour in Thai agriculture, the LFS wage rate is too high, and using it to calculate the labour income share would produce upwardly biased estimates. Following Tinakorn and Sussangkarn (1996) and Poapongsakorn (2006), the LFS average yearly wage is adjusted downwards by the wage payment of all agricultural workers, obtained from the social accounting matrix (SAM), developed by the Thailand Development Research Institute. This adjusted wage is the imputed wage based on the SAM, taking into account self-employed and unpaid labour.

For land income, the share is calculated as $S_N = (R \cdot N)/GDP_{FC}$, where S_N denotes the average land income share, R denotes the average land rent, N denotes land area, and GDP_{FC} again denotes GDP at factor cost, all in current prices. The rent for the agricultural sector, crops and livestock is estimated based on average rent per rai, derived by dividing total rent (actual and imputed) in the national accounts by the corresponding land area. The rent per rai is assumed to be the same for both crops and livestock. Total land rent and land area data are obtained from the National Economic and Social Development Board. It comprises the actual rent and the rent imputed for owner-occupied land.

After computing the factor income shares for labour and land, the share for *capital*, S_K, is computed as a residual. That is: $S_K = 1 - S_L - S_N$.

The resulting factor cost shares used in the calculation of TFP are described in Fig. 11.1. Table 11.2 summarizes their average values. The capital income shares reported are high, relative to those estimated by earlier studies for other developing countries, resulting (in the present study) in lower estimated levels of TFP growth. It is possible that by failing to adjust wage rates in agriculture for the prevalence of self-employed and unpaid family labour, earlier estimates of labour income have overestimated labour income shares and thereby underestimated the capital income shares derived as residuals, in turn overestimating TFP growth rates in agriculture.

11.4.2 The TFP determinants model

The TFP determinants model incorporates factors affecting the productivity of conventional inputs, X. A list of these determinants is equivalent to a list of the non-conventional inputs, Z. Our statistical analysis is based on a conceptual model in which the non-conventional inputs include agricultural research, domestic and foreign, as well as other economic and non-economic factors, such as extension services, infrastructure and weather.[4] Research lags are also incorporated, as discussed below. Other explanatory variables are explored in accordance with their potential connections with TFP in the Thai agriculture context.

In stylized form, the model is (with expected signs in parentheses):

$$TFP = g\,(R^p,\ R^f,\ E,\ I,\ RR,\ TO,\ W,\ D^e) \quad (11.7)$$

where TFP = total factor productivity, $R^p(+)$ = real public agricultural research expenditure, $R^f(+)$ = international agricultural research spillovers, $E(+)$ = real public

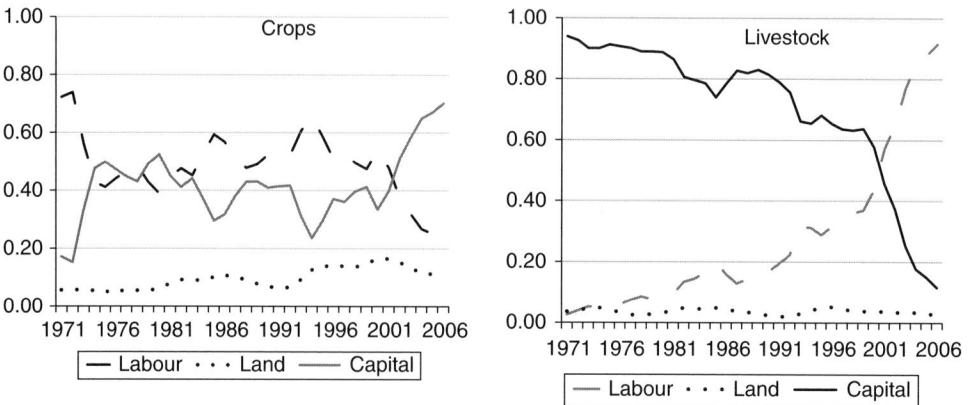

Fig. 11.1. Cost shares in crop and livestock sectors, 1971–2006. Source: Authors' calculations using quality-adjusted data described in the text.

Table 11.2. Summary of average factor cost shares.

1971–2006	Labour	Land	Capital
Crop	0.49	0.10	0.42
Livestock	0.27	0.04	0.70
Total agriculture	0.53	0.08	0.40

'Total agriculture' consists of crops, livestock, fisheries, forestry and agricultural services. See Fig. 11.1 for the time series behaviour of these cost shares as actually used in calculating total factor productivity growth in this study. *Source*: Authors' calculations.

agricultural extension expenditure, $I(+)=$ infrastructure (rural roads and irrigation), $RR(+)=$resource reallocation, $TO(+)=$trade openness, $W(+)=$weather or climate factors, $D^c=$case-specific dummy variables comprising: $D^{boom}(+)=$dummy variable for the world agricultural commodity boom of 1972–1974, $D^{bird}(-)=$dummy variable for the Avian Influenza outbreak of 2004.

Public agricultural research, within-country, is a potential source of technical change that raises productivity and sustains output growth (Ruttan, 1987). It increases the stock of knowledge, which either facilitates the use of existing knowledge or generates new technology. Hence, an increase in research expenditure within Thailand is expected to raise TFP.

International research spillovers are potentially important sources of productivity growth. But they have often been ignored in the literature on the impact of agricultural research, resulting in an omitted variable bias (Alston *et al.*, 1998b; Alston, 2002; Fuglie and Heisey, 2007). The model incorporates foreign research on crops and livestock relevant for Thailand, and it is expected to increase domestic TFP.

Agricultural extension involves a dissemination of research results to farmers through information distribution, training and demonstration. It may also indirectly influence the agricultural research process by conveying feedback from farmers to researchers that might improve future research. Effective agricultural extension should improve productivity.

Infrastructure is considered a fixed factor that contributes positively to agricultural growth and productivity (Evenson and Pray, 1991; Evenson, 2001). It is typically not included among the conventional inputs in growth accounting, and its effect on agricultural growth is thereby captured in the residual TFP.

Resource reallocation can raise TFP at the aggregate level by allowing factors to move from lower to higher marginal productivity sectors. For instance, movement of labour from the agricultural sector to a higher productivity sector such as manufacturing or services can increase TFP growth in the overall economy (Jorgenson, 1988). Within a sector, productivity growth can result from the reallocation of resources among subsectors and among commodities when their levels of TFP differ, and this does not necessarily require any new technology. Empirical evidence has shown that resource reallocation contributes significantly to TFP growth in Thailand (Chandrachai *et al.*, 2004; Warr, 2009).

Trade openness helps achieve economies of scale by expanding market size through export. Economies of scale bring about real cost reductions, thereby increasing productivity. It also enhances market competition through import and export. Competition influences technological development, thereby increasing TFP. More open economies and international trade are generally found to be favourable to TFP (Edwards, 1998; Urata and Yokota, 1994).

Weather or climate variation is considered a variable explaining changes in TFP under the conventional TFP decomposition framework (Evenson, 2001). Good weather, like more rainfall or less flooding and fewer droughts, should raise TFP relative to the opposite.

The world agricultural commodity boom of 1972–1974 raised the real price of internationally traded food commodities, thereby inducing additional production. That price boom has been shown to be one of the main driving forces behind Thailand's rapid agricultural growth in the early 1970s (Poapongsakorn, 2006). However, the increase in output might not have been fully reflected in the measured use of inputs. During a boom, farmers tend to utilize existing inputs more intensively, which does not necessarily show up in measured input growth. Measured productivity therefore rises, at least partly through measurement error.

Epidemic is represented by the 2004 outbreak of the Avian Influenza virus or Bird Flu. A dummy variable is used to capture the effect of the Bird Flu outbreak in the livestock productivity function. It should reduce TFP in the livestock sector.

Estimating research stocks using the perpetual inventory method

Because of the cumulative nature of the research process, a new finding is partially a result of previous findings. Constructing a research stock from past research expenditure has been a common practice. Creating a research stock has been adopted in many studies using various techniques to maintain a sufficient degree of freedom and mitigate multicollinearity (e.g. Evenson and Pray, 1991; Alston *et al.*, 1994; Huffman and Evenson, 1992, 2005; Hall and Scobie, 2006).

The perpetual inventory method is used to create a research stock from a flow of investment. This is a commonly used technique in calculating the capital stock in many countries, including Thailand (National Economic and Social Development Board, 2006). As today's innovation is partially built on past research, agricultural research can be considered as knowledge capital stock the effects of which may diminish over time but can also potentially last indefinitely (Alston *et al.*, 1998a; Griliches, 1998). The perpetual inventory method formula is (Eqn 11.8):

$$RS_t = R_t + (1-\delta)RS_{t-1} \qquad (11.8)$$

where RS_t is the research stock in year t, R_t is agricultural research expenditure in year t and δ is the depreciation rate. The initial stock (RS_0) is calculated as Eqn 11.9:

$$RS_0 = \frac{R_0}{g+\delta} \qquad (11.9)$$

where R_0 is agricultural research expenditure in the first year available and g is the average geometric growth rate of the research expenditure series during the first 20 years.[5] The difficult part is determining the depreciation rate. Previous studies have often assumed an annual depreciation rate of 5% (Coe and Helpman, 1995; Johnson *et al.*, 2005) and 15% (Hall and Scobie, 2006). This study employs three depreciation rates: 0%, 5% and 15%.[6] Using backward substitution, the research stock in Eqn 11.8 can be expressed as an infinite weighted sum of current and past research expenditure (Eqn 11.10):

$$RS_t = R_t + (1-\delta)\,R_{t-1} + (1-\delta)^2 R_{t-2} + \cdots (11.10)$$

The error correction estimation method

The error correction modelling (ECM) procedure of Hendry (1995) is employed as it permits investigation of both short-run and long-run determinants of TFP, while allowing dynamic and flexible forms of lags.[7] Another reason for using this approach is that it does not require the variables under consideration to have the same order of integration. Table 11.3 shows that the variables used in this study are a mixture of stationary series (or I(0)) and non-stationary series integrated of order 1 (or I(1)). Most of the variables are I(1) such as public research (R^p), extension (E), irrigation ($I^{irrigation}$) and rainfall (W^{rain}). Variables that are I(0) include foreign research (R^f), roads (I^{road}) and weather conditions ($W^{weather}$). This approach minimizes the possibility of estimating spurious relationships, while retaining long-run information without arbitrarily restricting the lag structure (Hendry, 1995). The ECM also provides estimates with valid t-statistics, even in the presence of endogenous explanatory variables (Inder, 1993).

The estimation procedure begins with an autoregressive distributed lag (ADL) specification of an appropriate lag order:

$$Y_t = \alpha + \sum_{i=1}^{m} A_i Y_{t-i} + \sum_{j=1}^{k} \sum_{h=0}^{n} B_{hj} X_{j,t-h} + \mu_t$$

$$(11.11)$$

Table 11.3. Augmented Dickey–Fuller test for unit roots, 1970–2006.

Variables	t-Statistics for level without time trend	t-Statistics for level with time trend	t-Statistics for first difference without time trend	t-Statistics for first difference with time trend
ln TFP_{crops}	−1.476(0)	−3.531(0)**	−5.036(1)*	−4.950(1)*
ln $TFP_{livestock}$	−4.370(0)*	−4.720(1)*	−6.245(0)*	−6.397(0)*
ln R^p_{crops}	−1.296(1)	0.240(0)	−3.887(0)*	−4.135(1)*
ln $R^p_{livestock}$	−2.018(0)	−1.612(0)	−5.737(0)*	−6.010(0)*
ln E_{crops}	−1.655(0)	−0.145(0)	−4.784(0)*	−5.003(0)*
ln $E_{livestock}$	−1.477(0)	−2.215(0)	−6.676(0)*	−6.732(0)*
ln R^f_{crops}	−6.505(1)*	−4.252(1)*	−4.149(0)*	−6.382(0)*
ln $R^f_{livestock}$	−3.032(1)*	−2.999(1)	−5.100(1)*	−5.038(1)*
ln TO	−2.030(0)	−1.496(0)	−7.998(0)*	−8.617(0)*
ln $I^{irrigation}$	−1.688(0)	−0.645(0)	−5.220(0)*	−5.936(0)*
ln I^{roads}	−0.992(1)	−3.829(5)*	−3.351(0)*	−3.386(0)*
ln RR	−1.532(0)	−1.674(0)	−5.187(0)*	−5.602(0)*
ln W^{rain}	−2.454(0)	−2.083(0)	−8.379(0)*	−8.717(0)*
ln $W^{weather}$	−6.198(0)*	−6.158(0)*	−10.070(0)*	−9.914(0)*

All variables are measured in natural logarithms. *and **denote the rejection of the null hypothesis implying the variable is stationary at the 5% and 10% level, respectively. Numbers in parentheses indicate the order of augmentation selected on the basis of the Schwarz criterion.

where α is a constant, Y_t is the endogenous variable at time t, $X_{j,t}$ is the jth explanatory variable at time t, and A_i and B_{hj} are parameters. The general ADL allows the initial lag length on all variables at two periods, except for the research variable where the lag length extends to four periods. The two-year lag is the established practice in modelling with annual data (Athukorala and Tsai, 2003).

Equation 11.9 can be rearranged by subtracting Y_{t-1} from both sides, yielding the explanatory variables in terms of differences, representing the short-run multipliers and the lagged levels of both the dependent and explanatory variables, capturing the long-run multipliers of the system (Eqn 11.12) (see equation 11.12 at bottom of page).

where $A^*_i = -[I - \sum_{i=1}^{m-1} A_i]$, $B^*_{hj} = \sum_{h=0}^{n-1} B_{hj}$, $C_0 = -\left[I - \sum_{i=1}^{m} A_i\right]$, $C_j = \left[\sum_{h=0}^{n-1} B_{hj}\right]$, I is the identity matrix and the long-run multipliers of the system are given by $C_0^{-1} C_j$.

Equation 11.12 forms the basis for the error correction mechanism representation of the model (Wickens and Breusch 1988;

Banerjee, et al. 1993; Hendry, 1995). Under the ECM, the long-run relationship is embedded within a sufficiently detailed dynamic specification, including both lagged dependent and independent variables, which helps minimize the possibility of estimating a spurious regression. The ECM can be estimated by the ordinary least squares method and the short- and long-run parameters can be separately identified. Equation 11.9 is the 'maintained hypothesis' for specification search. The full model is 'tested down' by dropping statistically insignificant lag terms using the standard testing procedure to obtain a parsimonious ECM. The final preferred model is required to satisfy standard diagnostic tests, including the Breush–Godfrey LM test for serial correlation in the regression residual, the Ramsey test for functional form mis-specification (RESET), the Jarque–Bera test of normality of the residual (JBN), Engle's autoregressive conditional heteroskedasticity test (ARCH) and the Augmented Dickey–Fuller test for residual stationarity (ADF).

$$\Delta Y_t = \alpha + \sum_{i=1}^{m-1} A^*_i \Delta Y_{t-1} + \sum_{j=1}^{k} \sum_{h=0}^{n-1} B^*_{hj} \Delta X_{j,t-h} + C_0 Y_{t-m} + \sum_{j=1}^{k} C_j X_{j,t-n} + \mu_t \qquad (11.12)$$

11.5 Data

The output and input data are time-series at an aggregate level, covering 37 years from 1970 to 2006. As TFP is computed for crops and livestock separately, the data sets are obtained for crop and livestock individually. Definitions and sources of data for TFP measurement are summarized in Table 11.4 and those for the TFP determinants are shown in Table 11.5.

11.6 Results

11.6.1 TFP Measurement: results from the growth accounting model

The general finding from the growth accounting analysis is that TFP makes an important contribution to its own sector's output growth. The details are summarized in Table 11.6. During the period 1971–2006, growth of the physical capital stock was the most important source of output growth in both the crop and livestock sectors, and TFP growth was second. Specifically, the average annual rate of TFP growth in the crop sector is estimated at 0.68%, accounting for 20.82% of crop output growth. Similarly, livestock TFP growth is estimated at 0.67%, accounting for 17.49% of livestock output growth. The patterns of crop and livestock TFP growth, as estimated by the above analysis, are shown in Fig. 11.2.[8] For agriculture as a whole, growth of the capital stock has accounted for 54.73% of the growth of value added, and TFP growth has accounted for a further 20.35%.

Table 11.4. Summary of the data used in TFP measurement, 1970–2006.

Variables	Definitions	Sources
Agricultural output	GDP at 1988 prices (value added)	National Income of Thailand, NESDB (1970–2006)
Agricultural labour	Number of employed persons aged 15 and above	Labor Force Survey, NSO (1971–2006); Poapongsakorn, 2006; and TDRI (1977–2003)
Agricultural land		
Crop land	Land used in crop production	Office of Agricultural Economics (1970–2006);
Livestock land	Grass and privately own area for livestock	Department of Livestock Development (1999–2006); Poapongsakorn, 2006 (1980–2003)
Agricultural capital	Net capital stock at 1988 prices	National Income of Thailand, NESDB (1970–2006)
Agricultural wage	Imputed wage of all workers, measured as private workers' wage adjusted by 1995 Social Accounting Matrix (SAM) wage to account for self employed and unpaid family labour	Labour Force Survey, NSO (1977–2006) Poapongsakorn, 2006 and TDRI (1977–2004) Coxhead and Plangpraphan, 1999 (1970–1976)
Land rent	Actual and imputed rent (rai)	NESDB
Labour quality-adjusted index	Qualitative changes in age, sex and education attainment of agricultural workers	TDRI (based on Labor Force Survey, NSO)
Irrigation	Accumulated irrigation area (rai), including small, medium and large scale irrigation projects	Office of Agricultural Economics (OAE)
Factor income share	Value of factor income divided by GDP at factor cost	NESDB (GDP at factor cost)

Table 11.5. Summary of the data used in TFP determinants model.

Variables	Abbreviation	Data sources	Years
Dependent variables			
Total factor productivity index (adjusted for input quality changes)[a]	TFP	Authors' calculation based on the growth accounting method	1971–2006
Explanatory variables			
Publicly funded, within-country research = real public research budget in the crop and the livestock sector	R^p	Bureau of the Budget, Office of Prime Minister	1961–2006
Foreign research spillovers	R^f		
For crops: = CGIAR funding to IRRI, CIAT and CIMMYT in US dollars[b]		CGIAR financial statements	1972–2006
For livestock: = import values of animal breeds as percentage share in livestock output		Office of Agricultural Economics (OAE)	1970–2006
Extension services = real public extension budget in the crop and livestock sector	E	Bureau of the Budget, Office of Prime Minister	1961–2006
Infrastructure			1970–2006
Irrigation = percentage share of irrigated area in total agricultural land area	$I^{irrigation}$	Office of Agricultural Economics	
Road = length of rural roads, unpaved roads and asphalt (km)	I^{roads}	Fan *et al.* (2004)	
Trade openness = agricultural export and import as percentage share in total agricultural output	TO	Office of Agricultural Economics	1970–2006
Resource reallocation (only available for crop model) = non-rice household as percentage share of total agricultural households	RR	Office of Agricultural Economics	1970–2006
Natural/case-specific factors		Office of Agricultural Economics	1970–2006
Rainfall = amount of rainfall in millimetre	W^{rain}		
Weather: drought or flooding = rice harvested as share in total rice planted area	$W^{weather}$		
Bird flu outbreak = dummy variable takes value 1 from 2004 and 0 otherwise	D^{bird}		
Agricultural commodity boom = dummy variable takes value 1 from 1972 to 1974	D^{boom}		

[a]TFP growth measure is converted into level of TFP index using 1971 as a base year, with the level of TFP set equal to unity for that year. [b]CGIAR stands for the Consultative Group of International Agricultural Research, IRRI is International Rice Research Institute, CIAT is International Center for Tropical Agriculture and CIMMYT is International Wheat and Maize Improvement Center. They are centres that have close collaborations with Thailand.

Table 11.6. Sources of growth in Thai agriculture.

Period	Output growth	Labour Unadjusted	Labour Adjusted for quality	Land Unadjusted	Land Adjusted for quality	Capital stock	Unadjusted TFPG	Adjusted TFPG
Crops								
1971–1980	3.98	1.43	1.43	0.27	0.32	1.91	0.37	0.33
1981–1990	4.79	1.28	1.33	0.12	0.34	0.85	2.54	2.27
1991–2000	2.92	−1.10	−0.99	−0.02	0.15	1.41	2.62	2.35
2001–2006	2.41	−0.90	−0.88	−0.02	0.19	6.22	2.91	3.13
1971–2006	3.27	0.20	0.25	0.10	0.25	2.09	0.88	0.68
Livestock								
1971–1980	5.84	0.43	0.34	0.08	0.08	1.72	3.62	3.70
1981–1990	4.56	1.02	1.05	0.06	0.06	2.54	0.94	0.91
1991–2000	1.34	0.43	0.79	−0.18	−0.18	0.48	0.60	−0.25
2001–2006	3.79	13.11	13.64	−0.01	−0.01	−7.56	−1.75	−2.28
1971–2006	3.83	3.00	3.00	0.17	0.17	0.17	0.88	0.67
Agriculture (crops, livestock, fisheries, forestry, agricultural services)								
1971–1980	3.32	1.63	1.62	0.17	0.24	0.87	0.65	0.59
1981–1990	4.28	1.27	1.37	0.09	0.25	0.94	1.99	1.72
1991–2000	2.74	−0.83	−0.62	0.00	0.10	2.72	0.85	0.54
2001–2006	2.55	−0.26	−0.07	−0.01	0.14	2.16	0.66	0.32
1971–2006	3.02	0.45	0.57	0.07	0.19	1.65	0.85	0.61

Source: Authors' calculations using data explained in the text.

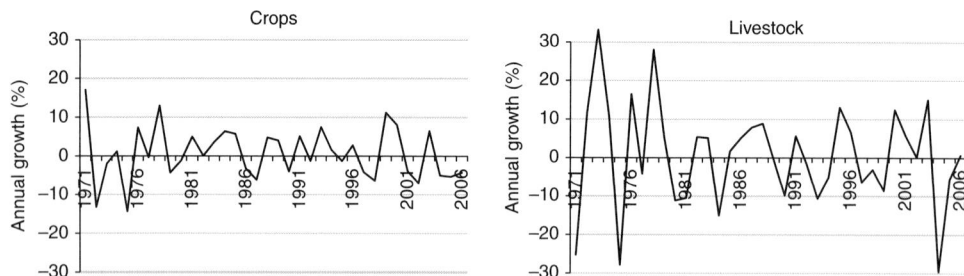

Fig. 11.2. TFP growth rates in crop and livestock sectors, 1971–2005. Source: Authors' calculations using data described in the text.

11.6.2 TFP determinants: results from the error correction models

In general, public agricultural research is the major factor positively influencing TFP in both the crop and livestock sectors. The positive and significant impact of public research is consistent with the theory and findings of studies of many countries (Evenson, 1993; Ruttan, 2002; Thirtle *et al.*, 2003). Other major determinants of TFP turn out to be different between the crop and livestock models. In Table 11.7, the results for crops are shown on the left-hand side and those for livestock on the right-hand side.[9]

The TFP determinant models in both the crops and livestock sectors are statistically significant at the 1% level in terms of the F test. Both equations pass all the standard diagnostic tests. The choice of dropping or keeping variables in the final models was statistical acceptance in the joint variable deletion tests against the maintained hypothesis. The error correction coefficient ($lnTFP_{t-1}$) has the expected negative sign

Table 11.7. TFP determinants in crop and livestock sectors – domestic research as a flow.

	Crop sector			Livestock sector	
	Dependent variable: $\Delta \ln TFP_t^{crop}$			Dependent variable: $\Delta \ln TFP_t^{livestock}$	
	Estimated coefficients (t-ratios)	Long-run elasticity		Estimated coefficients (t-ratios)	Long-run elasticity
Constant	−1.056 (−6.460)***		Constant	0.386 (2.246)**	
$\Delta \ln R_{t-3}^p$	0.155 (4.423)***		$\Delta \ln R_t^f$	0.012 (0.517)	
$\Delta \ln E_{t-1}$	0.137 (3.665)***		$\Delta \ln E_{t-1}$	0.119 (1.728)*	
$\ln R_{t-3}^p$	0.059 (1.876)*	0.067 (2.117)**	$\ln R_{t-3}^p$	0.128 (2.074)**	0.173 (2.111)**
$\ln R_{t-1}^f$	0.092 (2.955)***	0.105 (3.045)***	$\ln E_{t-1}$	−0.089 (−1.590)	−0.121 (−1.578)
$\ln I_{t-1}^{roads}$	0.033 (1.977)**	0.038 (1.962)**	$\ln R_{t-1}^f$	−0.003 (−0.168)	−0.004 (−0.167)
D^{boom}	0.127 (3.104)***	0.145 (3.189)***	D^{bird}	−0.165 (−2.720)***	−0.224 (−2.593)***
$\ln TFP_{t-1}$	−0.873 (−6.664)***		$\ln TFP_{t-1}$	−0.739 (−5.510)***	
N (observations)	34			35	
k (no. of parameters)	8			8	
Adjusted R^2	0.69			0.50	
F-statistic	11.31			5.93	
S.E. of regression	0.03			0.09	
Diagnostic tests:					
LM(1), F(1, N-k-1)	0.06 (p=0.79)			0.00 (p=0.99)	
LM(2), F(2, N-k-2)	1.42 (p=0.26)			1.47 (p=0.25)	
RESET, F(1, N-k-1)	0.89 (p=0.35)			1.80 (p=0.19)	
JBN, $\chi^2(2)$	0.77 (p=0.68)			0.86 (p=0.65)	
ARCH, F(1, N-2)	0.00 (p=0.98)			1.31 (p=0.26)	
ADF	−5.79 (p=0.00)			−4.89 (p=0.00)	

The level of statistical significance is denoted as: *, 10%, **, 5% and ***, 1%. All variables are measured in natural logarithms except the dummy variables. Long-run elasticities can be computed by dividing the estimated coefficients of the level terms by the positive value of the coefficient of the lagged dependent variable, $\ln TFP_{t-1}$. Short-run elasticities are coefficients of the variables expressed in rate of change terms, with delta (Δ) operator. Diagnostic tests are: *LM*, Breush–Godfrey serial correlation LM test; *RESET*, Ramsey test for functional form mis-specification; *JBN*, Jarque–Bera test of normality of residual; *ARCH*, Engle's autoregressive conditional heteroskedasticity test; *ADF*, Augmented Dickey–Fuller test for residual stationarity. Numbers in parentheses are *p*-values of the test statistics.

and is statistically significant at the 1% level. It indicates the speed of adjustment of TFP to exogenous shocks that cause the system to deviate temporarily from the steady state described by the long-run coefficients. The coefficients corresponding to $lnTFP_{t-1}$ are quite large (0.87 and 0.74), implying a high speed of adjustment to dissipate such shocks. Because all variables are measured in logarithms, the regression coefficients can be interpreted as elasticities, and the size of the coefficients also indicates the magnitude of their relative influence. Factors affecting TFP in each sector are discussed below.

Crops

In the crop sector, major factors affecting TFP are crop production research, both public and foreign, agricultural extension, infrastructure and the commodity boom. Public agricultural research (R^p) is statistically significant at the 1% and 5% levels in the short run and long run, respectively. In the short run, an increase in public agricultural research spending of 1% leads to an increase in TFP growth of 0.16%. The short-run effects operate with three-year lags. In the long run, a 1% increase in public research spending raises TFP by 0.07%. The larger short-run impact indicates that research produces an initial surge in TFP growth, which tapers off in the long run but does not vanish.

Foreign research spillovers (R^f) – measured as the Consultative Group on International Agricultural Research (CGIAR) spending on the International Rice Research Institute (IRRI), the International Maize and Wheat Improvement Center (CIMMYT) and the International Center for Tropical Agriculture (CIAT) – have a positive and significant impact on TFP in the long run.[10] A 1% increase in foreign research spending results in a steady-state (long-run) increase in TFP of 0.11%.

Agricultural extension (E) affects crop TFP only in the short run. The estimated coefficients of the change term of E are statistically significant at the 1% level and are positively signed. There is no evidence, however, that extension services significantly influence TFP in the long run.

Infrastructure, as represented by the rural roads variable, and case-specific factors, as represented by the agricultural commodity boom, are shown to have a positive and significant impact on TFP. This is consistent with the literature and with the general expectation that infrastructure improves agricultural productivity and that a commodity boom encourages farmers to grow more crops and to use existing inputs more intensively in order to reap the benefits of a world agricultural price surge, which in turn increases output and hence productivity. There is no evidence that other potential factors, such as resource reallocation, trade openness or weather conditions, are statistically significant.

Finally, it is possible to decompose the long-term relationship between productivity and its explanators, as follows. Consider the

estimated long-term relationship see equation 11.13 at bottom of page:

$$\ln TFP_t = \hat{a} + \hat{b}^1 \ln R_t^p + \hat{b}^2 \ln R_t^f$$
$$+ \hat{b}^3 \ln I_t^{roads} + \hat{b}^4 D_t^{boom} + u_t, \quad (11.13)$$

where the variables are defined as in Table 11.7 and $\hat{a}$ and $\hat{b}^i$, $i = 1,...,4$ denote the estimated parameters. Writing this equation in stylized form, for simplicity:

$$Y_t = \hat{a} + \sum_{i=1}^{4} \hat{b}^i X_t^i + u_t. \quad (11.14)$$

Because the estimated relationship must pass through the means of the data, it must be true that:

$$\bar{Y} = \hat{a} + \sum_{i=1}^{4} \hat{b}^i \bar{X}^i. \quad (11.15)$$

The mean value of the dependent variable is divided into two components: an 'unexplained' component, the estimated intercept term; and an 'explained' component, the sum of the means of the respective independent variables, each multiplied by its estimated coefficient. Rearranging,

$$\bar{Y} - \hat{a} = \sum_{i=1}^{4} \hat{b}^i \bar{X}^i. \quad (11.16)$$

That is, the mean of the dependent variable, net of the estimated intercept term, is equal to the sum of the estimated coefficients each multiplied by the means of its respective independent variable. The left-hand side of Eqn 11.16 is 'explained by' the right-hand side. This now makes it possible to partition the right-hand side of Eqn 11.16 into the proportions of the explained component of the mean dependent variable attributable to each of the independent variables.

The implied calculation implies that public R&D explains 35.4% of the explained portion of TFP (its mean minus the estimated intercept term); foreign R&D explains 34.7%; road infrastructure explains 29.8%; and the 1972–1974 price boom explains a trivial 0.1%.

Livestock

Major factors explaining livestock TFP are public agricultural research and the Avian

Influenza outbreak. Public research has a positive and significant impact only in the long run. The estimated long-run elasticity, statistically significant at the 5% level, suggests that a 1% increase in government research spending leads to a 0.17% increase in TFP.

The dummy variable representing the Bird Flu outbreak has a negative impact on TFP, as expected. Its coefficient is statistically significant at the 1% level. The commodity boom dummy variable is not significant, confirming that it is not directly relevant for livestock as it is in the case of crops. Other variables were tested from various experimental runs, but there is no evidence they are significant factors.

Again, applying the decomposition of the long-term relationship described above among the statistically significant explanatory variables, public R&D accounts for almost all of the mean value of the dependent variable.

11.6.3   Research as a flow or a stock

The results summarized in Table 11.7 treat the domestic research variable as a flow. Some earlier literature suggests it is more appropriate to treat it as a stock. Most such studies use the Almon polynomial lag structure and the perpetual inventory method to construct the stock (Alston *et al.*, 1998b; Hall and Scobie, 2006). These studies argue that it is the accumulated stock of knowledge, as represented by the accumulated stock of research output, that is relevant for productivity. This hypothesis is tested in Table 11.8, by substituting the accumulated stock of research expenditure, depreciated at 5% per annum (denoted $\ln R_{t-3}^{pS}$), for the flow variable (denoted $\ln R_{t-3}^{pS}$) as in Table 11.7. When the stock variable is used, the domestic research variable is insignificant at the 5% level of significance for both crops and livestock. The overall significance of the model, according to the F-test, is also lower than in the case of the flow model, especially in the case of crops.

A conjecture that might explain this finding is that domestic research acts as a catalyst for the adaptation and adoption of new research results coming from abroad, measured by the foreign research variable. But it is only new domestic research that can play this role. The accumulated stock of past domestic research results is of little value in facilitating the adoption of new research findings derived from foreign sources. According to this hypothesis, it is not possible to rely on past investments in agricultural research to facilitate the adaptation and adoption of new research results derived from external sources. Current investment in research is needed to perform this function.

11.7   Conclusions

This study estimates total factor productivity in the Thai crop and livestock sectors using the conventional growth accounting method. The findings confirm that TFP makes an important contribution to both crop and livestock output growth during the study period of 1970–2006. Specifically, TFP accounts for about 21% of crop output growth and 17% of livestock output growth. These TFP growth measures are converted into a TFP index level and used as the dependent variables in the subsequent TFP determinants models.

The error correction modelling technique of Hendry (1995) is employed in examining factors influencing the measured TFP. The models are estimated separately for the crop and livestock sectors. Results show that major factors influencing crop TFP are public investment in agricultural research, foreign research spillovers, infrastructure and the world commodity boom. For the livestock sector, major factors are public research and the Bird Flu outbreak.

The determinants of TFP are not confined just to agricultural research, but also include extension services, infrastructure, weather and case-specific factors, such as the commodity boom and the Bird Flu outbreak. Other factors left unexplained are probably due to measurement errors and unmeasured inputs. The degradation of environmental and natural resources associated with agricultural production could be an unmeasured negative input that has been ignored in this study, as in most such studies.

Table 11.8. TFP determinants in crop and livestock sectors – domestic research as a stock.

	Crop sector			Livestock sector	
	Dependent variable: $\Delta \ln TFP_t^{crop}$			Dependent variable: $\Delta \ln TFP_t^{livestock}$	
	Estimated coefficients (t-ratios)	Long-run elasticity		Estimated coefficients (t-ratios)	Long-run elasticity
Constant	−0.457 (−1.833)		Constant	0.242 (1.602)	
$\Delta \ln R^{pS}_{t-3}$	0.105 (0.587)		$\Delta \ln R^{f}_{t}$	0.016 (0.655)	
$\Delta \ln E_{t-1}$	0.032 (0.613)		$\Delta \ln E_{t-1}$	0.115 (1.482)	
$\ln R^{pS}_{t-3}$	−0.144 (−1.955)*	0.208 (−1.757)	$\ln R^{pS}_{t-3}$	0.102 (1.360)	
$\ln R^{f}_{t-1}$	0.194 (4.171)***	0.046 (4.182)***	$\ln E_{t-1}$	−0.085 (−1.070)	
$\ln I^{roads}_{t-1}$	0.084 (1.910)*	0.038 (1.826)	$\ln R^{f}_{t-1}$	0.006 (0.315)	
D^{boom}	0.163 (3.021)***	0.145 (3.257)***	D^{bird}	−0.222 (−3.107)***	−0.224 (−3.104)***
$\ln TFP^{crop}_{t-1}$	−0.693 (−4.576)***		$\ln TFP^{livestock}_{t-1}$	−0.716 (−5.140)***	
N (observations)	34			35	
k (no. of parameters)	8			8	
Adjusted R^2	0.43			0.46	
F-Statistic	4.64			5.16	
SE of regression	0.04			0.09	
Diagnostic tests:					
LM(1), F(1, N-k-1)	0.83(p=0.37)			0.21 (p=0.64)	
LM(2), F(2, N-k-2)	1.30(p=0.29)			1.62 (p=0.21)	
RESET, F(1, N-k-1)	0.05(p=0.82)			1.00 (p=0.33)	
JBN, χ^2(2)	1.67(p=0.43)			0.26 (p=0.87)	
ARCH, F(1, N-2)	0.18(p=0.67)			2.28 (p=0.14)	
ADF	−4.71(p=0.00)			−5.23 (p=0.00)	

The level of statistical significance is denoted as: *, 10%; **, 5%; and ***, 1%. All variables are measured in natural logarithms except the dummy variables. Long-run elasticities can be computed by dividing the estimated coefficients of the level terms by the positive value of the coefficient of the lagged dependent variable, ln TFP_{t-1}. Short-run elasticities are coefficients of the variables expressed in rate of change terms, with delta (Δ) operator. Diagnostic tests are: LM, Breush-Godfrey serial correlation LM test; RESET, Ramsey test for functional form mis-specification; JBN, Jarque-Bera test of normality of residual; ARCH, Engle's autoregressive conditional heteroskedasticity test; ADF, Augmented Dickey-Fuller test for residual stationarity. Numbers in parentheses are p-values of the test statistics.

Notes

[1] The fisheries subsector is not included because of the different nature of production and input types.
[2] The use of constant returns to scale (CRS) technology is sensible when dealing with aggregate country-level data (Coelli and Rao, 2003). For Thailand and the agricultural sector, the CRS technology is applied in all growth accounting studies: for example, Budhaka, 1987; Tinakorn and Sussangkarn, 1996; Kaipornsak, 1999; Chandrachai et al., 2004, Warr, 2009; and the National Economic and Social Development Board, 2006.
[3] Similar to the splicing technique, the pattern of land growth is converted to follow that of irrigated area. See Appendix 4B in Suphannachart (2009) for details.

[4] *The Handbook of Agricultural Economics* (Evenson, 2001) and other productivity studies (Alston *et al.*, 1998b; Evenson and Pray, 1991; Morrison Paul, 1999) have argued for the inclusion of case-specific and natural factors, such as major weather events, environmental degradation, epidemics and natural disasters.
[5] Twenty years was chosen following Hall and Scobie (2006) and Caselli (2003).
[6] No depreciation (0%) is added because today's research is always built up from past research, and research stock may not depreciate. A constant δ implies geometric decay of the stock of research knowledge.
[7] This method is used in many time-series studies but has apparently not yet been used in TFP determinants studies. It is also known as the London School of Economics Method or the General to Specific Method (GSM) developed by Hendry and his co-researchers (Davidson *et al.*, 1978; Hendry *et al.*, 1984; Hendry, 1995).
[8] Estimates of TFP growth and TFP indexes are shown in Appendix 11.1, Table A11.1.
[9] Before applying the ECM method, which does not impose pre-specified lag structure on the research variable, the commonly used Almon polynomial distributed lags were applied to both research expenditure data and research stocks. This model did not perform well in terms of statistical acceptance and did not fit the data well. Only the models expressed in rate of change terms, which only explain short-run variations of TFP, pass the standard diagnostic tests. Results are shown in Appendix Table A11.2. The research stocks were constructed based on current and past research expenditure, using both the Almon lag structure, assuming an inverted U-shape of research impacts, and the perpetual inventory method, assuming the geometric depreciation of research impacts. Suphannachart (2009) provides further details.
[10] The interaction term between public and foreign R&D was not statistically significant from various experimental runs and was therefore dropped out of the final parsimonious model.

References

Alston, J. (2002) Spillovers. *Australian Journal of Agricultural and Resource Economics* 46, 315–346.

Alston, J.M., Craig, B. and Pardey, P.G. (1998a) Dynamics in the Creation and Depreciation of Knowledge, and the Returns to Research. EPTD Discussion Paper No. 35. Environment and Production Technology Division, International Food Policy Research Institute, Washington, DC.

Alston, J., Norton, G. and Pardey, P. (1998b) *Science Under Scarcity: Principles and Practices for Agricultural Research Evaluation and Priority Setting*. CABI Publishing, Wallingford, UK.

Alston, J.M., Marra, M.C., Pardey, P.G. and Wyatt, T.J. (2000) Research return redux: a meta-analysis of the returns to agricultural R&D. *Australian Journal of Agricultural and Resource Economics*, 44, 185–215.

Athukorala, P. and Tsai, P.-L. (2003) Determinants of household saving in Taiwan: Growth, demography and public policy. *Journal of Development Studies* 39, 65–88.

Banerjee, A., Dolado, J., Galbraith, J. and Hendry, D. (1993) *Co-integration, Error Correction, and the Econometric Analysis of Non-Stationary Data*. Oxford University Press, Oxford, UK.

Budhaka, B. (1987) Thailand. In: Asian Productivity Organization (ed.) *Productivity Measurement and Analysis: Asian Agriculture*. Asian Productivity Organization, Tokyo.

Chandrachai, A., Bangorn, T. and Chockpisansin, K. (2004) National report: Thailand. Total Factor Productivity Growth: Survey Report. Asian Productivity Organization, Tokyo.

Chen, E.K.Y. (1997) The total factor productivity debate: Determinants of economic growth in East Asia. *Asian-Pacific Economic Literature* 11, 18–38.

Chockpisansin, K. (2002) Analysis of Total Factor Productivity Growth in Thailand: 1977–1999. M.A. Thesis (in Thai). Faculty of Economics. Chulalongkorn University, Bangkok.

Coe, D.T. and Helpman, E. (1995) International R&D spillovers. *European Economic Review*, 39, 859–887.

Coelli, T. and Rao, D.S.P. (2003) Total factor productivity growth in agriculture: A Malmquist index analysis of 93 countries, 1980–2000. Working Paper 02/2003, Centre for Efficiency and Productivity Analysis, University of Queensland, Brisbane, Australia.

Coelli, T., Rao, D.S.P., O'Donnell, C. and Battese, G. (2005) *An Introduction to Efficiency and Productivity Analysis*. Springer, New York, NY.

Coxhead, I. and Plangpraphan, J. (1999) Economic boom, financial bust, and the decline of Thai agriculture: Was growth in the 1990s too fast? *Chulalongkorn Journal of Economics* 11, 1–17.

Davidson, J., Hendry, D., Srba, F. and Yeo, S. (1978) Econometric modelling of the aggregate time-series relationship between consumers' expenditure and income in the United Kingdom. *Economic Journal* 88, 661–692.

Denison, E.F. (1967) *Why Growth Rates Differ: Postwar Experience in Nine Western Countries,* Brookings Institution, Washington, DC.

Edwards, S. (1998) Openness, productivity and growth: What do we really know? *The Economic Journal* 108, 383–398.

Evenson, R. (1993) Research and extension impacts on food crop production in Indonesia. *Upland Agriculture in Asia: Proceeding of a Workshop,* Bogor, Indonesia.

Evenson, R. (2001) Economic impacts of agricultural research and extension. In: Gardner, B. and Rausser, G. (eds) *Handbook of Agricultural Economics,* 1st edn, vol. 1, Elsevier, Amsterdam, The Netherlands, pp. 573–628.

Evenson, R. and Pray, C. (1991) *Research and Productivity in Asian Agriculture.* Cornell University Press, Ithaca, NY.

Fan, S., Jitsuchon, S. and Methakunnavut, N. (2004) The importance of public investment for reducing rural poverty in middle-income countries: The case of Thailand. DSGD Discussion Paper No.7, International Food Policy Research Institute, Washington, DC.

Fuglie, K. and Heisey, P. (2007) Economic returns to public agricultural research. *Economic Brief No.10.* Economic Research Service, U.S. Department of Agriculture, Washington, DC.

Griliches, Z. (1996) The discovery of the residual: A historical note. *Journal of Economic Literature* 34, 1324–1330.

Griliches, Z. (1998) R&D and productivity: the unfinished business. *Estudios de Economia. Department of Economics, University of Chile,* 25, 145–160.

Hall, J. and Scobie, G. (2006) The role of R&D in productivity growth: The case of agriculture in New Zealand: 1927 to 2001. New Zealand Treasury Working Paper 06/01, New Zealand Treasury, Wellington, New Zealand.

Hendry, D. (1995) *Dynamic Econometrics.* Oxford University Press, Oxford, UK.

Hendry, D., Pagan, A. and Sargan, J. (1984) Dynamic specification. In: Griliches, Z. and Intriligator, M. (eds) *The Handbook of Econometrics,* Vol. II, North-Holland, Amsterdam.

Huffman, W.E. and Evenson, R.E. (1992) Contributions of public and private science and technology to U.S. agricultural productivity. *American Journal of Agricultural Economics,* 74, 752–756.

Huffman, W. and Evenson, R. (2005) New econometric evidence on agricultural total factor productivity determinants: Impact of funding composition. *Working Papers Series No.03029.* Iowa State University, Department of Economics, Ames, IA.

Inder, B. (1993) Estimating long-run relationships in economics: A comparison of different approaches. *Journal of Econometrics* 57, 53–68.

Johnston, B. and Mellor, J. (1961) The role of agriculture in economic development. *American Economic Review* 51, 566–593.

Jorgenson, D. (1988) Productivity and postwar U.S. economic growth. *Journal of Economic Perspectives* 2, 23–41.

Jorgenson, D. (1995) *Productivity Volume 2: International Comparisons of Economic Growth,* MIT Press, Cambridge, MA.

Jorgenson, D. and Griliches, Z. (1967) The explanation of productivity change. *Review of Economic Studies* 34, 249–283.

Kaipornsak, P. (1999) Role of total factor productivity growth in Thai economy. *Thammasat Economic Journal* 16, 5–54 (in Thai).

Mahadevan, R. (2002) *New Currents in Productivity Analysis: Where to Now?* Asian Productivity Organization, Tokyo.

Morrison Paul, C. (1999) *Cost Structure and the Measurement of Economic Performance: Productivity, Utilization, Cost Economics, and Related Performance Indicators.* Kluwer Academic Publishers, Boston, MA.

Mundluk, Y. (1992) Agricultural productivity and economic policies: Concepts and measurements. OECD Development Centre Working Paper No. 75, OECD Development Centre, Issy-les-Moulineaux, France.

National Economic and Social Development Board (NESDB) (2006) *Capital Stock of Thailand,* 2006 edn, Office of the National Economic and Social Development Board, Bangkok, Thailand.

Oguchi, N. (2004) *Total Factor Productivity Growth: Survey Report.* Asian Productivity Organization, Tokyo, Japan.

Poapongsakorn, N. (2006) The decline and recovery of Thai agriculture: Causes, responses, prospects and challenges. In: *Rapid Growth of Selected Asian Economies: Lessons and Implications for Agriculture and Food Security.* FAO Regional Office for Asia and the Pacific, Bangkok, Thailand.

Pochanukul, P. (1992) An economic study of public research contributions in Thai crop production: a time and spatial analysis. PhD Thesis. Kyoto University, Japan.

Ruttan, V.W. (1987) *Agricultural Research Policy and Development*. Food and Agriculture Organization of the United Nations, Rome, Italy.

Ruttan, V.W. (2002) Productivity growth in world agriculture: Sources and constraints. *Journal of Economic Perspectives* 16, 161–184.

Solow, R. (1957) Technical change and aggregate production function. *Review of Economics and Statistics* 39, 312–320.

Suphannachart, W. (2009) Research and Productivity in Thai Agriculture. PhD Thesis, Australian National University, Canberra.

Suphannachart, W. and Warr, P. (2011) Research and productivity in Thai agriculture. *Australian Journal of Agricultural and Resource Economics* 55, 35–52.

Thirtle, C., Lin, L. and Piesse, J. (2003) The impact of research-led agricultural productivity growth on poverty reduction in Africa, Asia and Latin America. *World Development* 31, 1959–1975.

Tinakorn, P. and Sussangkarn, C. (1996) *Productivity Growth in Thailand,* Thailand Development Research Institute, Bangkok.

Urata, S. and Yokota, K. (1994) Trade liberalization and productivity growth in Thailand. *Developing Economies* 32, 444–458.

Warr, P. (2004) Globalization, growth, and poverty reduction in Thailand. *ASEAN Economic Bulletin,* 21, 1.

Warr, P. (2009) Aggregate and sectoral productivity growth in Thailand and Indonesia. *Working Papers in Trade and Development, 2009/10 Arndt-Corden Division of Economics*. Australian National University, Canberra, Australia.

Wickens, M. and Breusch, T. (1988) Dynamic specification: The long-run and the estimation of transformed regression models. *Economic Journal* 98, 189–205.

Appendix 11.1

Appendix Table A11.1. Public agricultural research expenditure and productivity growth in Thai agriculture.

	Agricultural R&D (million Bt)		Agricultural extension (million Bt)		TFP growth			TFP index		
	Crops	Livestock	Crops	Livestock	Crops	Livestock	Total agric.	Crops	Livestock	Total agric.
1961	60.05	3.04	n.a.	n.a.	n.a.	n.a.	-2.49	n.a.	n.a.	1
1962	95.09	3.66	n.a.	n.a.	n.a.	n.a.	5.18	n.a.	n.a.	1.05
1963	107.88	4.08	n.a.	n.a.	n.a.	n.a.	4.76	n.a.	n.a.	1.1
1964	134.45	4.23	n.a.	n.a.	n.a.	n.a.	4.03	n.a.	n.a.	1.15
1965	136.27	4.19	n.a.	n.a.	n.a.	n.a.	-5.94	n.a.	n.a.	1.08
1966	158.55	4.74	n.a.	n.a.	n.a.	n.a.	2.47	n.a.	n.a.	1.1
1967	200.47	9.31	n.a.	n.a.	n.a.	n.a.	-2.17	n.a.	n.a.	1.08
1968	242.49	8.25	n.a.	n.a.	n.a.	n.a.	3.82	n.a.	n.a.	1.12
1969	247.53	9.00	n.a.	n.a.	n.a.	n.a.	-0.06	n.a.	n.a.	1.12
1970	287.11	9.75	224.44	103.58	n.a.	n.a.	-3.72	n.a.	n.a.	1.08
1971	357.94	9.21	278.52	98.67	17.02	-25.27	-0.24	1	1	1.08
1972	272.21	8.8	226.24	102.56	-13.22	11.57	2.33	0.87	1.12	1.1
1973	230.05	10.44	182.05	160.17	-2	33.23	2.87	0.85	1.49	1.13
1974	193.92	10.65	186.2	100.81	1.23	10.95	0.93	0.86	1.65	1.14
1975	267.32	16.3	291.92	141.61	-14.35	-27.84	4.17	0.74	1.19	1.19
1976	303.64	21.48	310.13	189.17	7.38	16.48	-2.56	0.79	1.39	1.16
1977	306.89	19.41	387.1	214.02	-0.37	-4.16	-1.41	0.79	1.33	1.15
1978	309.45	28.35	391.27	259.71	13.01	27.98	4.84	0.89	1.7	1.2
1979	292.19	25.48	382.7	218.69	-4.28	5.19	4.58	0.85	1.79	1.26
1980	288.99	26.34	504.42	223.18	-1.16	-11.13	-6.55	0.84	1.59	1.17
1981	344.39	30.73	605.49	248.61	4.98	-10.51	5.87	0.89	1.42	1.24
1982	461.47	29.11	894.02	322.52	0.09	5.42	-1.92	0.89	1.5	1.22
1983	591.47	28.87	947.59	283.02	3.63	5.17	-3.12	0.92	1.58	1.18
1984	757.99	34.84	1310.05	424.21	6.41	-15.05	0.95	0.98	1.34	1.19
1985	961.33	42.74	1253.54	558.18	5.76	1.72	3.89	0.98	1.36	1.24
1986	957.29	43.54	1279.97	458.64	-3.15	5.13	2.42	1.03	1.43	1.27
1987	823.24	48.02	1023.53	380.07	-6.12	7.89	-6.62	1	1.55	1.18
1988	822.78	37.60	955.36	354.7	4.82	8.9	-3.22	0.94	1.68	1.15
1989	897.38	37.32	1056.43	407.79	4.05	-0.49	3.6	0.99	1.67	1.19
1990	1040.11	46.81	1399.67	736.95	-3.95	-9.85	3.52	0.98	1.51	1.23

1991	900.74	48.45	1488.25	715.56	5.17	5.66	1.04	1.6	2.87	1.26
1992	1253.08	90.60	1796.26	802	-1.25	-1.64	1.02	1.57	-2.34	1.23
1993	1869.40	111.71	2816.91	1375.53	7.53	-10.65	1.1	1.4	9.9	1.36
1994	1617.38	127.02	2657.63	1223.98	1.62	-5.06	1.12	1.33	-4.12	1.3
1995	1436.72	48.82	2319.15	1260.11	-1.27	13.07	1.1	1.51	-5.22	1.23
1996	1606.63	63.68	2903.44	1421.11	2.79	6.67	1.13	1.61	0.82	1.24
1997	1725.86	79.68	3145.04	1481.35	-4.09	-6.37	1.09	1.5	-2.49	1
1998	1514.99	70.86	2537.21	1289.62	-6.33	-3.09	1.02	1.46	5.18	1.05
1999	1740.38	55.11	2925.48	731.8	11.23	-8.58	1.13	1.33	4.76	1.1
2000	1861.34	85.59	3097.68	848.62	8.1	12.46	1.22	1.5	4.03	1.15
2001	1657.99	70.58	2752.4	644.51	-3.88	5.62	1.18	1.58	-5.94	1.08
2002	1438.41	56.41	2380.27	1018.92	-6.97	0.1	1.1	1.58	2.47	1.1
2003	911.76	55.73	2388.63	1218.69	6.54	15.03	1.17	1.82	-2.17	1.08
2004	620.99	23.84	2173.44	875.09	-4.99	-29.56	1.11	1.28	3.82	1.12
2005	545.89	16.67	1821.79	745.33	-5.27	-5.57	1.05	1.21	-0.06	1.12
2006	527.82	33.03	1555.11	1083.98	-4.21	0.69	1.01	1.22	-3.72	1.08

'Total agriculture' consists of crops, livestock, fisheries, forestry and agricultural services. 'n.a.' indicates data not available. *Source:* Research and extension spending are from the Bureau of the Budget, Office of the Prime Minister. They are deflated into real terms using GDP deflators.

Table A11.2. Results of TFP determinants models using Almon Lag structure.

Variables	Lag	Crops $\Delta \ln TFP_t^{crop}$ Coefficient	Livestock $\Delta \ln TFP_t^{livestock}$ Coefficient
Constant		−0.827	−1.740
		(−0.795)	(−0.606)
$\Delta \ln R_t^p$	0	0.023	−0.037
		(1.985)**	(−0.880)
	1	0.037	−0.055
		(1.985)**	(−0.880)
	2	0.042	−0.055
		(1.985)**	(−0.880)
	3	0.037	−0.037
		(1.985)**	(−0.880)
	4	0.023	
		(1.985)**	
Sum of lags		0.165	−0.185
		(1.985)**	(−0.880)
$\Delta \ln E_t$		0.056	
		(0.905)	
$\Delta \ln R_t^p \times \Delta \ln R_t^f$		−0.003	
		(−0.663)	
$\Delta \ln RR_t$		0.165	
		(2.334)**	
$\Delta \ln I^{irrigation}$			2.099
			(2.641)***
D^{bird}			−17.471
			(−1.673)*
N (no. of observations)		34	36
k (no. of parameters)		5	5
Adjusted R^2		0.17	0.17
F-statistics		2.74	3.39
S.E. of regression		5.43	12.38
Diagnostic tests:			
LM(1), F(1, N-k-1)		3.30 (p=0.08)	0.32 (p=0.57)
LM(2), F(2, N-k-2)		2.36 (p=0.11)	2.36 (p=0.11)
RESET, F(1, N-k-1)		0.07 (p=0.78)	3.33 (p=0.08)
JBN, $\chi^2(2)$		0.09 (p=0.95)	0.71 (p=0.70)
ARCH, F(1, N-2)		0.09 (p=0.76)	0.81 (p=0.37)
ADF		−5.40 (p=0.00)	−5.21 (p=0.00)

Numbers in parentheses underneath each coefficient are the t-ratio of the coefficient. The level of statistical significance is: *, 10%; **, 5%; and ***, 1%. Sum of lagged coefficient indicates long-run multiplier or long-run change in TFP growth given a permanent increase in the public research spending. Diagnostic tests are: LM, Breush-Godfrey serial correlation LM test; RESET, Ramsey test for functional form mis-specification; JBN, Jarque–Bera test of normality of residual; ARCH, Engle's autoregressive conditional heteroskedasticity test; ADF, augmented Dickey–Fuller test for residual stationarity. Numbers in parentheses are p-values of the test statistics. Source: Authors' calculations.

12 Constraints to Raising Agricultural Productivity in Sub-Saharan Africa

Keith O. Fuglie and Nicholas E. Rada

Economic Research Service, U.S. Department of Agriculture, Washington, DC

12.1 Introduction

Poverty and food insecurity are pervasive in sub-Saharan Africa (SSA).[1] In 2005, 51% of SSA's population earned less than PPP $1.25 per day (World Bank, 2010; PPP, international purchasing power parity) with a similar share of the population being food insecure (Shapouri *et al.*, 2010). A key, if not the principal, factor behind this disappointing record has been a lack of robust agricultural growth. It is this sector from which the majority of the region's population draws its livelihood, and their welfare is tied directly to the productivity of the resources at their disposal. The non-farm population also depends heavily on agriculture because a majority of their income is spent on food. Boosting agricultural productivity stimulates economic growth and poverty reduction through a number of avenues: it raises the income of farm households, increases availability and lowers the cost of food, frees resources – such as labour – for general economic development, saves foreign exchange, stimulates rural demand for non-farm goods and services, and creates a surplus for public and private investment (Johnston and Mellor, 1961).

That SSA was largely bypassed by the Green Revolution helps explain why the region has remained poor, but it does not explain why such productivity change has not come to Africa. Binswanger and Townsend (2000) noted that African agricultural productivity has historically remained low because of adverse resource endowments and poor governing institutions and policies, preventing sufficient capital accumulation for agriculture to be an engine of economic growth. They placed greater explanatory weight on institutional and policy factors over adverse resource endowments, and were optimistic that the structural adjustments (policy reforms) introduced in several countries throughout the 1980s and 1990s would improve agricultural growth. Although that hopefulness seemed to be premature, there has recently been some renewed interest in the potential for SSA's economic and agricultural growth. In a wide-ranging review of prospects for agricultural and rural development in the region, Binswanger-Mkhize and McCalla (2009) cited the reduction of armed conflict, improved macroeconomic management, the spread of democratic and civil society institutions, stronger regional organizations, and growing volumes of foreign aid. Moreover, since the early 1990s, SSA's rate of agricultural (and economic) growth has shown significant improvement (Ndulu *et al.*, 2007). If this agricultural growth is, however, primarily caused by improved terms of trade or expansion of resources rather than productivity growth, then prospects for accelerating or even sustaining it are limited. Productivity will, ultimately, need to grow for agriculture to make a sustained contribution to economic development and the reduction of poverty and food insecurity in Africa.

The purpose of this chapter is to assess agricultural productivity's record in sub-Saharan Africa and the role of research and development (R&D) and other policies in its growth. Although most of the recent acceleration in agricultural GDP growth seems *not* to be productivity led (it is primarily resource led), there does seem to be a handful of countries that have sustained modest agricultural total factor productivity growth for a number of decades. We find positive influences of international agricultural R&D and, at least for large countries, national agricultural R&D investments on productivity growth. We also find positive effects on agricultural productivity growth from structural adjustment/reduction

in agricultural taxation, improvements to labour-force schooling and reduction in armed conflict. The spread of HIV infection, however, has suppressed agricultural growth, whereas the relationship between infrastructure development and agricultural productivity growth is inconclusive.

12.2 Examining Agriculture's Record in Sub-Saharan Africa

This section discusses agriculture's place in the overall economic development of Sub-Saharan Africa and reviews agricultural production changes during the past 50 years.

12.2.1 Some basic indicators

As shown in Table 12.1, SSA's population in the first decade of the 21st century grew

Table 12.1. Development indicators for sub-Saharan Africa.

	1961–1970	1971–1980	1981–1990	1991–2000	2001–2008
Population, total (millions)	263	342	455	599	753.57
Population growth (annual %)	2.54	2.81	2.89	2.70	2.51
Regional GDP (billions of constant 2005 PPP dollars)	n.a.	n.a.	734	899	1300
Industry, value added (% of GDP)	31	33	34	29	30
Manufacturing, value added (% of GDP)	17	17	17	16	14
Services, etc., value added (% of GDP)	48	48	48	52	52
Agriculture, value added (% of GDP)	21	20	18	18	18
Trade (% of GDP)	48	55	53	57	64
Share of labour force employed in agriculture (%)	80	75	70	65	62
GDP per capita (constant 2005 PPP dollars)	n.a.	n.a.	1616	1501	1626
GDP per capita growth (annual %)	n.a.	n.a.	−1.07	−0.30	2.43
Percentage of population living on less than $2/day	n.a.	n.a.	75	77	74
Life expectancy at birth (years)	44	47	49	50	51
Adult literacy (percentage of population age 15 and over)	n.a.	n.a.	n.a.	57	n.a.

n.a. not available. Figures for 49 states in sub-Saharan Africa including South Africa. Source: World Bank, except for share of labour force in agriculture, which is from FAO.

by 2.5% per year, nearly the same as in the 1960s but slightly lower than the 1980s peak of 2.9%. The delayed demographic transition of SSA countries has meant that the proportion of young in the population, and therefore the dependency ratio (the ratio of the non-working to working population), has remained relatively high. This alone explains the low household saving rates observed in SSA compared with other developing regions (Ndulu *et al.*, 2007). During 2000–2009, about 60% of the labour force was employed in farming, but agriculture comprised only 16.7% of GDP (World Bank, 2010).

GDP per capita in the 2001–2008 period averaged $1626 (2005 PPP dollars) and grew by 2.43% per year, the best sustained performances of the past half century (World Bank, 2010). But African economies were still largely making up for lost ground from the 1980s and 1990s when GDP per capita had fallen because of languid economic growth. The higher economic growth of the 2000s has enabled poverty rates to decline (Sala-i-Martin and Pinkovskiy, 2010), although they remain substantially higher than other regions of the world.

Average life expectancy, another key indicator of well-being, saw some gradual improvement between the 1960s and 1980s but since then has stagnated at about 50 years (Table 12.1). In some countries, high HIV/AIDS infection rates have caused life expectancy to fall. Adult literacy seems to have improved somewhat between the 1980s and 2000s, with about six out of ten adults obtaining basic literacy skills.

Basic trends in agricultural production and resource use are given in Table 12.2. The Food and Agriculture Organisation's (FAO's) measure of gross agricultural output – measured in constant 2000 US dollars[2] – grew between 2.8% and 3.4% on average each year, except through the 1970s when growth averaged less than 1% annually. Agricultural GDP in constant dollars, on the other hand, grew on average more than 2% per year since 1970, and averaged 3.4% in the 2001–2008 period (World Bank, 2011). The difference between growth in real output and real GDP is primarily a terms-of-trade effect: changes

in real value-added include both changes in the real quantity of output and changes in the terms of trade between agricultural and non-agricultural goods and services. In other words, if agricultural prices rise (fall) faster than the general price level, real agricultural GDP will grow (decline) even if agricultural output remains constant. Thus, changes in the growth rates between agricultural output and agricultural GDP reflect the changes in agricultural prices relative to a general price index. Agriculture experienced increasing terms of trade (rising real prices) during the 1970s and again in the 2000s but declining terms of trade in the 1980s and 1990s.

12.2.2 Data deficiencies in measuring land and labour

Productivity measures output per unit of an input, some group of inputs or, in the case of TFP, an aggregation of all economic inputs employed in production. These inputs include land, labour, capital (farm machines, orchards, livestock, etc.) and various kinds of material inputs, such as fertilizer, fuel and animal feed. But productivity measures for SSA are heavily constrained by incomplete and poor quality data, particularly regarding production inputs. In assessing the area under crop cultivation, the most commonly used measure is the FAO estimate of arable land plus area under permanent crops. By this measure, SSA cropland increased from 130.5 million hectares (mHa) in 1961 to 207.3 mHa in 2008, or by an average rate of 0.98% per year. A second possible way of measuring cultivated land is to sum up the total area harvested for all crops (including annual and perennial crops). According to this measure, cultivated crop area in SSA increased from 83.1 mHa in 1961 to 188.0 mHa in 2008, or by 1.74% per year. If both measures are correct, it would imply a rapid increase in cropping intensity (the number of crops harvested from a hectare of land per year). Such an increase would require either a substantial increase in

Table 12.2. Agricultural indicators for sub-Saharan Africa.

	1961–1970	1971–1980	1981–1990	1991–2000	2001–2008
Gross agricultural output (billions of US dollars)	33.71	41.32	49.60	69.15	90.82
Crop share of agricultural output (% of total)	76.71	76.09	74.36	77.61	77.82
Livestock share of agricultural output (% of total)	23.29	23.91	25.64	22.39	22.18
Growth in real agricultural output (average annual %)	3.44	0.78	2.82	3.18	3.08
Growth in real agricultural GDP (average annual %)	n.a.	2.49	2.16	2.95	3.44
Arable land and land in permanent crops (million Ha)	136.66	146.46	155.42	171.25	194.00
Crop area harvested (million Ha)	92.73	99.79	109.73	146.74	175.08
Food crop share of total area harvested (% of total)	83.50	84.10	85.50	87.60	87.80
Cash crop share of total area harvested (% of total)[a]	16.50	15.90	14.50	12.40	12.20
Land in permanent crops (million Ha)	12.58	14.90	17.21	18.82	20.95
Land in permanent pasture (million Ha)	704.44	707.70	714.44	731.49	746.18
Agricultural labour force (millions)	80.57	96.52	118.42	148.82	179.91
Agricultural output per worker (US dollars per worker)	417.47	428.58	418.26	463.94	504.29
Growth in agricultural output per worker (average annual %)	1.56	−0.98	0.55	0.96	1.01
Crop yield (US dollars per hectare harvested)	278.39	315.21	336.26	365.25	403.47
Growth in crop yield (average annual %)	1.26	0.77	0.81	1.16	0.72
Head of livestock (millions of cattle equivalents)	171.64	206.15	240.90	283.31	352.32
Ruminant livestock (% of total)	97.41	96.84	96.23	95.63	95.32
Non-ruminant livestock (% of total)	2.59	3.16	3.77	4.37	4.68
Livestock yield (US dollars per cattle-equivalent)	41.51	43.91	48.47	50.53	53.34
Growth in livestock yield (average annual %)	0.31	1.04	0.47	0.34	0.36
Area harvested per worker (hectares)	1.15	1.04	0.93	0.99	0.97
Irrigated cropland (% of area harvested)	3.09	3.46	3.90	3.51	3.22
Fertilizer per area harvested (kg/ha)	3.04	7.07	9.95	8.57	7.62
Tractors per area harvested (units per 1000 ha)	0.69	0.94	0.99	0.86	0.87

n.a., not available. Monetary values in constant US dollars for the year 2000. Figures for 48 states in sub-Saharan Africa excluding South Africa; Ha, hectares; [a]Cash crops include cotton and other fibre crops, cocoa, coffee, tea, oil palm, rubber, tobacco and sugarcane. Source: FAO, except growth in agricultural value-added, which is from the World Bank.

year-round irrigation (to allow more than one crop per year) or a decline in fallow land. But irrigated area has remained at 3% to 4% of total cropland since the 1960s. And although land in long-term fallow has probably decreased because of population pressure (Pingali *et al.*, 1987), the FAO definition of arable land specifically excludes such land. Thus, such an increase in cropping intensity hardly seems plausible.

We suspect that, for some countries at least, the FAO arable land and permanent cropland series substantially under-estimate the growth in actual area under crop cultivation. Consequently, using this FAO cropland series would overstate productivity growth by attributing more of the observed increase in output to yield rather than area expansion. The difference between FAO's cropland and area-harvested measures are particularly large in the case of Nigeria (the largest country in SSA and responsible for about a quarter of the region's agricultural production). Figure 12.1 shows the trends in the two measures of cultivated land for Nigeria and for the rest of SSA. Nigeria's area harvested declined in the 1970s, then rapidly recovered and grew after 1981. Its FAO cropland measure, on the other hand, grew very gradually from 1961 to 1994, and then grew at a rate comparable to that of area harvested (Figure 12.16). The trend in crop output (not shown) more closely follows

the trend in area harvested than that of cropland. That the FAO area-harvested estimate has exceeded the FAO cropland measure since the early 1980s is troublesome because there is very little double-cropping in Nigeria. For the rest of SSA, however, the two series seem to be generally consistent, with cropping intensity rising slightly over time but remaining below 1.0 throughout the period (Fig. 12.1a).

Another problem emerges with the measurement of labour. FAO develops an estimate for the number of economically active adults engaged in agriculture for each country and year. These projections are derived from UN population estimates, International Labour Organization (ILO) labour force estimates and an assumption FAO introduces regarding long-term trends in the share of total labour 'economically active' in agriculture. Yet no adjustment is made for the intensity (hours worked) of that labour. This is significant because a large portion of agricultural work is seasonal

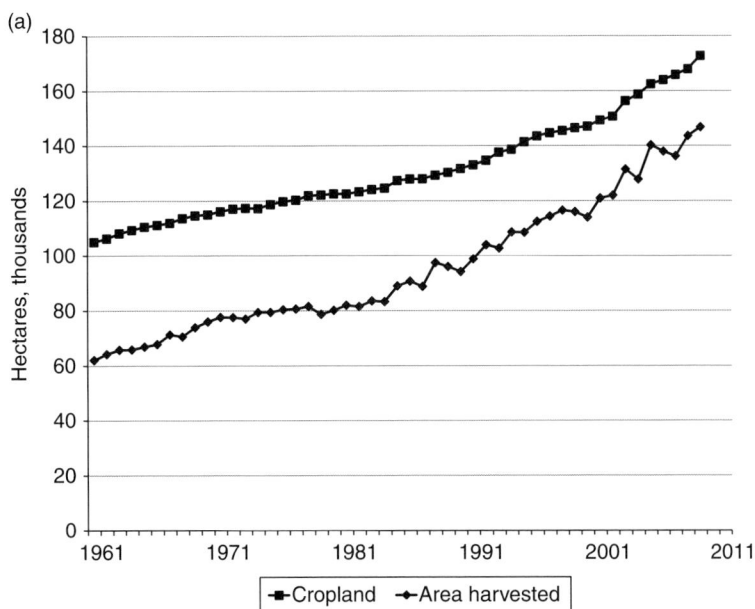

Fig. 12.1. Discrepancies in agricultural land and labour for sub-Saharan Africa. (a) Cropland measures for sub-Saharan Africa excluding Nigeria. (b) Cropland measures for Nigeria. (c) Cropland per worker and area harvested per worker for Nigeria.

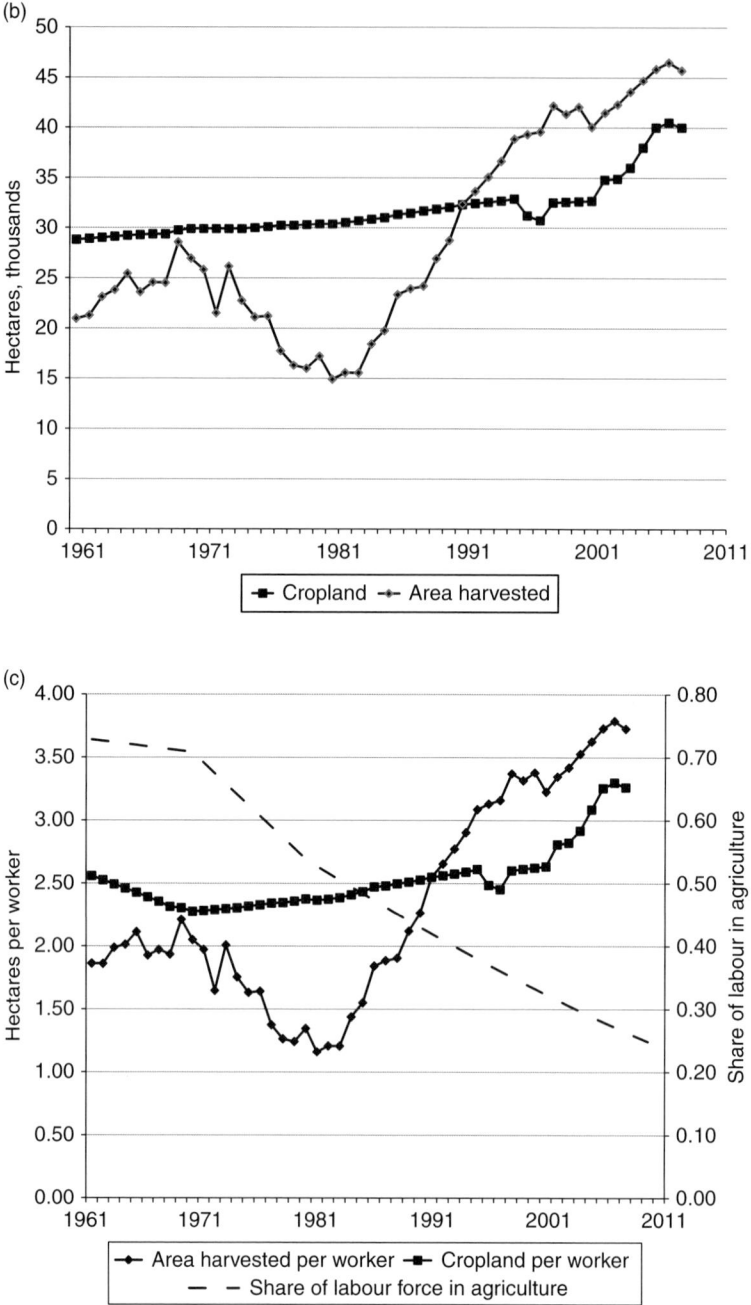

Fig. 12.1. Continued.

and part-time. Thus, employing the FAO's labour input in any productivity statistic adds the implicit assumption of constant work intensity, both spatially and tempo- rally. Recognizing this quality of the data helps us better understand and interpret observed differences in productivity that include this measure for labour.

For a number of SSA countries, there is considerable uncertainty regarding the share of total labour that is economically active in agriculture. Normally, this would be derived from either census or labour force surveys and updated over time as new survey information becomes available. Some SSA countries, however, have not conducted reliable population censuses or labour force surveys for several decades, if ever. Without reliable and up-to-date survey information, it is possible for major discrepancies to appear in the data, errors that might grow larger over time. Nigeria is an important instance where there is likely to be a large error in FAO's measure of agricultural labour. FAO's labour series shows that Nigeria's agricultural labour force peaked at 12.9 million persons in 1970 and has since declined in absolute number, even as the nation's total labour force grew by more than 2% per year. The implication is that FAO's estimate of the share of labour economically active in agriculture fell from 70% in 1970 to 28% by 2008. These numbers are difficult to reconcile with FAO's estimates of Nigeria's agricultural cropland, because they imply that crop area harvested per worker more than tripled between 1981 and 2007, from 1.2 ha per worker to 3.8 ha per worker. (Fig. 12.1c) These estimates seem especially unlikely given very little mechanization occurred in Nigeria over this period. For the rest of SSA, agricultural labour has grown by about 2% per year and the share of agricultural labour in total labour has declined more gradually.

12.2.3 Capital and material inputs

Two important forms of agricultural capital formation in SSA are the establishment of tree-crop plantations (primarily in forest zones) and the building-up of livestock herds (mainly in tsetse-free savannahs and highlands). Fixed capital – such as machinery, tools and structures – remains very low in SSA. For example, less than one tractor is used for every 1000 ha of crop area harvested. With respect to tree capital, area planted to perennial crops expanded from 12.6 mHa in the 1960s to 21.0 mHa during 2001–2008. About 60% of SSA's tree-crop area lies in West Africa, where cocoa and oil palm are the predominant species.

Farm holdings of livestock capital increased from 172 million head[3] in the 1960s to 352 million head during 2001–2008, ruminant species making up more than 95% of the total. The dominance of ruminants implies a reliance on pastures for feed and relatively little crop production diverted for feed to animals. FAO reports there are more than 700 mHa of permanent pastures in SSA, about four times the crop area. Most of the pastures are unimproved rangeland and lie in arid, semi-arid and savannah zones.

Transhumance and nomadic pastoralism are common practices in the region to address the seasonal availability of green pastures, and access to pastures and water holes is typically governed by the historical claims of kinship groups. In some areas, encroachment of cultivators on nomadic grazing lands is a recurring source of conflict (Oba and Lusigi, 1987). McIntire et al. (1992) showed, however, that integrated crop–livestock farming increased in relation to human population density on agricultural land. Cropland expansion, de-forestation and the removal of wildlife hosts reduce tsetse infestation and have enabled livestock husbandry to become more established in some areas, such as the sub-humid savannah belt of Nigeria (Bourn and Wint, 1994).

The use of synthetic fertilizer on cropland in SSA remains far below that of other developing regions. Average fertilizer rates peaked at around 10 kg of nutrients per ha in the 1980s and have since fallen to less than 8 kg/ha (Table 12.2). Fertilizer application rates fell sharply in some countries after subsidies were removed, for instance in Nigeria in the early 1990s. Elsewhere, such as in Kenya, fertilizer use has gradually increased over time and has averaged about 35 kg/ha since the mid-1990s. For the SSA region as a whole, fertilizer application rates are insufficient to sustain continuous cropping because total nutrients added to cropland from fertilizers, manure, fixation and other sources are estimated to be below the amount removed in the crop harvest (Stoorvogel and Smaling, 1990; Henao and Baanante, 2006). Thus, much of the cropland in SSA continues to require long fallow periods to allow nutrient levels to recover.

12.3   Measuring Agriculture's Performance

Total factor productivity measures the total resource cost of producing economic outputs. Unlike partial productivity measures (for example, labour productivity – output per worker or land productivity – or crop yield per hectare), TFP takes into account the contributions of all conventional inputs to production – land, labour, capital and materials. Although increases in labour or land productivity could be attributed to increasing the use of other inputs, increases in TFP reflect improvements in the efficiency of this aggregate bundle of inputs. As such, it is a more complete measure of productivity and more closely associated with the cost of producing outputs. Measuring TFP trends requires detailed information on all of the output and input quantities involved in agricultural production, plus information on prices and unit costs. This is an onerous task, even for countries with detailed agricultural data such as the USA. For the countries in SSA, data are incomplete and often of poor quality so indirect methods are required to derive approximate measures of TFP change.

12.3.1   Total factor productivity: a review

Summarized in Table 12.3 are findings from seven studies that have used various approaches to estimate long-term agricultural TFP growth in the SSA region. Despite each study employing different country aggregations, time periods and methods, seven of nine sets of results indicate a similar pattern of agricultural TFP growth: slow growth in the 1960s, negative growth in the 1970s, followed by recovery in the 1980s and subsequent decades, but overall low TFP growth – under 1% per year on average since 1961. The two exceptions are Lusigi and Thirtle (1997), whose sample also included five North African countries (probably explaining the higher average TFP growth reported), and Alene's (2010) estimate using a sequential Malmquist method. Interestingly, Nin-Pratt and Yu (Chapter 13, this volume) also used a sequential Malmquist method but did not find the unusually high TFP growth rates reported in Alene (2010). Nin-Pratt and Yu (Chapter 13, this volume) did attempt to improve the plausibility of the estimates by screening for outliers and forcing shadow values for land and labour to be positive. From Table 12.3, only Lusigi and Thirtle (1997) and Alene's (2010) sequential Malmquist results suggest SSA's TFP growth to be above 1% per annum. Given these circumstances, the collective evidence from the studies reported in Table 12.3 supports the view that true SSA agricultural TFP growth has averaged below 1% per year since 1961.

Each study in Table 12.3 found substantial variation in agricultural productivity growth across countries and over time. Some used multiple regression analysis to examine whether certain policies or country characteristics could explain these differences. Fulginiti et al. (2004) found a correlation between agricultural TFP growth and institutions, in that former British colonies and countries with higher levels of political rights generally performed better than other countries in the region. Nin-Pratt and Yu (Chapter 13, this volume) relate higher TFP growth to policy reform, especially the structural-adjustment policies that began to reduce net taxation of agriculture after the mid-1980s. Block (2010) investigated correlations between agricultural TFP growth and agricultural R&D, roads, the

Table 12.3. Review of estimates of agricultural total factor productivity growth for sub-Saharan Africa.

Study	Estimation method	Countries included (n = number of countries)	Period	TFP growth rate (Average annual percentage)					
				1960s	1970s	1980s	1990s	2000s	Whole period
Block (1995)	Production function	n = 39, excluding South Africa	1963–1988	0.78 (1963–73)	-0.24 (1973–83)	1.63 (1983–88)	n.a.	n.a.	0.54
Lusigi and Thirtle (1997)	Malmquist distance function	n = 47, including South Africa and 5 North African countries	1961–1991	n.a.	n.a.	n.a.	n.a.	n.a.	1.27
Fulginiti, Perrin and Yu (2004)	Stochastic frontier function	n = 41, including South Africa	1961–1999	0.68	-0.32	1.29	1.62	n.a.	0.83
Alene (2010)	Malmquist distance function (contemporaneous)	n = 47, including South Africa	1970–2001	n.a.	-0.90	1.40	0.50	n.a.	0.10
	Malmquist distance function (sequential)	n = 47, including South Africa	1970–2001	n.a.	1.40	1.70	2.10	n.a.	1.60
Nin-Pratt and Yu (Chapter 13, this volume)	Malmquist distance function (sequential and bounded)	n = 26, excluding South Africa	1961–2006		-1.33 (1961–1983)		1.37 (1984–2006)		0.20
Block (2010)	Production function	n = 47, including South Africa	1961–2007		0.14 (1961–1984)		1.24 (1985–2007)		0.61
Fuglie (2011)	Production function	n = 48, excluding South Africa	1961–2008	0.52	-0.19	1.14	1.34	1.00	0.75
	Production function (revised data for Nigeria)	n = 48, excluding South Africa	1961–2008	0.45	-0.20	0.81	1.26	0.83	0.63

n.a., not available. Source of data for all of these estimates is FAO. Output is the FAO measure of gross agricultural output aggregated using as fixed set of international prices for weights except Block (2010) who only includes crop output and aggregates using Africa-specific price weights. Inputs include agricultural land, labour, fertilizer, tractors, and head of livestock. All studies measure cropland as the FAO estimate of arable land plus land in permanent crops except Fuglie (2011), who uses area harvested for all crops as his cropland measure.

effects of civil war, and agricultural and macroeconomic policies. Because of data limitations, Block could only consider those factors one-at-a-time in a series of single-variable regressions. Those models exhibited positive and significant effects of R&D and policy reform on TFP but no significant relationship with road density or civil war.

The next section briefly reviews Fuglie's (2011) national and sub-regional findings on TFP growth in SSA and then examines the effects of government policy and other factors that might help explain why some countries have performed better than others in raising their agricultural productivity. Fuglie's (2011) TFP estimates are unique in that they attempt to reduce potential biases in TFP estimation by introducing some corrections to the agricultural input and output data, employ instrumental variables to avoid potential simultaneous equations bias, and include variables to control for land quality. Fuglie (2011) also introduces an alternative Nigerian dataset, testing the impact of the FAO data.

12.3.2 Regional and national indexes of agricultural TFP

Fuglie (2011) derives annual indexes of agricultural TFP during the 1961–2008 period for each SSA country, seven sub-regions, and for SSA as a whole. To do this, Fuglie first estimates a constant-returns-to-scale Cobb–Douglas production function (including variables to account for land quality differences across countries and over time).[4] The production elasticities from the regression are then used as factor weights for input aggregation.[5] Finally, agricultural productivity growth is computed using index numbers and growth accounting techniques, TFP change being the difference between gross agricultural output growth and aggregate agricultural input growth. Inputs include land (measured as total crop area harvested), labour (the number of economically active adults in agriculture), livestock capital (total animals, in cattle-equivalents), machinery

(the number of tractors in use) and material inputs (the quantity of fertilizer nutrients applied).

The estimated production elasticities or factor shares are: 0.25 for labour, 0.32 for land, 0.36 for livestock, 0.06 for fertilizer and 0.02 for tractors. Farm-supplied inputs of labour, land and livestock capital account for more than 90% of the total inputs used in agriculture. The elasticity for fertilizer implies that doubling fertilizer use from current levels (holding other inputs constant) would only boost output by about 6%. Getting a more powerful supply response from large increases in fertilizer usage would probably require complementary investment in irrigation, adoption of better soil and water management practices, plus access to improved crop varieties (Morris *et al.*, 2007).

The land quality coefficient estimated from the production function regression measures the relative productivity of irrigated cropland, favourable rainfed cropland (such as land high rainfall and gradual topology), and unfavourable rainfed cropland (low rainfall or steep topology) as defined by Sebastian (2007).[6] The coefficients for irrigated and favourable rainfed cropland for the full sample of countries are similar, 0.85 and 0.97, respectively, indicating that these lands produce nearly double the yield of unfavourable cropland. Because most irrigated cropland lies in areas that would otherwise be classified as unfavourable rainfed cropland, irrigation essentially brings this land up to the productivity level of high rainfall cropland. Most land equipped for irrigation in SSA has a cropping intensity of less than 1.0, meaning it is rarely cropped more than once and sometimes not at all (Fuglie, 2011).

Agricultural TFP growth trends for the entire sub-continent as well as its sub-regions are given in Fig. 12.2 and are provided for individual countries in Table 12.4. The indexes are set to 100 for 1961, changes thus reflecting output and TFP growth relative to that year. During the first 25 years of the post-independence period, agricultural TFP grew very slowly in SSA as a whole (0.2% per annum between 1961

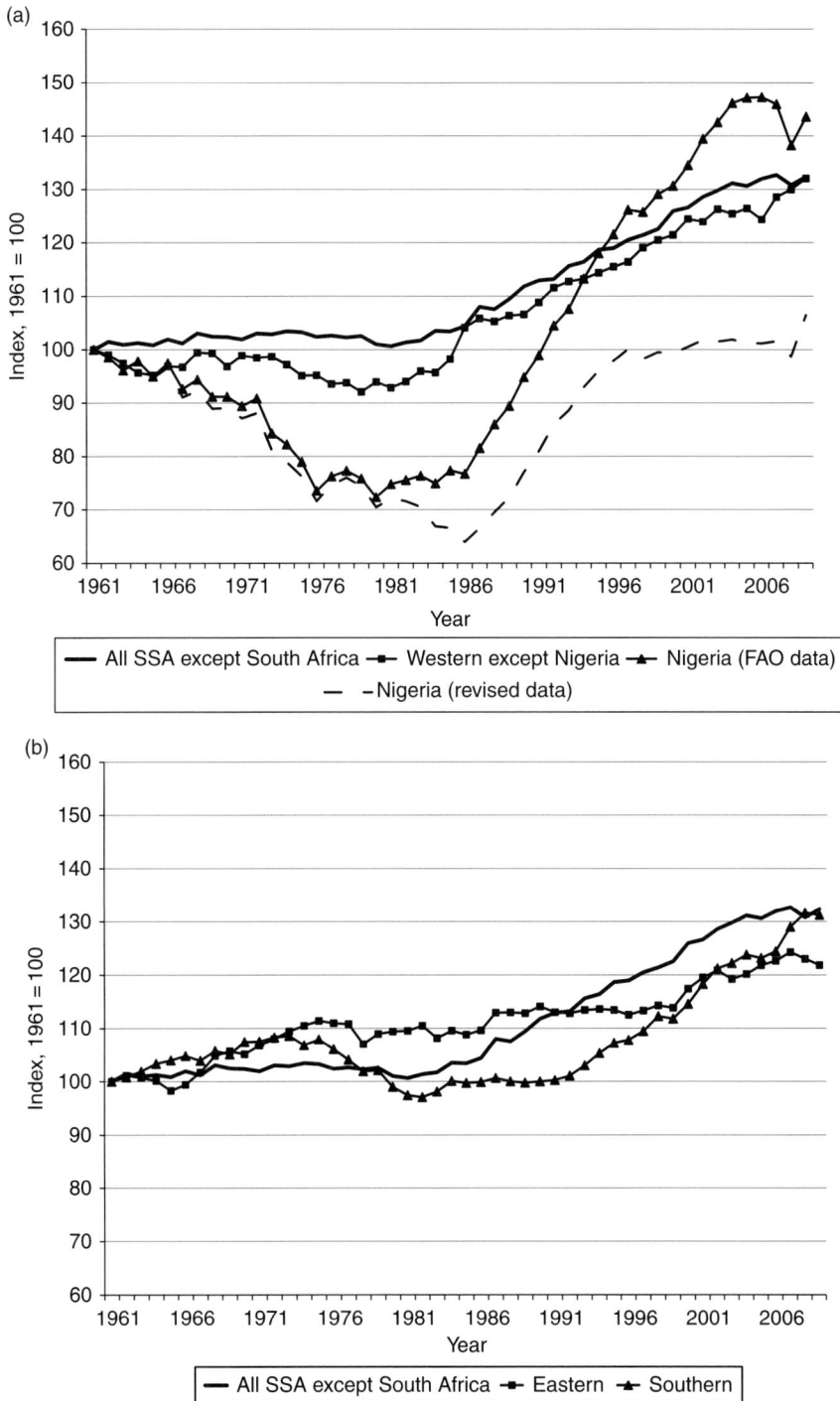

Fig. 12.2. Agricultural TFP indexes for sub-Saharan Africa, 1961–2009 (1961 = 100). (a) Western Africa and Nigeria. (b) Southern and Eastern Africa. (c) Sahel, Horn and Central Africa. Source: Fuglie (2011) with data updated through to 2009.

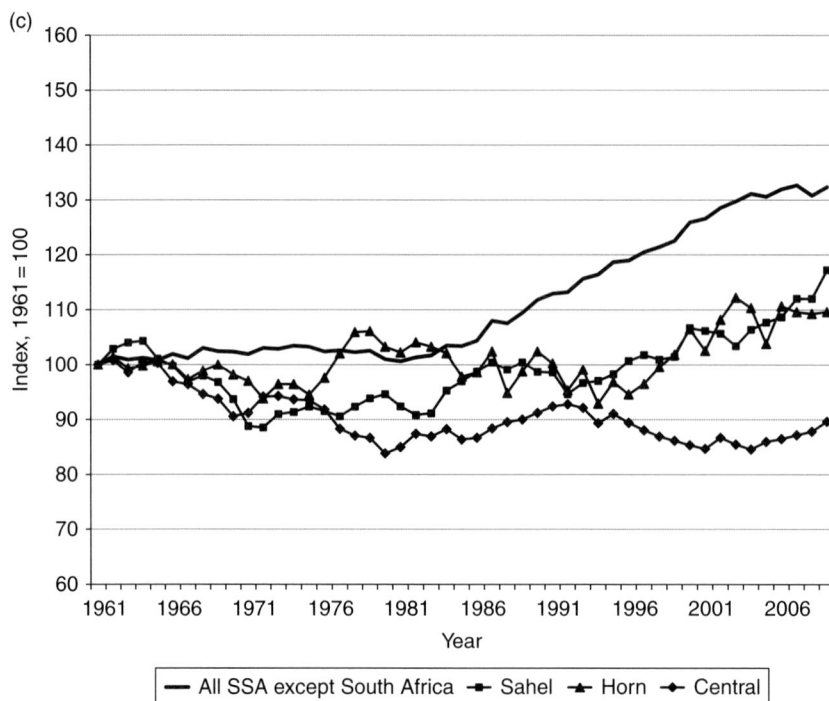

Fig. 12.2. Continued.

and 1984). That growth was dominated by productivity improvement in eastern and southern African countries. Between 1985 and 2008, there was a noticeable increase in the rate of TFP growth to 1.3% per year, western Africa and Nigeria leading all other regions. Reframing the results to focus on the post-1990s time period shows TFP improvements in southern Africa (principally led by Mozambique and Angola as peace was restored in those countries), the Horn of Africa (Ethiopia and Sudan), and the Sahel.

To counterbalance potential miscalculations in the FAO data, we also employ an alternative Nigeria dataset. This 'revised' dataset employs harvested area instead of the cropland series, assumes that farm labour grew by 2% per year rather than the FAO's assumed 0% per year, and uses Nigerian national data – collected by the International Institute of Tropical Agriculture (IITA) on crop production for root crops and cow peas, and by the USDA

on grains and oilseeds, rather than the FAO production data. An important difference of this Nigerian dataset is that its output data show less growth in yam production, the single largest agricultural commodity in FAO's estimation of gross value of production. This revision reduces the estimated growth rate in Nigeria's gross agricultural output. And, indeed, employing the revised Nigerian dataset and evaluating it during the 1961–2008 period shows a significant reduction in Nigeria's TFP growth, its mean rate falling from 0.97% to 0.21% per annum (Table 12.4). Interestingly, that substantial decline in Nigeria's TFP growth does not seem to weigh heavily on the regional TFP growth rate. Focusing strictly on the 1985–2008 period of TFP improvement, we find the revised Nigeria data reduces SSA's TFP growth rate to 1.1% annually, down from 1.3% when employing the FAO data.

Nin-Pratt and Yu (Chapter 13, this volume) attribute renewed post-1985 TFP

Table 12.4. Agricultural output and TFP indexes for countries and regions in sub-Saharan Africa.

| | Average Output | Gross Agricultural Output | | | | | Agricultural Total Factor Productivity | | | | | Avg TFP growth |
| | 2006–2008 | (1961 = 100) | | | | | (1961 = 100) | | | | | 1961–2008 |
	(billion US$)	1971	1981	1991	2001	2008	1971	1981	1991	2001	2008	%/year
Central Africa	**6.53**	**129**	**156**	**200**	**206**	**220**	**93**	**87**	**93**	**85**	**84**	**−0.34**
Cameroon	2.61	151	178	213	294	332	103	94	103	115	121	0.30
Cent. Afr. Rep.	0.67	136	172	208	296	336	91	84	94	110	110	0.37
Congo	0.24	121	138	158	203	248	82	89	85	89	101	0.06
Congo, DR	2.76	121	147	199	157	157	88	88	93	88	81	−0.14
Gabon	0.20	119	163	208	242	250	104	82	83	93	84	−0.35
Eastern Africa	**16.63**	**151**	**177**	**232**	**284**	**342**	**109**	**112**	**116**	**122**	**124**	**0.41**
Burundi	0.71	122	135	174	160	172	85	88	90	88	78	−0.22
Kenya	4.80	135	195	292	350	446	100	121	129	141	164	1.12
Rwanda	1.45	146	216	246	272	341	95	113	99	106	81	0.04
Tanzania	4.78	140	198	242	301	403	100	111	117	127	137	0.57
Uganda	4.88	177	152	200	263	277	128	129	129	127	107	0.19
Horn	**13.92**	**128**	**156**	**166**	**240**	**291**	**97**	**100**	**95**	**97**	**101**	**0.00**
Ethiopia, former	6.45	120	137	146	198	272	87	98	90	92	101	0.00
Somalia	1.23	142	185	180	204	209	109	104	110	121	121	0.26
Sudan	6.19	134	173	187	307	342	90	82	83	92	96	0.02
Sahel	**8.74**	**113**	**134**	**175**	**246**	**323**	**86**	**89**	**93**	**99**	**104**	**0.08**
Burkina Faso	1.80	129	158	289	436	557	85	79	95	120	105	0.36
Chad	1.13	104	111	147	215	237	86	97	93	94	89	0.01
Gambia	0.10	126	107	100	140	140	84	61	47	54	43	−1.85
Mali	2.01	124	167	225	299	394	79	93	107	110	116	0.61
Mauritania	0.34	109	121	141	169	189	86	99	95	93	94	−0.11
Niger	2.30	123	159	181	297	478	75	66	74	80	98	−0.27
Senegal	1.02	95	101	115	131	158	73	72	73	68	72	−0.63
Southern Africa	**10.62**	**138**	**145**	**165**	**214**	**252**	**106**	**97**	**100**	**116**	**128**	**0.32**
Angola	1.55	136	95	106	180	268	78	55	57	81	89	−0.32
Botswana	0.17	145	150	170	158	176	122	106	139	113	124	−0.22
Lesotho	0.09	112	122	125	143	123	86	93	95	82	81	−0.19
Madagascar	2.19	129	148	173	179	210	98	95	100	102	107	0.22

Continued

Table 12.4. Continued.

	Average Output 2006–2008	Gross Agricultural Output (1961 = 100)					Agricultural Total Factor Productivity (1961 = 100)					Avg TFP growth 1961–2008
	(billion US$)	1971	1981	1991	2001	2008	1971	1981	1991	2001	2008	%/year
Malawi	2.06	151	210	240	406	597	106	109	106	162	208	1.09
Mauritius	0.18	119	123	136	147	145	111	116	117	118	115	0.21
Mozambique	1.53	138	121	107	186	220	102	76	75	86	93	–0.38
Namibia	0.29	148	128	136	131	134	115	99	94	73	72	–1.07
Swaziland	0.19	152	218	261	243	267	143	180	191	208	226	1.52
Zambia	0.92	144	166	237	286	372	104	113	119	140	162	0.80
Zimbabwe	1.27	148	172	204	254	191	105	108	116	127	106	0.38
Western Africa	**13.34**	**136**	**162**	**234**	**348**	**423**	**99**	**93**	**109**	**124**	**127**	**0.66**
Benin	1.39	126	155	272	479	496	91	97	122	150	160	1.48
Côte d'Ivoire	4.52	162	254	352	484	543	100	103	107	130	132	0.61
Ghana	4.49	131	111	185	316	422	87	63	93	114	128	0.82
Guinea	1.36	119	138	184	257	338	100	107	116	106	107	0.30
Guinea-Bissau	0.19	76	105	143	201	243	78	68	91	86	95	0.34
Liberia	0.26	148	186	144	201	237	97	93	98	107	102	0.22
Sierra Leone	0.54	132	149	174	160	300	100	90	89	90	109	–0.24
Togo	0.58	125	140	198	277	297	91	82	75	83	78	–0.36
Nigeria	**27.85**	**132**	**124**	**238**	**361**	**467**	**89**	**73**	**97**	**130**	**153**	**1.10**
Nigeria (revised*)	**23.61**	**129**	**125**	**215**	**319**	**405**	**85**	**69**	**77**	**96**	**107**	**0.22**
All SSA	**97.61**	**134**	**150**	**204**	**278**	**341**	**102**	**100**	**111**	**124**	**132**	**0.59**
All SSA (revised*)	**93.37**	**133**	**150**	**199**	**269**	**329**	**101**	**99**	**107**	**118**	**124**	**0.44**

*Revised data for Nigeria uses alternative measure of output and agricultural labour. Output data uses USDA data for grains, oilseed and cash crops, national data on roots & tubers and legumes (reported in IFPRI, 2010) since 1994, and FAO data otherwise. The agricultural labor series uses FAO data for 1961 and assumes 2% annual growth for subsequent years.

Source: Fuglie (2011), revised using updated FAO data.

growth in SSA to structural adjustment policies, reducing agriculture's rate of effective taxation. But there is a growing body of evidence that suggests technological change is also playing a role in raising agricultural productivity. During the 1970s and 1980s, several eastern and southern African countries achieved significant improvements in maize productivity through a combination of improved varietal-adoption, government price support and fertilizer subsidization policies (Byerlee and Eicher, 1997). Maize is the dominant crop in this part of Africa, and its productivity growth could explain the better-than-average agricultural TFP growth observed in eastern and southern Africa during those decades. The success of maize faltered, however, when government support waned in the 1990s (Smale and Jayne, 2003), a story consistent with the lower TFP growth observed for these regions shown in Fig. 12.2. In West Africa (including Nigeria), some well-documented cases of successful technological innovations include maize (Smith *et al.*, 1994), cassava (Nweke, 2004) and rice (Dalton and Guei, 2003). All three are major crops in the region and the timing of the diffusion of productivity-enhancing innovations – post-1980 – corresponds with the increase in TFP growth observed for western Africa and Nigeria. In the 1990s, further technological improvements occur red in sorghum production in the Sahel-Sudan belt, where it is a major food staple (Deb and Bantilan, 2003), whereas the Sahel and Guinea savannah regions benefitted from gradual diffusion of animal traction (Smith *et al.*, 1994; Starkey, 2000) and improved natural resource management (Reij *et al.*, 2009). Such innovations – along with favourable weather trends in the Sahel (Olsson *et al.*, 2005) – are, again, consistent with the TFP patterns observed for these regions.

Among individual SSA countries, only a few seem to have been able to achieve sustained TFP growth over a long period, and several have shown productivity regression (Table 12.4). Kenya is one country (other than South Africa) that has sustained steady, long-term growth in agricultural TFP since the 1960s. Kenya's agricultural TFP increased by a total of 78% between 1961 and 2008, indicating that a given bundle of agricultural resources (land, labour, capital, materials, etc.) produced 78% more crops and livestock in 2008 than in 1961. Other countries that seem to have entered a sustained agricultural TFP growth path in the 1980s and 1990s include Benin, Cameroon, Ghana, Mali, Malawi, Tanzania, Zambia, Swaziland and possibly Nigeria. Each increased its TFP by at least 30% between 1980 and 2008 (the estimate of Nigeria's TFP growth was 58% using FAO data but only 10% using the revised data).

Several patterns of TFP growth are evident from the estimates in Table 12.4. A few countries seemed to be on a sustained TFP growth path but then saw productivity stagnate or decline. Côte d'Ivoire and Zimbabwe experienced positive TFP growth for several decades, but Zimbabwe suffered sharp productivity deterioration beginning around 1997, and Côte D'Ivoire's productivity stagnated after 2000. In both countries, the reversal in TFP growth correlated with periods of civil unrest and/or macroeconomic mismanagement. Another set of countries, notably Mozambique and Angola after 1991, showed strong TFP growth (or TFP recovery) after a prolonged period of TFP decline during protracted civil wars. Finally, a number of countries in SSA have shown no significant change in agricultural TFP during the past 50 years. Countries in Central Africa (other than Cameroon), the Horn of Africa, most small island states and scattered other countries fall into this 'no growth' category.

Although declining rates of agricultural productivity can sometimes be directly linked to periods of political unrest or poor governance, another potential source is natural resource degradation. Stoorvogel and Smaling (1990) and Henao and Baanante (2006) found that soil nutrient balances in most SSA countries have been negative since at least the 1980s (meaning that more nutrients were removed in crop harvests than were replenished from inorganic and organic sources). Negative nutrient balances, however, are not atypical of areas where swidden agriculture is commonly practised, as is the case in much of SSA. Moreover, the data used to estimate soil nutrient balances are of varying degrees of quality: data on inorganic fertilizer use, for example, are available only at the national level. The prevalence of significant, long-term land degradation is unknown, but the increases, albeit slow, in TFP reported in Table 12.4 for most regions do not suggest that long-term land degradation is pervasive in the sub-continent. Declining land quality is likely to be of greatest concern in regions of high population density and fragile soils, such as the Lake Victoria basin and the Central African and Ethiopian highlands.

12.3.3 Decomposing sources of growth

Fuglie's (2011) production function and TFP estimation allow for the decomposition of output growth into several components. A schematic for this growth decomposition is provided in Fig. 12.3. First, changes in value-added (sectoral GDP) are decomposed into a terms of trade effect and changes in real output. Growth in real output is then further decomposed into growth attributable to agricultural land expansion (extensification)

and growth attributable to raising yield per hectare (intensification). Finally, yield growth itself is decomposed into input intensification (i.e. more capital, labour and fertilizer per hectare of land), and TFP growth, where TFP reflects the efficiency with which all inputs are transformed into outputs. Improvements in TFP are driven by technological change and/or improved efficiency in resource allocation (such as what might come about when a country becomes more open to trade and farmers reallocate resources in response to new economic opportunities).[7] See Chapter 16 in this volume for a mathematical description of the growth decomposition illustrated in Fig. 12.3.

The decomposition of *output* growth into these components (Fig. 12.3) is both intuitively appealing and has some direct policy relevance: land expansion and input intensification are strongly influenced by changes in resource endowments and relative prices. Increasing population density or higher crop prices can induce more intensive use of existing farmland and investments in land improvement (Boserup, 1965). But in the short-run, the ability to raise yield through intensification is largely confined to existing technology. Changes in TFP, on the other hand, are driven by changes in technology

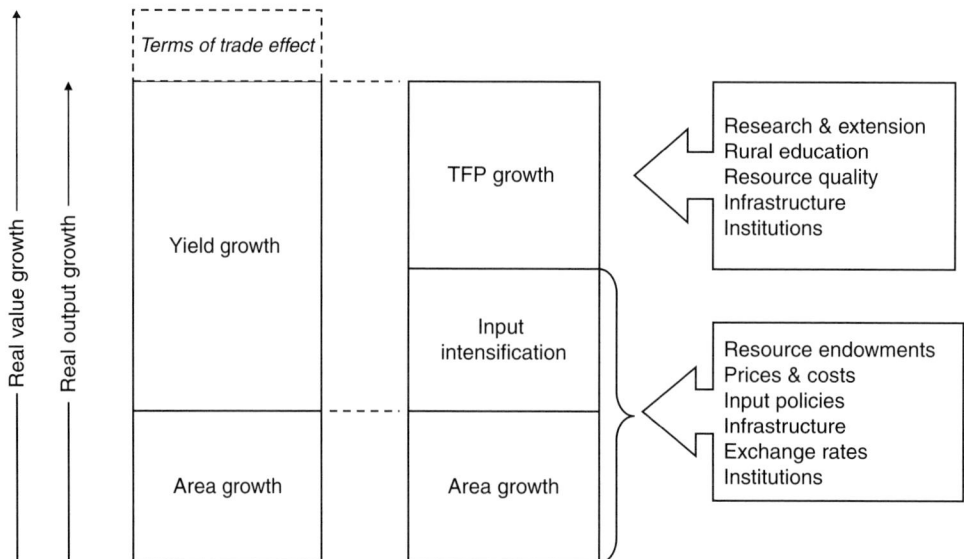

Fig. 12.3. Decomposing sources of agricultural growth.

and allocative efficiency. Yield growth resulting from incremental improvements to technology can be sustained over the long-run.

Table 12.5 shows SSA's output decomposition by decade since 1961. Real agricultural GDP growth averaged at least 2% per year over each decade, with growth accelerating in the 2001–2008 period to 3.4% per year. Over the whole 1961–2008 period, real output (absent terms-of-trade effects) grew at an average annual rate of 2.6%. Resource expansion accounted for 72% of that growth and improvements in TFP the other 28%. Agriculture's GDP growth acceleration during the 2001–2008 period, in comparison to the 1990s, was entirely a terms-of-trade effect because real output growth actually fell slightly between 1991–1999 and 2001–2008. Given the cyclical behaviour of commodity prices, current agricultural GDP growth rates are unlikely to be maintained or could even decline if real prices fall from the high levels seen since 2006.

Since the 1980s, there has been some improvement in the TFP growth rate, from less than 0.5% per year in the 1960s and 1970s to just over 1% per year in subsequent decades. This progress, though, has been partially offset by a declining rate of input intensification. Thus, yield growth has not accelerated, averaging less than 1.6% per year each decade since the 1960s. Sustaining agricultural growth over the long-run probably requires both greater input intensification and substantial improvement in TFP. The contribution of policy to agricultural TFP growth in SSA – especially national and international investments in agricultural research and development – is the focus of the remainder of the chapter.

12.4 Evaluating Agriculture's TFP Growth

We use multivariate regression analysis to examine how various types of policies might have contributed to agricultural TFP growth in SSA. We consider: (i) investments in research, not only in national agricultural research but also by the CGIAR system of international agricultural research centres; (ii) economic policy, including commodity price interventions, trade tariffs and input subsidies, as represented by the World Bank's measure of the nominal rates of assistance (NRA) to agriculture; (iii) human capital, both the education and health of the labour force; (iv) infrastructure, as represented by the extent of roads; and (v) political stability, or the lack thereof, as represented by the incidence of armed conflict.

Although investment in agricultural research provides an obvious mechanism for TFP growth through technical change, the other variables – economic policy, human capital, infrastructure and the absence of armed conflict – help to establish an enabling environment for economic growth. They facilitate farmer access to new technologies and markets, increase returns to savings and investments, and provide incentives for farmers to reallocate resources to the most profitable enterprises. Data limitations, however,

Table 12.5. Sources of agricultural growth in sub-Saharan Africa

	Real agricultural GDP	Terms of trade	Gross agricultural output	Cropland area	Yield	Inputs/ area	Total factor productivity
	A+B	A	B=C+D	C	D=E+F	E	F
	(average annual growth rate, %)						
1961–1970	n.a.	n.a.	3.03	2.26	0.78	0.59	0.19
1971–1980	2.49	1.45	1.04	−0.68	1.71	1.82	−0.10
1981–1990	2.16	−0.97	3.13	3.04	0.09	−0.97	1.06
1991–1900	2.95	−0.20	3.15	1.94	1.20	0.10	1.10
2001–2008	3.44	0.49	2.95	2.38	0.58	−0.22	0.79
1961–2008	n.a.	n.a.	2.53	1.63	0.90	0.31	0.59

n.a.= not available. Estimates are for all countries in SSA except South Africa and use FAO output and input data for Nigeria. Agricultural land area is crop area harvested adjusted for changes in quality. Source: Fuglie (2011), revised using updated FAO data.

constrain such an analysis. For many variables, data are either not available or incomplete for some countries. Other factors that might have an important role in promoting adoption of new technology, such as agricultural extension and credit services, are omitted because there are no good measures for them. Below, we describe in detail how we construct the variables that are included in our TFP growth analysis.

12.4.1 National agricultural research

Research affects productivity with a lag, but the effects can be long-lasting (Alston *et al.*, 1998). We treat research investments as creation of 'knowledge capital' and construct estimates of the agricultural research capital stock as the weighted sum of past annual investments in national agricultural research systems. The Agricultural Science and Technology Indicators (ASTI) project of the International Food Policy Research Institute provides data on annual spending and

the number of scientists employed in public agricultural research systems since 1981 for 33 SSA countries, including South Africa. We extend the series back to 1961 using data from Pardey *et al.* (1991). Annual research expenditures are measured in constant 2005 PPP dollars.[8] The number of scientists employed is a simple head count of full-time equivalent researchers holding a university degree.

The two measures of annual research effort – research spending and the number of scientists working on research – present different trends for agricultural research investment in SSA (Fig. 12.4). In constant 2005 PPP dollars, research spending in the region as a whole (excluding South Africa) stagnated at about $900 million around 1981 and remained at that level until 2000, after which research spending began rising again. The number of scientists employed, however, continued to grow throughout the period. A consequence of these trends is that research spending per scientist, measured in constant PPP dollars, declined by about half between 1981 and 2001, before stabilizing at around PPP$125,000 per scientist.

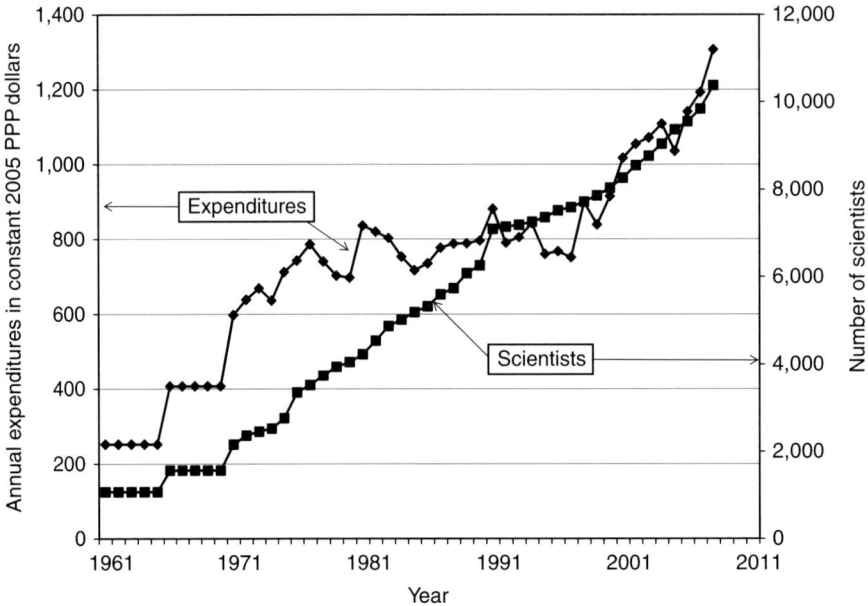

Fig. 12.4. Public agricultural R&D investment in sub-Saharan Africa. Figures show the sum total for 32 countries in sub-Saharan Africa, not including South Africa. Sources: Statistics for 1961–1980 are from Pardey *et al.* (1991); data from 1981–2008 are from IFPRI-ASTI.

Note that to put national research expenditures into a common unit (PPP dollars), they are first adjusted by a national price deflator (to put in real or constant local currency) and then converted to international dollars using the World Bank's PPP exchange rate (this adjustment was done by the authors of the data). Thus, research investments are adjusted according to variations in the price of a common basket of consumer goods over time and across countries. A preferable method would be to adjust research spending by indexes of the price of research goods (e.g. scientists' salaries, laboratory equipment, land for experimental plots, etc.), but such indexes are not available. If price trends in research goods are substantially different from price trends in consumer goods (for example, because scientist salaries – the principal research cost – are rising at a different rate than general price inflation), then the scientist count may be a better indicator of annual research effort than constant PPP expenditures.

Another issue in measuring research effort is whether to scale the data by country size. This decision depends in part on how spatially adapted new technologies emanating from this research effort are expected to be and whether there are economies of size in research systems. If an improved crop variety performs about as well on all of the area grown to that crop in the country, then one breeder working on one experiment station could serve 1000 ha as well as 10,000 ha. Countries with diverse ecological environments and farming systems might, however, need several research projects in order to develop and adapt technologies for different parts of the country. Small countries may simply not be able to afford research systems with sufficient scientific breadth (disciplinary, commodity and geographic coverage) or depth (scientists with advanced degrees), given the size of their agricultural sector. If a 'critical mass' of scientists is needed for a successful research system, it may be the

absolute size of the research system that is more relevant than research spending relative to agricultural output, cropland area, number of farms or another measure of size.

In our model, we test these hypotheses by including alternative measures of research capital. One is constructed from absolute spending levels on agricultural research. A second uses research spending per dollar of agricultural output. The third uses 'scientist-years'[9] rather than expenditures to measure research effort. Whether the absolute or relative measure performs better in explaining TFP growth can shed light on the 'small country problem' in agricultural research systems. If it is the absolute size of the research system that matters, then it would seem, at least in the African context, that there are significant economies of scale in agricultural research systems. And if scientist-years performs better than expenditures measured in constant PPP$, then the national price deflators and PPP exchange rates might not give internationally comparable estimates of R&D effort.

Research capital is treated like physical capital, in that it is an accumulation of past annual research investments. Like physical capital, research capital eventually depreciates through technology obsolescence but, unlike physical capital, research capital accumulates with a lag: it takes several years for the knowledge generated from research to be fully incorporated into higher farm productivity and output (Alston et al., 1998). To create research capital stocks from past research investments, we use the Almon or polynomial lag structure suggested by Alene and Coulibaly (2009). Each country c's research stock for year t $\left(NAR_{ct}^{stock}\right)$ is a weighted sum of the previous 16 years of research effort. The annual research effort (RE_{ct}) is alternatively measured as expenditures (in constant PPP$), expenditure per agricultural GDP and as a head count of scientists (Eqn 12.1) (see equation 12.1 at bottom of page):

$$NAR_{ct}^{stock} = 0.017(RE_{c,t}) + 0.034(RE_{c,t-1}) + 0.045(RE_{c,t-2}) + 0.057(RE_{c,t-3})$$
$$+ 0.068(RE_{c,t-4}) + 0.074(RE_{c,t-5}) + 0.080(RE_{c,t-6}) + 0.080(RE_{c,t-7})$$
$$+ 0.085(RE_{c,t-8}) + 0.080(RE_{c,t-9}) + \cdots + 0.017(RE_{c,t-16}) \tag{12.1}$$

That is, research expenditures begin to marginally affect productivity in the first year, their effects gradually rising, peaking in year eight, then diminishing and terminating in year 16 through technology obsolescence. Because of the 16-year time lag imposed on the annual R&D investment data (which began in 1961 – giving our research stock variable coverage over a 29-year period), we focus our multivariate analytical model of agricultural TFP growth determinants on the 1977–2005 time span.

12.4.2   International agricultural research

The CGIAR system of international agricultural research allocates 40–50% of its global research budget to SSA (CGIAR Annual Reports). Although SSA was largely bypassed by the Green Revolution of 1960s and 1970s, by the late 1990s nearly 20% of the crop area in SSA was sown to improved varieties developed by CGIAR centres (Evenson and Gollin, 2003). In addition to improved crop varieties, CGIAR has made significant impacts in SSA by developing and disseminating biological control agents for cassava pests (Zeddies et al., 2001). Maredia and Raitzer (2006) estimate that as much as 80% of the documented impact from CGIAR in SSA was due to cassava biological control.

It is possible to derive a research capital stock variable for CGIAR using data from CGIAR Annual Reports on its annual research spending for the SSA region. However, representing CGIAR's contribution to agricultural productivity in SSA in this way would only allow us to examine its impact on the region as a whole over time and not help to explain cross-country differences. Instead, we estimate the share of total crop area affected by CGIAR technologies for each country in SSA. Area affected by CGIAR technology includes area under improved crop varieties, area affected by biological control, and area under natural resource management technologies developed by CGIAR centres.[10] Total crop area is the sum of area harvested for all crops, including perennials.

The area affected by CGIAR research is a measure of technology dissemination, rather than research input. As such, it is likely to be affected by other variables in the model. To address this endogeneity problem, we use an instrumental variables approach in which area affected by CGIAR technology is modelled as a function of the other variables in the model, as well as a CGIAR research stock and the share of crop area planted to cassava. The CGIAR research stock is derived from past CGIAR investments for SSA[11] using the polynomial lag structure shown in Eqn 12.1. The share of total crop area planted to cassava captures the autonomous impact of the successful biological control programmes against the mealybug and green mite by IITA, one of the CGIAR centres. These biological control efforts involved mass rearing and release of insect pest predators by IITA and were self-sustaining once the pest predators were established in the field. As such, they did not involve any conscious adoption decision by farmers or even require much scientific or technical capacity in cooperating countries (Zeddies et al., 2001).

The model of CGIAR area impact $\left(CG_{ct}^{area}\right)$ is of interest in its own right because it allows us to test hypotheses about the complementarities between international and national agricultural research. Our modelling framework allows investments in national agricultural research to affect productivity independently, as well as by facilitating local adaptation and dissemination of CGIAR technologies. The coefficients on the NAR_{ct}^{stock} variable in the CG_{ct}^{area} and TFP models will indicate the importance of these two pathways for national agricultural research to have an impact on productivity.

12.4.3   Economic policies

The World Bank's nominal rate of assistance (NRA) to agriculture, reported annually for 18 SSA countries (including South Africa) through 2005 in Anderson and Masters (2009), provides a comprehensive measure of the price distortion caused by government commodity price interventions, input subsidies, trade policies, exchange rate policies and direct taxes on producers. The NRA gives the net effect of these policies on prices paid and received by farmers as a percentage of what prices would be in an undistorted market (Anderson and Masters, 2009).

For the SSA region as a whole, the average NRA has been consistently negative during the past several decades, meaning that the net effect of economic policies has been to tax agriculture. Structural adjustment policies implemented by some countries in the 1980s and 1990s seem to have reduced this bias against agriculture, even though in some cases these policies removed input subsidies. The mean NRA for the region rose from −22.0% in 1975–1979 to −11.9% in 2000–2004 (Anderson and Masters, 2009). The fact that NRA is negative means that the cumulative effect of government market interventions has been to tax agriculture in this region.

We expect that increases in NRA (lower effective taxation) will strengthen incentives for farmers to invest more in agriculture, adopt new technologies, shift resources to more profitable commodities and expand output generally. Such changes in resource allocation and technology utilization can raise agricultural TFP. Farmer response to policy reforms may not be immediate, however. To derive a predicted or expected value of NRA, we regress the NRA statistics against four of its own lagged values and then generate a predicted value, which we assume more accurately reflects how policy reform affects farmer decision making.

12.4.4 Human capital

Human capital of the labour force includes its skill level and health status. Barro and Lee (2010) have recently updated their internationally comparable average schooling level estimates for the working-age population, by country and over time. Their estimates, which are for the labour force as a whole and not just agricultural labour, show average schooling in SSA rose from

about two years to five years between 1970 and 2005. If more educated labour is more likely to migrate to non-farm or urban jobs, these estimates might overstate the average schooling level of farm labour. None the less, they should capture general tendencies (and differences among countries) of the importance given to general education, particularly since most labour in SSA continues to be employed in agriculture.

The spread of HIV/AIDS has had a major negative impact on the health status and depressed economic growth in several SSA countries, particularly in southern Africa. Dixon et al. (2002) estimate that HIV/AIDS reduced economic output in SSA by 2–4%. We expect HIV/AIDS to reduce agricultural productivity primarily through its effects on the labour supply. Not only are HIV/AIDS patients unable to work, other family members might have to reduce their farm labour supply in order to care for the sufferers. The fact that HIV/AIDS tends to disproportionally affect adults in their prime working years only exacerbates the effect on the labour force. Although other health problems, such as malaria and malnutrition, are also pervasive in SSA, this study models the health status of the labour force by the proportion of the population estimated to be infected with the HIV/AIDS virus. This may be the most significant *change* in the overall health status of the general population in the SSA region over the past several decades. There is also considerable variation over time and across countries in the incidence of the disease. Increased availability of antiretroviral therapy in many SSA countries has reduced these impacts in recent years, although most of this has occurred since 2005, after the period of our study (The Global Fund, 2011).

The World Bank reports the prevalence of HIV/AIDS infection as a percentage of a country's total population between the ages of 15 and 49, annually from 1990 to 2005. For most countries, HIV/AIDS prevalence was close to zero in 1990, and for those countries we assume it was zero prior to 1990. For countries with significant HIV/AIDS infection in 1990, we extrapolate infection rates back to 1977 by fitting a logistic epidemiology curve (assuming first infections occur in 1980) to create an HIV variable. By this procedure, HIV prevalence for the SSA region as a whole rose from zero in 1980 to 2.2% in 1990, and peaked at 4.8% in 2000 before falling to 4.3% by 2005.

12.4.5   Infrastructure

We measure transportation infrastructure by road density (km of roads/km² land area). Data on road density are available from the International Road Federation. For some countries with large, sparsely populated areas, this measure of road density might not reflect actual road density in populated or farmed areas. We experiment with alternative measures such as kilometres of road per square kilometres of crop area harvested (i.e. assuming roads are located primarily in farming areas). However, roads only capture one dimension of transportation and communication infrastructure. Besides railway and river transport, marketing costs will be affected by proximity to ports, availability of storage facilities, internal and cross-country restrictions on trade, availability of telecommunications for transmitting marketing and price information, and other factors. Road density is none the less a critical dimension of marketing infrastructure, and systematic measures of this variable are available for a number of SSA countries and over time.

12.4.6   Civil conflict

Civil conflict and war can destroy agricultural crops and livestock, disrupt trade, and displace large portions of a country's population. Mozambique's and Angola's prolonged civil wars in the 1980s displaced as much as one-third of the rural population. We account for civil conflict by employing the Uppsala Conflict Data Program armed conflict data set (Gleditsch et al., 2002).[12] The data consists of an indicator variable which takes on a value of 1 if a country experienced at least 25 battle-related deaths

in a given year. Roughly 18% of the time the countries in our dataset had at least this level of civil conflict, although five of these countries (Burundi, Mozambique, Rwanda, Sudan and Uganda) experienced conflicts about two-thirds of the years between 1977 and 2005. We expect the effects of conflict on growth to accumulate over time, so we measure the effect of conflict in year t as the cumulative number of years a country has experience such conflict since 1977. The coefficient on this variable will then measure the marginal effect of one additional year of conflict on the rate of productivity growth.

12.4.7   Data limitations

Although our indexes of agricultural TFP are available for 46 countries[13] in SSA for each year between 1961 and 2005, we are able to employ only subsets of this information in our TFP-growth determinants model. First, because of the lag structure created for national stocks of agricultural research, we are restricted to the 1977–2005 period. Second, the ASTI data on national agricultural research expenditures and scientist numbers cover only 32 countries in the region. Most of the excluded countries are either very small with populations under 1 million or are countries for which agricultural data are generally thought to be of poor quality (and therefore the TFP estimates are subject to a high degree of error). Countries that fall into the latter category include the Congo Democratic Republic (formerly, Zaire), Somalia, Angola, Chad, Liberia and Sierra Leone. Agricultural research data are also missing for Namibia, which did not gain independence from South Africa until 1990. Our estimates of cropland area affected by CGIAR technologies include all the 32 countries for which we have data on national agricultural research. If studies did not report any adoption of CGIAR technologies in a country, we assume it was zero.

We consider these 32 countries as our 'core' data set with measures of TFP, national agricultural research stock, and cropland area affected by CGIAR technology for the entire 1977–2005 period. Adding additional policy variables reduces the country and time coverage (Fig. 12.5). Including schooling levels with the core model reduces the coverage to only 27 countries (we lack data on schooling for Nigeria, Ethiopia, Burkina Faso, Guinea and Madagascar). For road density, only 12 of the 32 countries in our core sample have data for 1977–2005, whereas data for the other 20 countries are only available for more recent years. For economic policies, NRA data are available for only 18 of the 32 countries. Data on civil conflict and the incidence of HIV/AIDS (employing our estimation procedure for extrapolating HIV/AIDS prevalence for years prior to 1990) are available for all 32

core countries over the whole period. Finally, if we include all seven policy variables simultaneously in the model, we are left with only 14 countries in the sample (nine countries with complete data over 1977–2005 and five countries with complete data only for 2001–2005). See Table 12.6 for a complete list of the variables in the TFP determinants model, with definitions, sources and some summary statistics.

Our empirical strategy is to estimate a simultaneous equations model of crop area affected by CGIAR technology and agricultural TFP growth. The model is specified as follows (see equations 12.2a and 12.2b at bottom of page):

where subscripts c and t refer to country and year, respectively, and ε_{1t} and ε_{2t} are random and

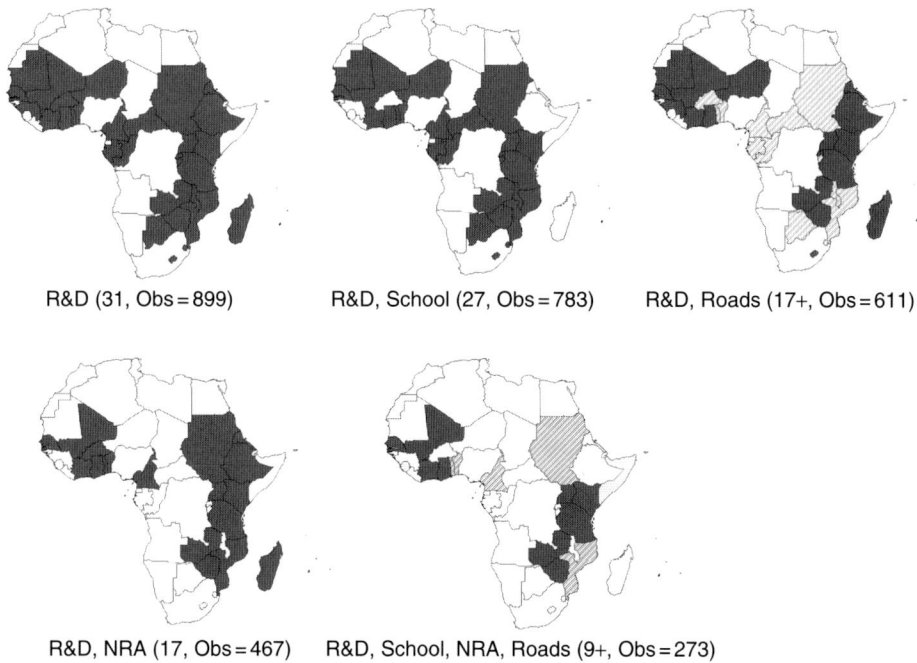

R&D (31, Obs = 899)   R&D, School (27, Obs = 783)   R&D, Roads (17+, Obs = 611)

R&D, NRA (17, Obs = 467)   R&D, School, NRA, Roads (9+, Obs = 273)

Fig. 12.5. Data coverage for policy variables during 1977–2005. Countries with complete data on listed variables during 1977–2005 are shaded and countries with data on variables for only some years are hatched. The first number in parentheses refers to the number of countries with data on these variables; the second number refers to the number of observations (countries × years of data per country).

$$CG_{ct}^{area} = \alpha_1 Ln\left(CG_t^{stock}\right) + \alpha_2 Ln\left(NAR_{ct}^{stock}\right) + \gamma_1 X_{kct} + \varepsilon_{1t} \qquad (12.2a)$$

$$Ln(TFP_{ct}) = \beta_1 Ln\left(CG_{ct}^{area}\right) + \beta\alpha_2 Ln\left(NAR_{ct}^{stock}\right) + \gamma_2 X_{kct} + \varepsilon_{2t} \qquad (12.2b)$$

Table 12.6. Description of variables in TFP determinants model.

Variable	Description	Units	Observations	Mean	SD	Min	Max	Data source
TFP	Index of agricultural total factor productivity	Base year (1977) = 100	1334	106.022	20.929	60.968	245.289	Fuglie (2011)
CG Area	Share of cropland impacted by CGIAR technologies	Percentage	1334	0.082	0.114	0.000	0.505	Compilation from Evenson and Gollin (2003), Zeddies et al. (2001), others
CG Stock	Stock of CGIAR research capital	2005 US dollars, millions	1334	92.742	43.866	16.273	153.991	CGIAR Annual Reports
NAR Stock	Stock of NAR research capital	2005 PPP dollars, millions	899	20.043	21.289	1.058	149.048	IFPRI-ASTI and Pardey et al. (1991)
NAR/GDP Stock	Stock of NAR/GDP research capital	PPP$ of R&D per PPP dollars of agricultural GDP	899	0.021	0.020	0.002	0.114	IFPRI-ASTI and Pardey et al. (1991)
SCI Stock	Stock of NAR research capital	Scientist-years	899	137.753	142.108	5.169	897.822	IFPRI-ASTI and Pardey et al. (1991)
Cassava	Cassava's share of total cropland harvested	Percentage	1334	0.071	0.100	0.000	0.477	FAO
HIV	Share of adult population infected with HIV/AIDs virus	Percentage	1334	0.025	0.052	0.000	0.289	World Bank
Civil war	Cumulative number of years since 1977 when there was at least 25 deaths in a year due to civil war or disturbance.	Years with civil disturbance since 1977	899	2.46	4.77	0.000	26.000	Gleditsch et al. (2002)
NRA	Nominal rate of assistance to agriculture	Percentage deviation from non-intervention policy	496	−0.117	0.139	−0.610	0.263	Anderson and Masters (2009)
Road	Road density	km roads per km² land area	725	0.151	0.202	0.007	0.990	International Road Federation
School	Average schooling of adult labour force	Years	928	3.790	1.922	0.508	9.270	Barro and Lee (2010)

independent error terms. The technology variables $\left(TFP_{ct}, CG_{ct}^{Area}, CG_{ct}^{stock}, \text{ and } NAR_{ct}^{stock}\right)$ form the 'core' of the model, X_{kct} is a vector in which k indexes the constant term, incidence of HIV (HIV_{ct}), prevalence of armed conflict ($CivilWar_{ct}$), economic policy (NRA_{ct}), road densities ($Road_{ct}$), education ($School_{ct}$) and cassava's share of total cropland ($Cassava_{ct}$). Given the missing observations of many of the variables in vector X_{kct}, as noted above, we run several regressions varying its composition. The values of the estimated parameters α_1, α_2, β_1 and β_2 allow us to derive a 'productivity' elasticity with respect to research, or the percentage change in productivity (or output) given a 1% change in the size of the *NAR Stock* or *CG Stock* variables. Taking the derivation of Eqn 12.2a and 12.2b with respect to $Ln\left(NAR_{ct}^{stock}\right)$ gives $\alpha_2\beta_1 + \beta_2$, or the total productivity elasticity with respect to national agricultural research. The first term ($\alpha_2\beta_1$) measures the impact of national agricultural research investment on helping to adapt

and disseminate CGIAR technologies within the country, whereas the second term (β_2) captures the direct effect of national research on productivity independent of CGIAR. The productivity elasticity with respect to CGIAR research is given by $\alpha_1\beta_1$. These productivity elasticities, together with the time structure of R&D impact specified in Eqn 12.1, allow us to estimate rates of return to national and international agricultural research in SSA (see Alston *et al.*, 1998, pp. 196–198, for a detailed description of these procedures).

12.5 Results on Determinants of Agricultural TFP Growth

Tables 12.7 and 12.8 present a selection of regression estimates of the TFP determinants model. These tables report results using absolute expenditures (i.e. un-scaled for country size) in the creation of the national

Table 12.7. Factors affecting area impacted by CGIAR technology in sub-Saharan Africa.

Variables	Model 1	Model 2	Model 3	Model 4	Model 5	Model 6
	CG Area	CG Area	CG Area	CG Area	CG Area	CG Area
Ln(CG stock)	0.0688***	0.0710***	0.0479***	0.0592***	0.0662***	0.0662***
	(16.15)	(11.08)	(9.675)	(11.85)	(8.559)	(6.667)
Cassava	0.645***	0.636***	0.899***	0.619***	0.512***	0.348***
	(26.11)	(11.92)	(18.98)	(23.41)	(7.831)	(3.082)
Ln(NAR stock)	0.0164***	0.00372	0.0180***	0.0122***	−0.0221***	−0.0435***
	(6.930)	(0.855)	(6.524)	(4.414)	(−3.486)	(−4.850)
HIV	0.222***	0.595***	0.268***	0.0988*	0.387***	0.320***
	(4.965)	(8.694)	(5.762)	(1.889)	(4.823)	(3.549)
Civil War	−0.00151***	−0.00333***	−0.000552	0.000275	−0.00214***	−0.00358***
	(−2.855)	(−5.709)	(−0.951)	(0.412)	(−2.870)	(−3.925)
NRA		0.0871***			0.0570**	0.0952***
		(3.642)			(2.090)	(2.763)
Ln(Road)			0.00595**			0.0336***
			(2.523)			(4.744)
School				0.0106***	0.0140***	0.00734*
				(6.128)	(5.383)	(1.901)
Constant	−0.298***	−0.249***	−0.205***	−0.283***	−0.186***	0.0267
	(−16.31)	(−8.629)	(−9.051)	(−13.60)	(−5.516)	(0.501)
Observations	928	496	640	783	389	273
R²	0.560	0.538	0.555	0.567	0.554	0.569
Adjusted-R²	0.557	0.533	0.551	0.564	0.546	0.556

A two-stage IV procedure is used to estimate the model using an annual panel of countries over 1977–2005. Because of missing observations on variables, the number of observations included in the estimation varies by model. t-Statistics in parentheses; significance tests indicated by ***$p<0.01$, **$p<0.05$, *$p<0.1$.

Table 12.8. Determinants of agricultural TFP growth in sub-Saharan Africa.

Variables	Model 1	Model 2	Model 3	Model 4	Model 5	Model 6
	Ln(TFP)	Ln(TFP)	Ln(TFP)	Ln(TFP)	Ln(TFP)	Ln(TFP)
CG Area	0.461***	0.815***	0.815***	0.521***	1.447***	2.038***
	(6.625)	(6.516)	(8.225)	(6.398)	(8.956)	(8.909)
Ln(NAR Stock)	0.0266***	0.0357***	0.0338***	0.0285***	0.0745***	0.0858***
	(4.891)	(3.388)	(4.765)	(5.010)	(5.851)	(4.511)
HIV	−0.171*	−0.847***	−0.495***	−0.262**	−1.264***	−1.672***
	(−1.810)	(−4.757)	(−4.635)	(−2.433)	(−6.790)	(−8.315)
Civil War	−0.00750***	−0.00864***	−0.00727***	−0.00865***	−0.0117***	−0.00860**
	(−6.766)	(−7.423)	(−5.713)	(−6.250)	(−8.296)	(−4.860)
NRA		0.338***			0.259***	0.124*
		(6.196)			(4.672)	(1.701)
Ln(Road)			−0.0297***			−0.0577***
			(−5.468)			(−3.921)
School				0.00596	−0.0102*	−0.0153*
				(1.540)	(−1.685)	(−1.884)
Constant	4.569***	4.584***	4.465***	4.535***	4.430***	4.203***
	(306.6)	(128.5)	(175.1)	(248.1)	(101.8)	(39.29)
Observations	899	467	611	783	389	273
R²	0.103	0.291	0.192	0.132	0.373	0.435
Adjusted-R²	0.0988	0.283	0.185	0.127	0.363	0.420

A two-stage IV procedure is used to estimate the model using an annual panel of countries over 1977–2005. Because of missing observations on variables, the number of observations included in the estimation varies by model. t-Statistics in parentheses; significance tests indicated by ***$p<0.01$, **$p<0.05$, *$p<0.1$.

agricultural research stock variable. The six model specifications reported in the tables contain alternative compositions of the vector X_{kct}, which, given the lack of country coverage for many of these variables, dramatically alters the sample size included in the regressions. For example, model 1 from Table 12.8, which contains only *HIV* and *CivilWar* in the X_{kct} vector, has 899 observations (31 countries over 29 years); when *NRA*, *Road* and *School* are also included (model 6, Table 12.8) the number of useable observations drops to 273.[14]

Additional results using alternative *NAR*-variable construction (i.e. scientist-years and research expenditures per agricultural GDP) are given in Appendix Tables A12.1 and A12.2. Using scientist-years to measure research investment (Appendix Table A12.1) gives very similar results to those shown in Tables 12.7 and 12.8, suggesting this modelling issue is not a concern. However, the model using *relative* research expenditures gives a very different result that is not robust across specifications (Appendix Table A12.2). The *NAR* variable

is negative or insignificant in five specifications and positive and significant in only one. It seems that small countries with (relatively) large research systems have not done better, and perhaps have done worse, than larger countries that have spent more total dollars on research but less per dollar of agricultural output. The better performance of the models using absolute spending on research supports the hypothesis – at least in the African context – that there are significant economies of size in agricultural research systems. We return to this question below and compare returns to research by country, using the results from Tables 12.7 and 12.8, to see whether, on average, larger countries have got higher returns from research than smaller countries.

Table 12.9 translates the findings from the econometric model into productivity elasticities with respect to national and international agricultural research. National agricultural research has a significant, direct effect on productivity (Table 12.8) and, to a lesser effect, facilitates the uptake of new technologies emanating from the CGIAR

Table 12.9. CGIAR and national agricultural research elasticities.

Variable effects	Coefficients	Average of models 1–4	Model 1	Model 2	Model 3	Model 4
CGIAR impact on CG Aarea	α_1	0.0617	0.0688	0.0710	0.0479	0.0592
NAR impact on CG Area	α_2	0.0126	0.0164	0.0037	0.0180	0.0122
CG area impact on TFP	β_1	0.6530	0.4610	0.8150	0.8150	0.5210
NAR direct impact on TFP	β_2	0.0312	0.0266	0.0357	0.0338	0.0285
NAR indirect impact on TFP through increasing CG Area	$\alpha_2\beta_1$	0.0082	0.0076	0.0030	0.0147	0.0064
R&D elasticities						
NAR total impact on TFP	$\alpha_2\beta_1 + \beta_2$	0.0394	0.0342	0.0387	0.0485	0.0349
CGIAR impact on TFP	$\alpha_1\beta_1$	0.0403	0.0317	0.0579	0.0390	0.0308

The coefficients from models 1–4 are taken from the econometric estimates reported in Tables 12.5 and 12.6. NAR, 'stock' value of research by national agricultural research system; CGIAR, 'stock' value of research by international agricultural research centres. CG area, share of cropland affected by technologies developed by CGIAR centres.

centres (Table 12.7). Through these two pathways, national agricultural research has a productivity elasticity of about 0.039, achieved by summing the mean impact of NAR in Tables 12.7 and 12.8 across models 1–4, which are the models in which we have the most confidence.[15] About 90% of the NAR impact is a direct effect, and about 10% is due to their collaboration with the CGIAR. The productivity elasticity, again averaged across models 1–4, of the CG Stock variable is somewhat lower, about 0.040. Recall, the CG Stock elasticity is generated by $\alpha_1\beta_1$ from Eqns 12.2a and 12.2b.

Using the estimates of the productivity elasticities with respect to research (Table 12.9) and the time path of research impact given by Eqn 12.1, we estimate the benefit stream over time from an initial $1 increase in research expenditure. From this cost–benefit stream we derive internal rates of returns (IRR) and cost–benefit ratios for different African countries. Rates of return to national agricultural research vary considerably across countries. Large countries with annual agricultural GDP greater than PPP$4 billion (this includes Nigeria, Ethiopia, Sudan, Kenya, Tanzania and Côte d'Ivoire) earned a mean IRR of 41%. Small countries (less than PPP$1 billion in agricultural GDP) earned a mean IRR of only 17%. Assuming a 10% real discount rate, this yields a benefit–cost ratio of 1.8 for small countries, compared

with 4.6 for large countries. For mid-sized countries (between PPP$1 and PPP$4 billion in agricultural GDP), the median IRR was 29%, giving a benefit–cost ratio of 2.9.

Having CGIAR technologies to draw from helped to raise returns to investments in NARS. For the average SSA country, returns to agricultural research without the CGIAR would have been about 23.8%, compared with 29.3% with the CGIAR.

We also derive an estimate of the IRR to CGIAR's historical investment in SSA (Table 12.10). Using the productivity-to-research stock elasticity of 0.040 and the same time pathway for research impacts (i.e. Eqn 12.1) implies that the CGIAR has yielded an IRR of 58%, or $5.3 in benefits for every $1 in expenditure. This is above the median 40% estimate of the IRR to international agricultural research in Africa and elsewhere reported by Alston et al. (2000) in a meta-analysis of studies on the returns agricultural research and far higher than the 8% IRR estimate by Maredia and Raitzer (2006), who conducted a meta-analysis of CGIAR impacts specifically in SSA. One reason why our estimate is so much higher than that of Maredia and Raitzer is that their study only included impacts that were carefully documented through ex-post field studies conducted before the year 2000. Since their review, there has been considerably more documented diffusion of CGIAR technologies in SSA (see Renkow and Byerlee, 2010).

Table 12.10. Returns to agricultural research in sub-Saharan Africa, 1977–2005.

Country	Ag GDP million 2005 PPP dollars	Ag R&D million 2005 PPP dollars	Returns to national agricultural research		
			IRR w/out CGIAR (%/year)	IRR with CGIAR (%/year)	Benefit–cost ratio (10% disc. rate)
Ethiopia	14,539	41	65.8	77.8	10.4
Sudan	14,110	39	50.0	59.2	7.1
Kenya	9,964	121	16.2	20.5	1.9
Tanzania	8,476	29	40.5	48.1	5.4
Ghana	7,211	31	43.3	51.4	5.9
Côte d'Ivoire	5,630	57	15.1	19.2	1.7
Zimbabwe	4,690	33	23.3	28.4	2.7
Cameroon	4,571	30	30.5	36.6	3.7
Uganda	4,399	42	21.1	26.0	2.4
Madagascar	3,182	17	23.8	29.1	2.8
Mali	2,900	25	24.9	30.2	2.9
Burkina Faso	2,326	24	19.2	23.8	2.2
Benin	2,219	11	31.7	38.0	3.9
Senegal	2,042	35	7.6	11.0	1.1
Rwanda	1,755	5	54.7	64.7	8.1
Malawi	1,746	22	14.5	18.6	1.7
Mozambique	1,732	15	18.9	23.5	2.2
Zambia	1,677	20	14.3	18.4	1.7
Niger	1,527	11	23.9	29.2	2.8
Guinea	1,357	12	18.7	23.2	2.1
Togo	1,149	11	18.8	23.4	2.2
Central African Rep.	1,125	4	32.2	38.6	4.0
Mauritania	1,081	9	27.0	32.7	3.2
Gabon	894	3	37.7	44.9	4.9
Burundi	839	10	14.8	18.9	1.7
Mauritius	600	16	5.7	8.9	0.9
Congo	519	6	16.0	20.2	1.8
Botswana	415	12	5.3	8.5	0.9
Swaziland	414	6	14.9	19.0	1.7
Gambia	358	4	17.6	22.0	2.0
Lesotho	108	11	–8.5	–6.4	0.2
Large countries	8,177	47	34.0	42.6	4.4
Mid-sized countries	1,844	16	23.6	28.9	2.6
Small countries	518	8	12.9	17.0	1.6
All countries	3,341	23	23.8	29.3	3.1
	m 2005 US$	m 2005 US$		IRR (%)	Benefit/Cost
CGIAR	38,386	133		57.7	6.2

IRR, internal rate of return to research (% per year). Countries with agricultural GDP >PPP$4 billion are defined as large; PPP$1–4 billion are mid-size and <PPP$1 billion are small. Sources: Agricultural GDP is annual average over 1980–2005 (World Bank); agricultural R&D is annual average spending over 1977–2005 by national agricultural research systems (ASTI) and the CGIAR for sub-Saharan Africa (CGIAR). The rate of return to NAR agricultural research with CGIAR assumes a research-to-output elasticity of 0.0394 and without CGIAR an elasticity of 0.0312. The CGIAR research-to-output elasticity is 0.0403 (see Table 12.5).

It is also possible, however, that the lag-structure for CGIAR impact is longer than what we have assumed in this study, which could lead to overestimating the returns to research. Although the evidence here suggests that international agricultural research has played an important role in raising productivity in African agriculture, more work is needed to better understand and quantify these impacts in the aggregate.

At the margin, it would seem that the highest payoff from additional R&D investment in SSA would come from strengthening the CGIAR system, followed by greater support for national agricultural research systems in large countries. Although returns to further expansion of mid-sized and small country research systems has lower returns compared with large-country and CGIAR research, the returns are none the less above the typical 'hurdle rates' of 10–12% used to evaluate development project investment decisions.

Below we draw some further conclusions from the econometric results presented in Tables 12.7 and 12.8. As above, we concentrate on the results from the first four specifications: models 1–4 in the tables.

Table 12.8's coefficient estimate of *CG Area* gives an indication of the average productivity improvement achieved from the diffusion of new CGIAR technologies. The estimate ranges from 0.46 to 0.85 in models 1–4, implying an average per hectare productivity gain of 46–85% on cropland affected by these technologies. This is consistent with yield impacts from the diffusion of improved varieties reported in Evenson and Gollin (2003) and the biological control of cassava pests described by Zeddies *et al.* (2001). Much of this yield improvement, according to those studies, came about from reduction in crop losses from biotic and abiotic stresses and did not involve increased use of external inputs or other changes in existing farming practices. This might explain why farmer schooling is influential on technology adoption (Table 12.7) but apparently not directly on productivity (Table 12.8), except through the adoption decision. Schultz (1975) argued that education confers cognitive skills that enable

farmers to adjust more quickly to the 'disequilibria' created when new technology is introduced. If all the gains (disequilibria) from new technology occur from initial adoption and not from subsequent changes in input use or other farming practices following adoption, then having more education would confer no further cognitive advantage other than to enable early adoption. This generalization certainly does not apply to all of the kinds of technologies being introduced and adopted by African farmers, but it may describe those that have achieved the widest area coverage (and economic significance) so far. The relatively low elasticity on the *School* variable is also consistent with what Lockheed *et al.* (1980) found in their survey of studies of the effects of farmer schooling on agricultural productivity in 'traditional' agricultural settings. In such settings, they found that 4 years of farmer schooling increased agricultural productivity by an average of about 1.3%; compared with 2.2% in our study.[16] This compares with an average of 9.5% in 'modernizing' agricultural settings, or ones undergoing significant technological or structural transformation (Lockheed *et al.*, 1980).

Policy reform (in the form of higher values of NRA) has had a direct effect on productivity, as well as an indirect effect by increasing the rate of CGIAR technology diffusion. The direct effect, however, accounts for about 86% of the total, suggesting that the primary way policy reform raised productivity was by providing stronger incentives for farm households to reallocate resources to more profitable crops and cropping practices. Anderson and Masters (2009) estimate that for SSA countries as a whole, the average value of the nominal rates of assistance to agriculture improved from –22.0% in 1975–1979 to –11.9% in 2000–2004 (i.e. net taxation of agriculture was reduced by 10.1 percentage points). The coefficient estimates from our model on the impact of NRA on productivity suggest that this magnitude of policy reform boosted productivity (or output) in SSA by about 4%. Further policy reform to raise the nominal rates of assistance to 0 (i.e. to reduce net taxation of agriculture from –11.9% to zero) would raise productivity by another 4.8%.[17]

The prevalence of HIV and the incidence of civil unrest suppressed agricultural productivity (Table 12.8). For every 1% rise in the population infected with HIV, farm productivity declined by a mean 0.44%. The mean increase in HIV from 0% to 5% in the SSA region over the study period implies that this disease has reduced regional agricultural output by at least 2%.[18] This is comparable to the estimate of a 2–4% loss in economic output from HIV/AIDS in Africa by Dixon *et al.* (2002). The increased availability of antiretroviral therapy, especially since 2004, has undoubtedly helped curb some of these economic losses. By 2009, approximately 36% of HIV/AIDs sufferers in SSA were receiving therapy (The Global Fund, 2011). Assuming a similar proportion of affected rural populations got access to antiretroviral therapy, the implied recovery of agricultural productivity would be on the order of 0.7%, or about US$640 million/year.

Armed conflict in many countries of the region was another cause of lost agricultural productivity. Every additional year of armed conflict resulted in a 0.9% decline in agricultural productivity (output), less than the economy-wide estimate of 2.3% per year of civil war by Collier (2007, p.27). One reason why our estimate is smaller might be that our sample excludes several countries most affected by civil war during our study period, such as Somalia, Congo DR, Angola, Liberia and Sierra Leone. Our estimate also does not include lost output from resource withdrawals from agriculture, so it is at best a lower-bound estimate.

The model results suggest that more road development encourages the diffusion of new agricultural technologies, although the estimated relationship between *Roads* and *TFP* is negative. Block (2010) found a similar negative relationship between road density and agricultural productivity growth in SSA and, like him, we believe this relationship is spurious. There is considerable evidence from Africa and other developing countries that improved rural road infrastructure encourages greater agricultural productivity: it lowers transportation costs and increases market access, which encourages farmers to devote more resources to commercial farming, increase their use of inputs, and shifts resources to more high-valued commodities (Zhang and Fan, 2004).

We believe our result reflects a limitation of the data. For most SSA countries we lack time-series data on roads, road quality and other dimensions of rural infrastructure. This means that there is not sufficient variation in the national road measures to assess impacts on productivity over time. It might be that to assess the economic impact of road infrastructure requires more detailed geospatial data A recent study by Dorosh *et al.* (2009) using geo-referenced data on agricultural production and road infrastructure in SSA found that reduced travel time to urban markets from more and better roads had a large and positive impact on agricultural production and stimulated more adoption of high-input/high productivity agricultural technologies. Most of the road infrastructure impacts they found, however, resulted from expansion of cultivated cropland in remote areas rather than productivity gains.

12.6   Conclusions

Despite some recent improvement, agricultural productivity growth in sub-Saharan Africa continues to lag behind every other region of the world. There are important data challenges in measuring agricultural productivity trends, but most studies agree that agricultural TFP in SSA was stagnant or declining in the 1960s and 1970s and turned positive by about the mid-1980s. Our results find that since the 1980s agricultural TFP in the region has been growing at roughly 1% annually, nearly half the 2% average growth rate for all developing countries (Fuglie, Chapter 16, this volume).West Africa seems to have had the strongest productivity growth in SSA in recent decades, especially Ghana, Benin and Nigeria, though the data on Nigeria are problematic. Kenya in East Africa has also done reasonably well in sustaining a modest rate of long-term productivity growth. Mozambique and Angola show rapid productivity gains since peace was restored to those countries in the 1990s, but this is mainly a case of recovering from productivity losses incurred during their long civil wars.

A number of factors seem to have contributed to the productivity growth renewal observed in recent decades. One driver is the accumulated knowledge capital from long-term national and international investments in agricultural R&D, which are gradually delivering improved technologies to farmers. We estimate that for large and mid-size African countries, agricultural R&D has generated reasonably high returns, on the order of $3 to $5 in benefits for every $1 spent on R&D. But for very small African countries, building comprehensive national agricultural R&D might not be economically viable because of economies of size in research systems. For these countries, tying into regional and international agricultural research networks and maintaining a policy environment that is receptive to technologies developed elsewhere seems to be crucial. In fact, the CGIAR system of international agricultural research centres has played an important role in raising agricultural productivity growth in SSA. Our results suggest that spending by CGIAR in SSA has generated an internal rate of return on the order of 58% per year, or about $6 in benefits for every $1 spent. Moreover, we find that national and international agricultural research in SSA are complementary – having a larger national

research system significantly improves the rate at which new technologies emanating from the international centres reach farmers. Despite these achievements, however, SSA agricultural research systems remain relatively weak and underfunded.

In addition to investing in agricultural research, strengthening the broader 'enabling environment' for farmers to access technology, markets and the necessary support services is crucial for raising agricultural productivity in SSA. Our results found that policy reforms that reduced the net taxation of agriculture stimulated new technology adoption and productivity growth, as did higher levels of labour force schooling. The spread of HIV/AIDS and widespread civil strife have, however, posed significant constraints to agricultural development in Africa.

Looking forward, there is reason to be cautiously sanguine about prospects for agricultural productivity growth in SSA. During the first decade of this century, both the CGIAR and national research systems have increased spending on agricultural research in SSA, the incidence of civil unrest has fallen, and greater availability of antiretroviral therapy and other measures have reduced the scourge of AIDS. If momentum on policy reform can be sustained, that too can continue to be a source of growth for African agriculture.

Notes

[1] Sub-Saharan Africa is defined here as the 47 developing countries that lie south of the Sahara Desert in Africa as they were constituted in 2005, excluding South Africa. For analytical purposes, we aggregate Ethiopia and Eritrea into 'former Ethiopia' because the two countries separated in 1993.

[2] Gross agricultural output measures change in quantities, holding prices fixed at constant 1999–2001 global average levels. Real agricultural output is based on the FAO gross agricultural production series. Using this fixed set of price weights, it is the aggregate value of 142 categories of crop outputs and 30 categories of animal outputs per year.

[3] Livestock capital is measured as the number of head of 'cattle equivalents'. This is the total sum of cattle, equine, camels, small ruminants, pigs and poultry on farms, each species weighted by its relative size according to the weights suggested by Hayami and Ruttan (1985).

[4] Fuglie (2011) uses a random-effects model, instrumenting for inputs to control for possible simultaneous-equations bias. His instruments include a measure of population-per-hectare of quality-adjusted agricultural land, global indexes of agricultural commodity prices, fertilizer prices, tractor prices and lagged values of the inputs.

[5] Under the assumptions that farmers maximize profits and markets are in long-run competitive equilibrium, production elasticities equal input cost shares.

[6] The main difference between favourable and unfavourable rainfall areas is the average length of the growing season, but even favourable rainfed areas are restricted to one crop per year in most of SSA. The main food crops in favourable rainfed areas include root crops, maize and rice. In unfavourable rainfall areas, the predominant cereal crops are more drought-resistant sorghum and millet.

[7] TFP growth may be decomposed into four parts: technical change, technical efficiency change, allocative efficiency and scale economies. The portions evaluated in any TFP evaluation are driven by data availability and quality, as well as by the method and assumptions a researcher imposes on that data.

[8] The Pardey et al. (1991) figures are adjusted to 2005 PPP dollars using the US implicit GDP price deflator.

[9] The total number of 'scientist-years' is simply the number of scientists working in a year, in full-time equivalents. Scientists are defined as research staff with at least a bachelor's degree.

[10] Douglas Gollin kindly provided a database of area sown to improved CGIAR crop varieties for each SSA country over the 1961–2000 period, a dataset he and Evenson derived from the commodity case studies presented in Evenson and Gollin (2003). We supplemented this with estimates of area affected by biological control of cassava pests and updated evidence of the adoption of improved crop varieties of rice, maize, beans and potatoes. See Renkow and Byerlee (2010) for a list of sources. For crops in which we do not have updated adoption estimates beyond 2000, we assume a constant share of adoption area for that crop for more recent years.

[11] CGIAR was formally established in 1971, but the first of the research centres that would later form the CGIAR opened in the Philippines in 1960. CGIAR Annual Reports give annual expenditures for the system and the share of expenditures by region. Using the SSA share of CGIAR expenditures from the 1980s, we extend CGIAR research spending for SSA back to 1968 (when the first CGIAR centre in Africa opened) and assume it is zero before that.

[12] We employed Version 4-2008 of the dataset, available at: http://www.prio.no/CSCW/Datasets/Armed-Conflict/USDP-PRIO/.

[13] Our data set excludes South Africa and aggregates Ethiopia and Eritrea into 'former Ethiopia' in order to consistently measure agricultural output and inputs for this country over the 1961–2005 period. For policies after the countries separated in 1993, we use data for Ethiopia. This leaves us with 46 country indexes of agricultural TFP.

[14] Some of the specifications of the CG Area diffusion model (Eqn 12.2a, Table 12.7) contain more observations because Nigeria is included in these regressions. Nigeria, however, is excluded from the TFP determinants model because of the uncertainty surrounding the TFP measurement for this country. In any case, the models' results are robust (little change in value and no change in signs or significance of the estimated coefficients) whether or not Nigeria is included.

[15] Specifications 1–4 give coefficient estimates that are all statistically significant, maintain the same signs, and are generally similar in value. In specifications 5 and 6, loss of observations and multicolinearity among variables lead to less robust results. For example, coefficient estimates for the technology variables are significantly higher than in the other specifications, and signs and/or significance of a few of the other variables change. For this reason we prefer to concentrate attention on the first four model specifications.

[16] The productivity impact of schooling in our model is found by multiplying the elasticity of schooling on technology diffusion (0.106 in model 4) by the effect of diffusion on TFP (0.5210 in model 4), to give an increase in TFP of 0.55% for each additional year of farmer schooling. Multiplying this by 4 years of schooling gives 2.2%. We have ignored the 'direct' effect of schooling on productivity given in the TFP determinants model because the schooling coefficient in this model is not statistically significant.

[17] While raising the NRA provides incentives to increase productivity of existing resources in agriculture, this is not the only way policy reform can affect growth. Reforms that improve the terms of trade between the agricultural and non-agricultural sectors can shift new resources into agriculture, causing further growth in the sector. Anderson and Masters (2009, pp. 46–47) show a modest improvement in the relative rate of assistance to agriculture (i.e. the ratio of the NRA to agriculture and the NRA to non-agricultural sectors, a good measure of how policies influence the agricultural terms of trade) from −25.2% to −17.9% over the same period for the SSA region. This terms-of-trade improvement probably provided additional output growth to SSA agriculture.

[18] Additional losses to output would come from the withdrawal of resources from agricultural production. Our estimate would, however, capture the output lost caused by a reduction in labour supply per capita from individuals still counted as part of the agricultural labour force. This might characterize many AIDS sufferers and their care givers.

References

Alene, A. (2010) Productivity growth and the effects of R&D in African agriculture. *Agricultural Economics* 41, 223–238.

Alene, A. and Coulibaly, O. (2009) The impact of agricultural research on productivity and poverty in sub-Saharan Africa. *Food Policy* 24, 198–209.

Alston, J., Norton, G. and Pardey, P. (1998) *Science Under Scarcity: Principles and Practices for Agricultural Research Evaluation and Priority Setting*. Cornell University Press, Ithaca, NY.

Alston, J., Chan-Kang, C., Marra, M., Pardey, P. and Wyatt, T. (2000) A meta-analysis of rates of return to agricultural R&D: Ex pede herculem? IFPRI Research Report 113, International Food Policy Research Institute, Washington, DC.

Anderson, K. and Masters, W. (eds) (2009) *Distortions to Agricultural Incentives in Africa*. World Bank, Washington, DC.

Barro, R. and Lee, J. (2010) A new data set of educational attainment in the world, 1950–2010. Working Paper No. 15902, April, National Bureau of Economic Research (NBER), Cambridge, MA.

Binswanger, H. and Townsend, R. (2000) The growth performance of agriculture in sub-Saharan Africa. *American Journal of Agricultural Economics* 85, 1075–1986.

Binswanger-Mkhize, H. and McCalla, A. (2009) *The Changing Context and Prospects for African Agricultural Development*. International Fund for Agricultural Development, Rome.

Block, S. (1995) The recovery of agricultural productivity in sub-Saharan Africa. *Food Policy* 20, 385–405.

Block, S. (2010) The decline and rise of agricultural productivity in sub-Saharan Africa since 1961. NBER Working Paper 16481. October. National Bureau of Economic Research, Cambridge, MA,

Boserup, E. (1965) *Conditions of Agricultural Growth: The Economics of Agrarian Change Under Population Pressure*. Aldine Publishing, New York, NY.

Bourn, D. and Wint, W. (1994) Livestock, land use and agricultural intensification in sub-Saharan Africa. ODI Network Paper No. 37a. Overseas Development Institute, London, UK.

Byerlee, D. and Eicher, C. (eds) (1997) *Africa's Emerging Maize Revolution*. Lynne Publishers, Boulder, CO.

CGIAR (various annual reports) *CGIAR Financial Report*. Consultative Group on International Agricultural Research, Washington, DC.

Collier, P. (2007) *The Bottom Billion: Why the Poorest Countries are Failing and What Can be Done About It*. Oxford University Press, New York, NY.

Dalton, T. and Guei, R. (2003) Ecological diversity and rice varietal improvement in West Africa. In: Evenson, R.E. and Gollin, D. (eds) *Crop Variety Improvement and Its Effects on Productivity*. CABI Publishing, Wallingford, UK, pp. 109–134.

Deb, U. and Bantilan, M. (2003) Impacts of genetic improvement in sorghum. In: Evenson, R. and Gollin, D. (eds) *Crop Variety Improvement and Its Effects on Productivity*. CABI Publishing, Wallingford, UK, pp. 183–214.

Dixon, S., McDonald, S. and Roberts, J. (2002) The Impact of HIV and AIDS on Africa's economic development. *BMJ* 324, 232–234.

Dorosh, P., Wang, H.G., You, L. and Schmidt, E. (2009) Crop production and road connectivity in sub-Saharan Africa: A spatial analysis. Africa Infrastructure Country Diagnostic Working Paper 19, February. World Bank, Washington, DC.

Evenson, R. and Gollin, D. (eds) (2003) *Crop Variety Improvement and Its Effects on Productivity*. CABI Publishing, Wallingford, UK.

FAO. FAOSTAT Database. Food and Agricultural Organization, Rome. Available at: http://faostat.fao.org/ (Accessed 1 April 2009).

Fuglie, K. (2011) Agricultural productivity in sub-Saharan Africa. In: Lee, D.L. (ed.) *The Food and Financial Crisis in Africa*. CAB International, Wallingford, UK (in press).

Fulginiti, L., Perrin, R. and Yu, B. (2004) Institutions and agricultural productivity in sub-Saharan Africa. *Agricultural Economics* 31, 169–180.

Gleditsch, N., Wallensteen, P., Eriksson, M., Sollenberg, M. and Strand, H. (2002) Armed conflict 1946–2001: A new dataset. *Journal of Peace Research* 39, 697–710. [Note: we used Version 4-2008 of the dataset.]

Henao, J. and Baanante, C. (2006) Agricultural production and soil nutrient mining in Africa: Implications for resource conservation and policy development, summary. Report prepared for the IFDC-An International Center for Soil Fertility and Agricultural Development, Muscle Shoals, AL.

Hayami, Y. and Ruttan, V. (1985) *Agricultural Development: An International Perspective*. Johns Hopkins University Press, Baltimore, MD.

International Road Federation (2006) *World Road Statistics.* International Road Federation, Geneva, Switzerland. Available at: http://www.irfnet.org/

Johnston, B. and Mellor, J. (1961) The role of agriculture in economic development. *American Economic Review* 51, 566–593.

Lockheed, M., Jamison, D. and Lau, L. (1980) Farmer education and farmer efficiency: A survey. *Economic Development and Cultural Change.* 29, 37–76.

Lusigi, A. and Thirtle, C. (1997) Total factor productivity and the effects of R&D in African agriculture. *Journal of International Development* 9, 529–538.

Maredia, M. and Raitzer, D. (2006) CGIAR and NARS partner research in sub-Saharan Africa: Evidence of impact to date. Commissioned Paper, CGIAR Science Council, Rome.

McIntire, J., Bourzat, D. and Pingali, P. (1992) *Crop–livestock Interaction in Sub-Saharan Africa.* World Bank, Washington, DC.

Morris, M., Kelly, V., Kopicki, R. and Byerlee, D. (2007) *Fertilizer Use in African Agriculture.* World Bank, Washington, DC.

Ndulu, B., Chakraborti, L., Lijane, L., Ramachandran, V. and Wolgin, J. (2007) *Challenges of African Growth: Opportunities, Constraints and Strategic Directions.* World Bank, Washington, DC.

Nweke, F. (2004) New challenges in the cassava transformation in Nigeria and Ghana. EPTD Discussion Paper No. 118, International Food Policy Research Institute, Washington, DC.

Oba, G. and Lusigi, W. (1987) An overview of drought strategies and land use in African pastoral systems. ODI Network Paper No. 23a. Overseas Development Institute, London, UK.

Olsson, L., Eklundh, L. and Ardö, J. (2005) A recent greening of the Sahel – Trends, patterns and potential causes. *Journal of Arid Environments* 63, 556–566.

Pardey, P., Roseboom, J. and Anderson, J. (eds) (1991) *Agricultural Research Policy: International Quantitative Perspectives.* Cambridge University Press, Cambridge.

Pingali, P., Bigot, Y. and Binswanger, H. (1987) *Agricultural Mechanization and the Evolution of Farming Systems in Sub-Saharan Africa.* Johns Hopkins University Press, Baltimore, MD.

Reij, C., Gray, T. and Smale, M. (2009) Agroenvironmental transformation in the Sahel: Another kind of 'Green Revoluion'. Discussion Paper 00914, November, International Food Policy Research Institute, Washington, DC.

Renkow, M. and Byerlee, D. (2010) The impacts of CGIAR research: A review of recent evidence. *Agricultural Economics* 35, 391–402.

Sala-i-Martin, X. and Pinkovskiy, M. (2010) African poverty is falling…much faster than you think! NBER Working Paper 15775, February. National Bureau of Economic Research, Cambridge, MA.

Schultz, T.W. (1975) The value of the ability to deal with disequilibria. *Journal of Economic Literature* 13, 827–845.

Sebastian, K. (2007) GIS/spatial analysis contribution to 2008 WDR: Technical notes on data and methodologies. Background Paper for the World Development Report 2008, World Bank, Washington, DC.

Shapouri, S., Rosen, S., Peters, M., Baquedano, F. and Allen, S. (2010) *Food Security Assessment, 2010–2020.* July. Economic Research Service, U.S. Department of Agriculture, Washington, DC.

Smale, M. and Jayne, T. (2003) Maize in Eastern and Southern Africa: 'Seeds' of success in retrospect. EPTD Discussion Paper No. 97, International Food Policy Research Institute, Washington, DC.

Smith, J., Barau, A., Goldman, A. and Mareck, J. (1994) The role of technology in agricultural intensification: The evolution of maize production in the Northern Guinea Savanna of Nigeria. *Economic Development and Cultural Change* 42, 537–554.

Starkey, P. (2000) The history of working animals in Africa. In: Blench, R. and MacDonald, K. (eds) *The Origins and Development of African Livestock: Archaeology, Genetics, Linguistics and Ethnography.* University College London Press, UK, pp. 478–502.

Stoorvogel, J. and Smaling, E. (1990) Assessment of soil nutrient depletion in sub-Saharan Africa: 1983–2000. Report no. 28. Vol. l-4. Winand Staring Center, Wageningen, The Netherlands.

The Global Fund (2011) Making a Difference: Global Fund Results Report 2011. The Global Fund to Fight Aids, Tuberculosis and Malaria, Geneva, Switzerland.

World Bank (2010) World Development Indicators Database, World Bank, Washington, DC. Available at: http://publications.worldbank.org/WDI/ (Accessed April 2011).

Zeddies, J., Schaab, R. Neuenschwander, P. and Herren, H. (2001) Economics of biological control of cassava mealybug in Africa. *Agricultural Economics* 24, 209–219.

Zhang, X. and Fan, S. (2004) How productive is infrastructure? A new approach and evidence from rural India. *American Journal of Agricultural Economics* 86, 492–501.

Appendix 12.1: Additional Regression Estimates of TFP Determinants Model

Table A12.1. Determinants of agricultural TFP using national agricultural scientist-years.

	Model 1	Model 2	Model 3	Model 4	Model 5	Model 6
Variables	Ln(TFP)	Ln(TFP)	Ln(TFP)	Ln(TFP)	Ln(TFP)	Ln(TFP)
CG Area	0.366***	0.719***	0.641***	0.434***	1.331***	1.717***
	(5.131)	(5.784)	(5.706)	(5.323)	(8.825)	(9.428)
Ln(Sci-Stock)	0.0305***	0.0484***	0.0381***	0.0273***	0.0772****	0.0623***
	(5.491)	(5.021)	(4.862)	(4.718)	(6.499)	(3.223)
HIV	−0.111	−0.895***	−0.381***	−0.203*	−1.218***	−1.537***
	(−1.193)	(−5.102)	(−3.498)	(−1.890)	(−6.682)	(−8.176)
Civil War	−0.00848***	−0.00988***	−0.00865***	−0.00929***	−0.0137***	−0.0115***
	(−7.437)	(−8.460)	(−6.604)	(−6.554)	(−9.417)	(−6.952)
NRA		0.336***			0.246***	0.158**
		(6.377)			(4.487)	(2.271)
Ln(Road)			−0.0284***			−0.0297**
			(−5.246)			(−2.123)
School				0.00580	−0.0184***	−0.0219**
				(1.499)	(−2.836)	(−2.382)
Constant	4.511***	4.468***	4.398***	4.497***	4.324***	4.311***
	(194.7)	(94.19)	(122.5)	(178.6)	(78.92)	(41.64)
Observations	899	467	611	783	389	273
R^2	0.111	0.312	0.197	0.131	0.386	0.447
Adjusted- R^2	0.107	0.304	0.190	0.125	0.376	0.433

A two-stage IV procedure is used to estimate the model using an annual panel of countries over 1977–2005. Because of missing observations on variables, the number of observations included in the estimation varies by model. t-Statistics in parentheses; significance tests indicated by ***$p < 0.01$, **$p < 0.05$, *$p < 0.1$.

Table A12.2. Determinants of agricultural TFP using national R&D expenditures scaled by agricultural GDP.

	Model 1	Model 2	Model 3	Model 4	Model 5	Model 6
Variables	Ln(TFP)	Ln(TFP)	Ln(TFP)	Ln(TFP)	Ln(TFP)	Ln(TFP)
CG Area	0.443***	0.792***	0.831***	0.398***	1.167***	1.660***
	(6.224)	(6.108)	(7.701)	(4.678)	(7.380)	(7.118)
Ln(NAR-Stock/Ag GDP)	−0.0273***	−0.0104	−0.00198	−0.0428***	−0.0146	0.00155
	(−3.918)	(−0.897)	(−0.235)	(−5.120)	(−1.199)	(0.100)
HIV	0.0125	−0.736***	−0.430***	−0.178*	−1.109***	−1.505***
	(0.126)	(−4.007)	(−3.538)	(−1.668)	(−5.820)	(−7.156)
Civil War	−0.00817***	−0.00924***	−0.00687***	−0.00921***	−0.0119***	−0.0110***
	(−6.958)	(−6.958)	(−4.760)	(−6.638)	(−7.702)	(−5.313)
NRA		0.301***			0.226***	0.203***
		(5.574)			(3.977)	(2.822)
Ln(Road)			−0.0290***			−0.0417***
			(−5.111)			(−2.907)
School				0.0169***	0.00288	−0.00682
				(3.983)	(0.508)	(−0.832)
Constant	4.518***	4.649***	4.545***	4.401***	4.586***	4.570***
	(152.9)	(95.75)	(123.7)	(108.6)	(86.68)	(54.14)
Observations	899	467	611	783	389	273
R^2	0.094	0.276	0.134	0.142	0.342	0.410
Adjusted- R^2	0.0900	0.269	0.127	0.137	0.332	0.395

A two-stage IV procedure is used to estimate the model using an annual panel of countries over 1977–2005. Because of missing observations on variables, the number of observations included in the estimation varies by model. t-Statistics in parentheses; Significance tests indicated by ***$p < 0.01$, **$p < 0.05$, *$p < 0.1$.

13 Agricultural Productivity and Policy Changes in Sub-Saharan Africa

Alejandro Nin-Pratt and Bingxin Yu
International Food Policy Research Institute, Washington DC

13.1 Introduction

In recent years, 'an improvement in economic indicators throughout Africa led some observers to argue that the region had finally solved its economic conundrums and could now expect sustained economic growth' (van de Walle, 2001). That optimism was fuelled by the end of several civil wars, a wave of democratization in several countries (which made possible the creation of the New Partnership for Africa's Development, or NEPAD, and a new agenda for development), the acceleration of economic growth, and significant improvements in the performance of the agricultural sector across Africa during the 1980s and 1990s. Output growth in sub-Saharan Africa (SSA) from 1964 to 1983 was on average 1.8%, with the worst performance occurring between 1972 and 1983, when output growth was less than 1%, below the rate of increase in the use of inputs in agriculture (1.2%). The recovery of SSA's agriculture resulted in output growth rates of 3.2% per year between 1984 and 2006. This recovery is also significant when compared with population growth. For the group of SSA countries in this study, population growth was more than 2.6% per year from 1964 to 2006. That high rate, together with the poor performance of SSA's agricultural sector,

resulted in negative growth rates in output per capita in the 1970s and early 1980s. This trend reverted after 1985, and by 2006 the level of output per capita in SSA was close to its level in the 1960s.

What are the factors behind the dynamism agriculture has shown in recent years? Is such growth related to structural adjustment and policy changes that occurred in the past 20 years? Can growth be sustained in the coming years? To answer these questions, it is necessary to analyse the determinants of output growth and establish the contribution of productivity to the improved performance of SSA's agricultural sector. Higher productivity results in a more efficient use of resources because it increases agricultural production per unit of input. A more efficient use of resources becomes increasingly important as countries begin to face resource constraints. A growing labour force in agriculture, together with land constraints, could result in diminishing returns to labour and a limited capacity to expand production. Under these circumstances, sustainable agricultural growth can be achieved only through increased total factor productivity (TFP), the amount of output per unit of total factors used in the production process. It is through increased productivity that the agricultural sector can make a substantial contribution to economic growth and development by increasing the welfare of agricultural workers and rural populations, allowing workers to move away from agriculture to more productive sectors, and generating surpluses that can be transferred to other sectors through prices, in particular at the early stages of economic development (e.g. Winters *et al.*, 1998).

Despite evidence of improved performance in the past 10 years, only a few studies have attempted to analyse SSA's agricultural productivity changes and the factors explaining those changes, but those studies have also shown evidence of agricultural productivity recovery in Africa. Estimates of how *much* productivity has increased vary depending on the study and on the time period analysed, ranging from 0.5% to 2% per year after 1985, a clear improvement from growth rates observed in the 1970s and early 1980s. Previous studies have also found similar patterns in TFP trends: low or negative TFP growth in the 1970s and 1980s and acceleration in the mid-1980s and 1990s. For examples, see Block (1995), Lusigi and Thirtle (1997), Fulginiti *et al.* (2004) and recently, Coelli and Rao (2005), Fuglie (2008), Avila and Evenson (2010) and Alene (2010).

This chapter updates the study by Nin-Pratt and Yu (2008), bringing new evidence of the contribution of policy changes in the late 1980s and early 1990s to the recovery of agriculture in SSA. To do this, we estimate a non-parametric Malmquist index and its components (efficiency and technical change), constraining the shadow input shares used in the estimation of distance functions to rule out the possibility of zero input shadow prices. We improve on our previous analysis of the impact of policy changes on agricultural TFP by using available data on distortions to agricultural incentives in SSA (Anderson and Masters, 2009).

We make three main contributions. First, we confirm the improved performance of SSA's agriculture since the mid-1980s, measured in terms of TFP growth. Second, we observe that policies applied by several SSA countries after independence imposed a heavy burden on agriculture, and that the structural adjustment implemented in the region brought a more favourable policy environment for agriculture. Finally, we determine that this more favourable policy environment resulted in a recovery of traditional export crops (coffee, cocoa and cotton) and other exported commodities, such as oil crops, tropical fruits, beef and sheep meat, that were negatively affected by policies in the 1970s. On the input side, the improved performance of agriculture in SSA can be related to an adjustment in the relative use of inputs in the production process that resulted in the reduction of fertilizer use in Nigeria, and an increase in the use of these inputs in the best-performing countries.

The chapter is organized as follows. The next section presents the methodology employed and the data used to estimate agricultural TFP. Section 13.3 presents productivity estimates and a discussion of those results, whereas the main findings from the literature on SSA policies in past decades are presented in section 13.4. This is followed in section 13.5 by results of the estimation of an econometric model relating agricultural TFP series with measures of rates of assistance. The final section summarizes main findings and conclusions.

13.2 Productivity Measures and Methodology

Productivity change is defined as the ratio of change in output to change in input. In the hypothetical case of a production unit using one input to produce one output, the measure of productivity is fairly simple to derive. Production units use several inputs, however, to produce one or more outputs and, under such circumstances, the primary challenge in measuring TFP stems from the need to aggregate the different inputs and outputs. The aggregation of inputs and outputs is both conceptually and empirically difficult. Several methods to aggregate inputs and outputs are available, resulting in different approaches to measuring TFP. Such methods can be classified into four major groups: (i) econometric production models; (ii) total factor productivity indexes; (iii) data envelope analysis (DEA); and (iv) stochastic frontiers (Coelli *et al.*, 1998).

The Malmquist index, pioneered by Caves *et al.* (1982) and based on distance functions, has become extensively used in the measurement and analysis of productivity after Färe *et al.* (1994). It showed that the index can be estimated using DEA, a non-parametric approach. The non-parametric Malmquist index has been especially popular because it is easy to compute and does not require information about input or output prices or assumptions regarding economic behaviour, such as cost minimization and revenue maximization. This is attractive in the context of African agriculture, where input market prices are either non-existent or insufficiently reported to provide any meaningful information for land, labour and livestock. In addition, the non-parametric approach can be applied in a multiple-input, multiple-output setting. Also important is its ability to decompose productivity growth into two mutually exclusive and exhaustive components: changes in technical efficiency over time (catching up) and shifts in technology over time (technical change).

13.2.1   The Malmquist TFP index

The Malmquist index measures the TFP change between two data points (e.g. those of a country in two different time periods) by calculating the ratio of the distance of each data point relative to a common technological frontier. Following Färe *et al.* (1994), the Malmquist index between period *t* and *t+1* is given by (Eqn 13.1) (see equation 13.1 at bottom of page):

This index is estimated as the geometric mean of two Malmquist indexes, one using as a reference the technology frontier in *t* (M^t), and a second index that uses the frontier in *t+1* as the reference (M^{t+1}). The distance function $D^t(x^t, y^t)$ measures the distance of a vector of inputs (x) and outputs (y) in period *t* to the technological frontier in the same period *t*. On the other hand, $D^{t+1}(x^t, y^t)$ measures the distance between the same vector of inputs and outputs in period *t*, but in this case to the frontier in period *t+1*. The other two distances can be explained in the same fashion.

Färe *et al.* (1994) showed that the Malmquist index could be decomposed into an efficiency change component and a technical change component, and that those results applied to the different period-based Malmquist indexes. As shown in Eqn 13.2 (see equation 13.2 at bottom of page), the ratio outside the square brackets measures the change in technical efficiency between period *t* and *t+1*. The expression inside the brackets measures technical change as the geometric mean of the shift in the technological frontier between *t* and *t+1* evaluated using the frontier at *t* and at *t+1*, respectively, as the reference. The efficiency change component of the Malmquist indexes measures the change in how far observed production is from maximum potential production between period *t* and *t+1*. Finally, the technical change component captures the shift of technology between the two periods. A value of the efficiency change component greater than one means that the production unit is closer to the frontier in period *t+1* than it was in period *t*: thus, the production unit is catching up to the frontier. A value of less than one indicates an efficiency regress. The same holds true for the technical change component of total productivity growth, signifying technical progress when the value is greater than one, and technical regress when the index is less than one.

$$M = \left[M^t \times M^{t+1}\right]^{1/2} = \left[\frac{D^t(x^{t+1}, y^{t+1})}{D^t(x^t, y^t)} \times \frac{D^{t+1}(x^{t+1}, y^{t+1})}{D^{t+1}(x^t, y^t)}\right]^{1/2}. \tag{13.1}$$

$$M = \frac{D^{t+1}(x^{t+1}, y^{t+1})}{D^t(x^t, y^t)} \times \left[\frac{D^t(x^{t+1}, y^{t+1})}{D^{t+1}(x^{t+1}, y^{t+1})} \times \frac{D^t(x^t, y^t)}{D^{t+1}(x^t, y^t)}\right]^{1/2}. \tag{13.2}$$

However, as in Nin *et al.* (2003a), the DEA approach used to estimate distances defines the frontier as a sequential frontier, ruling out the possibility of technical regress. The method has been extensively applied to the international comparison of agricultural productivity, as shown in the following examples: Bureau *et al.* (1995), Fulginiti and Perrin (1997), Lusigi and Thirtle (1997), Rao and Coelli (1998), Arnade (1998), Fulginiti and Perrin (1999), Chavas (2001), Suhariyanto *et al.* (2001), Suhariyanto and Thirtle (2001), Trueblood and Coggins (2003), Nin *et al.* (2003a), Nin *et al.* (2003b), Nin *et al.* (2008) and Ludena *et al.* (2007).

To define the input-based Malmquist index, it is necessary to define and estimate the distance functions, which requires a characterization of the production technology and of production efficiency. We do this in the next section by following Kuosmanen *et al.* (2004), formally defining technology and efficiency and relating that measure with allocative efficiency and an economic measure of performance. We chose this approach to highlight the importance of shadow prices in the non-parametric estimation of distance functions, and to be able to introduce new information in the estimation of distance functions to avoid the occurrence of zero shadow prices.

13.2.2 Technology and distance functions

We assume, as in Färe *et al.* (1998), that for each time period $t = 1,...., T$, the production technology describes the possibilities for the transformation of inputs x^t into outputs y^t, or the set of output vectors y that can be produced with input vector x. The technology in period t with $y^t \in R_+^m$ outputs and $x^t \in R_+^n$ inputs is characterized by the production possibility set (PPS) as follows in Eqn 13.3:

$$L^t = \{(y^t, x^t): \text{Such that } x^t \text{ can produce } y^t\}. \tag{13.3}$$

The technology described by the PPS L^t satisfies the usual set of axioms: closedness, non-emptiness, scarcity and no free lunch. The frontier of the PPS for a given output vector is defined as the input vector that cannot be decreased by a uniform factor without leaving the set.

Two different approaches have been used to define non-parametric distance functions: the envelope form and a dual equivalent approach that can be derived from the envelope or primal form (Kuosmanen *et al.*, 2004). The literature normally prefers the envelope approach to estimate distances. On the other hand, the dual form has the advantage of a more intuitive specification, offering an economic interpretation of the problem, as well. It also allows an explicit estimation of input and output shadow prices and the possibility of imposing bounds to those prices. We focus here on the dual form.

The dual linear programme measures efficiency as the ratio of a weighted sum of all outputs over a weighted sum of all inputs. The weights are obtained by solving the following problem in Eqn 13.4 (Coelli and Rao, 2001):

$$\max_{p,w} \sum_{k=1}^{m} p_k y_{ik} \Big/ \sum_{j=1}^{n} w_j x_{ij}. \tag{13.4}$$

subject to:

$$\sum_{k=1}^{m} p_k y_{ik} \Big/ \sum_{j=1}^{n} w_j x_{ij} \le 1 \quad i = 1,...,r$$

$$p_k, w_j \ge 0 \quad k = 1,...,m; j = 1,...,n$$

where the optimal weights p_k and w_j are, respectively, output k and input j shadow prices. Equation 13.4 shows clearly the intuition behind this approach to measure efficiency, but cannot be used as such because it has an infinite number of solutions. To solve that problem, we normalize the ratio by imposing $\sum_{j=1}^{n} w_j x_{ij} = 1$ (Coelli and Rao, 2001). With this new constraint, the dual problem is expressed as shown in Eqn 13.5 (with p and w different from ρ and ω):

$$\max_{\rho,\omega} \sum_{k=1}^{m} \rho_k y_{ik}$$

s.t.

$$\sum_{j=1}^{n} \omega_j x_{ij} = 1 \tag{13.5}$$

$$\sum_{k=1}^{m} \rho_k y_{ik} - \sum_{j=1}^{n} \omega_j x_{ij} \le 0 \quad i = 1,...,r$$

$$\rho_k, \omega_j \ge 0 \quad k = 1,...,m; j = 1,...,n$$

Kuosmanen *et al.* (2004) generalize the dual interpretation of the distance function to the case of closed, non-empty production sets, satisfying scarcity and no free lunch, showing that the distance has the following dual formulation (Eqn 13.6):

$$D_0^t(x^t, y^t) = \max\left\{ \frac{\rho y^t}{\omega x^t} : \frac{\rho y^t}{\omega x^t} \leq 1 \forall (y^t, x^t) \in L^t \right\}.(13.6)$$

They interpret this distance function as 'the return to the dollar'[1] at the 'most favourable prices, subject to a normalizing condition that no feasible input–output vector yields a return to the dollar higher than unity at those prices'. The optimal weights ρ_k and ω_j are, respectively, output k and input j shadow prices with respect to technology L^t. There exists a vector of shadow prices for any arbitrary input–output vector; however, these prices need not be unique. Kuosmanen *et al.* (2004) define the set of shadow price vectors as (Eqn 13.7) (see equation 13.7 at bottom of page):

They contend in the spirit of the theory of revealed preferences (Varian, 1984) that 'the observed allocation of inputs and outputs can indirectly reveal the economic prices underlying the production decision'. On the basis of this, they assume that decision-making units allocate inputs and outputs to maximize return to the dollar. Such prices are well defined and are observed by decision makers, but are not known by the productivity analyst. Assuming that decision-making units allocate inputs and outputs to maximize return to the dollar, Kuosmanen et al. (2004) define that the production vector (y_t, x_t) as allocatively efficient with respect to technology L and prices (ρ^t, ω^t) if and only if $(\rho^t, \omega^t) \in V^t(y^t, x^t)$. Allocative efficiency is a necessary but not sufficient condition for maximization of return to the dollar, given that it allows for technical inefficiency (production in the interior of the PPS). This dual approach to the problem of efficiency and input allocation is used below to analyse the plausibility of shadow prices obtained when estimating efficiency and eventually to correct those prices, introducing exogenous information into the linear programming problem.

13.2.3 Introducing bounds to shadow input shares

The lack of prior price information for inputs was pointed out as the prime motivation for estimating non-parametric Malmquist indexes for the analysis of TFP change in SSA. If we do not constrain the linear programming problem used in DEA to determine efficiency, we allow total flexibility in choosing shadow prices. Because of the lack of price information already mentioned, in most of the literature on efficiency and non-parametric TFP analysis, flexibility has been considered to be one of the major advantages of DEA when comparing it with other techniques used to measure efficiency or productivity (Pedraja-Chaparro *et al.*, 1997). Total flexibility for the weights has, however, been criticized on several grounds, given that the weights estimated by DEA can prove to be inconsistent with prior knowledge or accepted views on relative prices or cost shares.

Pedraja-Chaparro *et al.* (1997) stress two main problems with respect to allowing total shadow price flexibility. First, by allowing total flexibility in choosing shadow prices, inputs considered important a priori could be all but ignored in the analysis or could end up being dominated by inputs of secondary importance. Such is the case when, because of the particular shape of the PPS, linear programming problems assign a zero or close-to-zero price to some factors. Second, the relative importance attached to the different inputs and outputs by each unit should differ greatly. Although some degree of flexibility on the weights might be desirable for the decision-making units to reflect their particular circumstances, it may often be unacceptable that the weights should

$$V^t(y^t, x^t) = \left\{ (\rho, \omega) \in R_+^{n+m} : \frac{\rho y}{\omega x} = D^t(y, x); \frac{\rho y^t}{\omega x^t} \leq 1 \forall (y^t, x^t) \in L^t \right\}. \tag{13.7}$$

vary substantially from one decision-making unit to another. Another argument used against total flexibility of shadow prices (Kuosmanen *et al.*, 2006) is that, in some cases, a certain amount of information regarding the input and output prices or shares might be available. In that case, the analysis can be strengthened by imposing price information in the form of additional constraints that define a feasible range for the relative prices. A strong case therefore seems to exist for the analysis of shadow prices obtained from DEA when estimating efficiency and TFP, and eventually for considering the introduction of restrictions on shadow prices or cost shares, setting limits between which prices or shares can vary.

To define suitable limits to the value that input shares take, we set an upper and a lower bound (a_i,b_i) to the input share in Equation 13.5. We define the standard distance function where ρ and ω are, respectively, the output and input shadow prices and $\omega_i^t \times x_{io}^t$ (the input shadow prices multiplied by the input quantities) is equal to the implicit input shares, as shown in Coelli and Rao (2001) and Eqn 13.8:

$$D^t(y_k^t, x_k^t) = \underset{\rho,\omega}{Max} \sum_{r=1}^{s} \rho_r y_{ro}^t,$$

subject to:

$$\sum_{j=1}^{m} \omega_i^t x_{io}^t = 1, \qquad (13.8)$$

$$\sum_{r=1}^{s} \rho_r^t y_{rj}^t - \sum_{i=1}^{m} \omega_i x_{ij}^t \leq 0,$$

$$a_{io}^t \leq \omega_i^t x_{io}^t \leq b_{io}^t \quad i = 1,...,m,$$

$$\rho, \omega \geq 0.$$

Note that the introduction of bounds on shadow input shares constitutes additional constraints to the original formulation. Restricted and unrestricted models will provide the same results only if all the additional restrictions imposed are non-binding. In general, the narrower the imposed bounds, the larger the expected differences between the outcomes of each model.

To define the bounds for the input shares, we introduce information on the likely value of the shares of the different inputs, drawing

on Avila and Evenson (2010). In that paper, the authors estimated crop input cost shares for 32 SSA countries by adjusting carefully measured share calculations for India. Cost shares of SSA countries were calculated by scaling India's input shares, comparing India's input/cropland ratio to those ratios of the particular SSA country. Given that inputs used in the study by Avila and Evenson (2010) are similar to those used here, we use information from that study to determine the maximum and minimum share values for each input among all countries, and use those estimated shares as a rough reference to set the limits between which input shares in DEA estimates for SSA countries can vary. By setting these general limits for all countries, we allow input shares to vary, keeping flexibility and uncertainty about the true value of such shares and contemplating differences in circumstances among the individual countries. With the imposition of share bounds, the linear programming programme can no longer disregard the less favourable inputs, and we ensure that the most important outputs and inputs are given higher weights than those considered less important. TFP estimates, including bounds to input shares as constraints imposed on the linear programming problem used to estimate distances, will be referred to hereafter as 'constrained estimates'. Results from the standard DEA approach to estimate TFP will be referred to as 'unconstrained estimates'. A more thorough discussion of the bounds imposed and a comparison of the results of the constrained and unconstrained problems used in the estimation of distance functions can be found in Nin-Pratt and Yu (2010).

13.2.4   Data and countries included in this study

To estimate TFP growth in SSA, the only internationally comparable database available is that of FAO. We use national time-series data from 1961 to 2006 for the total quantity of different inputs and output volumes measured in constant 2000 international dollars.

We use one output (agricultural production) and five inputs (labour, land, fertilizer, tractors and animal stock) for 98 countries, including 26 SSA countries, to estimate TFP. Agricultural output is expressed as the quantity of agricultural production measured in millions of 1999–2001 international dollars, derived using a Geary–Khamis formula for the agricultural sector. This method assigns a single 'price' to each commodity, regardless of the country where it was produced. Agricultural land is measured as the number of hectares of arable and permanent cropland. Labour is measured as the total economically active agricultural population. Fertilizer is the metric tonnes of nitrogen, potash and phosphates used, measured in nutrient-equivalent terms. Livestock is the total number of animals (cattle, buffalo, sheep, goats, pigs and laying hens) measured in cow equivalents.

We combine a dissimilarity index developed by Fox *et al.* (2004) and a modification to the DEA model suggested by Andersen and Petersen (1993) to identify outliers that might influence our estimates of efficiency and TFP measures. According to Fox *et al.*, a 'mix outlier' is an observation with an unusual combination in terms of the size of vector elements relative to other observations. This means that even though its scale across all dimensions may not be different from the average, it might have a high variance in its scale relative to the average across individual dimensions. The dissimilarity index provides bilateral comparisons of the input–output vector of all countries, with a reference input–output vector defined as the mean of all countries, showing how different each country is from the mean. This method is capable of identifying outliers on the best-practice frontier and beneath the frontier, allowing measurement and other errors in all observations to become apparent.

The method by Andersen and Petersen (1993) uses a modification to the DEA model to allow ranking of ostensibly efficient observations. This modification involves removing the ith country from the constraint set and measuring technical efficiency for this country relative to all other countries in the sample. This efficiency measure is not bounded from above and can take values greater than one for the particular case of the ith country being efficient. The effect of this observation in overall efficiency results is measured as the change in average efficiency resulting from its exclusion from the PPS as one of the observations defining the frontier.

The use of a wide range of countries to define the production possibilities space in our study requires a careful interpretation of the results and of the meaning of the term 'inefficient' as applied to countries in this global context. The main question is: is it meaningful to compare agricultural productivity among this diverse set of countries? We believe the answer is 'yes', if we understand the actual nature of the comparison. First, the difference in efficiency among countries cannot be explained simply by differences in the adoption of an available technology, as could be the case in a comparison of farms producing in the same economic and natural environment. Efficiency of agriculture in international comparisons at the country level will depend on lags in international technological development, diffusion and adoption, as well as on the economic environment within the country (e.g. infrastructure, research, favourable policies and institutions, etc.) and on the quality of production factors (land, water availability and human capital), most of which we are not capturing with the information available. The implication is that there are no obvious recommendations to tackle the problem of inefficiency in this international context unless more information is considered in the analysis. In most cases, however, there is an essential role of adaptive research and of research and deveopment investments in reducing the gap between inefficient countries and the international frontier.

13.3 Agricultural TFP Growth, 1961–2006

13.3.1 TFP growth and decomposition

Agriculture's performance in SSA between 1961 and 2006 was poor. A simple average of TFP measures at the country level for a sample of 26 SSA countries shows that annual growth in that period was almost zero (0.18%). That average, however, hides significant variations across time, where two periods with contrasting results can be distinguished (Table 13.1). A first period of poor performance and decline, with negative productivity growth (−1.08% per year), stretches from the mid-1960s to the mid-1980s. Recovery starts in 1984–1985 and extends up to 2006, the last year for which information is available. During that period, TFP grew at an annual rate of 1.45%, with 1.06% growth in the first half of the period (1984–1995), and accelerated to 1.88% on average between 1996 and 2006.

The decomposition of SSA's TFP growth into efficiency and technical change shows that most TFP growth from 1985 to 2006 was the result of SSA catching up to the frontier after falling behind during the 1970–1983 period. It is clear from Table 13.1 that, between 1984 and 2006, the region was only catching up with its own efficiency levels of the early 1960s.

13.3.2   Changes in output structure and input use

Agricultural productivity is affected not only by the level of output per unit of input of the different crop and livestock activities, but also by the composition of outputs. This means that changes in the structure of production can alter the overall output/input ratio. Increased efficiency and accelerated output growth in SSA resulted from differential growth between subsectors. Looking at the evolution of the ratio between output of export crops (coffee, cocoa, cotton, tobacco, tropical fruits, oil crops, beet and sheep meat) and output of import activities (cereals, roots and tubers, milk and chicken meat), we find that the period of poor agriculture performance is associated with a reduction in the share of total output of export crops. During that period, the share in total output of commodities produced for domestic   consumption   and   normally

Table 13.1. TFP growth rates and growth decomposition for different periods.

	TFP	Efficiency	Technical change
	(Average annual percentage change)		
1961–2006	0.18	−0.43	0.61
1961–1983	−1.08	−1.63	0.56
1984–2006	1.45	0.78	0.67
1984–1995	1.06	0.60	0.46
1996–2006	1.88	1.55	1.64
By decade			
1961–1970	−1.43	−2.28	0.87
1971–1980	−0.88	−1.24	0.36
1981–1990	0.41	0.02	0.39
1991–2000	1.77	1.06	0.70
2001–2006	1.62	0.80	0.82

The growth rates in efficiency and technical change rates add up to the TFP growth rate. The product of the efficiency and technical change indices equals the TFP index for each year. Source: Authors estimation.

imported by most SSA countries increased their participation in total output. The period of accelerated output growth starting in the mid-1980s saw a recovery in production of exports and other competitive crops and products.

The most important changes on the input side were a reduction in the use of fertilizer and the increased use of labour and land between 1994 and 2006. The number of workers per hectare of arable land continued to grow in the second half of the 1990s, although at decreasing growth rates. This continuous growth in rural populations and labour in agriculture, together with slow growth in the number of head of animal stock and the number of tractors, resulted in a reduction in the use of capital per worker, whereas capital per hectare of land remained at levels similar to those at the beginning of the period.

Figure 13.1 shows the contribution of individual countries to regional TFP growth between 1984 and 2006 as the TFP growth rate of each country multiplied by the country's share in regional output. In the first ten years after the first positive results obtained from policy changes, four countries account for most of agricultural TFP growth: Nigeria, Sudan, Kenya and Ghana (Fig. 13.1a). Those countries contributed 75%, 12%, 7% and 5%,

(a)

(b)

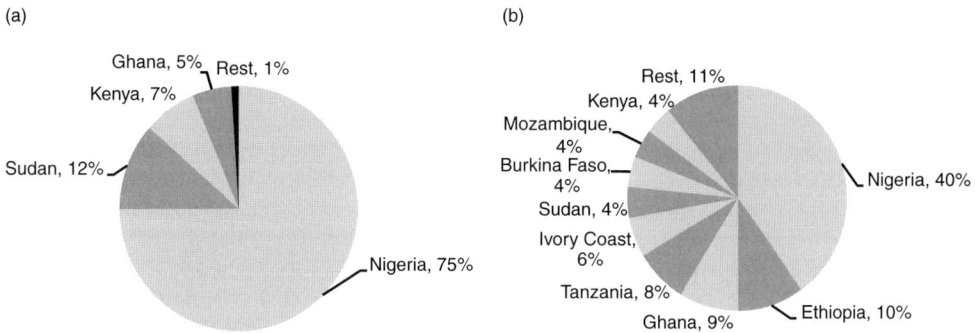

Fig. 13.1. Contribution of different countries to TFP growth in sub-Saharan Africa. (a) Country share of regional agricultural TFP growth, 1984–1993. (b) Country share of regional agricultural TFP growth, 1994–2006. Source: Authors' estimates.

respectively, of total TFP growth between 1984 and 1993, or 99% of total TFP growth in SSA during these years. The number of countries contributing to TFP growth increased significantly between 1994 and 2006, with nine countries accounting for 90% of TFP growth during that period. Nigeria remained the main contributor to TFP growth, but its contribution was reduced to 40% of total growth as TFP growth now extended to most SSA countries. Ethiopia contributed 10% of total TFP growth, followed by Ghana, Tanzania, Ivory Coast, Sudan, Burkina Faso, Mozambique and Kenya (Fig. 13.1b).

Looking at agricultural TFP growth rates for individual countries between 1984 and 2006, we observe that TFP grew more than 1.5% per year in ten of the 26 countries in our sample (Table 13.2). Angola, with 4% growth in TFP (average index of 1.04) is the country with the fastest growing agricultural TFP. Nigeria and Ghana follow, with average growth rates of 3.4% and 3%, respectively. Tanzania and Sierra Leone also show remarkable growth, with TFP growth rates of more than 2%.

This ranking changes if we consider the last years of the period, 1995–2006. As shown in the last six columns of Table 13.2, Angola still has the fastest growing agricultural TFP, but now Mozambique, Tanzania and Burkina Faso follow Angola as the countries with the best-performing agricultural sectors. Nigeria and Ghana

remain among the best SSA performers although, on average, TFP growth for those countries from 1995 to 2006 was lower than growth from 1984 to 2006. Mali, Zambia, Madagascar and Ethiopia also improved their performance significantly during that period.

Decomposition of TFP growth into its components in Table 13.2 shows that, in general, most of TFP growth is explained by efficiency gains, which corresponds to the fact that most countries are recovering from periods of negative productivity growth and reduction in efficiency. For instance, fast TFP growth in Angola is the result of catching-up after a long period of civil war, reflected in the zero growth rate of technical change (index = 1). As in Angola, TFP growth in Nigeria and Ghana is mainly explained by efficiency growth. In the case of other countries in coastal West Africa, only Benin shows a significant contribution of technical change to TFP growth. A similar result is obtained in East Africa. In southern Africa, the contribution of technical change to TFP seems to be important in the case of Swaziland and Zimbabwe, but agriculture's performance in these countries was poor because of growing inefficiency.

In terms of the changes in the use of inputs, we were interested in how such changes affect labour and land productivity, and, through them, overall TFP. Changes in those two partial productivity

Table 13.2. Ranking of countries by TFP growth performance, 1984–2006 and 1995–2006.

	1984–2006				1995–2006		
	TFP	Efficiency	Technical change		TFP	Efficiency	Technical change
	(Index for 2006 with base year 1984 = 1.000)				(Index for 2006 with base year 1995 = 1.000)		
Angola	1.040	1.040	1.000	Angola	1.063	1.063	1.000
Nigeria	1.034	1.034	1.000	Mozambique	1.043	1.043	1.000
Ghana	1.030	1.028	1.002	Tanzania	1.037	1.037	1.000
Tanzania	1.024	1.024	1.000	Burkina Faso	1.034	1.026	1.007
Sierra Leone	1.024	1.017	1.006	Sierra Leone	1.030	1.017	1.012
Togo	1.019	1.014	1.005	Nigeria	1.025	1.025	1.000
Kenya	1.018	1.013	1.005	Ghana	1.021	1.016	1.004
Sudan	1.017	1.017	1.000	Mali	1.021	1.019	1.002
Cameroon	1.016	1.011	1.005	Zambia	1.020	1.019	1.001
Chad	1.016	1.014	1.002	Madagascar	1.018	1.018	1.000
Mali	1.016	1.015	1.001	Ethiopia	1.016	1.015	1.001
Benin	1.016	1.000	1.016	Cameroon	1.014	1.005	1.009
Mozambique	1.014	1.014	1.000	Guinea	1.014	1.014	1.000
Burkina Faso	1.013	1.010	1.004	Zimbabwe	1.011	0.987	1.024
Zambia	1.013	1.013	1.000	Togo	1.010	1.004	1.006
Gabon	1.013	1.013	1.000	Kenya	1.010	1.001	1.009
Malawi	1.013	1.009	1.003	Ivory Coast	1.009	1.002	1.007
Ethiopia	1.010	1.010	1.001	Gabon	1.009	1.009	1.000
Ivory Coast	1.009	1.006	1.004	Malawi	1.008	1.001	1.007
Madagascar	1.004	1.004	1.000	Sudan	1.007	1.007	1.000
Guinea-Bissau	1.004	1.003	1.001	Guinea-Bissau	1.002	1.001	1.001
Zimbabwe	1.000	0.985	1.015	Benin	1.001	0.986	1.015
Senegal	1.000	0.996	1.004	Gambia	1.000	1.000	1.000
Guinea	0.999	0.999	1.000	Swaziland	0.999	0.942	1.061
Swaziland	0.993	0.957	1.037	Chad	0.998	0.994	1.004
Gambia	0.989	0.989	1.000	Senegal	0.977	0.970	1.006

Index = 1.000 in 2006 means zero growth between base year and 2006. The TFP index value is the production of the efficiency and technical change index values. Source: Authors' estimation.

measures are driven by changes in the labour–land ratio, which is affected by increases in rural populations and by the incorporation of arable land to crop production. If rural populations and the number of agricultural workers grow faster than yields, the result is a deterioration of rural living standards. Increased labour productivity is needed to increase agricultural workers' income, meaning yields need to increase faster than the number of workers per hectare (Block, 1995).

Considering growth in labour productivity together with TFP growth, we determine the best performing SSA countries during the period of improved performance (Table 13.3).

On average, these countries show high TFP growth, slow or negative growth of workers per hectare (with the exception of Angola), and increased labour and land productivity. These countries are more likely to have increased rural living standards through increased labour income in agriculture. A caveat to these results is that in many of these countries labour per hectare increased slowly because they were still able to incorporate more land into crop production to accommodate rural population growth. If the availability of land decreases in coming years, yields will need to increase faster to compensate for growth in rural populations and to improve rural income.

Table 13.3. Changes in labour/land ratios, labour and land productivity, and TFP in best-performing countries, 1995–2006.

	Labour productivity	Land productivity	Land–labour ratio	TFP
	(Average annual percentage change)			
Angola	2.63	4.5	−2.03	6.35
Nigeria	2.14	1.47	0.71	2.47
Ghana	2.09	1.46	0.51	2.06
Mozambique	1.83	1.94	−0.29	4.32
Guinea	1.69	0.71	−0.97	1.43
Cameroon	1.51	2.02	−0.57	1.43
Mali	1.32	1.7	−1.32	2.05
Zambia	1.17	1.76	−0.65	1.98
Ethiopia	1.16	0.42	0.61	1.64

Source: Authors estimation.

13.4   Policy Changes and Growth in Agriculture

According to Anderson and Masters (2009), most African countries gained independence in the 1960s at a time when central planning was widely seen as a promising strategy for economic development. In this environment, elected governments across Africa typically kept the marketing boards and other instruments for intervention that had been developed by previous administrations, thereby expanding their mandates and increasing public employment, in many cases as a means for electoral politics. In the 1970s, growing fiscal deficits, current account imbalances and overvalued exchange rates were supported by project aid and loans during a time of zero or negative real interest rates, as governments chose to ration credit and foreign exchange rather than expand the money supply. The result of growing government intervention, according to Anderson and Masters (2009), was political instability and weak market institutions.

African governments faced mounting pressures for public-sector reform with the rise in world real interest rates, combined with global recession that worsened Africa's terms of trade during the 1980s. These changes made it increasingly difficult for governments to finance the growing fiscal deficits associated with intervention. The World Bank, the IMF and USAID, as lenders of last resort, made their aid conditional on

devaluation, deregulation, privatization and retrenchment. As a result, trade policy reforms in the 1980s and 1990s were heavily influenced by structural adjustment programmes (SAPs) sponsored by the World Bank and the IMF. Loan conditions were often blamed for the economic stresses that accompanied them, but the actual implementation of reforms was typically slow and often subject to reversal or offsetting policy changes (Anderson and Masters, 2009).

As Anderson and Masters conclude, Africa's larger countries have had relatively interventionist governments, followed by periods of reform and a degree of recovery. Although the differences in the process of policy reform followed by those countries are frequently emphasized, there are also clear patterns across countries and clear trends in policy choices.

In 1994, the World Bank published a study intended to 'assess how much policy reform has taken place in Africa, how successful it has been, and how much more remains to be done'. That study (World Bank, 1994) concluded that progress had been made, but reforms remained incomplete. It also stressed the fact that the main factors behind the poor performance of SSA economies between the mid-1960s and the 1980s were: poor macroeconomic and sectoral policies that resulted in overvalued exchange rates; prolonged budget deficits; protectionist trade policies and government monopolies that reduced competition, negatively affecting productivity; and heavy taxation of agricultural exports. Food markets were controlled by state enterprises that also monopolized the import and distribution of fertilizers and other inputs, which were often supplied to farmers at subsidized prices and on credit. The prices farmers received were generally low because of taxation or high costs incurred by state enterprises. The negative impact of such policies on agricultural prices was particularly significant in the case of export crops. During this period, African governments followed a development strategy that prioritized industrialization, with a clear bias against agriculture (Kherallah *et al.*, 2000).

As emphasized by Kherallah *et al.* (2000), one of the most fundamental shifts in Africa's development strategy was to come to view agriculture not as a backward sector but as the engine of growth, an important source of export revenues and the primary means to reduce poverty. The idea behind the structural adjustment programmes was that reducing or eliminating state control over marketing would promote private-sector activity, and that fostering competitive markets would lead to increased agricultural production.[2]

Policy reforms have occurred in two major waves and been uneven across sectors and across countries. The first wave of reforms started in the 1980s and by the end of the decade almost two-thirds of African countries had managed to put better macroeconomic and agricultural policies in place. Improvements in the macroeconomic framework also enabled countries to adopt more market-based systems of foreign exchange allocation and fewer administrative controls over imports (World Bank, 1994).

The second wave of reforms came when many countries achieved major gains in macroeconomic stabilization, particularly since 1994. The devaluation of the CFA franc significantly improved the performance of the economy and of the agricultural sector in several West African countries. According to the World Bank (2000), by the end of the 1990s, the combination of sustained reforms and financial assistance was associated with better economic performance, at least at the aggregate level. Most

prices have been decontrolled and marketing boards eliminated (except in some countries for key exports, such as cotton and cocoa). Current account convertibility has been achieved: trade taxes have been rationalized from high average levels of 30–40% to trade-weighted average tariffs of 15% or less. Trade-weighted tariffs are now less than 10% in more open countries, such as Uganda and Zambia. Arbitrary exemptions, although still numerous, have also been rationalized.

In the case of agricultural reform, most policy changes took place after 1986–1987, and significant progress was attained. Most countries lowered export taxes, raised administered producer prices, reduced marketing costs (usually by deregulation and de-monopolization of export marketing) and depreciated the exchange rate of the domestic currency (Cleaver and Donovan, 1995). According to the World Bank (2008), the average net taxation of SSA agriculture was more than halved between 1980–1984 and 2000–2004. During the same period, agriculture-based countries (mostly African countries) lowered protection of agricultural importables, from a 14% tariff-equivalent to 10%, and reduced taxation of exportables from 45% to 19%. Most of the decline in taxation was the result of improved macroeconomic policies (World Bank, 2008).

As a result of these changes in the first years of reform, two-thirds of the adjusting countries were taxing their farmers less, and policy changes increased real producer prices for agricultural exporters. Most of the governments that had major restrictions on the private purchase, distribution and sale of major food crops before adjustment have withdrawn from marketing almost completely. Still, governments sold only a small share of their assets, although they have stopped expanding their public enterprise sectors (World Bank, 1994).

Market reforms were more comprehensive in food markets than in export crop or input markets. Kherallah *et al.* (2000) explain the progress in food market reforms by the losses those markets brought to governments, whereas, in contrast, the purchase and sale of export commodities brought considerable revenue to many governments. Also, major restrictions on the purchase and sale of agricultural commodities were eliminated in Benin (tubers), Ethiopia (teff, maize and wheat), Mali (millet and sorghum), Tanzania (maize), Malawi and Zambia partially (maize). There were no changes in Kenya and Zimbabwe (Kherallah *et al.*, 2000).

Anderson and Masters (2009) estimate nominal and relative rates of assistance (NRA and RRA, respectively) to measure the effects of government policies on returns to farmers in SSA.[3] We highlight here some of their main conclusions from the analysis of changes in rates of assistance to agriculture:

- At present, African governments have removed much of their earlier anti-farm and anti-trade policy biases, and most of these changes have come from reduced taxation of farm exports.
- Substantial distortions remain which still impose a large tax burden on Africa's poor. In constant (2000) US dollar terms, the transfers paid by farmers were reduced to an average of $41 per person working in agriculture, from a peak of $134 in the late 1970s. This lower amount is, however, appreciably larger than in other regions, given that in both Asia and Latin America, the average agricultural NRAs and RRAs reached zero by the early 2000s.
- Trade restrictions continue to be Africa's most important instruments of agricultural intervention. Domestic taxes and subsidies on farm inputs and outputs and non-product-specific assistance are a small share of total distortions to farmer incentives in Africa. As a result, policy incidence on consumers tends to mirror the incidence on producers, with fiscal expenditures playing a much smaller role than in more affluent regions.

13.5 Linking Policy Reforms with TFP Growth in Agriculture

Since the implementation of the structural adjustment, policymakers and academics have argued about the causes of and solutions to the African crisis, and about the impacts of the structural adjustment promoted by the international financial institutions in the 1980s and 1990s. For examples see Mosley *et al.* (1995), Bar-on (1997), Arndt *et al.* (2000), Boratav (2001), Kraev (2004) and Mkandawire (2005).

As discussed above, agricultural production in SSA was affected by distortions to agricultural incentives through both agricultural and non-agricultural policy measures. Importantly, the total effect of distortions on the agricultural sector depends not just on the size of the direct agricultural policy measures, but also on the magnitude of distortions generated by indirect policy measures altering incentives in non-agricultural sectors. It is relative prices, and hence relative rates of government assistance, that affect producers' incentives. We use here RRAs estimated by Anderson and Valenzuela (2008) to measure the effect of policy on agricultural TFP. This measure not only captures the overall delivered rates of distortion to domestic prices for agricultural and food products from policy interventions, but also takes into account assistance to producers of non-agricultural tradables.[4] Our model is expressed as (Eqn 13.9):

$$AgTFP_{it} = \beta_0 + \beta_{rra} * rra_{it} + \beta_i Dummy_i + \varepsilon_{it} \qquad (13.9)$$

where $AgTFP_{it}$ is the constrained and unconstrained agricultural TFP growth rate in year t of country I; β are parameters to be estimated; rra_{it} is the relative rate of assistance calculated by Anderson and Valenzuela (2008); $Dummy_i$ is the country-fixed effects with Togo as the control group; and ε_{it} is the error term. We expect the coefficient of RRA to be positive, because higher agricultural assistance could motivate producers to invest in new technology to improve productivity.

Before estimating the panel model, we conduct unit root tests for TFP and RRA indexes. The results of pooled panel tests for the null hypothesis of presence of a unit root are reported in Table 13.4. We can firmly reject the null hypothesis of non-stationarity in most cases, especially under the

Table 13.4. Panel unit root test results.

	Unconstrained cumulative TFP		Constrained cumulative TFP		Unconstrained TFP		Constrained TFP		Relative rate of assistance	
	Without trend	With trend	Without trend	With trend	Without trend	With trend	Without trend	With trend	Without trend	With trend
Levin-Lin-Chu test, Ho: Panels contain unit roots, Ha: All panels are stationary										
	0.836	0.022	0.264	0.003	1.000	1.000	1.000	1.000		
Harris-Tzavalis test, Ho: Panels contain unit roots, Ha: All panels are stationary										
	0.085	0.000	0.006	0.000	0.000	0.000	0.000	0.000		
Breitung test, Ho: Panels contain unit roots, Ha: All panels are stationary										
	0.329	0.994	0.117	0.883	0.000	0.000	0.000	0.000		
Im-Pesaran-Shin test, Ho: All panels contain unit roots, Ha: Some panels are stationary										
	0.078	0.000	0.001	0.000	0.000	0.000	0.000	0.000	0.000	0.000
Fisher-type test, Ho: All panels contain unit roots, Ha: At least one panel is stationary										
	0.000	0.000	0.000	0.000	0.000	0.000	0.000	0.000	0.000	0.000
Hadri LM test, Ho: All panels are stationary, Ha: Some panels contain unit roots										
	0.000	0.000	0.000	0.000	0.000	0.087	0.000	0.074	0.000	0.000

Source: Authors' calculation.

alternative hypothesis of partially stationary panel. This suggests that the series
considered as a panel is stationary and our
parameter estimates are valid.

We estimate the balanced panel data
linear model using feasible generalized least
squares. The dependent variables are the
unconstrained and constrained cumulative
TFP indexes from non-parametric analysis
and the explanatory variable is the RRA. We
assume heteroskedasticity across panels,
because it is common to have data on countries that have variations of scale. We also
assume that the error terms of panels are
correlated, in addition to having different
scale. That is, each country is assumed to
have errors that follow a different AR(1)
process. The effect of agricultural and non-
agricultural policy instruments on TFP
growth is examined, and the estimation
results in the presence of heterogeneous
AR(1) and cross-sectional correlation are
reported in Table 13.5. As expected, our
results show a positive effect of RRA on
agricultural TFP growth, showing the clear
influence that policy changes had in the
performance of SSA's agricultural sector.
This positive relationship between RRA
and TFP can also be seen in Fig. 13.2 for
cumulative agricultural TFP growth and
average RRA values for SSA.

Comments made about earlier versions
of this paper have suggested that the observed improved performance of SSA's agriculture did not correspond to the perception
that poverty in Africa has increased in past
decades, and that this increase in poverty
could have been a result of the implementation of SPAs. As reported by Sala-i-Martin
and Pinkovskiy (2010), a study from the
World Bank states that in sub-Saharan
Africa the extreme poverty rate rose from
47.4% in 1990 to 49% in 1999, with poverty
expected to rise in coming years (World
Bank, 2004). The UN Millennium Campaign
Deputy Director for Africa also stressed the
fact that poverty continues to intensify in
SSA because of the exclusion of groups of
people on the basis of class, caste, gender,
disability, age, race, religion and other status
(UN, 2009).

The study by Sala-i-Martin and Pinkovskiy (2010) shows that the perception
that poverty is increasing in SSA is wrong, a
result that fits the evidence of improvement
in the performance of SSA economies in
general and of agriculture production in
particular. Since 1995, according to Sala-
i-Martin and Pinkovskiy, African poverty
has been falling steadily, income distribution has become less rather than more une-
qual and growth has been accompanied by

Table 13.5. Panel regression results of the effect of policy (measured by the relative rate of assistance) on TFP growth.

	Unconstrained cumulative TFP		Constrained cumulative TFP		Unconstrained TFP		Constrained TFP	
	Coefficient	Standard error	Coefficient	Standard error	Coefficient	Standard error	Coefficient	Standard error
Relative rate of assist-ance (RRA)	0.142	(0.022)***	0.15	(0.023)***	0.067	(0.014)***	0.086	(0.014)***
Observations	557		557		557		557	
Number of countries	14		14		14		14	
Chi square	1658		1678		988.8		1387	

Unconstrained and constrained TFP refer to TFP estimates not imposing and imposing bounds to shadow prices, respectively, in linear programming models used to estimate distance functions. Cumulative TFP values are TFP levels relative to 1961's productivity levels for each country, allowing comparisons of own country's TFP levels across time but not across countries. The last four columns refer to results of annual changes in TFP and RRA. Source: Authors' calculation. ***refers to p-value <0.01

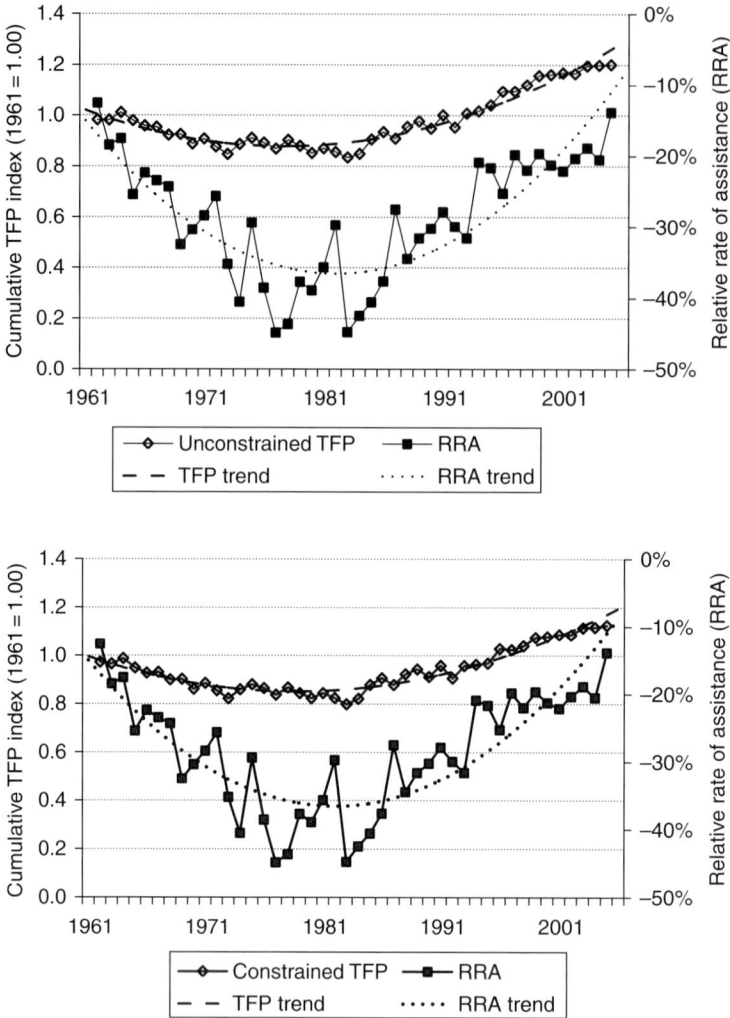

Fig. 13.2. Trend in agricultural TFP and the relative rate of assistance (RRA). Source: RRA from Anderson and Valenzuela (2008); TFP from authors' estimates. The trend lines are 2nd order polynomials.

a symmetric, sustained reduction in poverty that places SSA on track to achieve the Millennium Development Goal of halving poverty within a few years of 2015.

Figure 13.3 shows poverty estimates for SSA by Sala-i-Martin and Pinkovskiy (2010), together with the RRA to agriculture as a measure of policy change towards agriculture from Anderson and Masters (2009) and our estimates of cumulative growth of agricultural TFP. Poverty rates increased until the mid-1980s and stabilized after the start of the reforms, showing a clear decreasing

trend after 1995. The evidence from the agricultural sector, together with the improved performance of the economy and the reduction in poverty, all point to an improved performance of SSA economies.

13.6   Conclusions

In this study we analysed the evolution of SSA's agricultural TFP between 1961 and 2006, looking for evidence of recent changes

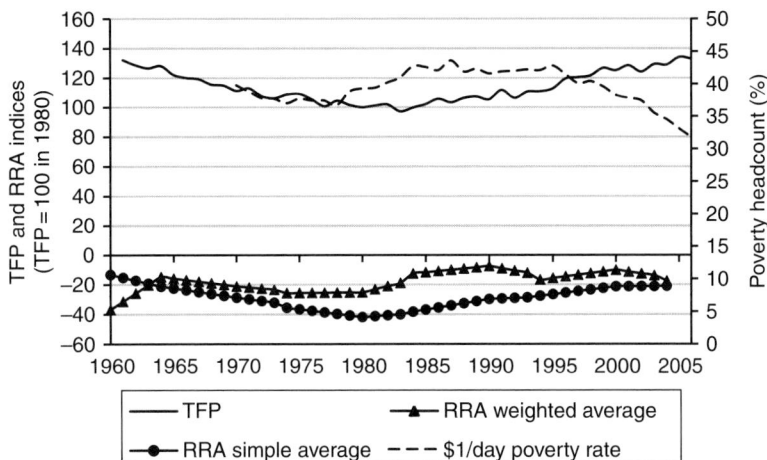

Fig. 13.3. Evolution of the cumulative index of agricultural TFP, gross agricultural subsidy equivalents and poverty in sub-Saharan Africa. RRA is the relative rate of assistance to agriculture. It is a measure of the gross agricultural subsidy equivalent. Source: RRA is from Anderson and Valenzuela (2008); poverty headcount is from the World Bank; TFP is the authors' estimation.

in growth patterns using a non-parametric Malmquist index and its components, efficiency and technical change indexes, for 26 SSA countries. Unlike previous studies using this methodology, we constrained the linear programming problem used to estimate distance functions for the Malmquist index to rule out the possibility of zero input shadow prices. We also looked at the contribution of different countries to total TFP growth in SSA and analysed changes in the composition of outputs and inputs. Finally, we estimated an econometric model relating TFP growth to global estimates of agricultural distortions.

The favourable impact of policy changes after 1985 shows that policies applied by several SSA countries after independence imposed a heavy burden on agriculture, and

Results of our TFP estimates show a remarkable recovery in the performance of SSA's agriculture, starting in the mid-1980s and early 1990s after a long period of poor performance and decline. Results of our econometric analysis point to policy changes conducted by SSA countries between the mid-1980s and the second half of the 1990s as among the major factors determining the agricultural sector's improved performance.

that the structural adjustment implemented in the region brought a more favourable policy environment for the primary sector. This more favourable policy environment resulted in improved allocation efficiency and increased production and, as a consequence, increased productivity.

On the output side, policy changes resulted in a recovery of traditional export crops (coffee, cocoa and cotton) and other export commodities, such as oil crops, tropical fruits, beef and sheep meat, which were negatively affected by policies in the 1970s.

On the input side, the improved performance of agriculture in SSA during 1984 to 2006 can be related to an adjustment in the relative use of inputs in the production process. The most important change is an absolute reduction in the total use of fertilizers. Most of this reduction is explained mainly by decreased fertilizer use in Nigeria. On the other hand, while Nigeria reduced the use of fertilizer, other countries increased it, with Ethiopia, Côted Ivoirè, Kenya, Benin, Cameroon and Mali accounting for 80% of growth in fertilizer use between 1983 and 2006.

Output growth and changes in the relative use of inputs resulted in a significant increase in output per hectare, after several years of little or no growth. Considering TFP growth together with balanced growth in land and labour productivity as indicators of good agriculture performance, we find nine countries (Angola, Nigeria, Ghana, Mozambique, Guinea, Cameroon, Mali, Zambia and Ethiopia) with relatively high TFP growth and sustained growth in labour and land productivity from 1995 to 2006. In most of these countries, growth in land and labour productivity can be explained by increased use of fertilizer per hectare and worker.

Despite improved agricultural performance between 1985 and 2006, several warning signs still exist, calling for more efforts to sustain TFP growth in the coming years. First, the decomposition of TFP growth into efficiency and technical change shows that most TFP growth from 1985 to 2006 is the result of SSA catching up to the frontier after falling behind between 1970 and 1984. Without increases in the rate of growth of technical change, TFP growth is expected to slow down in coming years as countries catch up with efficiency levels at the production frontier. According to our estimates, a slowdown in TFP growth is already apparent in the cases of Nigeria and Ghana, the leaders of SSA's agricultural recovery in the mid-1980s. Second, substantial distortions remain that still impose a large tax burden on Africa's poor, and are much larger than those in other developing regions (Anderson and Masters, 2009). Third, sustained growth in labour productivity faces the challenges of population growth and related increases in agricultural labour per hectare. In many countries, expansion of labour productivity was possible because those countries were still able to incorporate more land into crop production. If the availability of land reduces in coming years, yields will need to increase faster to compensate for growth in rural populations and to improve rural income.

Notes

[1] 'Return to the dollar' is an economic criterion to evaluate performance. It measures the ability of producers to attain maximum revenue-to-cost (introduced by Georgescu-Roegen, 1951, and referred to in Kuosmanen et al., 2004). The assumption of allocative efficiency depends on the specified economic objectives of the firms through the shadow price domain (Kuosmanen et al., 2004).

[2] The reforms included four types of measures, as summarized by Kherallah et al. (2000): (i) liberalizing input and output prices by eliminating subsidies on agricultural inputs and bringing domestic crop prices in line with world prices; (ii) reducing overvalued exchange rates; (iii) encouraging private-sector activity by removing regulatory controls in input and output markets; and (iv) restructuring public enterprises and restricting marketing boards to such activities as providing market information.

[3] NRA is defined as the percentage by which government policies have raised gross returns to farmers above what they would be without the government's intervention and are based on estimates of assistance to individual industries at the farm gate. As farmers are affected not only by the prices of their own outputs, but also by the incentives non-agricultural producers face affecting mobile resources engaged in other sectors, Anderson and Valenzuela (2008) also estimate NRAs for the non-farm sector to capture the effect policies have had on agriculture through their effect on non-farm activities. RRAs are then calculated as the ratio of farm and non-farm NRAs. See Anderson and Masters for details on these estimates.

[4] Defined as: $RRA = 100^*(100 + NRAag)/(100 + NRAnonag) - 1]$, NRAag and NRAnonag are nominal rates of assistance to agriculture and non-agriculture.

References

Alene, A. (2010) Productivity growth and the effects of R&D in African agriculture. *Agricultural Economics* 41, 223–238.

Andersen, P. and Petersen, N. (1993) A procedure for ranking efficient units in data envelopment analysis. *Management Sciences* 39, 1261–1264.

Anderson, K. and Masters, W. (eds) (2009) *Distortions to Agricultural Incentives in Africa*. World Bank, Washington, DC.

Anderson, K. and Valenzuela, E. (2008) Global estimates of distortions to agricultural incentives, 1955 to 2007. Database, World Bank, Washington, DC. Available at: ww.worldbank.org/agdistortions (Accessed 4 February 2010).

Arnade, C. (1998) Using a programming approach to measure international agricultural efficiency and productivity. *Journal of Agricultural Economics* 49, 67–84.

Arndt, C., Tarp Jensen, H. and Tarp, F. (2000) Stabilization and structural adjustment in Mozambique: An appraisal. *Journal of International Development* 12, 299–323.

Avila, A. and Evenson, R. (2010) Total factor productivity growth in agriculture: The role of technology capital. In: Pingali, P. and Evenson, R. (eds) *Handbook of Agricultural Economics,* vol 4. Elsevier, Amsterdam, The Netherlands, pp. 3769–3822.

Bar-on, A. (1997) Assessing Sub-Saharan Africa's structural adjustment programmes: The need for more qualitative measures. *Journal of Social Development in Africa* 12, 15–27.

Block, S. (1995) The recovery of agricultural productivity in Sub-Saharan Africa. *Food Policy* 20, 385–405.

Boratav, K. (2001) Movement of relative agricultural prices in Sub-Saharan Africa. *Cambridge Journal of Economics* 25, 395–416.

Bureau, C., Färe, R. and Grosskopf, S. (1995) A comparison of three nonparametric measures of productivity growth in European and United States agriculture. *Journal of Agricultural Economics* 45, 309–326.

Caves, D., Christensen, L. and Diewert, W. (1982) The economic theory of index numbers and the measurement of input, output, and productivity. *Econometrica* 50, 1393–1414.

Chavas, J. (2001) An international analysis of agricultural productivity. In: Zepeda, L. (ed.) *Agricultural Investment and Productivity in Developing Countries*. Food and Agriculture Organization, Rome, Italy, pp. 21–38.

Cleaver, K. and Donovan, W. (1995) Agriculture, poverty, and policy reform in Sub-Saharan Africa. World Bank Discussion Papers, Africa Technical Department Series, No. 208, World Bank, Washington, DC.

Coelli, T. and Rao, P.D.S. (2001) Implicit value shares in Malmquist TFP index numbers. Working Paper No. 4/2001, Centre for Efficiency and Productivity Analysis, University of Queensland, Brisbane, Australia.

Coelli, T. and Rao, P.D.S. (2005) Total factor productivity growth in agriculture: A Malmquist index analysis of 93 countries, 1980–2000. *Agricultural Economics* 32, 115–134.

Coelli, T., Rao, P.D.S. and Battese, G. (1998) *An Introduction to Efficiency and Productivity Analysis*. Kluwer Academic Publishers, Boston, MA.

FAO. FAOSTAT Database. Food and Agriculture Organization of the United Nations. Available at: http://www.fao.org/. (Accessed 5 May 2007).

Färe, R., Grosskopf, S., Norris, M. and Roos, P. (1998) Malmquist productivity indexes: A survey of theory and practice. In: Färe, R., Grosskopf, S. and Russell, R. (eds), *Index Numbers: Essays in Honor of Sten Malmquist,* Kluwer Academic Publishers, Boston, MA, pp. 127–190.

Färe, R., Grosskopf, S., Norris, M. and Zhang, Z. (1994) Productivity growth, technical progress, and efficiency change in industrialized countries. *American Economic Review* 84, 66–83.

Fox, K., Hill, R. and Diewert, W. (2004) Identifying outliers in multi-output models. *Journal of Productivity Analysis* 22, 73–94.

Fuglie, K. (2008) Is a slowdown in agricultural productivity growth contributing to the rise in commodity prices? *Agricultural Economics* 39, 431–441.

Fulginiti, L. and Perrin, R. (1997) LDC agriculture: Nonparameric Malmquist productivity indexes. *Journal of Development Economics* 53, 373–390.

Fulginiti, L. and Perrin, R. (1999) Have price policies damaged LDC agricultural productivity? *Contemporary Economic Policy* 17, 469–475.

Fulginiti, L., Perrin, R. and Yu, B. (2004) Institutions and agricultural productivity in Sub-Saharan Africa. *Agricultural Economics* 4, 169–180.

Kherallah, M., Delgado, C., Gabre-Madhin, E., Minot, N. and Johnson, M. (2000) The road half traveled: Agricultural market reform in Sub-Saharan Africa. IFPRI Food Policy Report, International Food Policy Research Institute, Washington, DC.

Kraev, E. (2004) Towards adequate analysis and modeling of structural adjustment programs: An analytical framework with application to Ghana. Working Paper, School of Public Policy, University of Maryland, College Park, MD.

Kuosmanen, T., Post, T. and Sipiläinen, T. (2004) Shadow price approach to total factor productivity measurement: With an application to Finnish grass-silage production. *Journal of Productivity Analysis* 22, 95–121.

Kuosmanen, T., Cherchye, L. and Sipiläinen, T. (2006) The Law of One Price in data envelopment analysis: Restricting weight flexibility across firms. *European Journal of Operational Research* 127, 735–757.

Ludena, C., Hertel, T., Preckel, P., Foster, K. and Nin, A. (2007) Productivity growth and convergence in crop, ruminant, and non-ruminant production: Measurement and forecasts. *Agricultural Economics* 37, 1–17.

Lusigi, A. and Thirtle, C. (1997) Total factor productivity and the effects of R&D in African agriculture. *Journal of International Development* 9, 529–538.

Mkandawire, T. (2005) Maladjusted African economies and globalization. *Africa Development* 30, 1–33.

Mosley, P., Subasat, T. and Weeks, J. (1995) Assessing adjustment in Africa. *World Development* 23, 1459–1479.

Nin, A., Arndt, C., Hertel, T. and Preckel, P. (2003a) Is agricultural productivity in developing countries really shrinking? New evidence using a modified non-parametric approach. *Journal of Development Economics* 71, 395–415.

Nin, A., Arndt, C., Hertel, T. and Preckel, P. (2003b) Bridging the gap between partial and total factor productivity measures using directional distance functions. *American Journal of Agricultural Economics* 85, 937–951.

Nin-Pratt, A. and Yu, B. (2008) An updated look at the recovery of agricultural productivity in Sub-Saharan Africa. IFPRI Discussion Paper 00787, International Food Policy Research Institute, Washington, DC.

Nin-Pratt, A. and Yu, B. (2010) Getting implicit shadow prices right for the estimation of the Malmquist index: The case of agricultural total factor productivity in developing countries. *Agricultural Economics* 41, 349–360.

Pedraja-Chaparro, F., Salinas-Jimenez, J. and Smith, P. (1997) On the role of weight restrictions in data envelopment analysis. *Journal of Productivity Analysis* 8, 215–230.

Rao, P.D.S. and Coelli, T. (1998) Catch-up and convergence in global agricultural productivity, 1980–1995. CEPA Working Papers, No. 4/98, Centre for Efficiency and Productivity Analysis, University of Queensland, Brisbane, Australia.

Sala-i-Martin, X. and Pinkovskiy, M. (2010) African poverty is falling…much faster than you think! NBER Working Paper 15775, National Bureau of Economic Research, Cambridge, MA.

Suhariyanto, K. and Thirtle, C. (2001) Asian agricultural productivity and convergence. *Journal of Agricultural Economics* 52, 96–110.

Suhariyanto, K., Lusigi, A. and Thirtle, C. (2001) Productivity growth and convergence in Asian and African agriculture. In: Lawrence, P. and Thirtle, C. (eds) *Asia and Africa in Comparative Economic Perspective.* Palgrave, London, pp. 258–274.

Trueblood, M. and Coggins, J. (2003) Intercountry agricultural efficiency and productivity: A Malmquist index approach. Working Paper, Department of Applied Economics, University of Minnesota, St Paul, MN.

United Nations (2009) United Nations Millenium Campaign. Available at: http://www.endpoverty2015.org/africa/news/unmillennium-campaign-petitions-african-governments-address-inequality-andunemployment/07/may/09 (Accessed 4 February 2010).

Van de Walle, N. (2001) *African Economies and the Policy of Permanent Crisis, 1979–99.* Cambridge University Press, Cambridge.

Varian, H. (1984) The nonparametric approach to production analysis. *Econometrica* 52, 579–597.

Winters, P., de Janvry, A. Sadoulet, E. and Stamoulis, K. (1998) The role of agriculture in economic development: Visible and invisible surplus transfers. *Journal of Development Studies* 34, 71–97.

World Bank (1994) *Adjustment in Africa: Reform, Results, and the Road Ahead. A Policy Research Report.* Oxford University Press, New York, NY.

World Bank (2000) *Can Africa Claim the 21st Century?* World Bank, Washington, DC.

World Bank (2004) Millennium Development Goals. Available at: http://web.worldbank.org/WBSITE/EXTERNAL/EXTABOUTUS/0,,contentMK:20104132~menuPK:250991~pagePK:43912~piPK:44037~theSitePK:29708,00.html (Accessed 4 February 2010).

World Bank (2008) *World Development Report 2008: Agriculture for Development.* World Bank, Washington, DC.

World Bank. World Development Indicators Database, World Bank, Washington, DC. Available at: http://publications.worldbank.org/WDI/

14 South African Agricultural Production and Investment Patterns*

Frikkie Liebenberg
University of Pretoria

14.1 Introduction

Since the establishment of the Union of South Africa in 1910,[1] its agricultural sector has seen significant shifts in structure that have been accompanied by ongoing institutional reorientation in its agricultural technical support services. Starting with the Land Act of 1913, followed by other legislation, such as the Cooperative Act of 1920, the Agricultural Marketing Act of 1937, and subsequent Apartheid-era legislation and policies, ethnically based disparities in access to land, markets, credit, technical support services and so on were entrenched. These policy measures established a dualistic structure of South African agriculture. Although the number of black farmers dropped from 1.04 million in 1921 to 437,807 in 1951, the number of white-owned farms increased from 81,432 to 118,186 over the same period (Bureau of Census and Statistics, 1960). Today the sector is dominated by large white-owned 'commercial agriculture' that is highly mechanized and dependent on hired labour, alongside communal or subsistence orientated farm households for much of the native population living in the former homelands. Most of the data and discussion of agriculture in South Africa refer to 'commercial agriculture,' or seem to, because during the Apartheid-era little effort was made to record the agricultural activities of black farmers as thoroughly as for white-owned farms.

This study highlights the shifts in agricultural production and input use during the 1910–2010 period and discusses certain salient features of public-sector investment in agriculture during that time. It then discusses certain issues regarding measurements of land use, labour and capital use in South African agriculture in order to assess more accurately long-term trends in agricultural productivity.

14.2 Agricultural Trends in South Africa

In 1910, agriculture (as measured by GDP or value-added) accounted for 19.3% of total economic output.[2] The agricultural

* The research supporting this chapter was funded principally by the South African Department of Agriculture and the Agricultural Research Council, with additional support from the International Food Policy Research Institute, InStePP, the University of Pretoria, and the Bill and Melinda Gates Foundation. The author thanks Philip Pardey, Colin Thirtle and Dirk Blignaut for their assistance in the analysis. The chapter itself draws on two previous research reports done in collaboration with Philip Pardey.

share of total economic output declined steadily throughout the 20th century, to just 2.2% by 2010. The absolute size of the agricultural sector grew almost every decade until the 1970s at an overall average annual rate of 3.66%, from US$1.6 billion (Rand 10.1 billion) in 1910 to US$10.2 billion (R 64.9 billion) in 1974 (both measured in 2005 prices).[3] From 1910 to 1928, real agricultural output grew by 3.56% per year. After the global economic depression and severe droughts of the early 1930s, the agricultural economy experienced a period of strong growth in conjunction with expanded farmer settlement and agricultural development support. Agricultural GDP reached US$7.9 billion (R 50.0 billion, in constant 2005 prices) in 1951, an increase of 8.95% per year for the 1934–1951 period. During the period 1951–1974, output growth slowed to an average of 2.27% per year. The agricultural sector then declined to a low point of US$5.8 billion (R 36.9 billion at 2005 prices) in 1992, reflecting in part the effects of another severe drought in the 1991 and 1992 cropping seasons. Thereafter, agricultural output rebounded to a peak of US$8.2 billion (R 52.4 billion at 2005 prices) in 2002, after which international market pressures, changing domestic agricultural policies, economy-wide influences and adverse weather conditions drove a period of decline.

14.2.1   Agricultural output

Prior to 1995, commodity-specific data on agricultural output reported in the *Abstract of Agricultural Statistics* excluded production originating from the former homelands (Directorate of Agricultural Statistics, 1995). Since 1995, crop-specific output statistics were revised to include production originating from these areas (Directorate Agricultural Statistics, 2011). The *Abstract of Agricultural Statistics* only reports the past 30–40 years of statistics, and data for the years prior to this were obtained from the Directorate of Agricultural Statistics, which now includes production from the former homelands/

reserves (Directorate of Agricultural Statistics, 2011). A lesser known feature of the statistics reported by the Directorate on the aggregate of production (gross value of production) is that this measure has always included production originating from the former homeland areas.[4]

Estimates by Liebenberg and Pardey (2010) show that even though corn production originating from the homelands today accounts for less than 3% of total national corn output, it was a very significant 27.2% in 1918. As for sorghum production, the share of national output from the homelands dropped from 74.3% in 1918 to less than 1% by 2002. In 1950, the homeland shares of production were 18.8% for corn and 46.4% for sorghum. This emphasizes the need to account for homeland production if productivity analysis for the country is disaggregated to the commodity level, i.e. if output indexes are constructed from commodity-specific price and quantity data. Past studies that have utilized the national gross value of production as the output statistic do, however, account for production from the homelands.

The mix of agricultural output changed markedly over the years, as shown in Table 14.1. In 1911, about 55% of the value of South African agricultural output was livestock products, with wool (20%), dairy (19%), cattle (15%) and sheep (15%) accounting for 69% of the value of livestock production. By 2010, the livestock share had shrunk to 51% of agricultural output by value (with poultry production now accounting for 46% of the total). The field crops share was 34% in 1911, grew to 47% in 1971 (largely because of an expansion of cereals and sugarcane production), declined significantly to 28% in 2004, regained some market share to reach 33% in 2008, but declined to 24% in 2010. A reduction in corn and wheat production accounted for most of the post-1971 decline. The share of horticultural output expanded steadily for the entire period since 1910, starting at 10% that year and increasing to 26% by 2010. Up until the late 1980s, the growth in the value of horticultural output averaged 2.9% per year. After a brief downturn in output growth from 1989 to

Table 14.1. Gross value of agricultural production and industry share, 1910–1911 to 2009–2010.

Period	Total	Field crops	Horticulture	Livestock
	Constant 2005 Rand, million	Output value share (%)		
1910–1911	14,013	33	12	55
1915–1916	14,969	39	11	50
1920–1921	14,593	34	14	52
1925–1926	17,799	33	11	55
1930–1931	15,355	36	14	50
1935–1936	21,066	39	16	46
1940–1941	25,366	40	13	47
1945–1946	30,727	36	16	48
1950–1951	56,361	34	11	55
1955–1956	56,431	39	14	47
1960–1961	61,860	43	15	43
1965–1966	68,565	38	17	45
1970–1971	74,289	47	17	36
1975–1976	68,383	34	18	47
1980–1981	68,383	34	18	47
1985–1986	76,251	40	19	41
1990–1991	71,140	35	22	43
1995–1996	69,934	36	23	40
2000–2001	68,402	34	25	41
2005–2006	73,129	24	26	50
2009–2010	88,317	24	26	51
		Annual growth rates (%)		
1910–1947	2.39	0.40	0.75	−0.49
1947–1971	2.30	0.65	0.34	−0.84
1971–1989	0.63	−0.94	0.04	1.00
1989–2010	0.24	−2.25	1.75	0.68
1947–2010	1.15	−0.82	0.75	0.22
1910–2010	1.84	−0.32	0.75	−0.08

Livestock aggregate includes 11 commodities, field crops, 22 commodities, and horticulture, 12 commodities. Annual nominal expenditures adjusted for inflation using the GDP implicit price deflator from South African Reserve Bank (2011). *Source:* Directorate of Agricultural Statistics (2011).

1992, the sector resumed growing at impressive annual rates, especially in the wine sector (4.64% from 1992 to 2004), deciduous fruit (5.29%) and citrus fruit (7.37%). This growth was partly in response to improved access to international markets, because global trade sanctions against South Africa were eliminated following the end of Apartheid policies in 1994.

The quantity of total agricultural output grew at an average annual rate of 1.66% from 1911 to 2010. From 1911 to 1945, output grew by only 1.95% per year, accelerating to 3.92% annually over the following three decades, then slowing to just 1.43%

per year for the period 1982–2000. Since 2000, output growth has increased to 1.66% per year. Since 1911, growth in horticultural output (fruit and vegetables) outpaced that of field crops and livestock by almost 0.7% per year. Field crop production kept pace with livestock output from 1911 until the mid-1960s; during the subsequent two decades, it actually grew at a faster rate than the livestock sector. However, during the period 1982–2010, field crop production grew by only 0.22% per year, lagging behind the corresponding growth in annual livestock output of 1.85%. Since 2000, growth in field crop production has fallen substantially behind the corresponding growth in live-stock output, which increased by 2.4%.

Thus, the overall growth in South Africa's total agricultural output has been largely driven by strong growth in the horti-cultural sector, with the comparatively slower growth in field crop output during more recent decades acting as a drag on the overall pace of growth. Moreover, the rate of growth in agricultural output – and espe-cially field crop production, which includes staple food crops such as wheat, corn and grain sorghum – has fallen below the rate of population growth. South Africa's popula-tion grew by 2.43% per year from 1982 to 2008, compared with 1.51% per year for overall agricultural output and just 0.36% per year for field crops. (Although the rate of population growth has slowed in more recent years – to 1.35% per year since 2000 – field crop production has slowed even more dramatically to just 0.31% per year during the same period.) Notably, the slowdown in both total output and crop output in South Africa in recent decades parallels similar trends in the USA, where total output grew by 1.63% per year during the 1980s (com-pared with 2.22% per year for the previous decade), slowing to 1.28% per year from 1990 to 2002 (Alston *et al.*, 2010).

Agricultural trade constituted 2.7% of South Africa's GDP in 2006, with agricul-tural exports accounting for about 6.9% of total exports (Directorate of Agricultural Statistics, 2011). This is significantly below the share in 1932, when agriculture accounted for 78.4% of total South African exports. Since then, agricultural exports as a share of the country's total exports has declined steadily, to a low of 6.5% in 1993, after which the agricultural share grew to an average of 8.2% for the period 1994–2007. South Africa has always been a net exporter (by value) of agricultural prod-ucts. In 1975, agricultural exports exceeded imports by R 20.7 billion but the lingering effects of trade sanctions imposed because of Apartheid policies, combined with a fail-ure to remain internationally competitive in agriculture, have left the country barely able to sustain its net agricultural exporter status in recent years.

14.2.2   Agricultural input use

Up until the 1950s, input use by farmers in the reserves, as reported in agricultural cen-sus reports, was limited to estimates of area planted and was excluded in the data repor-ted in the *Abstract of Agricultural Statistics*. Since 1995, this series was revised to include it. For labour use, no records were kept of employment by farmers in the homeland areas. An effort is made in this study to sys-tematically include family and proprietor labour, as well as seasonal workers, and to exclude domestic servants (the latter being included in the farm worker population reported in the *Abstract of Agricultural Statis-tics*) for commercial agriculture.[5] The effect of this departure from the labour data used in earlier studies is discussed below.

With capital inputs, the situation is more problematic. Only information on the stock of capital items was taken into account in census years, and then was limited to the more important capital items, such as trac-tors and ploughs, but not infrastructure. No indication was given as to whether or not the machinery and equipment were newly purchased. Since 1975, the situation has changed and is discussed in greater detail below, but it is safe to say that the data on capital inputs still mostly reflect (or are dominated by) the trend in the 'commercial sector' and, since 1975, that of homeland farmers to an uncertain extent. Information on material inputs – largely purchased from off-farm sources – is sourced directly from wholesalers and suppliers and is generally perceived to be inclusive of total use. A comparison between the data published in the *Abstract of Agricultural Statistics* and the data reported on current expenditure in the censuses shows that this is not the case. For example, expenditure on fertilizer reported in the *Abstract of Agricultural Statistics* matched that of the census until 1979/1980, but thereafter was reported to be at levels well below that of the census. More analysis is required, however, into the nature of the differences.

Data on land use is based on the author's own estimates derived from census and sur-vey reports, with land allocated to timber plantations and other uses (residential, etc.) excluded because the forestry industry is excluded from estimates of agricultural output in South Africa. It also excludes agricultural land use in the homelands because no reliable information is available, except for a few sporadic observations in census reports. Estimates on land in farming made by the FAO generally exceed those used locally in South Africa by about 15 million hectares (ha), which presumably include land in agriculture in the former homelands, of which 2.5 million ha are considered to be cropland (FAO, 2009). In 1991, the former homelands had an estimated 14.5 million ha of agricultural land worked by about 1.3 million households (Directorate of Agricultural Statistics, 2011).

Figure 14.1 indicates the significant structural changes in farmland use in South African agriculture since 1910. Total farmed

area by commercial agriculture grew to a peak of 91.8 million ha in 1960. It then declined steadily to 82.2 million ha in 1996, a level at which it has more or less stabilized. The total number of farms followed a similar pattern, peaking in 1953 at 119,600 and declining at an average rate of 1.23% per year thereafter, so that by 2007 the number of farms had dropped to less than half the number that existed five decades earlier. The interplay between changing farm numbers and the total area in farms means that average farm size declined during the first half of the 20th century (from 1019 ha in 1910 to 730 ha in 1952) and increased during the second half of the century, to average 1640 ha in 2000. Average farm size has continued to grow: in 2002, it was 1833 ha/farm. Preliminary results from the 2007 Agricultural Census indicate a continuing increase in average farm size, to

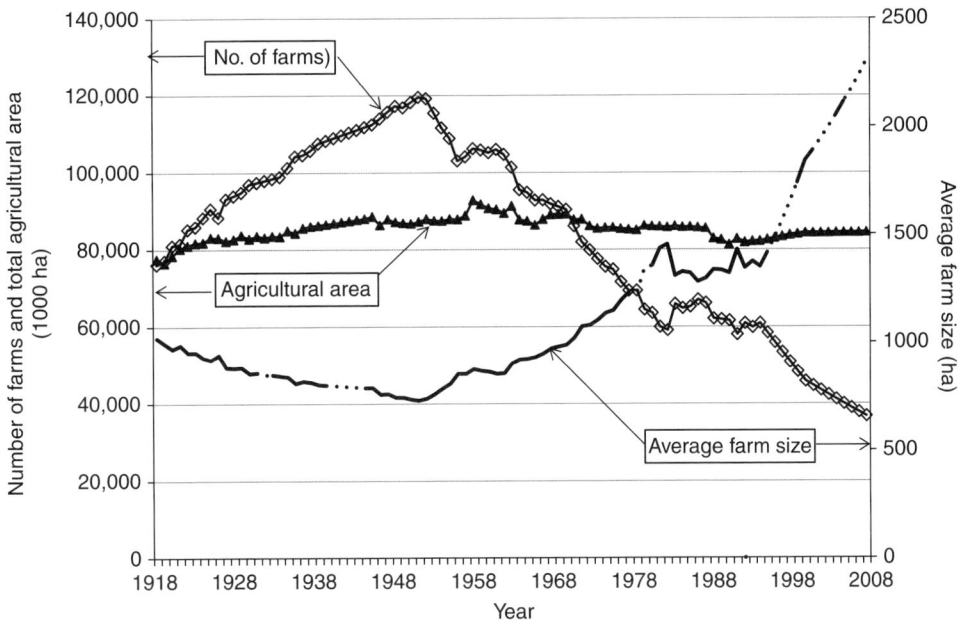

Fig. 14.1. Number, total area and average size of farms in South Africa, 1918–2010. Dashed sections of farm size plot indicate estimates (via interpolation). Farm numbers are a total count of farms. Statistical definitions of a farm changed over time. For example, Union of South Africa (1948) defined a farm as all 'occupied farms in rural areas', as well as any occupied holding greater than 0.86 ha in urban areas that was producing agricultural output for sale. Presently, only commercial farms or business entities registered for value added tax and/or income tax are designated as farms for the purpose of statistical compilations. Sources: Statistics South Africa (2011) and Directorate of Agricultural Statistics (2011).

about 2099 ha/farm, and a continuing decline in farm numbers, to 39,966 (Statistics
South Africa, 2011).

Table 14.2 shows period averages beginning in 1947–1948 to 1949–1950, with
5-year averages thereafter, in the total cost
shares of four agricultural input categories:
labour, land, capital and materials. Material
inputs have claimed an increasing share of
total costs, around 25.3% in 1947–1950 to
an average of 46.9% over the past 5 years.
Reported capital costs have fluctuated at
around a 30% share of total costs over the
same period, whereas labour inputs have
steadily declined as a share of total costs,
from almost 32.4% in 1947–1950 to less
than half of that (14.6%) by 2005–2010. At
the beginning of the period, land costs
accounted for 10.4% of total costs, growing
to 13.1% by the mid-1980s, then shrinking
to just 6.3% of total costs by 2005–2010.
Notably, Alston et al. (2010) reported US
land cost shares that followed a similar
trend, starting at 17% of total cost in 1949,
growing to 20% during the late 1970s and
early 1980s (when land prices soared), then
falling to 15% by 2002. According to these
data, however, land cost shares are uni-
formly lower in South Africa compared with

the USA, perhaps reflecting a much smaller
share of cropped to total land in South Africa
versus the USA, along with a smaller share
of that cropped land under irrigation.[6]

Looking in more detail at land costs, in
nominal terms it grew by 5.7% per year during the period 1947–1959, when the total
area under cultivation was still increasing.
Thereafter it increased by 9.95% per year,
whereas land in farming declined by 0.18%
per year. Total labour costs fluctuated at
around an average of R 160.3 million during
the 1960s, but then declined in real terms
by 1.08% per year until 1985. They increased
during the period 1985–1996 by 2.04% per
year, but then began to decline and have
continued to do so through 2009–2010.

According to Thirtle et al. (1993), during the period from 1947 to 1970, the cultivated corn area in the summer rainfall
region expanded as oxen were increasingly
replaced by tractors. This spurred the
expansion of average farm size (as measured
by area per farm; see Fig. 14.1) along with
labour use, as well the use of chemical fertilizers and higher-yielding seed varieties
(Payne et al., 1990). After 1970, the increased
mechanization of crop harvesting activities
through the use of combine harvesters began

Table 14.2. Input cost shares, 1947–1948 to 2009–2010.

	Labour	Land	Materials	Capital
Period		Input cost share (%)		
1947/48–1949/50	32.37	10.42	25.34	31.87
1950/51–1954/55	33.42	13.75	25.56	27.27
1955/56–1959/60	29.36	10.72	28.93	30.99
1960/61–1964/65	26.07	9.96	29.87	34.11
1965/66–1969/70	25.62	8.00	32.40	33.97
1970/71–1974/75	23.88	9.78	34.03	32.31
1975/76–1979/80	20.51	12.47	39.00	28.02
1980/81–1984/85	16.70	13.06	39.24	31.00
1985/86–1989/90	19.64	13.69	39.52	27.15
1990/91–1994/95	21.77	11.17	37.32	29.75
1995/96–1999/00	21.39	9.20	40.41	29.01
2000/01–2004/05	17.22	7.19	43.53	32.07
2005/06–2009/10	14.63	6.32	46.92	32.13

Labour includes proprietor and family labour, seasonal and full-time labour – domestic servants are excluded.
Land refers to all land in farming, but excludes timber plantations and other uses (farm yards and buildings).
Source: Calculated from data obtained from Directorate of Agricultural Statistics (2011).

to alleviate a peak demand for labour at harvest time, thus contributing to a decline in overall labour use.

Agriculture's general pattern of labour, land and machinery use in summer and winter rainfall areas evolved in parallel. The overall expansion of the cultivated area was largely complete by 1947, with machinery increasingly substituting for labour throughout South African agriculture (Thirtle *et al.*, 1993). The Pass Laws Act of 1952[7] might have accelerated this ongoing factor substitution effect, especially during the late 1960s when the conditions of the Act were severely applied. Other polices probably had a bigger effect, however. Farmers were given access to cheap credit, which for periods of time involved negative real interest rates, and tax breaks that allowed capital equipment to be written off within the first year after purchase. By the end of the 1981–1983 drought, the credit and tax concessions were largely gone; the price of gold had plummeted; and the Rand was drastically devalued. These events combined to make domestic inputs, especially labour, much cheaper than capital items (which were largely imported), causing a dramatic reversal of the historical trend in labour's share of

total costs. During the late 1980s and early 1990s, labour use increased considerably as a substitute for relatively expensive capital. Since then, new post-Apartheid legislation regarding the security of land tenure for agricultural labour tenants working on large farms and the stipulation of minimum wages has again caused the sector to shed labour.

14.3 Public Spending on Agriculture

Government spending on agriculture has shown highly varying degrees of importance since the establishment of the Union government in 1910 (Fig. 14.2). During the first two decades after the Union's establishment, Department of Agriculture expenditure fluctuated at fairly low levels. Support for agricultural development took the form of subsidized imports of breeding stock, steam ploughs, and a strong focus on training students both locally and abroad. Agricultural spending levels grew strongly during the multi-year drought that followed the onset of the depression in the early 1930s. Drought relief programmes were introduced, along with a strong drive to

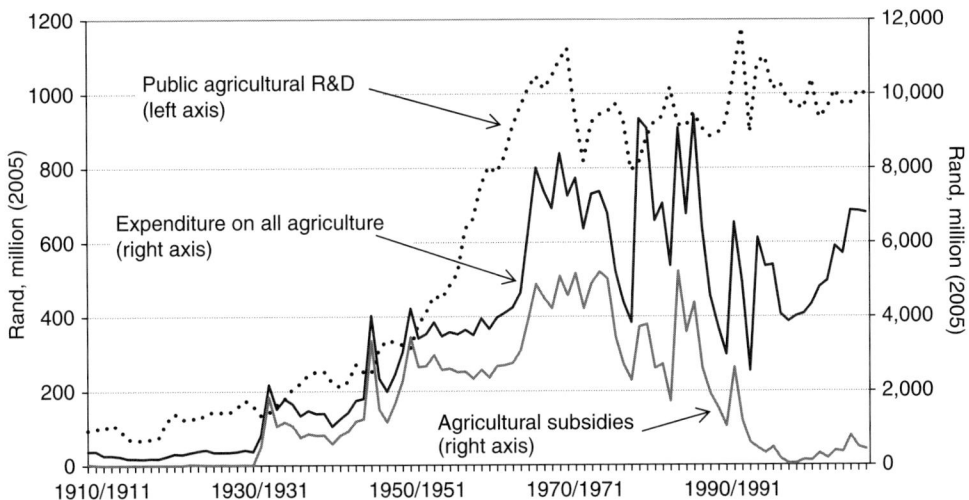

Fig. 14.2. Public expenditure on agriculture in South Africa, 1910–2009. Expenditure on agricultural R&D. Sources: Union of South Africa (1910–1959), Republic of South Africa (1960–2011), and Agricultural Research Council (1993–2010). Annual expenditures adjusted for inflation using the GDP implicit price deflator from South African Reserve Bank (2011).

curb soil and veldt erosion, while a programme to 'Train Rural Unemployed People'
began to take shape. The latter was to
become the government's main farmer settlement drive for the next 30 years. Credit
support by means of soft loans through the
State Advances and Recoveries Office under
the governance of the Farmers Assistance
Board – later to become the Agricultural
Credit Board – was largely administered
through the farmer settlement programme.

Support programmes established in the
1930s were in aid of purchases of land,
infrastructure development in a broad sense
and production credit. From the 1960s
through the 1980s, industry support subsidies, such as stockpiling of butter and government funding of export losses, began to
exhibit a strong influence on the trends in
government support, in addition to the support provided through the Credit Board
loans. Prior to this, public investment in
agricultural research and development
increased substantially from the mid-1950s
and continued to grow until the mid-1970s,
when it began to stagnate.

The ability to maintain high levels of
support payments to farmers came under
pressure in the mid-1970s as a result of
increased expenditure on domestic security
services and tough international economic
conditions at the time. In the early 1980s,
there was another increase in support to
farmers, as a result of another severe drought
and the introduction of a 'land use conversion' initiative to shift marginal cropland
out of crop production and into livestock
farming on planted pastures. Thereafter, all
subsidy support to farmers was phased out
in the 1990s, and the increase seen in the
expenditure on agriculture was a result of
implementing the Land Reform and Restitution initiatives beginning in 1994.

14.3.1   Investment in agricultural
research and development

In inflation-adjusted terms, South Africa
invested just US$16.3 (R 63.0) million on
public agricultural research and development (R&D) in 1911. Real public agricultural

R&D spending grew steadily, by an average
of 5.1% per year until 1952 (Fig. 14.2). The
pace of growth accelerated to 7% over the
subsequent 19 years to a total of US$290.1
(R 1122.6) million by 1971. From 1971, spending on public agricultural R&D then declined
by an average of 2.9% per year, in inflation-
adjusted terms, to US$204 (R 793) million in
1980 and thereafter recovered somewhat to
reach US$268 (R 1038) million in 2003.
Notably, real public spending on agricultural
R&D failed to grow significantly after 1972,
except for a brief jump to US$304 (R 1177)
million in 1993, brought about by structural
adjustment payments during the establishment of the Agricultural Research Council
(ARC). In fact, if external income generated
by the ARC is excluded, public agricultural
R&D spending for every year in the entire
1971–2007 period was less than the inflation-
adjusted 1971 amount of R 1123 million.[8] In
2007, with the external income generated by
the ARC excluded, direct public investment
in agricultural R&D was equal to just 70% of
the corresponding 1971 figure. Several of
the switching points, when the growth rate
in public agricultural R&D spending changed, coincide with changes in the administrative structure of public agricultural research
agencies. Others relate to changes in science
policy more generally.

The changing rates of growth in R&D
spending track the changes in the institutional organization and governance of the
public-sector research and extension services. In 1952, research and extension were
reorganized with a much stronger regional
focus, and all public-sector research and
extension capacity within the agrological
region, inclusive of the faculties of agriculture, was focused on solving the productive
constraints of the region as part of a multi-
disciplinary team. In 1971 the focus shifted,
and the responsibility for university agriculture faculties was transferred to the
National Department of Education, followed
by the establishment of autonomous national
commodity institutes.

In the international context, South
African agricultural GDP shrank as a share of
overall GDP throughout the 20th century, as
it did for Australia and the USA. The trends

(and value) of agricultural GDP to the overall GDP ratio in South Africa and Australia are similar, but the corresponding ratio for the USA declined at a faster rate and was generally considerably below the South African figure. Notwithstanding the Australian and South African similarities in the agricultural shares of their respective economies, surprisingly South Africa invested more intensively in agricultural research than did Australia (and the USA) for the first three quarters of the 20th century. In the early 1970s, the relativities changed, with South Africa generally falling below Australia (and well below the USA) in terms of public agricultural R&D intensity, as the pace of R&D investment faltered as did South African agricultural economic growth.

Despite South Africa's recent poor intensity performance relative to Australia and the USA, in 2000 South Africa's intensity of commitment to agricultural R&D per unit of agricultural GDP (US$2.50 of research spending per $100 of agricultural output) was on a par with the corresponding average of US$2.36 for high-income countries (Pardey *et al.*, 2006). South Africa, however, spends about half as much on agricultural R&D per capita of the general population and about a fifth as much per capita of the economically active agricultural population, compared with the corresponding average intensity ratios of the high-income countries. South Africa's performance on this score has, according to Flaherty *et al.* (2010), declined to US$2.10 of research spending per $100 of agricultural output.

14.3.2 The number of agricultural scientists

A total of 120 researchers was engaged in public agricultural R&D in South Africa in 1911, about half of them employed by the Department of Agriculture and the other half by the university agriculture faculties and regional experiment stations. This number grew steadily to a total of 503 researchers in 1940, declined briefly to 445 researchers during World War II and then resumed growing.

In the two decades following World War II, the total number of researchers increased from 618 in 1949 to 903 in 1976, an average annual rate of growth of 1.8%. The total number of agricultural researchers continued to grow for the following 20 years (at a rate of 2% per year), peaking at an estimated 1322 researchers in 1996. From 1997 to 2003, voluntary retrenchments and net attrition in the public and semi-public sectors saw the number of full-time-equivalent researchers decline to 1055, a contraction of 3.1% per year for an overall loss of 20% of the country's total scientific research capacity in the agricultural sciences. The number of full-time-equivalent scientists working for ARC peaked in 1996 at 761, dropping precipitously to 460 in 2004, with small increases thereafter to 471 in 2007, and a decline to 433 in 2010. Recent estimates suggest that growth in the total number of full-time-equivalent researchers working for public agricultural R&D agencies in South Africa stalled in the mid-1980s, declining to 784.3 in 2008, 64.7% of the 1213 researchers employed in 1985 (Liebenberg *et al.*, 2010; Flaherty *et al.*, 2010. See also Kahn *et al.* (2004) for a discussion of the exodus of R&D personnel from South Africa during this time).

14.4 Productivity Trends

14.4.1 Land and labour productivity

From 1911 to 1924, land productivity grew at an average annual rate of 1.41%, four times that of the corresponding rate of labour productivity growth, which averaged 0.32% per year (Fig. 14.3). Throughout the 20th century, there were three phases of distinct growth patterns in these two partial productivity measures. During the pre-World War II years (1911–1940), land productivity grew by 1.86% per year, whereas labour productivity declined by 0.23% per year. The rate of growth of both land and labour productivity picked up over the subsequent four decades following the war (1947–1981), averaging an impressive 4.71%

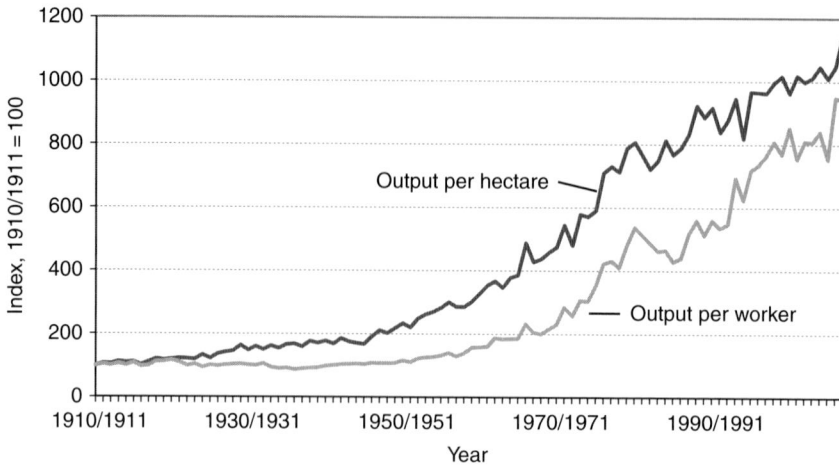

Fig. 14.3. Agricultural labour and land productivity in South Africa, 1910–2010. Labour data were adjusted to consistently include proprietor and family labour, seasonal and full-time labour. Domestic servants are excluded. These series are Tornqvist quantity indexes. Sources: see text for details.

per year for labour productivity and 4.23% per year for land productivity. Since then, productivity growth rates for both land and labour have slowed considerably, down to 2.33% per year for labour and only 1.34% per year for land productivity.

14.4.2   Total factor productivity

Several studies have measured rates of total factor productivity (TFP) growth for South African agriculture, with no apparent consensus or pattern emerging from or evident in the different measures. Some of these differences might be attributable to differences in the range of years covered by each study, but differences in data coverage and treatment no doubt play a role, too, making a collective assessment of these studies problematic.

In 2010, Liebenberg and Pardey extended the aggregate input, output and TFP measures first reported by Thirtle *et al.* (1993) for the period 1947–1991 and updated in Schimmelpfennig *et al.* (2000) to 1997. Thirtle *et al.* (1993) indicate their aggregate output measure consists of a Divisia aggregation of three pre-aggregated groups of outputs (perceived as representative of commercial agriculture only): namely, crops, horticulture and livestock products, drawn from annual

issues of the *Abstract of Agricultural Statistics* (Directorate of Agricultural Statistics, 2011). The input index consists of an aggregation of measures of land, labour, intermediate inputs (packing fuel, fertilizer, dips and sprays, and other non-farm items), and capital inputs (fixed improvements, machinery and livestock capital). Liebenberg and Pardey (2010) updated this to cover the period 1947–2008. The present study extends the period to 2010 and uses the most recent data available from the Directorate of Agricultural Statistics, which has since revised its capital data series with specific changes to its estimates of rent paid on land and a revised livestock inventory.

Figure 14.4 shows the results of the most recent update to the South African agricultural TFP index. According to this measure agricultural TFP grew on average by 1.58% per year from 1947 to 2010. As with previous studies, the highest rate of TFP growth was achieved in the 1970s and 1980s was an impressive (and perhaps questionable) 4.22% per year. This is substantially higher than the 0.98% per year rate reported here for the immediate post-World War II decades. Notably, TFP growth slowed to 1.2% per year for the period 1981–2010 and, depending on the beginning point selected, could even have declined by

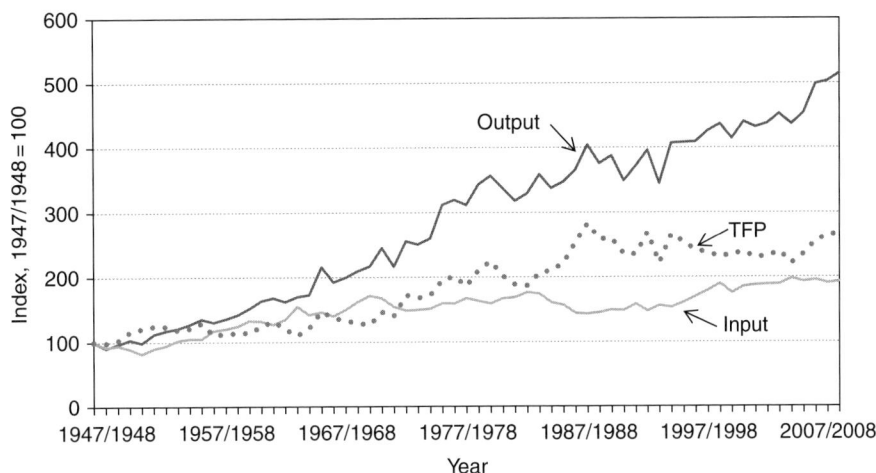

Fig. 14.4. Output, input and TFP indexes of South African agriculture, 1947/1948 to 2009/2010. Source: Author's estimates. See text for details.

0.25% per year since 1989. This slowdown or even decline in agricultural TFP growth is apparently owing to a much stronger growth in the rate of input use.

Recent studies by Conradie *et al.* (2009a,b) extend the earlier methods used by Thirtle *et al.* (1993) to compile regional estimates of aggregate input, output and TFP growth for the Western Cape region. That region has distinctive agro-climatic attributes. Specifically, it is the only region within the country that experiences winter rainfall, so its agricultural output is thus dominated by deciduous fruit and wine grapes, whereas output in the rest of the country consists mainly of field crops and livestock products.

Conradie *et al.* (2009b) estimate that during the period 1952–2002, TFP in the Western Cape grew on average by 1.22% per year. This is about the same rate as overall TFP growth at the national level, but the regional-versus-national pattern of growth over comparable time periods was considerably different. Whereas the reported TFP growth at the national level stretches back to 1965, the growth in TFP for the Western Cape begins in 1971. From 1971 to 2002, the Western Cape region saw productivity grow at 0.89% per year, less than half the corresponding rate of growth at the national level, which Conradie *et al.* (2009b) put at

more than 2% per year). Again, differences in data sources and treatment may account for some of the disparities, but it is also likely that differences in the composition of output and inputs and other factors play a role in these regional differences, as they do in the considerable national-versus-state differences in productivity patterns reported for the USA by Alston *et al.* (2010).

Contrasting the growth phases in TFP observed here with the timing of the major institutional reorganizations of the agricultural R&D services referred to previously, it is interesting to note that the TFP growth rates appear to be connected with this with a 14- to 17-year lag. Although many other factors play a role in explaining this, the trends in agricultural R&D service provision undoubtedly had an effect, as shown by an internal rate of return to investment on the order of 44%, as estimated by Schimmelp-fennig *et al.* (2000).

14.5 Limitations in the Statistics Used for Productivity Analysis

The Liebenberg and Pardey (2010) study on agricultural productivity trends in South Africa was developed by extending the 1947–1997 series developed by Schimmelp-fennig *et al.* (2000) to 2008, and in so doing sought to faithfully deploy the same methods, data types and sources used in the earlier compilations by these authors. The data used by Schimmelpfennig *et al.* (2000) and the earlier study by Thirtle *et al.* (1993) were largely taken from the *Abstract of Agricultural Statistics*, published by the Directorate of Agricultural Statistics, and supplemented by historical material and some greater detail from the Department of Agriculture. The annual issues of the *Abstract*, however, rarely report data for periods longer than 30 to 40 years and, as a result, earlier databases were constructed by splicing data from several issues of the *Abstract* to cover the period beginning 1947–1948. Liebenberg and Pardey used more detailed underlying data from past agricultural census reports, juxtaposed against the data from the *Abstract*, which reveal inconsistencies in the specification of the aggregates reported in the *Abstract* (or in the way they have been interpreted) over time, some of which were adjusted for by Liebenberg and Pardey (2010). In the present study, the variables for which major adjustments were made include land in farming and agricultural employment, the significance of which is discussed below.

Further analysis since the Liebenberg and Pardey (2010) study reveals fewer apparent caveats regarding the data taken from different *Abstract* reports. Given the country's history of racial segregation, which can be traced back to its colonial origins, it is little wonder that its statistical records were affected by that situation. The *Abstract of Agricultural Statistics* essentially provides a summary of the more important variables included in estimating agriculture's contribution to the national income accounting system. As mentioned by Thirtle *et al.* (1993), the accounting conventions are not always readily apparent. During the Apartheid era of racial segregation, and especially during the 1975–1994 period when the homelands were established, the contribution of the economies of homeland areas was separated out from the national income accounts. The reintegration of those territories back into national statistics after 1994 remains problematic in the absence of a comprehensive agricultural census that includes farmers from those territories. Thus, at a disaggregated level, agricultural statistics generally continue to focus mainly on the commercial farming sector. However, as will be shown in the rest of this section, inconsistent inclusion of factors, such as production and input use by farmers in the former homelands and changing definitions (or composition) of land use and labour statistics, do have an important bearing on the data that is eventually used in analysing agricultural productivity trends.

14.5.1   Agricultural labour

Statistics on on-farm employment in the *Abstract of Agricultural Statistics* do not provide a detailed description of the composition of the data beyond whether seasonal and casual labour was included.[9] Prior to August 1965, seasonal labour was generally not included in the *Abstract*. It was sporadically reported in the censuses, but not explicitly repeated in the *Abstract*. Historically, commercial agriculture has always relied heavily on hired labour, both seasonal and full-time, and the issue of ready access to a cheap labour force by settler farmers formed much of the political debates and labour policies as it had in other former colonies (Gilliomee, 2007; Binswanger *et al.*, 1995). Expressed as a ratio to full-time farm workers, the amount of seasonal labour in South Africa varied between 60.24% (1937) and 96.69% (1958), and sometimes exceeded the size of the full-time labour force, as in 1971.

In those years where data on proprietor and family labour as well as domestic servants were enumerated in the census, they are included in the statistics on farm employment in the *Abstract* but not listed separately (leading to significant inconsistencies in terms of composition of this data series). The estimated number for proprietor and family labour in 1953/54 was 154,271, or 14.42% of the combined total for full-time farm workers actually involved in farming activities (excluding domestic servants).

Their share declined to 8.77% of the aggregate of the combined total of full-time farm workers by 1993.

Owner and family labour, including relatives of the farm owner or occupier, were for the first time reported as 50,996 in the 1945–1946 Agricultural Census. Because of the influence of World War II, when many rural adult males were still in military service, this can be regarded as abnormally low. In 1937, the whites employed as full-time farm labour amounted to 205,261, with the non-white component of the labour force totalling 764,945. This changed to 15,460 white labourers in 1946 (706,733 non-white), or 7.53% of the regular labour force recorded in 1937. In 1958, the occupier and relatives category was enumerated again for three years, then standing at 132,560 and climbing to 158,475 in 1960. Thereafter, it reappeared in the 1985–1986 agricultural survey as 82,861 and was enumerated as proprietor (64,042) and family labour (17,473). In 2007, owner and family labour declined to 47,570. As a proportion of the full-time labour force, this segment varied between 8.68% in 1946 and 18.95% in 1958, and by 2007 was around 11.5%.

Figure 14.5 shows the different trends in farm worker population between this study, the study by Schimmelpfennig et al. (2000), and the statistics reported in the Abstract. This study differs from earlier studies in a number of ways. Earlier studies accounted for seasonal and casual labour using fixed proportions relative to the regular labour force according to specific data points available from various agricultural censuses. Liebenberg and Pardey (2010) used linear projections in seasonal and casual labour between census years, juxtaposed against trends in the regular labour force, using a more detailed set of census statistics. This is not ideal, as a different set of factors – such as area planted, prevailing production conditions and the stage of mechanization – determine the size of this labour force. Earlier studies estimated seasonal labour by adding a fixed ratio of the regular labour force, whereas this ratio has varied significantly over time. In the present study, the number of seasonal workers was estimated by using (i) where available, the trend in salaries paid to casual and seasonal workers and (ii) the trend in the volume of production in the absence of salary data.

Accounting for the inconsistent inclusion of domestic servants – and by excluding it from the data – and adding estimates for owner and family labour show that these

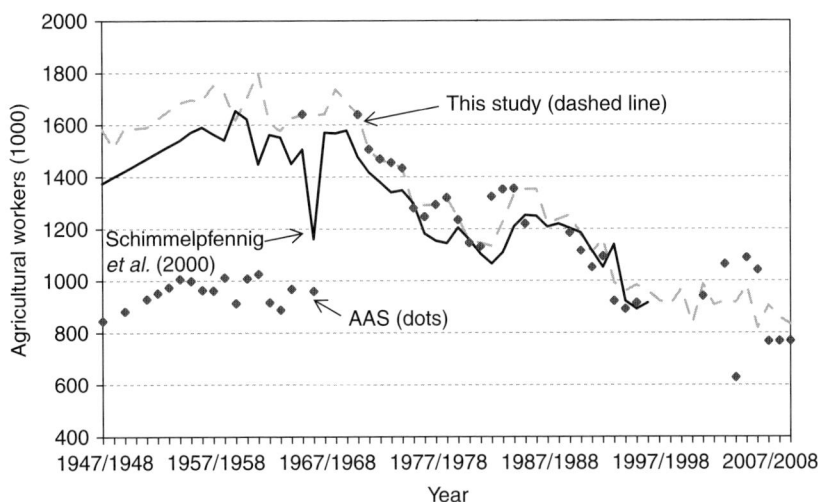

Fig. 14.5. Agricultural labour in South Africa as measured from different sources. Units are 1000s of agricultural workers. Source: See text for details.

adjustments in aggregate have a significant effect on the labour force estimates for the years prior to 1954, compared to the data reported in the *Abstract*.

14.5.2 Land use

This and previous TFP studies use the total area in farming in calculating the land index. Coverage on this in the *Abstract* is limited to census years with no inter-census extrapolation, and the most recently published data available are for 1991 (Directorate of Agricultural Statistics, 2011). The *Abstract* reports this according to the financial year specified in the census – a practice which is rarely highlighted and could easily lead to misinterpretation by the end-user. Schimmelpfennig *et al.* (2000) and earlier studies by Thirtle *et al.* (1993) seem to have suffered for that reason. This study deviates from earlier studies by using the aggregated cultivated area from the underlying data on area planted, as reported in the agricultural censuses and survey reports, to estimate the inter-census years observations and it assigns the data to the year in which the land use pattern was enumerated. The results differ slightly from the Schimmelpfennig study, as shown in Fig. 14.6a, for the period 1947–1948 to 2009–2010. Prior to 1976, the trend is largely the same, but it follows a different allocation of the data by one year. Since 1976, the trend is the same until 1986. The deviation since 1986 arises from a change in the way land use is reported, with land under forestry now being included in the *Abstract*. However, it continues to be excluded in this study to maintain consistence over time.

In addition, as opposed to data on agricultural output, land use data (area planted, etc.) published in the *Abstract* represent that of white-owned farms only in regions considered to be in South Africa as defined at the time. With the establishment of the various homelands and self-governing territories (SGT) between 1975 and 1982, land use statistics would have excluded white-owned farms in these areas until these regions were reincorporated into South Africa in 1993. Hence, the 1992–1993 Agricultural Census shows an increase in land in farming (when these areas were again treated as part of South African territory). The effect of this would be a downward bias in land area and an upward bias in TFP estimates for the 1975–1993 period. Assuming that the share of white-owned farmland in the homeland and SGT were relatively small, the net effect could be assumed to be relatively small.

The inclusion of production from the non-white farmers in the native reserves deserves attention in terms of the effect of the previous governments' racial segregation policies on the level of agricultural activity of black farmers in general. This is partly reflected in Fig. 14.6a, where the trend in area planted in homelands (this author's estimates) is shown against that of white commercial farmers. The area planted by native farmers remained above 2 million ha until a decade after World War II. Then it began to decline gradually until 1960–1961 and then sharply over the subsequent 14 years. By the mid 1970s it was only 13.6% of its 1923–1924 peak. Yields obtained by black farmers were also reported to be typically much lower than those of commercial farmers. The suppressing effect that the discriminatory policies increasingly had on the contribution of this section of the farming community is significant enough to explain the low estimates of productivity growth in the South African agricultural sector, as measured by past studies of the period 1947–1968.

In existing studies, the share weights for land are calculated from estimates of the values of land rent and the percentage of area rented (Wiebe *et al.*, 2001). With 68.6% of the total land area of South Africa being suitable only for extensive livestock farming, one suspects that reported aggregate rental rates would largely reflect market trends in the livestock industries (Directorate of Agricultural Statistics, 2011). Using the aggregate of land in farming as the measure of land productivity could thus potentially hide important responses to market signals by farmers. The trends in rain-fed

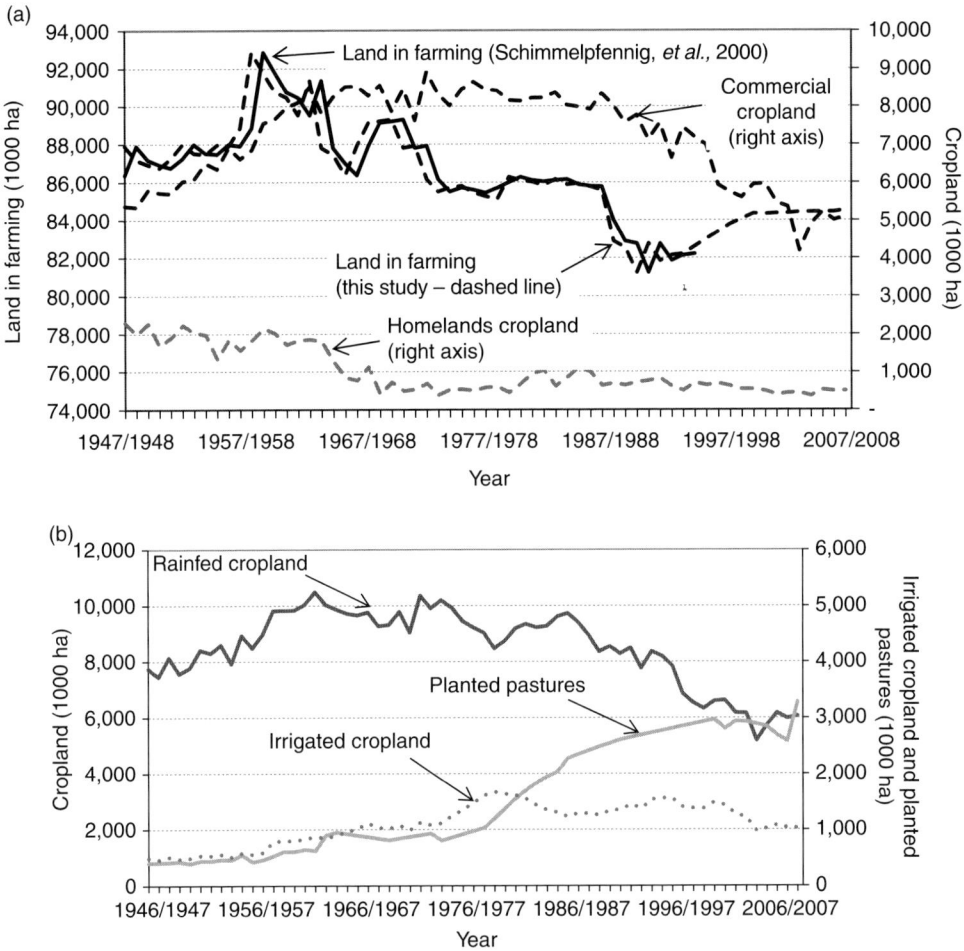

Fig. 14.6. Estimates of agricultural land use in South Africa. (a) Farmland in agriculture, 1947/1948 to 2007/2008. (b) Rainfed cropland, irrigated cropland and planted pastures, 1910/1911 to 2009/2010. Planted pastures are pastures that have been sown with improved forages. Not shown in the figure are unimproved range lands. Source: Agricultural Census Reports, Statistics South Africa (2011).

cropland, irrigated cropland and planted pastures, based on the revised aggregate land in farming series (net of forestry and indigenous forests), are shown in Fig. 14.6b. This reveals very different trends in each of these categories over time. The area of rainfed cropland peaked at 10.49 million ha in 1963–1964 and hovered at a slightly lower level until 1976, after which it decreased by 3.71% per annum to 8.47 million ha in 1981. After that, it increased by 2.11% per year until 1988, and decreased annually thereafter by 2.45%, to 5.75 million ha in 2008. That figure is 3.99 million ha lower than its 1988 level and represents 74.3% of the cultivated rainfed area in 1947–1948.

The cultivated area under irrigation (largely horticultural production) grew by 3.63% per annum from its 1947–1948 level of 0.49 million ha to 0.87 million ha in 1963–1964. In contrast with rainfed cropland, it thereafter grew by 2.01% per year until 1976, then increased by 8.12% annually over the next 5 years, to peak at 1.67 million ha in 1981. Then it declined by 4.41% per year until 1988, after which it increased by a moderate 1.12% per year to reach an estimated 1.53 million ha in 2008. In contrast, the area of planted pastures increased throughout the whole period, growing at 2.43% per year for the 28 years since 1947–1948 to 0.81 million ha in 1975–1976. In response to the subsidized land conversion programme to convert marginal cropland to planted pastures, its annual growth then increased to 8.66% per annum until 1987–1988 and continued to grow by 2.37% per annum until 2005, levelling off at about 2.97 million ha.

The contrasting trends shown in Fig. 14.6 emphasize that market forces faced by farmers involved in the different agricultural sub-sectors faced different incentives in the use of cultivated land. Hence, different

rental values would have been attached to the use of the cultivated land in these sub-sectors, contrary to the imputed average based on the total area in farming. This presents an opportunity for further refinement of the analysis in future work.

14.5.3 Capital accounts

As mentioned previously, data on capital input use published in the *Abstract* is mostly representative of the commercial farming sector in South Africa. Figure 14.7a and b

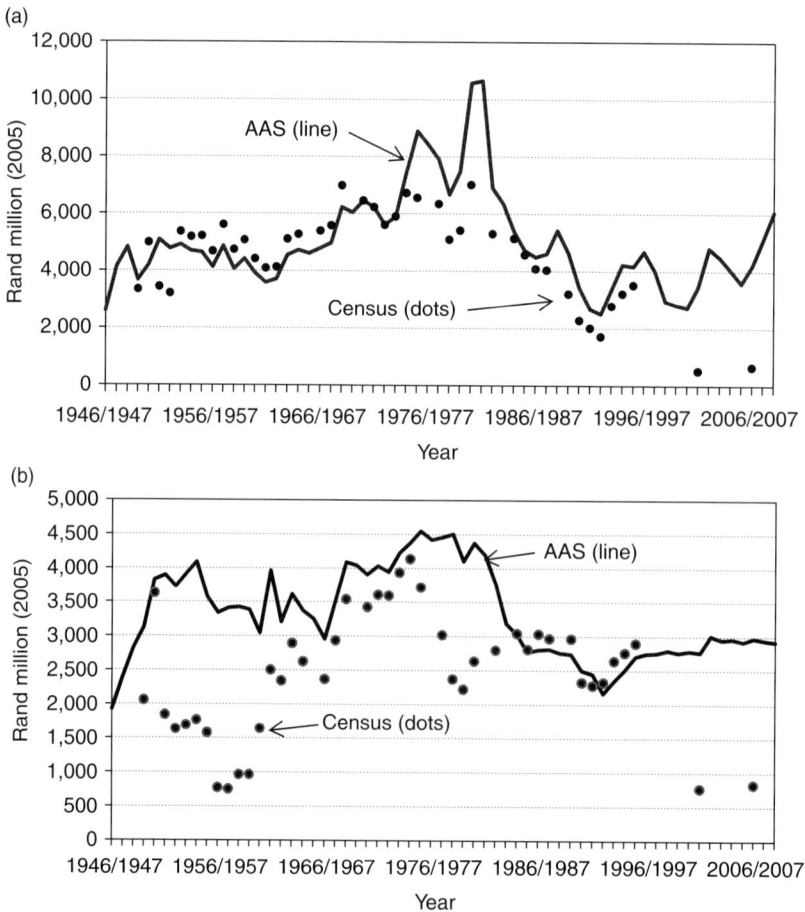

Fig. 14.7. Capital expenditures on farm machinery and fixed improvements reported by the *Agricultural Census* and *Abstract of Agricultural Statistics*. (a) Expenditures on farm machinery and equipment. (b) Expenditures on fixed improvements. Sources: Estimates from the Directorate of Agricultural Statistics (2011), Agricultural Census estimates from Statistics South Africa (2011).

compare the reported census statistics with the estimates reported in the *Abstract* on machinery and equipment (Fig. 14.7a) and fixed improvements (Fig. 14.7b). Up to 1975–1976, the data on capital formation of machinery published in the *Abstract* precisely matched the reported expenditure on machinery and equipment in the agricultural censuses since 1949–1950 when this was first enumerated in the agricultural census. Thereafter, estimates of capital formation on machinery and equipment reported in the *Abstract* (used as the basis to subsequently estimate the value of farm assets, interest and depreciation) began to deviate from estimates reported in the censuses. Information obtained from the Survey of Agricultural Machinery: Sales Statistics (Retail) (SAMS) about retail purchases of farm machinery and equipment became the preferred source from which the capital formation accounts were derived. The possible motivation for this change was the perceived greater accuracy of the SAMS survey.

The SAMS survey was conducted annually from 1968 to 1994 by the Directorate of Agricultural Engineering of the Department of Agriculture. In 1994, when the research arm of this directorate was transferred to the responsibility of the South African Agricultural Research Council, estimates on retail sales were sourced from the South African Agricultural Machinery Association through its AGFACTS information service (J. Rankin, Edenvale, Secretary SAAMA, 2010, personal communication). Both the value and sales volume of essentially all the categories of machinery and implements consistently exceeded estimates obtained through the agricultural censuses. In the case of tractors, the reported expenditure in the census varied between 36.0% and 80.5% of the SAMS survey from 1975–1976 through 1995–1996, the last year in which reliable data on tractor purchases could be obtained from the agricultural census. Data for the SAMS survey were collected from the suppliers of agricultural machinery and equipment and, though imports and re-exports were accounted for, the survey did not account for the possibility of sales to either the homelands or to non-agricultural equipment users,

such as municipalities, construction companies, etc. It is also uncertain whether, in accounting for trade, exports to Southern African Development Community member states were taken into account. (SADC is an inter-governmental organization that promotes economic integration among its 15 member states.)

Prior to 1962, the agricultural census estimates of capital did not include investments in farm residential property, although this was included in *Abstract* estimates. This explains the difference of up to 140% between *Abstract* and census statistics on capital investment for this period. From 1963, the capital formation of fixed assets reported in the *Abstract* and the census results show the same trend, as shown in Fig. 14.7b. Differences in magnitude result from adjustments made by the Directorate of Agricultural Statistics to account for investments in some kinds of fixed assets not included in the censuses, such as irrigation and electrical equipment. Inter-census-year estimates on capital formation in fixed improvements are indexed against the trend in the price index of building materials, with adjustments made in depreciation rates to again 'closer reflect' observations obtained from subsequent censuses. As the last comparatively reliable census was that of 1992–1993, the concern arises that current estimates of investment in fixed improvements reported in *Abstract* might have become unreliable. The author's own estimates of construction are derived from investment in new construction – a subcomponent of the investment in fixed improvements. The approach of using the price index of building materials as the basis to project past investments in fixed improvements might, in time, lead to an overestimate of current levels of investment in fixed improvements.

In calculating interest flows on capital inputs, all studies used a low rate of 2% per annum to reflect the 'considerable credit subsidies that were available to commercial farmers over most of the period' (Wiebe *et al.*, 2001). As shown in Fig. 14.2, the subsidy payments to commercial farmers in South Africa were completely phased out by the late 1990s. Currently, the producer support estimate for South African agriculture is on the order of 5% of farm receipts (Organisation for Economic Co-operation and Development, 2006), which is well below that of the USA. For consistency's sake, this study used the same interest rate, even though the motivation for doing so might no longer be entirely valid.

14.6 Conclusion

The rate of growth in agricultural output has slowed since the 1980s, largely as a result of a slowdown in the rate of growth in field crop production. Indeed, agricultural output growth in South Africa – and, for that matter, southern Africa – has in recent decades lagged behind the rest of Africa, even though the country's agricultural productivity growth has historically outpaced productivity growth elsewhere in the continent. This might be due to the stagnation of public support for agriculture, especially in R&D. The composition of agricultural outputs in South Africa has also changed, with higher-valued horticultural crops gaining market share at the expense of staple food crops and livestock products.

The composition of input use has changed, too. Notwithstanding high levels of rural unemployment, the evidence reported in this chapter indicates that South African agriculture has substantially increased its use of material inputs and continued to invest significantly in capital inputs, whereas the use of labour in agriculture has declined.

Over time, the country has made huge investments in agricultural R&D, and for the earlier years it even outperformed other ex-colonial developed countries on this score, such as the USA and Australia. Support to farmers through, initially, farmer settlement programmes, soil conservation and rural and market infrastructure development was particularly emphasized in past agricultural policy, as well. These earlier investments, however, seem to have been poorly correlated with improvements in productivity during the earlier phase of the development of the country's agricultural sector.

South African agriculture seems to have sustained a competitive edge during the decades prior to the late 1980s, with strong growth in agricultural exports and more muted, but still pronounced, growth in net agricultural trade surplus. Its agricultural exports and net trade balances have, however, declined precipitously in more recent years. These trade trends are loosely concordant with changes in the pattern of TFP growth for South Africa, which grew at much slower rates in more recent years compared with earlier decades.

More detailed analysis of the underlying data, however, reveals – in this study and the past studies it builds on – that measured productivity growth might have been influenced by inconsistent measurement of the role of homeland farmers in the national agricultural income accounts. This could have significantly offset and distorted the causal relationship that this and past studies have found between the research investments and productivity growth. This calls for a revisit of the empirical analysis of South African agricultural productivity growth. Future work should try to merge both sectors to establish a more complete picture of South African agricultural sector performance.

Notes

[1] The Union of South Africa was formed in 1910 out of the former British colonies of Cape Colony, Natal, and the former Boer republics of Transvaal and Orange Free State. In 1960 the country was renamed the Republic of South Africa.

[2] Agricultural GDP does not include GDP from the (processed) food sector. The combined output of the farm and agribusiness sectors – including food and fibre processors, distributors, and relevant parts of the beverage industries like wine and beer, all of which are reported in the national accounts as part of the manufacturing

sector – would almost double the sectoral share, such that the combined food and agricultural industries would constitute about one-third of total GDP.

[3] Here, and throughout this chapter, 'R' denotes rand, the local currency unit of South Africa.

[4] This fact is not spelled out in the explanatory notes to the data tables in the various issues of the *Abstract of Agricultural Statistics* since it first appeared in 1958.

[5] The classification of females in the homelands has changed considerably from earlier population censuses. In 1921 and 1936 all 'Bantu' women in the reserves (later to become the homelands) were classified as peasant farmers under the group 'Agriculture' (1.3 million and 1.6 million, respectively). In 1946, only those who were returned as peasants were classified as such (a total of 306,146), all others being classified as 'dependants under the not economically active group'. In 1951, only those who returned themselves as farmers were classified as peasants (9846), the remainder being classified as dependants (Bureau of Census and Statistics, 1960).

[6] According to the Directorate of Agricultural Statistics (2011), 13.7% of South Africa's total land area is potentially arable and around 69% of its total land area is only suitable for grazing. Moreover, a large share of the grazing area is in the semi-arid Karooveld that dominates much of the western half of the country.

[7] The Pass Laws Act of 1952 was part of a historical series of such Acts that, in its earliest incarnation in 1797, sought to exclude all 'natives' from the Cape Colony. The 1952 Act made it compulsory for all black South Africans over the age of 16 to carry a 'pass book' at all times. An employer was defined under the law and could only be a white person. The pass also documented permission requested – and denied or granted – to be in a certain region and the reason for seeking such permission. Under the terms of the law, any government employee could strike out such entries, basically cancelling permission to remain in the area.

[8] By way of comparison, in 2007 the USA spent $3.77 billion on public agricultural R&D, equivalent to $1.45 billion (2000 prices), more than it did in 1971 despite a slowdown in the average annual rate of growth during the 1970–2007 period, compared with the rate of growth during the previous 50 years (Alston *et al.*, 2010).

[9] This should not be confused with the estimates of the number people economically active in agriculture as reported in population census reports. This is typically higher as it includes employment in agriculturally related industries in, for instance, the food processing, forestry and game industries, which are not included in the definition of agriculture used by the Directorate of Agricultural Statistics.

References

Agricultural Research Council (2010) ARC Financial Database. Agricultural Research Council, Pretoria, South Africa. (Accessed 16 July 2010).

Alston, J., Babcock, B. and Pardey, P. (eds) (2010) *Shifting Patterns of Agricultural Production and Productivity Worldwide*. Midwest Agribusiness Trade and Research Information Center, Iowa State University, Ames, IA.

Binswanger, H., Deininger, K. and Feder, G. (1995) Power, distortions, revolt and reform in agricultural land relations. In: Behrman, J. and Srinivasan, T. (eds) *Handbook of Development Economics*, Vol. III. Elsevier, Amsterdam, The Netherlands.

Bureau of Census and Statistics (1960) *Union Statistics for Fifty Years: 1921–1960. Jubilee Issue*. Pretoria, Union of South Africa.

Conradie, B., Piesse, J. and Thirtle, C. (2009a) District-level total factor productivity in agriculture: Western Cape Province, South Africa, 1952–2002. *Agricultural Economics* 40, 265–280.

Conradie, B., Piesse, J. and Thirtle, C. (2009b) What is the appropriate level of aggregation for productivity indices? Comparing district, regional and national measures. *Agrekon* 48, 1, 9–20.

Department of Finance (1911–1959 annuals). *Estimates of Expenditure to be Defrayed from the National Income Account*. Department of Finance, Pretoria, Union of South Africa.

Directorate of Agricultural Statistics (annual issues) *Abstract of Agricultural Statistics*. Department of Agriculture, Pretoria, Republic of South Africa.

FAO. FAOSTAT Database. Food and Agriculture Organization of the United Nations. Rome, Italy. Available at: http://faostat.fao.org/ (Accessed 13 August 2009).

Flaherty, K., Liebenberg, F. and Kirsten, J. (2010) South Africa. ASTI Country Note. International Food Policy Research Institute, Washington, DC.

Giliomee, H. (2009) *The Afrikaners: Biography of a People*. 2nd edn. Tafelberg, Cape Town, South Africa.

Kahn, M.W., Blankley, R., Maharajh, T.E., Pogue, V., Reddy, Cele, G. and du Toit, M. (2004) *Flight of the Flamingos: A Study on the Mobility of R&D Workers*. HSRC Press, Cape Town, South Africa.

Liebenberg, F. (2011) South African agricultural production, productivity and research performance in the 20th century. PhD Dissertation, University of Pretoria, South Africa.

Liebenberg, F. and Pardey, P. (2010) South African agricultural production and productivity patterns. In: Alston, J., Babcock, B. and Pardey, P. (eds) *The Shifting Patterns of Agricultural Production and Productivity Worldwide*. Midwest Agribusiness Trade and Research Information Center, Iowa State University, Ames, IA.

Liebenberg, F., Pardey, P. and Khan, M. (2010) South African agricultural research and development: A century of change. Staff Paper P10-1. Department of Applied Economics, University of Minnesota, St. Paul, MN.

Organisation for Economic Co-operation and Development (2006) OECD Review of Agricultural Policies: South Africa. OECD Publishing, Paris, France.

Pardey, P., Beintema, N., Dehmer, S. and Wood, S. (2006) Agricultural research – A growing global divide? Food Policy Report. International Food Policy Research Institute, Washington, DC.

Payne, N., van Zyl, J. and Sartorius von Bach, H. (1990) Labour-related structural trends in South African commercial grain production: A comparison between Summer and Winter rainfall areas, 1945–87. *Agrekon* 29, 4, 407–415.

Republic of South Africa (1960–2009 annuals) *Estimates of National Expenditure*. Department of Finance, Pretoria, South Africa.

Schimmelpfennig, D., Thirtle, C., van Zyl, J., Arnade, C. and Khatri, Y. (2000) Short and long-run returns to agricultural R&D in South Africa, or will the real rate of return please stand up? *Agricultural Economics* 23, 1, 1–15.

South African Reserve Bank (2009) Long run Gross Domestic Product (GDP) and Agricultural Contribution to GDP Statistics. Unpublished data provided upon request. South African Reserve Bank, Pretoria, South Africa.

South African Reserve Bank (2011) Long run Gross Domestic Product (GDP) and Agricultural Contribution to GDP Statistics. Unpublished data provided upon request. South African Reserve Bank, Pretoria, South Africa.

Statistics South Africa (2011) Agricultural Census and Survey Reports obtained from Statistics South Africa. Pretoria, South Africa.

Thirtle, C., Sartorius von Bach, H. and van Zyl, J. (1993) Total factor productivity in South Africa, 1947–1991. *Development Southern Africa* 10, 3, 301–318.

Union of South Africa (1911-1959 annuals) *Estimates of Expenditure to be Defrayed from the National Income Account*. Department of Finance, Pretoria, South Africa.

Wiebe. K., Schimmelpfennig, D. and Soule, M. (2001) Agricultural policy, investment and productivity in sub-Saharan Africa: A comparison of commercial and smallholder sectors in Zimbabwe and South Africa. In: Zepeda, L. (ed.) *Agricultural Investment and Productivity in Developing Countries*. FAO Economic and Social Development Paper 148. Rome, Italy.

15 Measures of Fixed Capital in Agriculture*

Rita Butzer,[1] Yair Mundlak[2] and Donald F. Larson[3]
[1]*University of Chicago;* [2]*Hebrew University of Jerusalem;*
[3]*Development Research Group, World Bank*

15.1 Introduction

Data on sectoral investment and capital stocks are essential for empirical research in sectoral productivity, yet cross-country panels are rare for countries outside of the Organisation for Economic Co-operation and Development (OECD). Crego *et al.* (1998) introduce a database on the capital stock in the agricultural and manufacturing sectors for 57 developed and developing countries for the years 1967–1992. We have updated the agricultural component of this database to the year 2000 for a subset of 30 countries. We construct three capital sub-components series: treestock, livestock and fixed capital in agriculture. We modify a commonly used methodology for integrating investment to obtain the fixed capital stock. This methodology can also be used to compute comparable fixed capital stocks for other sectors and for the economy as a whole, in order to facilitate comparative analyses. An analysis of capital stocks in manufacturing and agriculture,

as well as for the economy as a whole for 60 countries for 1967–1992 can be found in Larson *et al.* (2000). Butzer (2011) updates the data series on capital stocks in manufacturing and for the whole economy to 2000 for the subset of 30 countries.

After describing selective characteristics of the updated agricultural capital data set, including the evolution of aggregate capital stocks over time and the changing composition of agricultural capital, we revisit an earlier analysis of agricultural productivity that utilizes our data set on agricultural capital. We then show how the frequently used practice of employing farm machinery as a proxy for agricultural fixed capital leads to substantially different results.

15.2 Background

Measures of agricultural capital are fundamental to two important and related fields of empirical study. The first concerns the

* This chapter relies heavily on two earlier papers. Larson *et al.* (2000) was funded by the World Bank's Research Support Budget (RPO 680–50). Mundlak *et al.* (2008) was funded by the World Bank's Research Support Budget under the research project 'The Contributions of Governance to Growth in Agriculture' (RPO 94759). We benefitted greatly from the input of participants at the workshop, *Causes and Consequences of Global Agricultural Productivity,* co-sponsored by Farm Foundation, NFP and the USDA's Economic Research Service. Keith Fuglie graciously shared data for the analysis. We appreciate Isabel Tejedo's research assistance with variable construction. The authors would also like to thank Polly Means for greatly improving the presented figures and Keith Fuglie and an anonymous reviewer for helpful comments on an earlier draft.

determinants of agricultural productivity and growth; the second concerns the structural transformation of developed and developing countries. Both areas of study are important for understanding economic development, because agriculture remains the largest source of employment in many poor countries today. For example, the World Bank's *World Development Report* (2008) on agriculture estimates that 75% of the world's poor live in rural areas and that most poor rural households depend on agriculture as their primary source of income. The report classifies 21 developing countries as agricultural and an additional 19 as transitional.

In general, as economies grow, labour flows from agriculture to other sectors. As a consequence, the shares of labour and gross domestic product (GDP) in the agricultural sector decline as part of this process; however in economies that perform well, the sector continues to grow through a build-up of physical and human capital and through the adoption of more productive technologies (Larson and Mundlak, 1997; Mundlak *et al.*, 1998; Mundlak, 2001).

Measured stocks of capital are required to characterize the transformation process and to distinguish growth that is due to the accumulation of input factors from that due to changes in factor productivity. Moreover, in order to compare this transformation process among countries, a compatible measure is needed across countries and through time. In general, this has been lacking, especially for panels that include developing countries.[1] By necessity, therefore, empirical studies have looked to proxies for agricultural capital, most often measures related to on-farm machines, livestock and orchards, or have omitted capital entirely.[2] As argued below, the composition of capital varies as the prevalent technologies change as part of the structural transformation of economies. Consequently, the roles played by on-farm machines or other proxies for a general measure of capital change. In general, machinery capital becomes less relevant as the stock of capital in agriculture grows; this creates problems for studies of agricultural productivity and growth.

Mundlak *et al.* (1999) and Coelli and Rao (2005) list a number of early cross-country studies of agricultural productivity, including important studies by Bhattacharjee (1955), Hayami

and Ruttan (1970, 1971), Evenson and Kislev (1975), Nguyen (1979), Mundlak and Hellinghausen (1982), Kawagoe and Hayami (1983, 1985), Antle (1983), and Lau and Yotopoulos (1989). Varied capital measures, based on combinations of farm machinery, tractors, livestock, orchards, and, less frequently, data on irrigation and farm structures, were used. Recent econometric studies are less uniform in their methodology, but many still include capital proxies (most often in the form of livestock and machinery) in estimated production functions. Examples include Fulginiti and Perrin (1993, 1998), Block (1994), Craig *et al.* (1997), Wiebe *et al.* (2003), Cermeño *et al.* (2003), Lio and Liu (2008), and Cermeño and Vázquez (2009).

Increasingly, agricultural productivity studies have also drawn on non-parametric or semi-parametric approaches. These are related to index number theory, and frequently employ data envelope analysis (DEA) programming methods to estimate either input or output distance functions from which productivity measures are defined. In the context of country panels, these studies most often entail decomposing changes in total factor productivity, via a Malmquist index fitted to measured inputs and outputs (see Färe *et al.*, 1994, for an early application). In some instances, a second-stage regression is used to explain variations in calculated productivity in terms of a set of determining state variables – for example, past research expenditures or the strength of political institutions.

Still, although the approaches applied to the problem of estimated agricultural productivity have become more varied, the underlying measures have not, and the alternative methods often build on the same data and the same proxies for capital as earlier studies. See, for example, Lusigi and Thirtle (1997), Arnade (1998), Coelli and Rao (1995), Fleming (2005), Fuglie (2008), Alene (2010), Headey *et al.* (2010), Nin-Pratt and Yu (2010) and Fulginiti (2010).

15.3 Measuring Agricultural Capital Stock

The series of agricultural capital stock presented here consists of three components: fixed capital, livestock and treestock. In this sense, we distinguish between stores of capital that arise through investments and privately held productive natural endowments, including land and water resources. National accounts report fixed capital investment, which does not wholly include livestock and treestock; we therefore compute each component separately. Data sources along with the computer program used to calculate the capital series are documented in Crego *et al.* (1998).[3]

We construct the fixed-capital series on the basis of national account investment data, using a modification of the perpetual inventory method. The method requires integration of the investment data to obtain capital stocks. For livestock, the initial data are the number of animals. We need only calculate the values of the individual herds and then aggregate these values to obtain the total for the full stock of animals. For treestock, we use the estimated present value of future income derived from the area planted in orchards.

15.3.1 Fixed capital

The data series on fixed capital is based upon national accounts data on fixed investment, which includes structures, equipment, machinery, and so on. The mapping from fixed investment to fixed capital follows the methodology of Ball *et al.* (1993). The capital stock is represented as a weighted sum of past investments, where the sequence of relative efficiencies of capital of different ages serves as the weights. The following function of physical depreciation is chosen to describe the relationship between the efficiency of an asset and its age. Let L be the lifetime of the capital good, and β be a curvature parameter bounded from above by one in order to restrict productivity to be nonnegative. Thus (Eqn 15.1):

$$s_j = (L - j)/(L - \beta j), \quad 0 \le j \le L$$
$$s_j = 0, \quad j \ge L \qquad (15.1)$$

The asset is discarded at age L, at which time its relative productivity becomes zero.

To analyse this expression, we note that $ds_j/dj = L(\beta - 1)/(L - \beta j)^2 < 0$, for $0 \le j < L$, indicating that the productivity falls with age (use). The speed of the change in the depreciation with age depends on the sign of curvature parameter, β:

$$d^2 s_j /dj^2 = 2L\beta(\beta - 1)/(L - \beta j)^3 \qquad (15.2)$$

When β is positive but less than unity $d^2 s_j/dj^2 < 0$, the physical depreciation accelerates with time (use), and the productivity curve is concave. Conversely, when β is negative, the productivity curve is convex. The lifetime of the asset is taken as a random variable with a normal distribution truncated at two standard deviations on both sides. Figure 15.1 illustrates the dependence of the productivity paths of buildings and agricultural machinery on the parameters in question.[4] The data sources used here do not provide any information on the separate components of fixed capital, so we choose a single set of parameters.[5] Thus, the capital stock in any given year is the sum of the relative efficiency for that year of all past investments.

Data on the value of gross fixed capital formation in agriculture (in local prices) were obtained from national accounts data. If the data series on investment is not sufficiently long, information on the initial capital stock is also needed to construct the capital series. This information is rarely available. Researchers use several different techniques for fixing, or seeding, the initial value. They are often forced to choose among competing seeding techniques on the basis of criteria such as whether the methods generate negative initial values (e.g. Nehru and Dhareshwar, 1993). This choice would not be necessary if the investment series were sufficiently long (large T) because the productivity of old capital goods is low and their contribution to the current stock is small. Therefore, our approach is to generate lengthier investment time series when they do not exist. We do this by regressing the logarithm of the investment–output ratio on time for the study period. We then use this regression to estimate past values of the investment–output ratio and

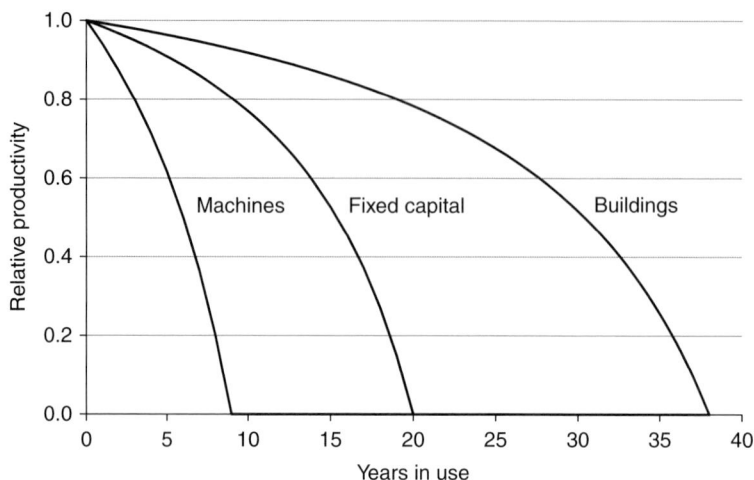

Fig. 15.1. Examples of relative productivity paths.

apply them to the published output data to generate the needed missing investment values. If the output values are not available, we can estimate them from a regression of output on time.[6]

The data series on fixed-capital stock are constructed in constant local prices. To facilitate cross-country comparisons, the series are converted to a common currency (US dollars) by dividing by the real exchange rate, e_t, defined as the nominal exchange rate, E_t, multiplied by the relative price of the US dollar deflator to the domestic deflator, P^{US}/P.[7]

15.3.2 Livestock capital

A considerable amount of agricultural capital is embodied in livestock, which are used to perform various roles – such as breeding and draft power. According to the United Nations' (UN) accounting practices, animals that are not used for slaughter are included as fixed capital investments.[8] After closely examining the data, we find, however, that this is not the case for many countries. Furthermore, changes in livestock used for slaughter are included as 'increases in stocks', not as fixed capital. Thus we construct a separate data series on livestock capital, recognizing that there might be some overlap,

but also that livestock accounts for a considerable share of agricultural capital and should not be ignored. We return to this issue in Appendix 15.1.

Conceptually, the calculation of the livestock is fairly straightforward. The Food and Agriculture Organization (FAO) reports the quantities of all farm animals – cattle, sheep, pigs, poultry, and so on. We aggregate the value of these individual components to obtain the livestock. Ideally, we would use market prices of live animals to value local herds, but these data were not consistently available. In their place, we use regional export unit values, based on FAO trade data, to value domestic herds. We calculate separate prices for each region by dividing regional dollar export values by regional export quantities. These unit prices are then applied to national herd statistics for each category of livestock. We convert the aggregate values to constant (1990) dollars by using the US GDP deflator for the agricultural sector.

15.3.3 Treestock

Standing orchards, plantations and smallholder trees represent another important category of investment in agriculture. According to the UN accounting practices, the value of investments in treestock should be included along with other land improvements in national accounting systems. A close examination of country data suggests, however, that in practice such stocks might go unaccounted (see Appendix 15.1). We therefore construct a direct estimate of the value of treestock.

The available information for constructing the treestock consists of FAO data on the area harvested by crops, production and output prices. We begin by using the condition for long-run equilibrium, in which the cost of investment in the orchard equals the present value of the expected future income generated by the orchard. The income from the orchard is value of output less the costs of production. There are no published data on production costs; thus we construct our estimates under the assumption that production costs account for 80% of gross revenues.

We derive the yield (output per hectare) from the data on output and area. The yield depends on the age of the trees, but the necessary information for estimating the yield curve is not available. We therefore calculate the present value of the orchard under the assumption that the orchards are halfway through their assumed lifetime.[9] For the expected price, we calculate a 5-year moving average of actual domestic producer prices, converted to nominal dollars, centred on the current year (two periods forward and two periods lagged). The income per hectare is imputed forward in time (with discounting) for each crop and then aggregated. We use the US interest and inflation rates to calculate a real interest rate, which serves as the discounting factor. Finally, we convert the result to constant (1990) dollars by applying the US agricultural GDP deflator.

15.3.4 Data[10]

The economies included in the agricultural investment data set are shown in Table 15.1, as categorized by income level.[11] For most countries included in the data set on agricultural capital stocks, the series on the sectoral

Table 15.1. Income classification of sample countries.

Income category	Number of countries	Countries
Low income	6	India, Indonesia, Kenya, Malawi, Pakistan, Tanzania
Lower middle income	6	Egypt, Morocco, Peru, Philippines, Sri Lanka, Tunisia
Upper middle income	4	Republic of Korea, Mauritius, Turkey, Uruguay
High income	14	Australia, Austria, Canada, Cyprus, Denmark, Finland, France, Greece, Italy, Netherlands, Norway, Sweden, UK, USA
Total	30	

breakdown of fixed investment begin in the 1960s, thus sufficiently complete to estimate capital stocks beginning in 1967. All series on livestock and treestock begin in 1961. Looking forward, there is potential for the expansion of the data set on agricultural fixed capital to include more countries, but with a later starting date for the series.[12]

15.4 Evolution of Capital Stocks

Having a data series on agricultural capital that is comparable across countries and over time allows us to examine the evolution of capital stocks and gain insight on the process of agricultural growth. In Figs 15.2 and 15.3, we present the frequency distribution of growth rates of the capital components for the 30 countries in our sample.[13] Capital accumulation in agriculture since 1970 has been positive in all of the countries (Fig. 15.2). The median growth rate of agricultural capital is 5.4%. There is a similar pattern for each of the components of agricultural capital. Fixed capital and treestock have median growth rates of 5.9% and

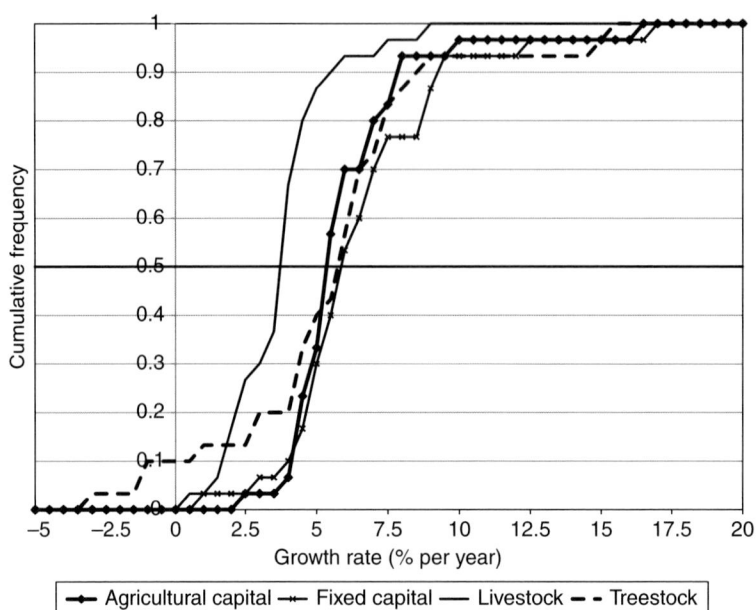

Fig. 15.2. Growth of capital stocks, 1970–2000. The graph shows the frequency distribution of the growth rates in agricultural capital stocks. Growth in all components of agricultural capital was positive except for a few cases where treestock capital declined. The median growth rate of fixed capital was the highest at 5.9%, followed by treestock at 5.7%, whereas the median growth rate of livestock capital was the slowest at 3.6%. Source: Authors' estimates.

5.7%, respectively. Although livestock grew in all countries, the growth was slower than for the other two components (median of 3.6%, with the distribution lying almost completely to the left of the others). This changing composition of agricultural capital is even more apparent in Fig. 15.3. In more than 90% of the countries in our sample, the share of livestock in total agricultural capital fell over the period, with treestock and fixed capital becoming increasingly important inputs to agricultural production, depending upon the geographical constraints faced by the country.

We can gain insight on the dynamics of growth in agriculture by examining the changing composition of agricultural capital over time. The patterns of capital accumulation for each component vary over time. Table 15.2 presents the average annual growth rates for the sample for each decade. Fixed capital grew at a rapid pace of 8.1% in the 1970s. The slowdown in fixed investment

began in the 1980s and continued into the 1990s, although fixed capital was still growing by 3.6%. Livestock grew at a similar pace in the 1970s and 1980s (3.9% and 4.0%, respectively), but then experienced slower growth in the 1990s. Treestock achieved a growth rate for the entire period of slightly less than that of fixed capital, but with a very different pattern over the decades. Treestock grew more slowly than fixed capital and livestock in the 1970s (3.2%) but achieved nearly double-digit growth rates in the 1980s. Although the growth slowed in the 1990s, treestock continued to grow faster than fixed capital and livestock. Thus the decreasing share of livestock in total agricultural capital was due in part to the rapid growth in treestock in the 1980s and 1990s, as well as to the slowdown in the accumulation of animal stocks in the 1990s.

In addition to the time dimension, there is much cross-country variability in agricultural

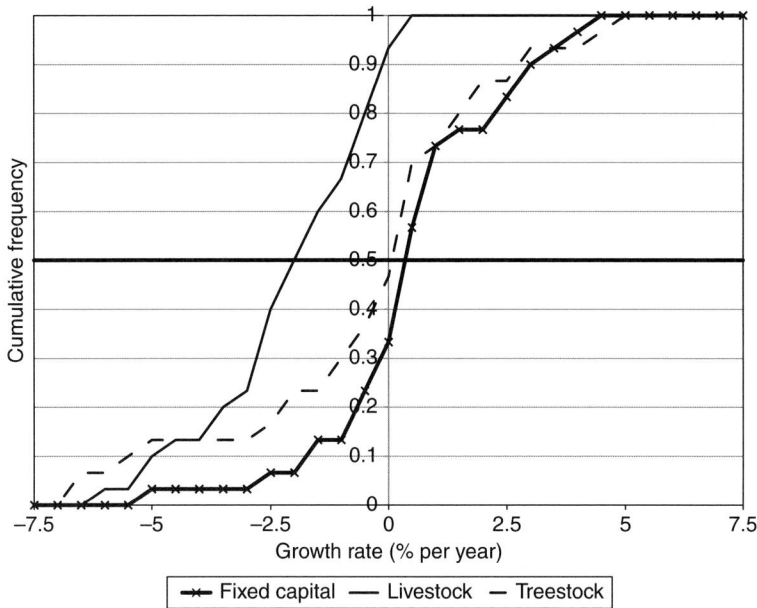

Fig. 15.3. Growth in component shares of agricultural capital, 1970–2000. The graph shows the frequency distribution of the growth in the component shares of agricultural capital stock. Livestock capital had a declining share of total capital stock (negative median growth rate), whereas the component shares of treestock and especially fixed capital increased. Source: Authors' estimates.

Table 15.2. Growth in agricultural capital over time.

	Full sample	Decades			Country classification	
	1970–2000	1970s	1980s	1990s	High income	Middle and low income
	Average annual growth rate (%)					
Capital component						
Fixed capital (Structures and equipment)	6.39	8.05	4.81	3.75	6.26	6.51
Livestock	3.60	3.93	4.02	2.21	3.34	3.83
Treestock	5.37	3.18	9.93	5.89	4.01	6.57
Tractors	3.45	5.84	3.25	1.59	0.84	5.79

capital stocks. For purposes of comparison, we divide the sample into high-income countries versus middle- and low-income countries. Fixed capital grew slightly faster in middle- and low-income countries (6.5% versus 6.3%), whereas the growth of treestock was considerably faster in middle- and low-income countries (6.6% versus 4.0%).

The growth rate of livestock was only slightly higher in middle- and low-income countries (3.8% versus 3.3%).

These growth patterns led to interesting changes in the composition of capital for each income group as shown in Table 15.3. Almost half of agricultural capital in high-income countries was composed of fixed

Table 15.3. Component shares in total agricultural capital.

	Weighted				Unweighted			
	1970	1980	1990	2000	1970	1980	1990	2000
High-income countries								
Fixed capital	0.46	0.55	0.54	0.46	0.59	0.68	0.70	0.68
Livestock	0.21	0.19	0.11	0.09	0.20	0.15	0.12	0.11
Treestock	0.32	0.26	0.35	0.45	0.21	0.17	0.18	0.20
Middle- and low-income countries								
Fixed capital	0.27	0.32	0.35	0.36	0.27	0.32	0.33	0.31
Livestock	0.28	0.32	0.28	0.17	0.31	0.29	0.27	0.21
Treestock	0.45	0.36	0.37	0.47	0.42	0.38	0.40	0.48

capital throughout the study period. In 1970, a third was in treestock, and 20% was in livestock. By 2000, the share of treestock was nearly as high as the share of fixed capital, whereas livestock accounted for less than 10%. In middle- and low-income countries, 45% of agricultural capital was in treestock in 1970, whereas fixed capital and livestock had nearly equal shares. The share of fixed capital rose to over a third, whereas the share of livestock fell to 17%. Looking at the unweighted averages of the country shares on the right-hand side of Table 15.3, the differences between the two subsets of countries are even more pronounced.

15.5 Comparison of Data on Agricultural Fixed Capital with FAO Data on Tractors

The new database on agricultural fixed capital has several advantages, in particular the sound theoretical basis for the construction of the series, as well as the broad coverage of fixed capital to include structures and all types of equipment and machinery. From the UN System of National Accounts (2008), fixed assets are defined as 'produced assets that are used repeatedly or continuously in production processes for more than one year'. Tractors represent one component of fixed capital.

The obvious drawback to the new database is the limited availability of national accounts data on gross fixed-capital formation needed to construct the stock of fixed capital.[14] In the absence of cross-country data sets on agricultural fixed capital, the most commonly used proxy is FAO data on the numbers of tractors in use. This is a partial measure, in that it does not include buildings, irrigation systems, local infrastructure, as well as other types of machinery, nor does it take into account the varying quality and horsepower of tractors.[15] We compute the correlation between the FAO data series on the numbers of tractors in use and our data on agricultural fixed capital (of which tractors are one component) for each country; the average of the country correlations is 0.49. The median correlation is 0.84, so for half of the countries in our sample, the data series are somewhat correlated. Only 30% of the sample have correlations greater than 0.9, whereas the correlation was actually negative for six countries.[16] Figure 15.4 shows a plot of the data on fixed capital in agriculture against the FAO data on tractors.[17] The prevalence of vertical clusters suggests that the data on tractors were more stagnant than the data on the value of agricultural fixed capital.

Looking at Table 15.2, we see from the decade growth rates that the data series for tractors show the same pattern of slowing growth as fixed capital. Although the growth rates in fixed capital are similar across income groups, there are, however, remarkable differences between the growth rates of tractors. In high-income countries,

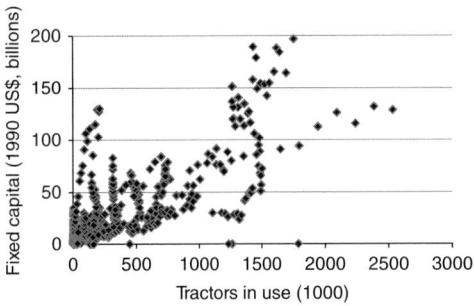

Fig. 15.4. Tractors in use versus agricultural fixed capital, 1967–2003. Source: tractors in use from FAO; fixed capital stock in constant 1990 US dollars from authors' estimates.

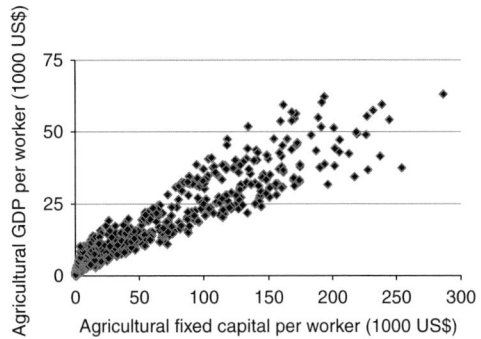

Fig. 15.5. GDP per worker versus fixed capital per worker in agriculture, 1967–2003. Source: Agricultural GDP in constant 1990 US dollars from the World Bank; number of agricultural workers from FAO; agricultural fixed capital in constant 1990 US dollars from authors' estimates.

there was very little growth in tractors, whereas fixed-capital stocks grew at more than 6%. It is clear from these comparisons that tractors are not a convincing proxy for agricultural fixed capital and, by inference, of total agricultural capital.

We compare the data on agricultural capital to other economic variables to obtain a sense of its relevancy for economic growth. Figure 15.5 shows the relationship between average labour productivity (output–labour ratio) and the fixed capital–labour ratio for agriculture.[18] This scatter diagram traces the production function in terms of capital intensity, without allowing for the effects of other pertinent variables. Plotting tractors per worker against agricultural GDP per worker (Fig. 15.6) shows a weaker and much more volatile relationship, however.

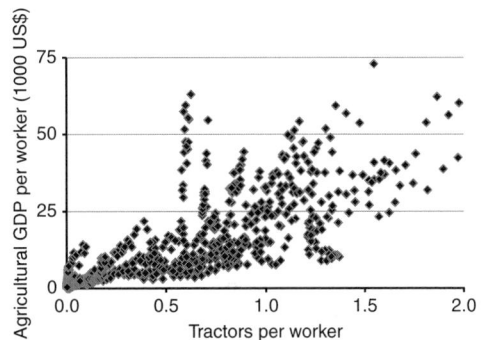

Fig. 15.6. Agricultural GDP per worker versus tractors per worker, 1967–2003. Source: agricultural GDP in constant 1990 US dollars from the World Bank; number of agricultural workers and tractors in use from FAO.

15.6 Agricultural Production Function

Mundlak *et al.* (1999) first utilized the panel data set on agricultural capital in an analysis of agricultural productivity for 37 countries for the time period 1970–1990. They presented a model of production under heterogeneous technology, where the implemented technology is chosen jointly with the level of inputs. The empirical formulation allows for the dependence of parameters of the function, as well as the inputs, on a set of state variables, where state variables are defined to be a set of exogenous variables that characterize the environment in which the production decisions are made. Examples of state variables might include culture, geography, institutions and market integration.

Mundlak *et al.* (2011) provide an in-depth discussion of panel data analysis, and in particular the dominance of the within estimator, utilizing an updated version of the data set on agricultural capital stock to extend the analysis to 2000.[19] The authors draw comparisons with the between estimators as well, noting that, although the between estimators are biased, they still relay valuable information for understanding the underlying processes of growth. The authors explore the role of the country and time effects in accounting for the variability in productivity, and the ability of the state variables to capture some of these effects.

Using this prior work as a staging ground, we show the consequences of using an incomplete measure of capital on the estimates of production elasticities by replacing our measure of agricultural fixed capital with the FAO data on tractors. Before doing so though, some background information on the underlying approach is required.

15.6.1   Theory

We estimate agricultural production, y, as a function of the inputs, x, and state variables, s, for a panel of countries, i, over time, t. The inputs and output, as well as the implemented technology, are determined by maximizing the expected profit conditional on the state variables. The solution of the system yields the production where output and inputs are expressed in logs (Eqn 15.3): [20]

$$y_{it} = x_{it}\beta(s) + s_{it}\gamma + m_{oit} + u_{oit} \qquad (15.3)$$

where $u_{oit} \sim$ IID $(0, \sigma_{00})$, $u_{jit} \sim$ IID $(0, \sigma_{jj})$, $E(u_{oit}u_{jit}) = 0$, $E(u_{oit}x_{jit}) = E(u_{oit}s_{jit}) = 0$ for all j,i,t,t'. Output and input prices are included in the set of state variables. Note that state variables not included as regressors add to the equation disturbance, and because they also affect the inputs they constitute a source for simultaneity bias. With panel data, the term m_{oit} is decomposed to a country effect and a time effect $m_{oit} = m_{oi} + m_{ot}$.

The effect of the state variables is assumed to be linear; they serve as shifters in the production function. We replace $\beta(s)$ with $\beta(S)$ where S is the set of s values in the sample, $S = \{s_{it}: i = 1,...n; t = 1,...,T\}$. Then $\beta(S)$ is the estimated constant associated with S. As the set S does not converge, the estimates obtained for $\beta(S)$ with different samples are not repeated estimates of an overall population constant β. In fact,

the underlying premise is that such a constant does not exist. We refer to the technology represented by $\beta(S)$ as the core technology associated with S.

15.6.2   Data

Output and inputs

Output is measured as agricultural GDP in 1990 US dollars.[21] Inputs to agricultural production include land, capital, labour and fertilizers. The construction of the agricultural capital data has already been described in depth. Hectares of agricultural area are used for the measure of land. This includes arable land, land under permanent crops and permanent pastures. Agricultural labour is defined as the economically active population in agriculture. Fertilizer consumption is measured in metric tonnes. It is often viewed as a proxy for the whole range of chemical inputs.[22]

Note that the dependent variable is value added; in a competitive economy the elasticity of fertilizer, the cost of which is allowed for in the computation of value added, should be nearly zero (an outcome of the envelope theorem). Whether the fertilizer variable is relevant is determined empirically. The inference is, therefore, that when the elasticity is different from zero, there is a gap between the official price and that used by producers.

State variables[23]

State variables are meant to represent the economic environment related to the choice of the implemented techniques. Because the motivation for including these variables has already been discussed in full in the previous studies, we review them only briefly here.

Technology

Three variables are included to account for differences in technology. The most commonly used variable to represent the level of technology in a country is a measure of human capital, which can be proxied by schooling years of the labour force (aged 15+).[24] The peak yield variable represents a measure of the frontier of implemented technologies. It is measured by country-specific Paasche indices, constructed from peak commodity yields weighted by land areas. The development indicator variable measures the overall level of development in a country, which would incorporate the effects of various public goods, such as infrastructure, public health, research and institutions. It is measured as the ratio of the country per capita output (GDP) to that of the USA. The hypothesis is that higher levels of technology would have a positive effect on agricultural productivity.

Institutions

We use two variables obtained from Freedom House to measure the influence of institutions on agricultural productivity. The measure of political rights reflects the electoral process, political pluralism and participation, and functioning of the government. The civil liberties measure includes aspects of freedom of expression and belief, associational and organizational rights, rule of law, and personal autonomy and individual rights. The hypothesis is that the physical, legal, and regulatory infrastructure and institutions support overall development, including agricultural; however, it is possible that these effects might already be captured by the development indicator variable.

Incentives

Three measures of incentives are included in the estimation. The effects of these variables would be over and above their indirect effect through resource allocation. The mechanism for the direct effect is through the choice of technique. Relative prices reflect the terms of trade between the agricultural sector and the overall economy. Price variability, calculated as a moving standard deviation of the relative prices from the three previous periods, is a measure of the market risk faced by agricultural producers. In addition to the sector-specific risk, there is an economy-wide market risk, that of price volatility for the economy as a whole, measured by the rate of inflation. Inflation may also serve as an indicator of overall macroeconomic instability, which is thought to inhibit growth.

Physical environment

Agricultural production depends on the physical environment or natural conditions. We represent the environment by using two variables: potential dry matter (PDM) and a factor of water availability (FWA). (The measures are based on Buringh et al., 1979, and were used in Mundlak and Hellinghausen, 1982, and Binswanger et al., 1987.) The first variable is intended to measure the theoretical potential production of dry matter. The production of dry matter requires moisture. Arid areas might have a large value for PDM, but actual production is small owing to water deficit. The relative water availability is measured by the ratio of actual transpiration to potential transpiration. These two variables are country specific and do not vary with time.

15.6.3   Sample description

The sample was determined by data availability. It consists of annual data from 30 countries for a 29-year period (1972–2000). The information conveyed by the sample is summarized in Table 15.4. The first column shows the average annual growth rates of the variables for the whole sample period. Agricultural output grew at a rate of 5.3%. All three components of agricultural capital grew faster than the other inputs. Agricultural fixed capital grew fastest, at a rate of 6.4%. Treestock grew at a rate of 5.8%. Livestock also grew during this period, but at a slower rate of 3.6%. Led by increasing rates of off-farm migration, agricultural labour declined at the average rate of 0.6%. Fertilizer grew on average at the rate of 1.9%. There was little change in the agricultural area. The terms

Table 15.4. Growth rates and the decomposition of the sum of squares for production function variables.

	Average annual growth rate	Decomposition of the sum of squares		
		Between time	Between country	Within time-country
	(%)	(Percentage of total)		
Output				
GDP	5.29	8.68	89.77	1.55
Inputs				
Structures and equipment	6.40	8.06	89.90	2.04
Capital of agricultural origin	4.94	5.50	91.94	2.55
Livestock	3.59	3.09	95.72	1.18
Treestock	5.77	4.13	93.35	2.53
Tractors	3.26	1.51	94.12	4.36
Agricultural area	0.01	0.00	99.93	0.07
Labour	−0.60	0.07	99.03	0.90
Fertilizer	1.87	1.01	96.47	2.53
Technology				
Schooling	1.67	7.32	88.05	4.63
Peak yield	1.41	79.37	6.30	14.33
Development indicator		1.10	95.46	3.44
Institutions				
Civil liberties		1.57	79.28	19.16
Political rights		1.17	82.67	16.16
Prices				
Relative prices	−1.39	18.55	32.54	48.91
Price variability		5.20	18.56	76.25
Inflation		3.12	8.50	88.37

Estimates are derived from a panel of 30 countries during 1972–2000. Source: Mundlak *et al.* (2011).

of trade of agriculture declined at the average rate of 1.4%. The technology measures show a growth rate of schooling of 1.7% and 1.4% for peak yield.

The data are subject to much variability over time and across countries. To describe the variability further, we decompose the total sum of squares to the three orthogonal components: within-country-time, SSW(it), between-country, SSB(i), and between-time, SSB(t). Let $x_{..}$, $x_{i.}$ and $x_{.t}$ denote the overall mean, the country mean, and the time mean, respectively. Thus, SS total $= SS(x_{it} - x_{..})$ is decomposed to SSW(it) $= SS(x_{it} - x_{i.} - x_{.t} + x_{..})$, SSB(i) $= SS(x_{i.} - x_{..})$, and SSB(t) $= SS(x_{.t} - x_{..})$, where, for any variable z, we use the notation: $SS(z) = \sum_i \sum_t z_{it}^2$.

The last three columns of Table 15.4 show the percentage of each component in the total sum of squares. The between-country differences account for most of the variability in output (nearly 90%), with the between-time variability accounting for almost another 9%. Thus, a regression that allows for country and time effects, without any quantitative variables, would yield an R^2 of 0.985, so that the unexplained residual accounts for less than 2% of the total sum of squares of output. The between-country variability accounts for an even higher percentage of the total variability in inputs.

Looking at the state variables reveals that schooling, development and the institution measures are dominated by between-country variability. The peak yield measure varies primarily over time, reflecting the changes in the implemented technology. Price variability and inflation have a considerable amount of within variability, reflecting more transitory effects.

15.7 Empirical Results

The importance of capital, in particular fixed capital, was shown in Mundlak *et al.* (1999, 2011), the results of which are summarized in Table 15.5. We organize the empirical results of the model into three blocks where the variables in each block are transformed. The first block presents the within-time-country estimates. The 'within estimator' focuses on factors driving changes within countries over time, while ignoring differences across countries and systematic changes over time. The second block presents the between-time estimates, representing the time-series component, common to all countries, and as such captures the impact of changes over time in the available technology. The last block presents the between-country estimates, summarizing the between-country variability. The estimates are based on the locus of points that go across the different techniques implemented by the countries that, in principle, operate under the same available technology. Because of the orthogonal structure of the regressors, it is possible to estimate the three blocks separately.

As expected, the different transformations yield different coefficients. The spread in productivity across countries is a different economic process than the spread in productivity for a country through time, thus the factors explaining the variability *should* differ. The within estimates represent a stable production function conditional on the state variables, whereas the between estimates are subject to the jointness effect and

Table 15.5. Econometric estimates of an international agricultural production function.

	Independent block regressions					
	Within time-country		Between time		Between country	
Variable	Estimate	t-Score	Estimate	t-Score	Estimate	t-Score
Inputs						
Structures and equipment	0.30	15.25	0.63	46.43	0.19	7.29
Capital of agricultural origin	0.07	3.40	0.25	19.89	0.11	12.08
Agricultural area	0.54		−1.17	−2.93	0.05	3.16
Fertilizer	0.05	2.26	−0.19	−7.97	0.44	21.78
Labour	0.04	0.99	−1.51	−15.29	0.16	11.88
Sum of estimates	1.00				0.94	
Technology						
Schooling	−0.07	−1.50	−0.32	−5.35	0.14	3.31
Peak yield	0.25	2.28			0.23	0.62
Development indicator	0.58	8.41	0.69	20.14	−0.19	−2.38
Institutions						
Civil liberties	−0.02	−1.85	−0.03	−3.46	0.03	0.81
Political rights	0.00	0.15	−0.02	−1.35	0.03	1.19
Prices						
Relative prices	0.21	9.48	0.19	8.47	0.97	5.97
Price variability	−0.27	−6.32	−0.68	−13.13	−1.84	−3.35
Inflation	−0.01	−2.64	−0.01	−2.42	−0.07	−3.99
Environmental						
Potential dry matter					−0.66	−12.31
Factor of water availability					0.22	4.96
Summary statistics						
Panel R-squares	0.493			0.999		0.977
Durbin–Watson statistic			1.997			

N = 870 (panel of 30 countries over 1972–2000). *Source*: Mundlak *et al.* (2011).

will be biased. The between estimates do not present pure input elasticities. Thus we impose constant returns to scale on the inputs for the within estimation only.

15.7.1   Inputs

The sum of the within elasticities of both types of capital is 0.37, with fixed capital accounting for most of the impact on productivity. Turning to the between regressions, the elasticity on fixed capital is particularly high (0.63) in the between-time regression. This suggests that the pace of the implementation of changes in the available technology was strongly constrained by the level of the fixed-capital stock in agriculture. The between-time estimate of capital of agricultural origin is significantly positive, though more modest in value (0.25). Although the between-time coefficient on fixed capital is robust to different specifications of the model, the other between-time coefficients are difficult to interpret in light of the limited time series (29 years) and the concern over multicollinearity.

The between-country estimate of fixed capital is smaller than the within estimate, whereas the between-country coefficient on livestock and treestock is larger (0.11). These estimates are more similar to those obtained from cross-country studies, in which the impact of capital on the changes in technology over time is not considered.[25]

15.7.2   State variables

State variables are included in the analysis to eliminate the bias of the estimated input coefficients that is caused by the jointness property of choosing inputs and implemented technology simultaneously. Looking first at the within estimation, the technology variables give mixed results. The development indicator is quite robust, indicating that the more productive the economy is as a whole, the higher is the productivity of agriculture. Peak yield has a significantly positive impact on productivity. Schooling

has an unexpected negative coefficient, though not significant.[26] As expected, the price coefficient is positive, and that of the price variability is negative. Inflation has a significantly negative effect. The institutional measures do not capture any effect beyond what was already reflected by the development indicator.

The between-country results reflect the determinants of cross-country variability at a given point in time, where countries implement different technologies, but presumably operate under the same set of available technology. Schooling is positively significant, suggesting that education was conducive to the techniques used by the more productive countries. Peak yield and the institutional measures are not significant. Development has changed to being significantly negative.[27] The price variables and inflation continue to have the expected signs. Looking at the country-specific environmental variables, water availability increases agricultural productivity. Potential dry matter is significantly negative.[28]

15.7.3   Estimates utilizing FAO data on tractors

As discussed above, in the absence of cross-country data sets on agricultural fixed capital, researchers often used FAO data on number of tractors as a proxy. We showed that the data series on tractors are not a convincing proxy for agricultural fixed capital data. To demonstrate the implications for econometric analysis, we estimate agricultural productivity once again, with the FAO data on the number of tractors in use replacing the data on agricultural fixed-capital stocks. The results are presented in Table 15.6. The coefficient on tractors is not significant in the within estimations. Due to the omission of the fixed capital input, the other input elasticities change as well: the elasticities of capital of agricultural origin, land, and fertilizer increase; whereas the labour elasticity becomes negative. Omitting fixed capital from the empirical production function leads to unrealistic results.

Table 15.6. Agricultural production function estimates using tractors as a proxy for fixed capital.

| | Independent block regressions | | | | | |
| | Within time-country | | Between time | | Between country | |
Variable	Estimate	t-Score	Estimate	t-Score	Estimate	t-Score
Inputs						
Tractors	0.01	0.53	−1.60	−15.71	0.10	7.41
Capital of agricultural origin	0.14	6.14	0.80	71.67	0.12	9.83
Agricultural area	0.85		−10.42	−19.28	0.01	1.01
Fertilizer	0.11	3.80	−0.73	−22.41	0.54	37.69
Labour	−0.11	−2.69	4.77	22.79	0.17	11.44
Sum of estimates	1.00				0.94	
Technology						
Schooling	0.00	0.05	5.57	21.64	0.29	6.84
Peak yield	0.38	3.16			0.89	2.32
Development indicator	0.92	12.48	1.29	25.50	−0.14	−1.69
Institutions						
Civil liberties	−0.03	−3.05	0.10	5.73	−0.03	−0.82
Political rights	0.02	2.00	−0.36	−17.92	0.16	6.20
Prices						
Relative prices	0.16	6.35	−0.28	−8.89	1.24	7.28
Price variability	−0.33	−6.87	−0.59	−6.93	−4.82	−7.99
Inflation	0.00	−1.43	0.05	13.06	0.07	1.60
Environmental						
Potential dry matter					−0.42	−6.71
Factor of water availability					0.49	8.98
Summary statistics						
Panel R-squares		0.355		0.996		0.975
Durbin–Watson statistic			2.473			

N = 870 (panel of 30 countries over 1972–2000).

Interestingly, the input elasticities from the between-country estimations using the tractor data are somewhat similar to those using the agricultural fixed capital data (compare the last two columns of Tables 15.5 and 15.6). The coefficient on tractors is 0.10, whereas the coefficient on fixed capital is 0.19. The choice of capital series (tractors versus fixed capital) has little effect on the between-country elasticities of capital of agricultural origin and labour. The reason for the similar results from the two regressions lies in the correlations of the transformed capital data. Examining the transformations of the capital data series, the correlations of the between-country transformations are much higher than those of the within transformations (0.85 versus 0.48). This is because the differences in the composition of capital and the declining importance of tractors are lost as the time-dimension of the data is collapsed into country averages. Thus, if the capital data series were compared on a cross-country basis, one might erroneously conclude that tractors are a valid proxy for agricultural fixed capital. It is crucial to view both the between and within transformations of the data series to fully capture its evolution.

15.8 Conclusions

This chapter reports a time series of capital stock data for agriculture for 30 countries for the period 1970–2000. The capital stocks for agriculture consist of three components: fixed capital, livestock and treestock. The data suggest that, as economies grow, agricultural capital stocks accumulate and the composition of agricultural capital changes. In particular, livestock declines as a share of total agricultural capital, and capital from treestock and fixed investments in machinery, irrigation and buildings becomes increasingly important. Moreover, the degree of the change differs between high-income and middle- and low-income countries.

We review a study on agricultural productivity that highlights the important role of agricultural capital, particularly fixed capital. This finding differs from earlier studies that used incomplete measures of agricultural capital and different methodologies of estimation. We show explicitly that our findings are the result of both the improved data set on agricultural capital stocks and estimations that focus on the within transformation of the data. The results clearly show that data on tractors are poor proxies for agricultural fixed capital.

Notes

[1] Ball *et al.* (2004, 2008) construct panel data sets of agricultural capital stocks for various OECD countries starting in the early 1970s. The Groningen Growth and Development Centre (GGDC) houses the EU KLEMS database on output and input growth at the industry level for 25 European Union member states, as well as the USA and Japan from 1970–2007 (http://www.euklems.net).

[2] Two-sector models of structural adjustment require comparable measures of capital in and out of agriculture. The lack of a consistent measure has prompted some researchers to exclude direct measures of agricultural capital from their analyses (see, for example, Restuccia *et al.* (2008); Bah and Brada (2009); and references therein). In an examination of sectoral differences in total factor productivity (TFP), Caselli (2005) uses assumptions on factor shares to derive the sectoral (agricultural versus non-agricultural) capital stocks necessary to compute TFP measures comparable across sectors.

[3] The paper can be downloaded from the World Bank at http://go.worldbank.org/IJ0CWCLVR0. More recent agricultural investment series were obtained from updates of the original sources, as well as from United Nations National Accounts data. When available, the authors choose to rely on the quality of the National Accounts data and to use its investment data. In some instances, country sources on investment are used. The quality of these sources can vary by country and even over time. Although internal checks on these data have been conducted, one must keep these limitations in mind when using the data.

[4] The *s* curvature parameters (β) and lifetime parameters (*L*) in Fig. 15.1 are taken from Ball *et al.* (1993). For buildings, $\beta=0.75$ and $L=38$ years; for agricultural machinery, $\beta=0.50$ and $L=9$ years.

[5] Judging the available evidence, the authors used the following parameters: curvature of 0.70, mean service life of 20 years and standard deviation of 8 years.

[6] An alternative method of estimating earlier investment data in agriculture was necessary for Indonesia because the previously mentioned method resulted in an unrealistic data series. Researchers at the FAO are introducing a more sophisticated method of estimating missing investment data, where they account for the possibility of structural breaks in the GDP and investment series (Daidone and Anriquez, 2011).

[7] We convert to comparable units in three steps. First, the series are valued in current prices in local currency using national GDP deflators. Then using 'market exchange rates' from the International Monetary Fund (IMF), the values are converted to current US dollars. Finally, the series are then deflated by the US agricultural GDP deflators to obtain the capital stock series in constant (1990) US dollars. Using this method allows comparability in the conversion of the three components of agricultural capital. Alternatively, one could convert the data on fixed-capital stocks using exchange rates in the base year. For more discussion on the conversion and deflation methods used, see Appendix A in Larson *et al.* (2000).

[8] Based on the System of National Accounts (SNA) used by the UN, 'Gross fixed-capital formation includes outlays on reclamation and improvement of land and development and extension of timber tracts, mines, plantations, orchards, vineyards etc., and on breeding and dairy cattle, draft animals, and animals raised for wool' (United Nations, 1991, p. xiv). In the 1993 version of the SNA, cultivated assets are noted to be part of gross fixed-capital formation; 'Cultivated assets consist of livestock or trees that are used repeatedly or continuously over periods of time of more than one year to produce other goods or services. Thus, livestock that continue to be used in production year after year are fixed assets. They include, for example, breeding stock, dairy cattle,

sheep reared for wool and draught animals. On the other hand, animals raised for slaughter, including poultry, are not fixed assets. Similarly, trees (including shrubs) that are cultivated in plantations for the products they yield year after year – such as fruit trees, vines, rubber trees, palm trees, etc. – are fixed assets. On the other hand, trees grown for timber that yield a finished product once only when they are ultimately felled are not fixed assets, just as cereals and vegetables that produce only a single crop when they are harvested cannot be fixed assets.' In the most recent (2008) version of the SNA, the terminology changes from 'cultivated assets' to 'cultivated biological resources', while the description remains similar.

[9] Estimated lifetimes are assigned specifically for each tree crop, ranging from 20 years for coffee to 200 years for olives (see Appendix 15.1). For a discussion of vintage in the evaluation of orchards, see Akiyama and Trivedi (1987).

[10] The data set on agricultural capital stocks is available online at: http://www.agproductivity.org

[11] The income classifications are based upon the World Development Indicators of the World Bank in 2000.

[12] Since the dissolution of the Soviet Union, Czechoslovakia and Yugoslavia in the early 1990s, many of the newly independent countries report data on agricultural gross fixed capital formation.

[13] Average annual growth rates reported in this chapter were obtained from trend regressions (natural log of variable on time).

[14] As mentioned previously, the raw data needed to construct the livestock and treestock measures are available from the FAO for a larger number of countries.

[15] In some studies, the data on tractors is converted into horsepower equivalents in an attempt to control for the quality differences in tractors (e.g. Craig et al., 1997).

[16] For more details on the distribution of the correlations, see Appendix 15.1.

[17] For ease of display, data from the USA were excluded in Fig. 15.4. Including the USA would not alter the finding of the lack of a strong relationship between the two series.

[18] Since livestock and treestock are included in agricultural GDP, it would be more appropriate to examine the relationship between agricultural GDP and *total* agricultural capital, which includes these components. We compromise here for the purpose of a more direct comparison with Fig. 15.6. Both fixed capital and capital of agricultural origin (livestock and treestock) are included in the productivity analysis, which is reviewed in the next sections.

[19] This is a revised version of Mundlak et al. (2008). See Mundlak et al. (2008, 2011) for a more thorough econometric discussion on heterogeneous technology and the role of the state variables. To summarize, the identification of the production function is achieved through allocation error. With heterogeneous technology, state variables (including prices) appear in both the production function and the factor demand equation, and thus cannot serve as instrumental variables.

[20] The full empirical framework is presented in Mundlak et al. (2008).

[21] Various sources of National Accounts data were used to compile longer time series, including the United Nations, World Bank, OECD, IMF and, if necessary, country-specific sources. Using agricultural GDP provides internal consistency to our estimation because the investment data are also taken from the national accounts. We apply real exchange rates to convert agricultural GDP to a common currency. Alternatively, one could use a gross agricultural output series based on a common international price set, such as reported by the FAO.

[22] Data on agricultural area, agricultural labour and fertilizers were downloaded from the FAOSTAT database.

[23] The list of possible state variables is vast. The variables presented in this analysis were chosen on the basis of our prior knowledge of what are important determinants of agricultural productivity, as well as for the availability of the data. This list is not meant to be exclusive, but rather suggestive. Indeed, there must be other determinants as the state variables in our model account for about half of the country effect (Mundlak et al., 2011).

[24] Total education data from Barro and Lee (2001) are reported for every 5 years up to 2000 through the World Bank website (http://go.worldbank.org/8BQASOPK40). Data for other years are obtained through linear interpolations.

[25] The early literature on cross-country studies on agricultural production was reviewed in Mundlak et al. (1999).

[26] In Butzer (2011) and Mundlak et al. (2011), the reduced form estimation (state variables as regressors, omitting the inputs) is shown. In the reduced form, the within coefficient on rural schooling is positive. The introduction of inputs to the regression led to the change in sign. This suggests that the significantly positive coefficient of schooling is confounded in the inputs and there is not enough variability to sort out the effects.

[27] Butzer (2011) shows results from separate estimations for the two subsets of countries. The impact of the development indicator is much larger in middle- and low-income countries than for the high-income

countries (within estimates of 0.80 and 2.52 respectively). Policies and institutions may have varying impacts depending on the current level of economic development. In countries where institutions are not well developed, the marginal impact of a change in the institutions (proxied here by the development indicator variable) would be greater than in countries which have advanced institutions. This differential impact across subsets of countries is being picked up by the between-country estimator in Table 15.5.

[28] In a previous study where agricultural capital entered the estimation as a composite variable, potential dry matter had a significantly positive effect. Once agricultural capital was disaggregated into its various components, the coefficient on potential dry matter lost significance and even turned negative (Mundlak *et al.*, 1999).

References

Akiyama, T. and Trivedi, P. (1987) Vintage production approach to perennial crop supply: an application to tea in major producing countries. *Journal of Econometrics* 36, 133–161.

Alene, A. (2010) Productivity growth and the effects of R&D in African agriculture. *Agricultural Economics* 41, 223–238.

Antle, J. (1983) Infrastructure and aggregate agricultural productivity: International evidence. *Economic Development and Cultural Change* 31, 609–619.

Arnade, C. (1998) Using a programming approach to measure international agricultural efficiency and productivity. *Journal of Agricultural Economics* 49, 67–84.

Bah, M. and Brada, J. (2009) Total factor productivity growth, structural change and convergence in the new members of the European Union. *Comparative Economic Studies* 51, 421–446.

Ball, V.E., Bureau, J.-C., Butault, J.-P. and Witzke, H. (1993) The stock of capital in European Community agriculture. *European Review of Agricultural Economics* 20, 437–450.

Ball, V.E., Bureau, J.-C., Butault, J.-P. and Nehring, R. (2001) Levels of farm sector productivity: An international comparison. *Journal of Productivity Analysis* 15, 5–29.

Ball, V.E., Butault, J.-P. and Mesonada, C. (2004) Measuring real capital input in OECD agriculture. *Canadian Journal of Agricultural Economics* 52, 351–370.

Ball, V.E., Lindamood, W., Nehring, R. and Mesonada, C. (2008) Capital as a factor of production in OECD agriculture: Measurement and data. *Applied Economics* 40, 1253–1277.

Barro, R. and Lee, J.-W. (2001) International data on educational attainment: Updates and implications. *Oxford Economic Papers* 53, 541–563.

Bhattacharjee, J. (1955) Resource use and productivity in world agriculture. *Journal of Farm Economics* 37, 57–71.

Binswanger, H., Mundlak, Y., Yang, M.-C. and Bowers, A. (1987) On the determinants of cross-country aggregate agricultural supply. *Journal of Econometrics* 36, 111–131.

Block, S. (1994) A new view of agricultural productivity in sub-Saharan Africa. *American Journal of Agricultural Economics* 76, 3, 619–624.

Buringh, P., van Heemst, H. and Staring, G. (1979) Computation of the absolute maximum food production of the world. In: Linneman, H., De Hoogh, J., Keyzer, M. and van Heemst, H. (eds) *MOIRA, Model of International Relations in Agriculture*. North-Holland, Amsterdam.

Butzer, R. (2011) The role of physical capital in agricultural and manufacturing production. PhD Thesis. University of Chicago, Chicago, IL.

Caselli, F. (2005) Accounting for cross-country income differences. In: Aghion, P. and Durlauf, S. (eds) *Handbook of Economic Growth*. Elsevier, Amsterdam, The Netherlands.

Cermeño, R. and Vázquez, S. (2009) Technological backwardness in agriculture: Is it due to lack of R&D, human capital, and openness to international trade? *Review of Development Economics* 13, 673–686.

Cermeño, R., Maddala, G. and Trueblood, M. (2003) Modeling technology as a dynamic error components process: the case of the inter-country agricultural production function. *Econometric Reviews* 22, 289–306.

Coelli, T. and Rao, P.S.D. (2005) Total factor productivity growth in agriculture: A Malmquist index analysis of 93 countries, 1980–2000. *Agricultural Economics* 32, 115–134.

Craig, B., Pardey, P. and Roseboom, J. (1997) International productivity patterns: Accounting for input quality, infrastructure, and research. *American Journal of Agricultural Economics* 79, 1064–1076.

Crego, A., Larson, D. Butzer, R. and Mundlak, Y. (1998) A new database on investment and capital for agriculture and manufacturing. Working Paper Number 2013. World Bank, Washington, DC.

Daidone, S. and Anriquez, D. (2011) An extended cross-country database for agricultural investment and capital. ESA Working Paper No. 11–16, FAO, Rome.

Evenson, R. and Kislev, Y. (1975) *Agricultural Research and Productivity*. Yale University Press, New Haven, CT.

Färe, R., Grosskopf, S., Norris, M. and Zhang, Z. (1994) Productivity growth, technical progress, and efficiency change in industrialized countries. *American Economic Review* 84, 66–83.

Fleming, E. (2007) Agricultural productivity change in Pacific island countries. *Pacific Economic Bulletin* 22, 32–47.

Fuglie, K. (2008) Is a slowdown in agricultural productivity growth contributing to the rise in commodity prices? *Agricultural Economics* 39, supplement, 431–441.

Fulginiti, L. (2010) What comes first, agricultural growth or democracy? *Agricultural Economics* 41, 15–24.

Fulginiti, L. and Perrin, R. (1993) Prices and productivity in agriculture. *Review of Economics and Statistics* 75, 471–482.

Fulginiti, L. and Perrin, R. (1998) Agricultural productivity in developing countries. *Agricultural Economics* 19, 45–51.

Hayami, Y. and Ruttan, V.W. (1970) Agricultural productivity differences among countries. *American Economic Review* 60, 895–911.

Hayami, Y. and Ruttan, V.W. (1971) *Agricultural Development: An International Perspective*. Johns Hopkins University Press, Baltimore, MD.

Headey, D., Alauddin, M. and Rao, D.S.P (2010) Explaining agricultural productivity growth: An international perspective. *Agricultural Economics* 41, 1–14.

Kawagoe, T. and Hayami, Y. (1983) The production structure of world agriculture: An intercountry cross-section analysis. *Developing Economies* 21, 3, 189–206.

Kawagoe, T. and Hayami, Y. (1985) An intercountry comparison of agricultural production efficiency. *American Journal of Agricultural Economics* 67, 1, 87–92.

Larson, D. and Mundlak, Y. (1997) On the intersectoral migration of agricultural labor. *Economic Development and Cultural Change* 45, 2, 295–319.

Larson, D., Butzer, R., Mundlak, Y. and Crego, A. (2000) A cross-country database for sector investment and capital. *World Bank Economic Review* 14, 371–391.

Lau, L. and Yotopoulos, P. (1989) The meta-production function approach to technological change in world agriculture. *Journal of Development Economics* 31, 2, 241–269.

Lio, M. and Liu, M.-C. (2008) Governance and agricultural productivity: A cross-national analysis. *Food Policy* 33, 504–512.

Lusigi, A. and Thirtle, C. (1997) Total factor productivity and the effects of R&D in African agriculture. *Journal of International Development* 9, 4, 529–538.

Mundlak, Y. (2001) Production and supply. In: Gardner, B.L. and Rausser, G.C. (eds) *Handbook of Agricultural Economics, Agricultural Production*. Elsevier, Amsterdam.

Mundlak, Y. and Hellinghausen, R. (1982) The intercountry agricultural production function: Another view. *American Journal of Agricultural Economics* 64, 664–672.

Mundlak, Y., Larson, D. and Crego, A. (1998) Agricultural development: Issues, evidence, and consequences. In: Mundlak, Y., Bruno, M. and Cohen, D. (eds) *Contemporary Economic Issues: Proceedings of the Eleventh World Congress of the International Economic Association, Tunis. Volume 2. Labour, Food, and Poverty*. St. Martin's Press, New York, NY.

Mundlak, Y., Larson, D. and Butzer, R. (1999) Rethinking within and between regressions: The case of agricultural production functions. *Annales d'Economie et de Statistique* 55–56, 475–501.

Mundlak, Y., Butzer, R. and Larson, D. (2008, revised 2011) Heterogeneous technology and panel data: The case of the agricultural production function. World Bank Policy Research Working Paper 4536. World Bank, Washington, DC.

Nehru, V. and Dhareshwar, A. (1993) A new database on physical capital stock: Sources, methodology and results. *Revista de Analisis Economico* 8, 37–59.

Nguyen, D. (1979) On agricultural productivity differences among countries. *American Journal of Agricultural Economics* 61, 565–570.

Nin-Pratt, A. and Yu, B. (2010) Getting implicit shadow prices right for the estimation of Malmquist indices: The case of total factor productivity in developing countries. *Agricultural Economics* 41, 349–360.

Restuccia, D., Yang, D. and Zhu, X. (2008) Agriculture and aggregate productivity: A quantitative cross-country analysis. *Journal of Monetary Economics* 55, 234–250.

United Nations (1991, 1993, 2008) *National Accounts Statistics: Main Aggregates and Detailed Tables.* United Nations, New York, NY.

Wiebe, K., Soule, M., Narrod, C. and Breneman, V. (2003) Resource quality and agricultural productivity: A multi-country comparison. In: Wiebe, K. (ed.) *Land Quality, Agricultural Productivity, and Food Security: Biophysical Processes and Economic Choices at Local, Regional, and Global Levels.* Edward Elgar Publishing, Northampton, MA.

World Bank. World Development Indicators Database. World Bank, Washington, DC.

World Bank (2008) *World Development Report 2008: Agriculture for Development.* World Bank, Washington, DC.

Appendix 15.1

Livestock and Treestock as Components of Fixed Capital

To understand the extent to which fixed capital includes treestock and livestock, we calculate the ratio of these two components to fixed capital. In many countries the capital in livestock and treestock is considerably larger than that in fixed capital (Table A15.1). The magnitude of the difference raises doubts as to whether fixed capital includes these components. We cannot, however, conclude that this is the case for all countries, and we cannot answer the question of coverage of fixed costs with our data. Within the agricultural sector, the growth of fixed capital generally exceeds that of total capital, indicating a smaller growth rate of livestock and treestock.

Lifetimes of Tree Crops

To compute the value of treestock, we assign the following lifetimes for the trees: coffee, 20 years; bananas/plantains, cocoa, oranges and other citrus, apples, peaches/nectarines, papayas, lemons/limes, mangos, pears, currants, dates, figs and grapes, 25 years; oil palm, rubber and cherries, 30 years; cashews, 40 years; almonds, 50 years; coconuts, 75 years; tea, 100 years; and olives, 200 years.

Correlations of Data on Fixed Capital with FAO Data on Tractors

Although we have shown that the FAO dataset on tractors is not a suitable proxy for the data on agricultural fixed capital for all countries, in light of the lack of national accounts investment data for a larger sample of countries, it is desirable to examine to what extent the alternative data might be useful. To this end, in Table A15.2, we report the correlation coefficients for data on fixed capital and tractors for each country. The table is sorted by the correlation values so that the countries

with the highest correlations are at the top. It is interesting to note that of the 11 countries with correlations less than 0.78, eight are high-income countries.

The average correlations for each income group are reported at the bottom of Table A15.2. The correlation for the high-income countries is much lower than for the other three income groups. This suggests that in the more developed countries, tractors and machinery account for a smaller component of fixed capital. This finding corresponds to the large difference between the growth rates for fixed capital and those of tractors in high-income countries, as seen in Table 15.2.

Table A15.1. Average ratio of treestock and livestock to agricultural fixed capital.

Country	Treestock	Livestock
Australia	0.058	0.684
Austria	0.045	0.131
Canada	0.008	0.198
Cyprus	3.399	0.158
Denmark	0.008	0.231
Egypt	0.501	0.314
Finland	0.002	0.090
France	0.264	0.246
Greece	24.537	0.256
India	0.671	1.353
Indonesia	3.009	0.487
Italy	2.883	0.095
Kenya	6.353	3.578
Korea, Republic of	0.072	0.057
Malawi	4.350	1.805
Mauritius	0.616	0.073
Morocco	10.866	2.239
Netherlands	0.021	0.229
Norway	0.009	0.076
Pakistan	0.280	1.573
Peru	0.464	1.122
Philippines	6.876	1.275
Sri Lanka	4.588	0.297
Sweden	0.004	0.137
Tanzania	2.658	5.914
Tunisia	5.845	0.234
Turkey	3.131	0.596
UK	0.015	0.348
USA	0.072	0.477
Uruguay	0.737	8.575

Average ratios for 1970–2000.

The high correlations for the low-income countries (five of which have correlations greater than 0.82) offer the hope for the use of tractor data as an alternative. However, further research must be done to explore this possibility because these are correlations for the entire series. Corresponding to the analysis on panel data in Mundlak *et al.* (2011), it would be necessary to consider the correlations for the within transformations of the data series.

Table A15.2. Correlations of data on fixed capital with FAO data on tractors.

Country	Correlation	Income category
Mauritius	0.96	UMI
Italy	0.95	HI
Indonesia	0.95	LI
Korea, Repubic of	0.95	UMI
Canada	0.94	HI
Greece	0.93	HI
Tunisia	0.93	LMI
India	0.92	LI
Turkey	0.92	UMI
Pakistan	0.90	LI
Cyprus	0.90	HI
Kenya	0.88	LI
Norway	0.86	HI
Peru	0.85	LMI
Egypt	0.85	LMI
Morocco	0.83	LMI
Malawi	0.82	LI
Uruguay	0.80	UMI
Austria	0.78	HI
Sri Lanka	0.45	LMI
Finland	0.41	HI
Netherlands	0.33	HI
UK	0.21	HI
France	0.06	HI
Philippines	−0.28	LMI
Tanzania	−0.30	LI
Sweden	−0.55	HI
Denmark	−0.75	HI
USA	−0.82	HI
Australia	−0.91	HI

Income categories from the World Bank. HI, high income; UMI, upper middle income; LMI, lower middle income; LI, low income.

16 Productivity Growth and Technology Capital in the Global Agricultural Economy

Keith O. Fuglie

Economic Research Service, US Department of Agriculture, Washington, DC

16.1 Introduction

The chapters of this volume have presented some of the latest and most comprehensive assessments of productivity growth for agriculture in various countries and regions of the world. As reviewed in the introduction to this volume, the global story is a mixed one. Industrialized countries have generally sustained relatively strong rates of total factor productivity (TFP) during the past several decades, although Australia and South Africa show signs of productivity stagnation. In transition countries there has been a fairly robust productivity recovery after more than a decade of economic reforms that forced a sharp contraction on the agricultural sectors of these countries. But just as the reform process has been uneven across these countries, so has the pace of their agricultural recovery. Among developing countries, several, most notably Brazil and China, have achieved remarkable productivity gains over the past several decades. Others, especially those in sub-Saharan Africa, continue to lag far behind the kind of productivity growth most other countries are achieving.

What does all this add up to? In this closing chapter I have two principal objectives. First, I extend my previous work (Fuglie, 2008, 2010b) on measuring in a consistent and comparably fashion agricultural TFP growth for various countries and regions and for the world as a whole. Second, I re-examine the model in Evenson and Fuglie (2010) on the correlation between national capacities in research and extension with long-run agricultural productivity growth with these updated estimates. I use the national 'technology capital' indexes described in Evenson and Fuglie (2010) to test whether developing countries that invested more in technology capital achieved faster growth in agricultural productivity. This work continues a long line of research on the technological determinants of agricultural growth, dating from Hayami and Ruttan (1971, 1985), Evenson and Kislev (1975), Craig *et al.* (1997), Wiebe *et al.* (2003) and Avila and Evenson (2010), that seeks to understand better the role of agricultural science and technology in improving food security and economic welfare around the world.

In the next section of this chapter I outline a practical, 'growth accounting' approach for measuring changes in agricultural TFP across a broad set of countries given limited international data on production outputs, inputs and their economic values. Considerable attention is given to data and measurement issues. Like in my previous work (Fuglie, 2008, 2010b), I adjust agricultural land area for the quality differences among rainfed and irrigated cropland and pastures. Applying the lessons from other chapters in this volume, I use alternative measures (from FAO) for cropland in sub-Saharan Africa (Fuglie and Rada, Chapter 12), agricultural labour in transition countries (Swinnen *et al.*, Chapter 6) and Nigeria (Fuglie and Rada, Chapter 12), and agricultural machinery capital globally (Butzer *et al.*, Chapter 15). Although the measure for farm machinery I develop here – which includes a broader set of capital stock than a simple count of tractors in use – is an improvement over previous studies, it still probably falls short of the comprehensive measures described in Butzer *et al.* (Chapter 15). Getting more complete, global measures of agricultural capital stock is probably the most pressing challenge in improving our ability to decipher the rate and direction of global agricultural productivity growth.

16.2   Methods and Data

16.2.1   Measuring TFP growth and its causes

Total factor productivity

Here, I sketch out the procedures used to construct internationally comparable measures of agricultural TFP growth relying primarily on FAO data on agricultural inputs and outputs, and supplementary information on production costs from other studies. Zhao, Sheng and Gray (Chapter 4, this volume) present a thorough discussion of growth accounting methods for assessing changes in agricultural TFP and the reader is referred to this chapter for a more comprehensive conceptual treatment of the subject.

If total factor productivity (TFP) is defined as the ratio of total output to total inputs in a production process and total output is given by Y and total inputs by X, then TFP is simply (Eqn 16.1):

$$TFP = \frac{Y}{X}. \qquad (16.1)$$

Changes in TFP over time are found by comparing the rate of change in total output with the rate of change in total input. Expressed as logarithms, changes in Eqn 16.1 over time can be written as (Eqn 16.2):

$$\frac{d\ln(TFP)}{dt} = \frac{d\ln(Y)}{dt} - \frac{d\ln(X)}{dt} \qquad (16.2)$$

which simply states that the rate of change in TFP is the difference in the rate of change in aggregate output and input.

Agriculture is a multi-output, multi-input production process, so Y and X are vectors. When the underlying technology is represented by a constant-returns-to-scale Cobb–Douglas production function and where (i) producers maximize profits so that the output elasticity with respect to an input equals the cost share of that input; and (ii) markets are in long-run competitive equilibrium so that total revenue equal total cost, then Eqn 16.2 can be written as Eqn 16.3:

$$\ln\left(\frac{TFP_t}{TFP_{t-1}}\right) = \sum_i R_i \ln\left(\frac{Y_{i,t}}{Y_{i,t-1}}\right)$$
$$- \sum_j S_j \ln\left(\frac{X_{j,t}}{X_{j,t-1}}\right). \qquad (16.3)$$

where R_i is the revenue share of the ith output and S_j is the cost-share of the jth input. Total output growth is estimated by summing over the growth rates for each commodity weighted by its revenue share. Similarly, total input growth is found by summing the growth rate of each input, weighted by its cost share. TFP growth is just the difference between the growth of total output and total input.

One difference among growth accounting methods is whether the revenue and cost share weights are fixed or vary over time. Paasche and Laspeyres indexes use fixed weights, whereas the Tornqvist–Thiel and other chained indexes use variable weights. Allowing the weights to vary reduces potential 'index number bias'. Index number bias arises when producers substitute among outputs and inputs depending on their relative profitability or cost. In other words, the growth rates in Y_i and X_j are not independent of changes R_i and S_j. For example, if labour wages rise relative to the cost of capital, producers are likely to substitute more capital for labour, thereby reducing the growth rate in labour and increasing it for capital. For agriculture, index number bias in productivity measurement appears to be more likely for inputs than outputs. Cost shares of agricultural capital and material inputs tend to rise in the process of economic development, whereas the cost share of labour tends to fall. Commodity revenue shares, on the other hand, seem to show less change over time.

To reduce potential index number bias in TFP growth estimates, I vary cost shares by decade whenever such information is available. For outputs, however, base year prices (or equivalently, base year revenue shares) are fixed because these depend on FAO's measure of constant, gross agricultural output (described in more detail below). The base period for output prices is 2004–2006.

A key limitation in using Eqn 16.3 for measuring agricultural productivity change is a lack of representative cost share data for most countries. Many types of agricultural inputs (such as land and labour) might not be widely traded and heterogeneous in quality, making price or cost determination difficult. Some studies have circumvented this problem by estimating a distance function, such as a Malmquist index, that measures productivity using data on output and input quantities alone (see Nin-Pratt and Yu, Chapter 13 in this volume for a description of this method). But this method is sensitive to the dimensionality problem: results of the model are sensitive to the number of outputs, inputs and countries included in estimation (Lusigi and Thirtle, 1997). Coelli and Rao (2005) have also observed that the input shadow prices derived from the estimation of this model vary widely across countries and over time and in many cases are zero for major inputs such as land and labour, which is not plausible. Instead, I compile estimates from previous studies of input cost shares or production elasticities for individual countries or regions and apply these to Eqn 16.3.

For countries for which I lack data on cost shares, I approximate these by applying cost shares from a 'like' country. The section below on 'input cost shares' provides details on the data sources and assumptions. This is similar to the approach used by Avila and Evenson (2010), who applied agricultural input cost shares from Brazil and India to other developing countries, except that I use a richer set of information on cost shares and include industrialized and transition countries in the analysis.

The framework outlined above provides a simple means of decomposing the relative contribution of TFP and inputs to the growth in output. Using a dot above a variable to signify its annual rate of growth, the growth in output is simply the growth in TFP plus the growth rates of the inputs times their respective cost shares:

$$\dot{Y} = \dot{TFP} + \sum_{j=1}^{J} S_j \dot{X}_j. \tag{16.4}$$

I call Eqn 16.4 an *input cost decomposition* of output growth because each $S_j \dot{X}_j$ term gives the growth in cost from using more of the jth input to increase output.[1] It is also possible to focus on a particular input, for example, land (which I will designate as X_1), and decompose growth into the component due to expansion in this resource and the yield of this resource (Eqn 16.5):

$$\dot{Y} = \dot{X}_1 + \left(\frac{\dot{Y}}{X_1}\right) \tag{16.5}$$

This decomposition corresponds to what is commonly referred to as *extensification* (land expansion) and *intensification* (land yield growth). We can further decompose yield growth into the share owing to TFP and the share owing to using other inputs more intensively per unit of land:

$$\dot{Y} = \dot{X}_1 + \dot{TFP} + \sum_{j=2}^{J} S_j \left(\frac{\dot{X}_j}{X_1}\right). \tag{16.6}$$

I call Eqn 16.6 a *resource decomposition* of growth because it focuses on the quantity change of a physical resource (land) rather than its contribution to changes in cost of production. See Fig. 12.3 in Chapter 12 of this volume for a graphical depiction of the growth decomposition described in Eqns 16.5 and 16.6.

TFP and technology capital

Although the growth decomposition described above is useful for illustrating the role of productivity change and resource utilization in expanding output, it does not explain why these trends are occurring. The transition from resource-led to productivity-led growth was a major 20th-century development in world history (Hayami and Ruttan, 1971, 1985). But the speed at which various countries have made this transition has varied widely, and for some countries hardly at all. Hayami and Ruttan (and others since them) attributed the different rates of productivity growth to differences in their accumulation of human capital, which they took especially to mean formal institutions conducting agricultural research and development (R&D). Hayami and Ruttan (1971, 1985) lacked sufficient data to characterize R&D investments, however, and proxied for this using labour force education. Since their work, much data has been accumulated on national capacities in R&D as well as agricultural extension and general education, which Robert Evenson developed into indexes of 'technology capital' (Evenson and Fuglie, 2010; Avila and Evenson, 2010). I use these indexes of national technology capital to explore whether they can explain differences in agricultural productivity performance among countries. My approach is similar to that used in Evenson and Fuglie (2010) and Avila and Evenson (2010), in which estimates of long-run average TFP growth are regressed against indexes of national technology capacities. These technology capital indexes, one measuring a nation's ability to invent and innovate new agricultural technology and a second a nation's ability to extend new technologies to farmers, are briefly presented here and described in more detail in Evenson and Fuglie (2010) and Avila and Evenson (2010).

To represent the capacity to develop or adapt new agricultural technology, an 'Invention–Innovation' (II) index is con-structed from two indicators, the number of public-sector agricultural scientists per thousand hectares of arable land (Pardey et al., 1991, and updated from Agricultural Science and Technology Indicators) and industry research and development as a percentage of GDP (UNESCO). Agricultural scientists per crop area represent capacity to breed and adapt appropriate varieties and agronomic practices for the crops and environments in a country. The UNESCO indicator captures a country's capacity to adapt and manufacture appropriate industrial inputs for agriculture. Similarly, the capacity to extend and adopt agricultural technology is represented by an index of 'Technology Mastery' (TM). The TM index is also a composite of two indicators, the number of extension workers per thousand hectares of arable land and the average years of schooling of males over 25.[2] Values for the II and TM indexes are constructed for a set of 87 developing countries for two points in time: the average capacity scores over 1970–1975 and 1990–1995. Each index ranges in value from 2 to 6, with 2 representing countries with minimal or no capacity (i.e. no formal research, no extension service and a largely illiterate population) and 6 countries that have acquired capacities comparably with those of developed countries (Evenson and Fuglie, 2010).

To examine the relationship between technology capital and productivity growth, technology capital in period t is hypothesized to influence long-run average TFP growth over subsequent years. Because the technology capital indexes have been constructed for two periods, we effectively have a two-period panel dataset. We let the II and TM levels in 1970–1975 explain average annual TFP growth during 1971–1990 and II and TM levels in 1990–1995 explain TFP growth during 1991–2009. Causality between technology capital and productivity growth is established through the lag structure of the model (i.e. present technology capital affects future growth performance) and the panel structure of the data (through a difference-in-difference model, described in Eqn 16.8 below).

The first estimating equation examines the interaction between research and extension. It is often contended that a lot of technology, often imported, is 'on the shelf' but has not diffused because of poor extension services or low farmer schooling. Others maintain that agricultural technology requires innovation and adaptation to local conditions before it can be successfully adopted, and therefore local research capacity is the limiting factor. We examine this question by comparing the productivity performance between countries that have given relatively more or less emphasis to research versus extension and education. These factors enter the equation as a series of indicator variables describing different combinations of II and TM capacities. This estimating equation is given by (Eqn 16.7):

$$\overline{TFP}_p \equiv \frac{\sum\limits_{k=0}^{19}\left(T\dot{F}P_{c,p+k}\right)}{20}$$

$$= \sum\limits_{i=2}^{6}\sum\limits_{j=2}^{6}\delta_{i,j}Dij_{c,p}. \qquad (16.7)$$

where $T\dot{F}P_{c,t} = \ln(TFP_{c,t}/TFP_{c,t-1})$ is the growth rate in country c's agricultural TFP in year t and $Dij_{c,p}$ is an indicator variable for the country's II and TM capacities in the base period p (p=1970 and 1990).[3] $Dij_{c,p}$ takes on a value of 1 if both $II_{c,t} = i$ and $TM_{c,t} = j$, and 0 otherwise. The dependent variable $\overline{TFP}_p$ is the average annual TFP growth rate over the 20-year period subsequent to when technology capacities (the Dij indicator variables) are observed. Because $II_{c,t}$ and $TM_{c,t}$ each have five levels (i.e. they take on values from 2 to 6), there are potentially 25 different combinations of II and TM capitals. Thus Eqn 16.7 could have as many as 25 Dij indicator variables, although only 19 such combinations are present in the data. The indicator variable coefficients $\delta_{II,TM}$ measure the average long-run TFP growth rate for all the countries with this II and TM combination. Looking at productivity growth in the years after II and TM are measured accounts for the lag between when research is done and when new technology resulting from that research is likely to be adopted by farmers.

Note that the model structure in Eqn 16.7 is a very flexible form – productivity growth for any II and TM combination is independent of productivity growth of any other combination. Another advantage of the model is that it allows us to examine the marginal effects of changes in the one type of technology capital given some level of the other. Holding II (research capacity) at some level J and then examining how the coefficients $\delta_{J,2}...\delta_{J,6}$ vary allows us to examine how marginal increases in TM (agricultural extension and schooling) affect TFP growth. Similarly, examining the values of coefficients $\delta_{2,K}...\delta_{6,K}$ allow us to say something about the marginal effect of research capacity holding TM fixed at some level K.

One limitation of the model in Eqn 16.7 is that it does not control for other factors that might be correlated with both TFP growth and technology capital. The panel structure of the data allows for a more rigorous test of the relationship between technology capital and TFP growth by estimating a 'difference in difference' model. By taking first differences of the variables, we can assess whether countries that *increased* their technology capital between 1970 and 1990 were able to *accelerate* productivity growth in agriculture compared with countries that did not. The estimating equation for this version of the model is given by:

$$\left(\overline{TFP}_{p+1} - \overline{TFP}_p\right) = \delta_{II}\left(II_{c,p+1} - II_{c,p}\right)$$
$$+ \delta_{TM}\left(TM_{c,p+1} - TM_{c,p}\right). \qquad (16.8)$$

In Eqn 16.8, the dependent variable is the change in the average TFP growth rate between the two periods (1971–1990 and 1991–2009). The explanatory variables are the changes in II and TM capitals between 1970 and 1990. The coefficients δ_{II} and δ_{TM} indicate the average rate by which TFP growth changed as countries increased (or decreased) their II and TM capacities by one unit between the two periods. Equation 16.8 is estimated using data for all the developing countries in the sample as well as separately for three regional groups of countries (Latin American, sub-Saharan African and Asia) to see whether there might be systematic differences across regions.

It is important to consider whether the estimates of Eqns 16.7 and 16.8 might suffer from omitted variable bias. In addition to technology and human capital, TFP growth could be affected by errors in measurement, 'left-out' factors of production, infrastructure, weather fluctuations, civil disturbances, economies of scale, gains in allocative efficiency from market liberalization and other variables. Although the 'difference in difference' model removes some country-specific factors that might influence TFP growth, it does not control for changing circumstances within countries. Several of these omitted variables are, however, probably not relevant to our model because of the long period over which we measure TFP growth (i.e. we take average TFP growth over 20 years). Thus, short-run fluctuations to output or TFP owing to natural or civil disturbances will tend to be averaged out. Regarding scale economies, Hayami and Ruttan (1985) and Binswanger *et al.* (1995) find little evidence that farm size explains productivity differences among developing countries. For infrastructure, Evenson and Fuglie (2010) included a road density variable in their model, but this was not significant in explaining TFP growth across countries so is excluded here. Market liberalization and institutional reforms that improve allocative efficiency will also cause TFP to grow, although the effect might only be temporary because once resources have been reallocated to realize the efficiencies, growth will again stagnate unless improved technology is forthcoming. For productivity growth to be sustained over the long run, it is difficult to conceive of factors other than science and technology that could explain major differences across countries.

16.2.2 Data

FAO's 1961–2009 annual time series of crop and livestock commodity production and land, labour, livestock capital, fertilizer and machinery resources are the primary source for agricultural outputs and inputs used to construct the national and global productivity measures. In some cases these are modified or supplemented with data from other sources (national statistical agencies, mostly) where alternative data are considered to be more accurate or up-to-date, as described below.

Output

For agricultural output, FAO publishes data on annual production of 198 crop and livestock commodities by country since 1961, aggregates this into a measure of the gross production value using a common set of commodity prices from 2004–2006 and expresses this in constant 2005 international dollars. FAO excludes production of animal forages but includes crop production that is used for animal feed and seed in estimating gross production value. The FAO also provides a measure of output net of domestic production used for feed and seed. The net production measure does not, however, exclude imported grain that may be used as feed or seed, or grain that is exported and used in another country for these purposes.

Because current (or near current) prices are fixed to aggregate quantities and measure changes in real output over time, the FAO gross production value is equivalent to a Paasche quantity index. The set of common commodity prices is derived using the Geary–Khamis method. This method determines an international price p_i for each commodity which is defined as an international weighted average of prices of the i-th commodity in different countries, after national prices have been converted into a common currency using a purchasing power parity (PPP_j) conversion rate for each j-th country. The weights are the quantities produced by the country. The computational scheme involves solving a system of simultaneous linear equations that derives both the p_i prices and PPP_j conversion factors for each commodity and country. The FAO updates these prices every 5 years and recalculates its index of gross production value back to 1961 using its most recent set of international prices. See Rao (1993) for a thorough description and assessment of these procedures.

I use the FAO value of gross agricultural output in constant 2005 international dollars as the basis for a consistent measure of output for each country and the world. However, owing to the influence of weather and other factors, agricultural production is exceptionally volatile from year to year, and it can be difficult to disentangle short-run fluctuations from long-term trends. To relieve the data of some of these fluctuations, I smooth the output series for each country using the Hodrick–Prescott filter (setting λ = 6.25 as recommended for annual data by Ravn and Uhlig, 2002). Figure 16.1 illustrates the effect of this smoothing technique on gross agricultural output for Zambia and Jordan.[4] It is evident that even with smoothing there is still considerable curvature in the output series, although much of the year-to-year fluctuation in output has been removed from the data. I assume that the smoothed series provides a better indi-

cator of productivity trends and that annual variation around this trend is primarily due to short-term disturbances such as weather.

Inputs

For agricultural inputs, FAO publishes data on cropland (total and irrigated), permanent pasture, labour employed in agriculture, animal stocks, the number of tractors in use and inorganic fertilizer consumption. I supplement these data with better or more up-to-date data from national or industry sources when available. For fertilizer consumption, the International Fertilizer Association has more up-to-date and accurate statistics than FAO on fertilizer consumption by country, except for small countries. For agricultural statistics on China, a relatively comprehensive dataset is available from the Economic Research Service (b) with original data from the National Bureau of Statistics of China.

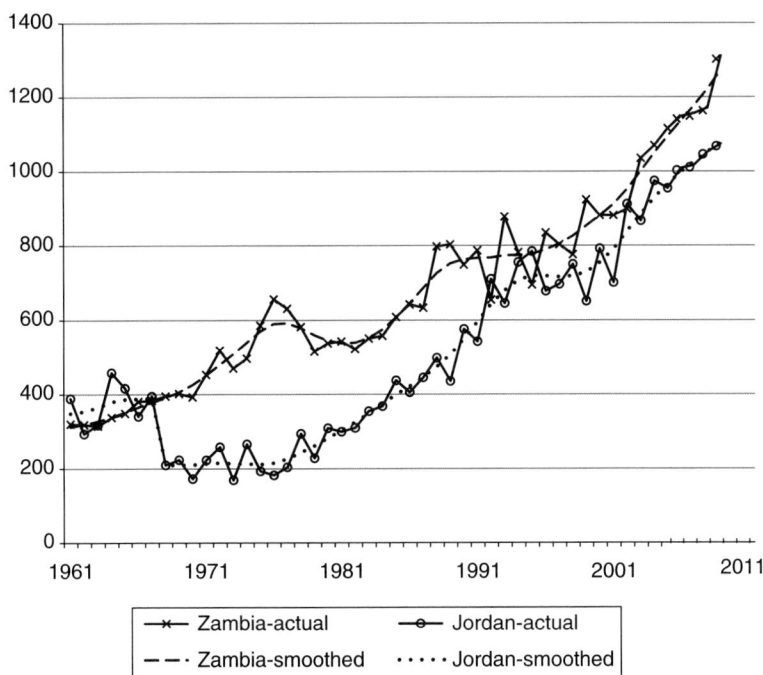

Fig. 16.1. The effects of smoothing on gross agricultural output measures. The dashed curves are output series that have been smoothed using the Hodrick–Prescott filter. This is meant to remove some of the annual fluctuations in output caused by weather and other short-run disturbances but preserve sufficient curvature to capture productivity trends.

For Brazil, I use results of the recently published 2006 Brazilian agricultural census (IBGE) and for Indonesia, I compiled improved data on agricultural land and machinery use (Fuglie, 2010a). For Taiwan, I use statistics from the Council of Agriculture. For the countries of the former Soviet Union, FAO reports data only from 1991 and onward. I extend the time series for each of the former Soviet Socialist Republics (SSRs) back to 1965 from Shend (1993). Also, because FAO labour force estimates for former SSRs and Eastern Europe are not reliable for the post-1990 years (Lerman et al., 2003; Swinnen et al., 2005), sources I use for agricultural labour data are EUROSTAT for the Baltic states and Eastern Europe, CISSTAT for Russia, Belorussia and Moldova, the International Labor Organization's LABORSTA for Ukraine, and national data reported by the Asian Development Bank for Asiatic former Soviet republics.

Inputs are divided into five categories. *Farm labour* is the total economically active adult population (males and females) in agriculture. *Agricultural land* is the area in permanent crops (perennials), annual crops, and permanent pasture. Cropland (permanent and annual crops) is further divided into rainfed cropland and cropland equipped for irrigation. For agricultural cropland in sub-Saharan Africa, however, I use total area harvested for all crops rather than the FAO series on arable land (see Fuglie and Rada in Chapter 12 of this volume for a discussion of why this series seems to be a better measure of agricultural land in this region). For China I use sown crop area for cropland in that country, given unreasonably discontinuities in both the FAO and Economic Research Service's arable land series for China.[5] I then aggregate rainfed cropland, irrigated area and permanent pasture into a quality-adjusted measure that gives greater weight to irrigated cropland and less weight to permanent pasture in assessing agricultural land changes over time (see the next section on 'land quality'). *Livestock* is the aggregate number of animals in 'cattle equivalents' held in farm inventories and includes cattle, camels, water buffalos, horses and other equine

species (asses, mules and hinnies), small ruminants (sheep and goats), pigs and poultry species (chickens, ducks and turkeys), with each species weighted by its relative size. The weights for aggregation are based on Hayami and Ruttan (1985, p. 450): 1.38 for camels, 1.25 for water buffalo and horses, 1.00 for cattle and other equine species, 0.25 for pigs, 0.13 for small ruminants and 12.50 per 1000 head of poultry. *Fertilizer* is the amount of major inorganic nutrients applied to agricultural land annually, measured as metric tonnes of N, P_2O_5 and K_2O nutrients. *Farm machinery* is an aggregation of four-wheel riding tractors, two-wheel pedestrian tractors, and power harvester-threshers in use, adjusted by the average metric horse-power for each type of machine. The FAO reports time series data for only four-wheel tractors and harvest-threshers; it recorded information on two-wheel tractors in the 1970s then discontinued this series until recommencing it again in 2002. For interim years I collected national farm machinery statistics on two-wheel tractors for the following Asian countries: China, Japan, South Korea, Taiwan, Thailand, Philippines, Indonesia, India, Bangladesh, Pakistan and Sri Lanka. These are the main countries where pedestrian tractors are widely employed. For aggregation purposes, I assume the following average metric horsepower (CV) per machine: 40 cv for four-wheel tractors, 12 cv for two-wheel tractors, and 25 cv for power combines.[6]

Although these inputs account for the major part of total agricultural input usage, there are a few types of inputs for which complete country-level data are lacking, namely, use of chemical pesticides, seed, prepared animal feed, veterinary pharmaceuticals, energy and farm structures. More detailed input data are, however, available for several of the countries from which I have data on input cost shares. To account for these inputs, I assume that their growth rate is correlated with one of the five input variables just described and include their cost with the related input. For example, services from capital in farm structures as well as irrigation fees are included with the agricultural land cost share; the cost of chemical pesticide and seed is included with the fertilizer cost share; costs of animal feed and veterinary medicines are included in the livestock cost share, and other farm machinery and energy costs are included in the tractor cost share. So long as the growth rates for the observed inputs and their unobserved counterparts are similar, then the model captures the growth of these inputs in the aggregate input index.

Land quality

The FAO agricultural database provides time-series estimates of agricultural land by country and categorizes this as either cropland (arable and permanent crops) or permanent pasture. It also provides an estimate of area equipped for irrigation. The productive capacity of land among these categories and across countries can be very different, however. For example, some countries count vast expanses of semi-arid lands as permanent pastures even though these areas produce very limited agricultural output. Using such data for international comparisons of agricultural productivity can lead to serious distortions, such as significantly biasing downward the econometric estimates of the production elasticity of agricultural land (Peterson, 1987; Craig et al., 1997).

In this study, because I estimate only productivity growth rather than productivity levels, differences in land quality across countries is less of a problem. The estimates depend only on changes in agricultural land and other inputs over time. A bias might arise, however, if changes occur unevenly among land classes. For example, adding an acre of irrigated land would probably make a considerably larger contribution to output growth than adding an acre of rainfed cropland or pasture. To account for the contributions to growth from different land types, I derive weights for irrigated cropland, rainfed cropland, and permanent pastures on the basis of their relative productivity and allow these weights to vary regionally. So as not to confound the land quality weights with productivity change itself, the weights are estimated using country-level data from the beginning of the period of study (i.e. using average annual data from 1961–1965). I first construct regional indicator variables ($REGION_i$, i = 1,2,...5, representing developed and former Soviet countries, Asia-Pacific, Latin America and the Caribbean, West Asia and North Africa, and sub-Saharan Africa), and then regress the log of agricultural land yield against the proportions of agricultural land in rainfed cropland ($RAINFED$), permanent pasture ($PASTURE$), and irrigated cropland ($IRRIG$). Including slope indicator variables allows the coefficients to vary among regions (Eqn 16.9) (see equation 16.9 at bottom of page):

The coefficient vectors α, β and γ provide the quality weights for aggregating the three land types into an aggregate land input index. Countries with a higher proportion of irrigated land are likely to have higher average land productivity, as will countries with more cropland relative to pasture. The estimates of the parameters in Eqn 16.9 reflect these differences and provide a ready means of weighting the relative qualities of these land classes. Because of the limited amount of irrigated cropland in some regions, the coefficient on $IRRIG$ was held constant across all developing country regions.

Coefficient estimates for each region were divided by $α_i$. Thus, the normalized β and γ coefficients indicate the productivity of pasture and irrigated land relative to rainfed cropland (the normalized α coefficients equal 1). The regression estimates show that, on average, one hectare of irrigated land was between two and three times as productive as rainfed cropland, which in turn was 10–20 times as productive as permanent pasture,

$$\ln\left(\frac{Ag\ output}{Cropland + Pasture}\right) = \sum_i \alpha_i \left(RAINFED * REGION_i\right) + \sum_i \beta_i \left(PASTURE * REGION_i\right)$$
$$+ \sum_i \gamma_i \left(IRRIG * REGION_i\right). \tag{16.9}$$

with some variation across regions (see the lower part of Table 16.1 for the normalized land quality coefficients for each region). The results give plausible weights for aggregating agricultural land across broad quality classes. Indeed, this approach to account for land quality differences among countries is similar to one developed by Peterson (1987), who derived land quality weights by regressing average cropland values in US states against the share of irrigated and un-irrigated cropland and long-run average rainfall. He then applied these regression coefficients to data from other countries to derive an international land quality index. The advantage of my model is that it is based on international rather than US land yield data and provides results for a larger set of countries.

The effects of this land quality adjustment on global land-use change are shown in Table 16.1. When summed up using unadjusted data, between 1961 and 2009 total global agricultural land expanded from 4437 million ha to 4880 million ha or by about 10%. When adjusted for quality, 'effective' agricultural land expanded by 31%, or three times the rate of growth in raw area. The reason is that irrigated area expanded much faster than other types of land and, when weighted for its greater productivity, it implies a much greater expansion in 'effective' agricultural land. For the purpose of TFP calculation, accounting for the changes in the quality of agricultural land over time increases the growth rate in total agricultural inputs and commensurately reduces the estimated growth in TFP.

This adjustment for changes in different classes of land allows us to refine further the resource decomposition of output growth in Eqn 16.6 to isolate the contribution of irrigation apart from expansion in cropland area to output growth. Letting X_1 be the quality-adjusted quantity of (rainfed cropland equivalent) land, a change in X_1 is given by:

$$\Delta X_1 = \Delta(Cropland) + \beta\Delta(Pasture)$$
$$+ (\gamma - 1)\Delta(Irrigated\ area). \quad (16.10)$$

The first two terms indicate the expansion in land area (with growth in pasture area adjusted for quality to put it on comparable terms with cropland expansion). The third term isolated the contribution to growth from irrigation expansion: $(\gamma - 1) * 100\%$ gives the percent augmentation to yield by equipping an acre of cropland with supplemental irrigation. Dividing Eqn 16.7 by X_1 converts the expression into percentage changes so that it shows the respective contributions of changes in rainfed cropland, pasture area and irrigation to output growth. Combined with Eqn 16.6, the resource decomposition expression shows the contributions to agricultural growth from changes in agricultural land, water resource use, other inputs per hectare of land and TFP.

Input cost shares

The FAO (and supplementary) quantity data allow us to calculate the growth rates for five categories of production inputs (land, labour, machinery capital, livestock capital and material inputs represented by fertilizer), but to combine these into an aggregate input measure requires information on their cost shares or production elasticities. For this I draw upon other productivity studies that have compiled relatively complete measurements for selected countries and then assign these as 'representative' input cost shares for different regions of the world. Table A16.2 in Appendix 16.1 shows the input cost shares or production elasticities compiled from 14 studies (eight from developed countries, six from developing countries and two from transition countries or regions) and the regions to which these were applied for the purpose of input aggregation. For instance, the cost shares for Brazil were applied to South America, West Asia and North Africa, the cost shares for India were applied to other countries in South Asia and the cost shares for Indonesia were applied to developing countries in South-east Asia and Oceania. These assignments were based on judgements about the resemblance among the agricultural sectors of these countries. Countries assigned to the cost shares from Brazil tended to be middle-income countries having relatively large livestock sectors, for example.

Table 16.1. Global agricultural land use changes between 1961 and 2009.

Total agricultural land (millions of hectares)

Region	Rainfed cropland			Irrigated cropland			Permanent pasture			Total agricultural land		
	1961	2009	% change	1961	2009	% change	1961	2009	% change	1961	2009	% change
Developed countries	391	371	−5	28	47	66	886	767	−13	1277	1139	−11
Transition countries	283	246	−13	11	25	123	358	378	6	641	624	−3
Developing countries	666	938	41	100	233	132	1853	2180	18	2519	3117	24
World	1340	1555	16	140	305	118	3097	3325	7	4437	4880	10

Total agricultural land in quality-adjusted units (millions of hectares of 'rainfed cropland equivalents')

Region	Rainfed cropland			Irrigated cropland			Permanent pasture			Total agricultural land		
	1961	2009	% change	1961	2009	% change	1961	2009	% change	1961	2009	% change
Developed countries	391	371	−5	61	101	66	84	72	−13	535	544	2
Transition countries	283	246	−13	28	61	123	10	11	6	320	318	−1
Developing countries	666	938	41	215	501	132	175	205	18	1056	1644	56
World	1340	1555	16	304	662	118	268	289	8	1912	2506	31

Land quality adjustment factors

	World	DC	LDC	SSA	LAC	WANA	Asia LDC
Rainfed cropland	1.00	1.00	1.00	1.00	1.00	1.00	1.00
Irrigated cropland	2.13	2.15	2.50	1.74	1.01	1.45	2.99
Permanent pasture	0.03	0.09	0.03	0.02	0.03	0.02	0.06

DC, developed and transition countries; LDC, less developed countries; SSA, sub-Saharan Africa; LAC, Latin America & Caribbean; WANA, West Asia and North Africa. *Source:* Agricultural land area from FAO, with adjustments made for Indonesia and China. Cropland includes FAO's measure of arable land and land under permanent crops except for sub-Saharan Africa, where cropland equals total area harvested. Cropland for China is total sown area. Land quality adjustments reflect the average productivity of different land types relative to rainfed cropland and are derived from regressions (see text).

Although the assignment of cost shares to countries lacking input cost data is unfortunate, an argument in favour is that there is a significant degree of congruence among the cost shares reported for the country studies shown in Table A16.2. For the developing-country cases (India, Indonesia, China, Brazil, Mexico and sub-Saharan Africa), the cost shares indicate that traditionally farm-supplied inputs (land, labour and livestock capital) dominate the agricultural production process. These three input classes accounted for between 60% and 98% of total resources in production, whereas inputs supplied by industry (machinery, or fixed capital, and purchased materials such as fertilizers), accounted for a far smaller share of resources. The cost share of inputs supplied by industry rises with the income of a country, and accounts for a third or more of total costs in the more highly industrialized countries. The use of modern inputs in transition countries, on the other hand, fell sharply after reforms were initiated in the early 1990s, and this is reflected in the cost shares for these countries.

Country and regional productivity

The methodology and data described above allow me to calculate agricultural TFP indexes for nearly every country of the world on an annual basis since 1961. Some countries have dissolved or are too small to have complete data, however. For the purpose of estimating long-run productivity trends, I aggregate some national data to create consistent political units over time. For example, data from the nations that formerly constituted Yugoslavia are aggregated to make comparisons with productivity before Yugoslavia's dissolution; data were aggregated similarly for Czechoslovakia, Ethiopia and the former Soviet Union (I also construct TFP series for individual SSRs beginning in 1965). Because some small island nations have incomplete or zero values for some agricultural data, I constructed three composite 'countries' by aggregating available data for island states in the Lesser Antilles, Micronesia and Polynesia. The countries included in the analysis account for more than 99.7% of FAO's global gross agricultural

output. The only areas not included in the analysis that have significant agricultural production are the West Bank and Gaza.

In addition to individual countries, I aggregate the data and construct TFP indexes at the regional level. Input and output quantity aggregation is straightforward because they are all measured in the same units (although not adjusted for quality differences in the inputs). To obtain cost shares at the regional level, I take the weighted averages of the cost shares for the countries composing that region. The weights are the country's share of total costs (or revenue) within the region. In this way, I obtain TFP indexes for 'North America', 'Transition countries of the former Soviet bloc', 'the Sahel', etc. Table 16.2 provides a complete list of countries included in the analysis and their regional groupings.

16.3 Results

16.3.1 Growth rates for agricultural total factor productivity

Table 16.3 provides productivity measures for the global agricultural economy as a whole. The figures show average annual growth rates by decade since 1961. Output growth has remained remarkably consistent over time, 2.7% per year in the 1960s and between 2.1% and 2.5% per year every decade since then. The source of output growth, however, shifted from being primarily input driven to productivity driven. Annual growth in total inputs fell from 2.5% in the 1960s to 0.7% in the 2000s (it was even lower in the 1990s but this was affected by a sharp contraction in the agricultural sector of the former Soviet bloc countries). Annual TFP growth, meanwhile, rose from 0.2% in the 1960s to about 1.7% since 1990.

Labour productivity growth has tended to lag growth in land productivity (because the number of workers in agriculture has been expanding faster than agricultural land area), but labour productivity growth accelerated after the 1980s and was growing at about 2.3% during 2001–2009.

Table 16.2. Countries and regional groupings included in the productivity analysis.

Sub-Saharan Africa (SSA)

Central	Eastern	Horn	Sahel	Southern	Western	Nigeria
Cameroon	Burundi	Djibouti	Burk. Faso	Angola	Benin	
CAR	Kenya	Ethiopia[a]	C. Verde	Botswana	Côte d'Ivoire	
Congo	Rwanda	Somalia	Chad	Comoros	Ghana	
Congo, DR	Seychelles	Sudan	Gambia	Lesotho	Guinea	
Eq. Guinea	Tanzania		Mali	Madagascar	G. Bissau	
Gabon	Uganda		Mauritania	Malawi	Liberia	
São Tomé &			Niger	Mauritius	Sierra Leone	
Principe			Senegal	Mozambique	Togo	
				Namibia		
				Réunion		
				Swaziland		
				Zambia		
				Zimbabwe		

Latin America & Caribbean (LAC)

Northeast	Andes	S. Cone	C. America	Caribbean	N. America	Africa, Developed
Brazil	Bolivia	Argentina	Belize	Bahamas	Canada	South Africa
Fr. Guiana	Colombia	Chile	Costa Rica	Cuba	USA	
Guyana	Ecuador	Paraguay	El Salvador	Dom. Rep.		
Suriname	Peru	Uruguay	Guatemala	Haiti		
	Venezuela		Honduras	Jamaica		
			Mexico	Les. Antilles[b]		
			Nicaragua	Puerto Rico		
			Panama	Trin. & Tob.		

Asia / Former Soviet Union

Developed	NE Asia, LDC	SE Asia	South Asia	Baltic	E. Europe	CAC
Japan	China	Brunei	Afghanistan	Estonia	Belarus	Armenia
Korea, Rep.	Korea, DPR	Cambodia	Bhutan	Latvia	Kazakhstan	Azerbaijan
Taiwan	Mongolia	Indonesia	Nepal	Lithuania	Moldova	Georgia
Singapore		Laos	Sri Lanka		Russia	Kyrgyzstan
		Malaysia	Bangladesh		Ukraine	Tajikistan
		Myanmar	India			Turkmenistan
		Philippines	Pakistan			Uzbekistan
		Thailand				
		VietNam				

Europe / West Asia & North Africa / Oceania

Northwest	Southern	Transition	West Asia	North Africa	Developed	Developing
Austria	Cyprus	Albania	Bahrain	Algeria	Australia	Fiji
Belgium–Lux.	Greece	Bulgaria	Iran	Egypt	N. Zealand	Micronesia[b]
Denmark	Italy	Czechoslovakia[a]	Iraq	Libya		N. Caledonia
Finland	Malta	Hungary	Israel	Morocco		PNG
France	Portugal	Poland	Jordan	Tunisia		Polynesia[b]
Germany	Spain	Romania	Kuwait			Solomon Is.

Continued

Table 16.2. Continued.

	Europe			West Asia & North Africa		Oceania	
Northwest	Southern	Transition	West Asia	North Africa	Developed	Developing
Iceland		Yugoslavia[a]	Lebanon			Vanuatu
Ireland			Oman			
Netherlands			Qatar			
Norway			S. Arabia			
Sweden			Syria			
Switzerland			Turkey			
UK			UAR			
			Yemen			

LDC, developing countries. CAC, C. Asia & Caucasia. [a]Statistics from the successor states of Ethiopia (Ethiopia and Eritrea), Czechoslovakia (Czech and Slovak Republics) and Yugoslavia (Slovenia, Croatia, Bosnia, Macedonia, Serbia and Montenegro) were merged to form continuous time series from 1961 to 2009. [b]Composite countries composed of several small island nations.

Table 16.3. Productivity indicators for world agriculture.

Period	Gross output	Total input	Total factor productivity	Output per worker	Output per hectare	Cereal yield
			Average annual growth rate (%)			
1961–1970	2.74	2.55	0.18	1.13	2.45	2.88
1971–1980	2.30	1.70	0.60	1.58	2.09	2.08
1981–1990	2.12	1.50	0.62	0.62	1.75	1.88
1991–2000	2.21	0.55	1.65	2.00	2.16	1.57
2001–2009	2.49	0.65	1.84	2.80	2.64	1.80
1971–1990	2.25	1.53	0.72	1.11	1.97	2.25
1991–2009	2.29	0.70	1.59	1.97	2.27	1.42
1961–2009	2.23	1.28	0.95	1.19	2.00	1.99

Gross output: FAO gross production value in constant 2004–2006 international dollars. Total input: author's aggregation of agricultural land, labour, capital and material inputs (see text). TFP: the difference between output growth and total input growth, based on the author's estimation. Output per worker: FAO gross production value divided by number of persons working in agriculture. Output per hectare: FAO gross production value divided by total arable land and permanent pasture. Cereal yield: global production of maize, rice and wheat divided by area harvested of these crops. The average annual growth rate in series Y is found by regressing the natural log of Y against time, i.e. the parameter B in $\ln(Y) = A + Bt$.

Growth in agricultural output per total agricultural land area (total yield) has mimicked the trends in output growth, remaining fairly steady around an average of 2.1% per year during the past 50 years. The growth rate in cereal yield, however, showed signs of slowing after 1990. Global cereal yield was increasing by about 2.5% per year in the 1970s and 1980s but by only 1.3% per year during 1991–2009. The decline in cereal yield growth does not, however, seem to be representative of agriculture as a whole. It has been offset by productivity improvements elsewhere – rising yield growth in other commodities and greater intensification of land use – to keep total output per hectare of agricultural land rising at historical rates. Note that growth in global agricultural TFP is generally lower than growth in both land productivity and labour productivity. This reflects an intensification of capital improvements and material inputs in agriculture, which raise land and labour productivity but are removed from growth in TFP.

The decomposition of global output growth into contributions from inputs and TFP is depicted in Fig.16.2. Figure 16.2a shows the contributions of various inputs to growth according to their share of total costs (see Eqn 16.4), and the residual (output growth above total input growth) that we define as TFP. The height of each column gives the average annual rate of growth in output over the period. The first column shows the average over the entire 1961–2009 period and the following columns show growth by decade. During this 48-year period, total inputs grew about 60% as fast as gross agricultural output, implying that improvement in TFP accounted for about 40% of output growth. The contribution of TFP to output growth, however, grew over time, and by the most recent decade (2001–2009), TFP accounted for 74% of the growth in global agricultural production.

Fig. 16.2a shows the changing composition of input growth over time. Growth in material inputs, especially fertilizers, was a leading source of agricultural growth in the 1960s and 1970s, when green revolution cereal crop varieties became widely available in developing countries. Fertilizer use also expanded considerably in the Soviet Union during these decades, where they were heavily subsidized. The exceptionally low rate of input growth in global agriculture during the 1990s was due primarily to the rapid withdrawal of resources from agriculture in the countries of the former Soviet bloc. By the early 2000s agricultural resources in this region had stabilized and there was a recovery in the rate of global input growth compared with the 1990s. Growth in agricultural labour tends to follow population growth rates in low income countries but turns negative through structural transformation when countries become richer (see Binswanger-Mkhize and d'Souza, Chapter 9 of this volume). By the most recent decade (2001–2009), the global agricultural labour probably peaked, as declining agricultural employment in developed countries, transition countries, Latin America and China offset

rising agricultural employment in other developing countries, most notably in sub-Saharan Africa and South Asia.

Figure 16.2b decomposes the sources of global agricultural growth slightly differently. Instead of by input cost, it shows the relative contribution of land and irrigation expansion, input intensification on land and TFP (see Eqns 16.6 and 6.10). The rate of expansion in natural resources (land and water) has diminished over time, whereas the rate of growth in resource yield has risen. The source of yield gain has, however, shifted markedly from input intensification to improvement in TFP.

The estimates of global agricultural output and TFP growth are disaggregated among regions and sub-regions in Table 16.4 (results for specific countries are given in Appendix Table A16.2). The regional results reveal that the global trend is hardly uniform, with three general patterns evident:

1. **In developed regions**, total agricultural inputs have been declining since the 1980s (output growth is less than TFP growth) and at an increasing rate; TFP growth offset the declining resource base to keep output from falling and has remained robust (above 1.5% per year in all regions except Oceania (Australia and New Zealand).

2. **In developing regions**, productivity growth doubled between the 1960s–1980s and the 1990s–2000s, from less than 1% to more than 2% per year. Input growth has been slowing each decade but still expanding enough to keep output growing at more than 3% annually for each of the last three decades. Two large developing countries in particular, China and Brazil, have sustained exceptionally high TFP growth. Several other developing regions, including South-east Asia, North Africa, Central America and the Andean region, also registered accelerated TFP growth in the 1990s or 2000s. The major exception is the developing countries of sub-Saharan Africa where long-run TFP growth remained below 1% per year.

3. **In transition countries**, the dissolution of the Soviet Union in 1991 imparted a major shock to agriculture as these countries made a transition from centrally planned to

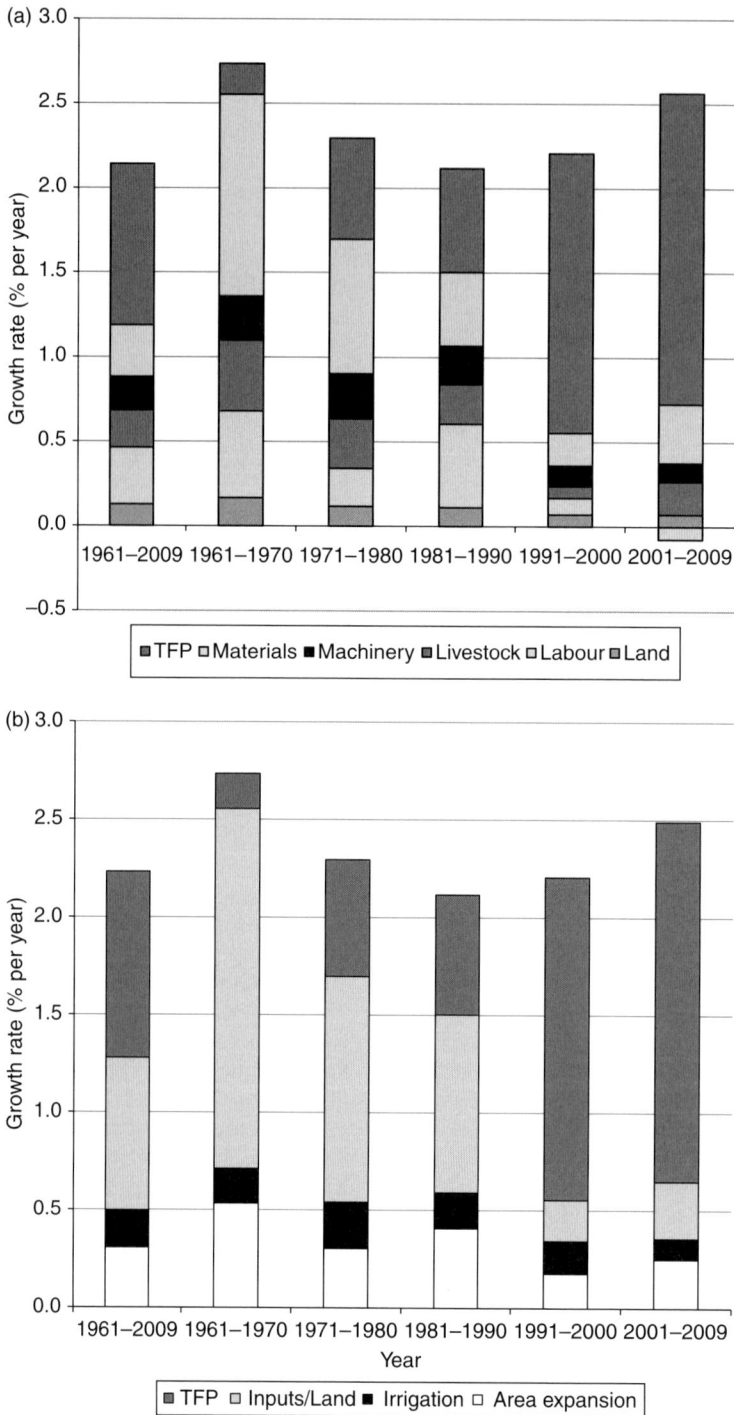

Fig. 16.2. Sources of global agricultural growth. (a) Input cost decomposition. (b) Resource decomposition. The height of the bar shows the average annual growth rate in gross agricultural output during the period specified. The shaded components of the bar show the contribution of that component to total output growth.

Table 16.4. Agricultural output and productivity growth for global regions by decade.

Region	Agricultural output growth (annual %)					Agricultural TFP growth (annual %)				
	1961–1970	1971–1980	1981–1990	1991–2000	2001–2009	1961–1970	1971–1980	1981–1990	1991–2000	2001–2009
All developing countries	3.15	2.97	3.43	3.64	3.34	0.69	0.93	1.12	2.22	2.21
Sub-Saharan Africa	2.95	1.19	2.82	3.05	2.69	0.17	-0.05	0.76	0.99	0.51
Latin America & Caribbean	3.05	3.31	2.26	3.14	3.41	0.84	1.21	0.99	2.30	2.74
Caribbean	1.70	1.97	0.68	-0.73	-0.18	-1.00	0.57	-0.26	-0.55	-0.16
Central America	4.63	3.72	1.36	2.95	2.24	2.83	1.95	-1.69	3.05	2.33
Andean countries	2.97	2.75	2.77	3.08	3.19	1.49	1.18	0.55	2.12	2.60
Northeast (Brazil, mainly)	3.56	3.86	3.41	3.65	4.44	0.25	0.60	3.02	2.62	4.03
Southern Cone	1.80	2.87	1.13	3.15	2.79	0.58	2.56	-0.82	1.61	1.29
Asia (except West Asia)	3.26	3.10	3.67	3.78	3.41	0.91	1.17	1.42	2.73	2.78
Northeast (China, mainly)	4.79	3.32	4.49	5.17	3.39	0.94	0.67	1.71	4.10	3.05
Southeast Asia	2.63	3.92	3.31	2.89	4.45	0.57	2.10	0.54	1.69	3.29
South Asia	2.02	2.66	3.31	2.65	3.32	0.63	0.86	1.31	1.22	1.96
West Asia & North Africa	2.87	3.05	3.64	2.82	2.35	1.40	1.66	1.63	1.74	1.88
North Africa	2.62	1.58	4.53	3.34	3.57	1.32	0.48	3.09	2.03	3.04
West Asia	2.98	3.65	3.29	2.60	1.77	1.21	2.21	0.95	1.70	1.34
Oceania	2.53	2.34	1.58	2.07	2.29	-0.14	0.47	-0.73	0.54	1.33
All Developed Countries	2.05	1.93	0.72	1.37	0.58	0.99	1.64	1.36	2.23	2.44
USA & Canada	2.06	2.29	0.68	1.96	1.41	1.25	1.67	1.31	2.18	2.24
Europe (except FSU)	1.96	1.60	0.42	0.24	-0.16	0.58	1.44	1.43	1.25	1.98
Europe, Northwest	1.56	1.36	0.51	0.34	-0.09	0.85	1.48	1.55	1.80	2.75
Europe, Southern	2.11	1.96	0.69	1.32	-0.42	1.97	2.03	1.30	2.42	3.04
Australia & New Zealand	2.90	1.68	1.48	3.21	-0.22	0.72	1.53	1.35	2.62	1.09
NE Asia, developed	3.31	2.23	1.23	0.18	-0.24	2.34	2.46	1.74	2.23	2.07
Transition Countries	3.27	1.32	0.85	-3.51	1.96	0.57	-0.11	0.58	0.78	2.28
Eastern Europe	2.67	1.73	-0.04	-1.35	0.04	0.54	0.59	0.81	0.79	0.78
Former Soviet Union (FSU)	3.59	1.10	1.30	-4.69	2.96	0.53	-0.51	0.63	0.59	3.29
Baltic[a]	3.56	0.93	1.09	-6.01	2.10	2.11	-0.49	0.58	0.82	2.20
Central Asia & Caucasia[a]	3.41	4.71	0.56	0.08	4.33	-0.36	2.02	-0.89	0.65	2.45
Eastern Europe FSU[a]	3.16	0.76	1.39	-5.39	2.70	0.89	-0.85	0.86	0.92	4.00
World	2.74	2.30	2.12	2.21	2.49	0.18	0.60	0.62	1.65	1.84

[a]Data for former Soviet republics covers 1965–2009 only. The average annual growth rate in series Y is found by regressing the natural log of Y against time, i.e. the parameter B in ln(Y) = A + Bt. Source: Author's estimates. See Table 16.3 for list of countries in each regional group.

market-oriented economies. In the 1990s, agricultural resources sharply contracted and output fell. Total agricultural inputs were still declining in 2001–2009 but at a much slower rate than during 1991–2000. Productivity growth, which was minimal during the USSR era, took off in 2001–2009. As a result, output growth again turned positive. Gross agricultural output in 2009 was, however, below Soviet-era levels in every region except Central Asia and Caucasia (CAC).

The strong and sustained productivity growth described here is broadly consistent with results of the detailed country and regional case studies presented in the other chapters of this volume. Among industrialized countries, agricultural TFP growth has remained at historical levels in the USA (Wang *et al.*, Chapter 2, this volume), Canada (Cahill and Rich, Chapter 3, this volume) and Western Europe (Wang *et al.*, Chapter 5, this volume), but has fallen in Australia (Zhao *et al.*, Chapter 4, this volume) and South Africa (Liebenberg, Chapter 14, this volume). The case studies found evidence that these patterns were correlated with the rate of growth in public investments in agriculture, particularly in R&D.

For transition countries, Swinnen *et al.* (Chapter 6, this volume) provide an explanation for the renewed but uneven recovery of agricultural productivity in this region. They find it to be related to the pace of economic reforms implemented since the collapse of the Soviet Union, especially in the institutions governing land and labour relations and in the functioning of agricultural markets. Similarly to what happened earlier (and more smoothly) in China, moving from collective and state-owned corporate farming responding to state mandates to privately (especially family-) operated farms responding to market incentives brought significant gains in efficiency (Rozelle and Swinnen, 2004). Once the initial gains from institutional reform were realized, China was able to sustain productivity growth through technological change (Tong *et al.*, Chapter 8, this volume). Whether this pattern will also be followed in the countries of the former Soviet Union and Eastern Europe remains to

be seen; it will likely depend on their policies governing the development of and access to new agricultural technology.

For developing countries, the robust growth in agricultural TFP over the past one to three decades measured for Brazil (Gasques *et al.*, Chapter 7, this volume), China (Tong *et al.*, Chapter 8, this volume), and Indonesia (Rada and Fuglie, Chapter 10, this volume) is consistent with the results presented here, as is the result of relatively low TFP growth for sub-Saharan Africa (Fuglie and Rada, Chapter 12; Nin-Pratt and Yu, Chapter 13, this volume). The Indian productivity trend reported by Binswanger-Mkhize and d'Souza (Chapter 9, this volume) is drawn directly from my estimates. India represents a middle case of moderate TFP growth of about 1.3% per year since the 1970s–1990s, although in 2001–2009 it seemed to also accelerate to more than 2% per year. Binswanger-Mkhize and d'Souza argue that India will need to achieve strong agricultural TFP growth if the sector is to be a major source of employment generation and poverty reduction if the country is to be a major source of employment generation and poverty reduction if the country. Finally, for Thailand, my results track the TFP growth estimates of Suphannachart and Warr (Chapter 11, this volume) closely for 1961–1993 but then diverge. For the years after 1993 I find continued TFP improvement while they find falling TFP. The principal reason for this difference appears to be a higher input cost share that Suphannachart and Warr give to agricultural capital stock, which in turn results in a higher rate of growth in total inputs. As Butzer *et al.* explain in Chapter 15 (this volume), internationally comparable measures of capital stock and the cost of capital services have been lacking for agriculture, and this can confound analyses of productivity and growth. More complete and comparable data on agricultural capital is one of the most pressing needs to improve our ability to assess long-term trends in global agricultural productivity.

16.3.2   Technology capital and TFP growth

What explains the apparent acceleration in agricultural TFP growth in developing countries, or at least in many of them? The case studies in this volume identified institutional and economic reforms as an important source of productivity growth, at least in the medium term, and research and development for sustaining productivity growth over the long term. The model described above on technology capital and TFP growth examines this question for a group of 87 developing countries during a 40-year period.

Table 16.5 shows the econometric estimates of Eqn 16.7, where long-run average TFP growth rates for 87 developing countries are regressed against combinations of innovation–invention and technology–mastery capitals. The regression coefficients in Table 16.5 are arrayed in a matrix corresponding to the II and TM combinations to which they refer. The coefficient estimates reflect the average annual TFP growth rate (percentage) for all countries having technology capital in that II and TM class. The numbers in parentheses below the coefficients indicate the number of observations that fell in that class. For example, there were 18 countries that were characterized as having little or no technology capital (II class=2 and TM class=2). These countries as a group achieved a mean annual TFP growth of 0.41%, which was not significantly different from zero. At the other end of the technology capital scale there were two countries with II class=6 and TM class=6, and these achieved an average annual TFP growth rate of 3.29%. These countries are Brazil and China, large countries that have invested heavily in agricultural research and extension. Figure 16.3 plots out these coefficients visually. There is a clear progression to higher TFP growth as countries increase II and TM technology capital. Countries needed, however, a minimal capacity in both research and extension-schooling in order to sustain significant productivity growth. When either II capital or TM capital were at very low levels (class 2), mean TFP growth rates were not significantly different from zero. With one exception, technology capitals of (II,TM) combinations of (3,3) and higher were all associated with positive and significant TFP

Table 16.5. Technology capital and agricultural TFP growth.

		Invention–Innovation (II) class (Agricultural research and industry R&D)					Marginal effect of II holding TM fixed
		2	3	4	5	6	
Technology mastery (TM) class (Agricultural extension + schooling)		Coefficients show average annual TFP growth rate (percentage) (number in parentheses is number of observations with II–TM combination)					
	2	0.41 (n=18)	0.64 (n=14)	0.42 (n=8)	0.42 (n=1)		$F_{(3,155)}=$ 0.10 ns
	3	−0.01 (n=9)	1.03*** (n=25)	1.44*** (n=15)	1.20* (n=2)		$F_{(3,155)}=$ 2.48 ^
	4	0.35 (n=4)	0.76** (n=12)	1.34*** (n=29)	2.07*** (n=8)	1.14* (n=2)	$F_{(4,155)}=$ 1.79 ^
	5		0.21 (n=2)	1.44** (n=7)	1.93*** (n=9)	2.03 (n=2)**	$F_{(3,155)}=$ 1.10 ns
	6				1.15** (n=5)	3.29*** (n=2)	$F_{(1,155)}=$ 3.99 ^^

F-test of marginal effect of TM holding II fixed

	$F_{(2,155)}=$ 0.32 ns	$F_{(3,155)}=$ 0.48 ns	$F_{(3,155)}=$ 1.33 ns	$F_{(4,155)}=$ 0.79 ns	$F_{(2,155)}=$ 1.42 ns

*,**,*** indicate coefficients are significant from zero at 10%, 5% and 1% significance level, respectively. ^,^^ indicate rejection of hypothesis that all coefficients in row or column are equal at 10% and 5% significance level and 'ns' indicates cannot reject hypothesis of equal coefficients. Data sample: 87 developing countries over two periods.

Number of obs =	174	$F_{(18,155)}=$	2.06	Prob > F =	0.010
R-squared =	0.193	Adj R-sqr =	0.100	Root MSE =	0.013

Source: Author's estimates of equation (16.7).

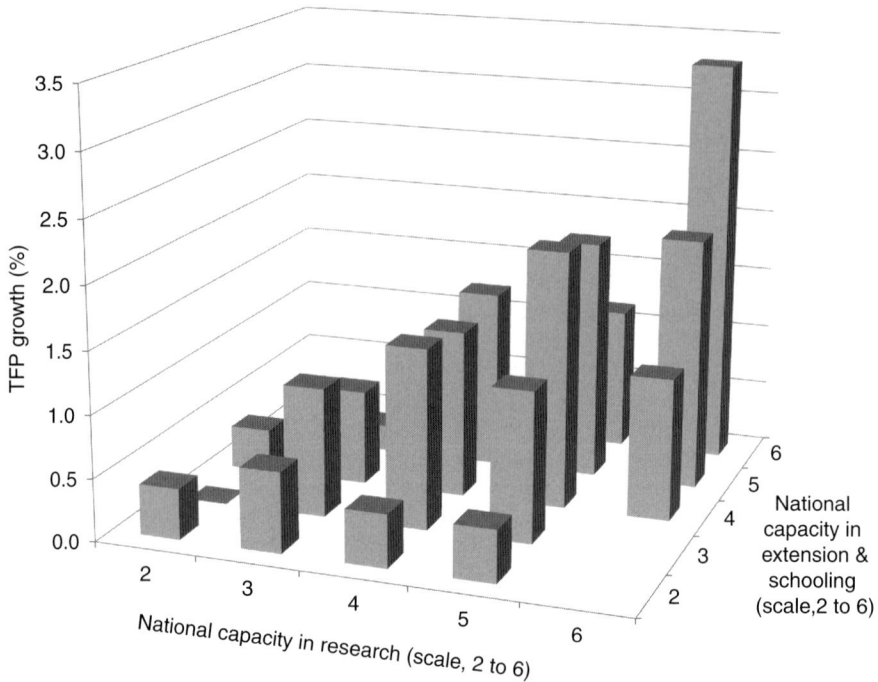

Fig. 16.3. Technology capital and agricultural TFP growth. Source: Author's estimates.

growth. The exception is (II,TM) class (3,5) that consists of only two countries – Panama in 1971–1990 and Zimbabwe in 1991–2009. Both of these countries suffered from political instability and poor macroeconomic performance over these periods, which might account for their low agricultural productivity growth (0.21% per year on average) despite significant levels of extension-schooling and some research capacity.

The F-statistic tests reported in the final column and row of Table 16.5 examine the marginal effects of research and extension holding the other fixed. Casual observation indicates that TFP growth rates tended to rise at higher levels of either II or TM capital (holding the other fixed), but the F-statistic tests the hypothesis that all of the row (or column) coefficients are equal. In other words, it tests the hypothesis that there was no significant increase in TFP growth with a marginal increase in one of the types of technology capital. Neither II capital (research) or TM capital (extension and schooling) was effective at raising agricultural TFP growth

without at least a minimal capacity in the other. But in the case of research, TFP growth rose significantly with marginal increases in II capital when TM capital was held constant at level 3, 4 and 6 (TFP growth also rose when TM capital was held fixed at 5 but the increase in TFP growth was not statistically significant). On the other hand, in no case did a marginal increase in TM capital significantly increase TFP growth when II capital remained constant. In other words, agricultural extension and schooling do not seem to be substitutes for research and development capacity. Improved capacity to invent and adapt new technology to country-specific conditions was a requisite for sustaining long-run TFP growth in agriculture.

What the above estimates demonstrate is that countries with higher levels of II and TM capitals experienced more rapid agricultural TFP growth. But it could be that unobserved characteristic of the countries might be influencing both variables, undermining casual inference. The difference-in-differences model (Eqn 16.8), on the other hand, tests whether countries that *increased* their II or TM capitals between 1970–1975 and 1990–1995 also saw an *increase* in their average TFP growth rates between 1971–1990 and 1991–2009. The results find that countries that increased their II capital between 1970–1975 and 1990–1995 achieved more rapid agricultural TFP growth in the decades following, whereas an increase in TM capital did not. Increasing II capital by one unit on the index scale raised the average annual TFP growth rate by 0.46% (Table 16.6). The evidence is strongest for Latin America, where an increase in II capital was associated with an increase in the TFP growth rate of 0.76%. The effect of II capital on TFP growth in Asian countries was also positive and significant (0.48%), whereas for sub-Saharan Africa it was positive but not statistically significant. The evidence presented earlier in this volume (Fuglie and Rada, Chapter 12; Nin-Pratt and Yu, Chapter 13) provide insights into why research capacity in sub-Saharan Africa did not seem to have had much impact on growth in the region: small countries might not have been able to achieve sufficient scale in their national R&D systems, economic and trade policies have reduced incentives to agricultural producers, the AIDS/HIV epidemic has reduced the health of the population, civil disturbances and war have been widespread, and poor infrastructure reduces access to markets.

16.4 Conclusions

The framework outlined in this chapter provides a means for viewing agricultural productivity trends at the global level. It draws together the major available data series on national agricultural outputs and inputs to estimate growth in TFP in a consistent fashion by country, region and the world as a whole. The principal innovations introduced in this chapter compared with my earlier work using this approach (Fuglie, 2008, 2010b) are (i) chain-indexing the total input index using variable factor shares; (ii) decomposing growth into *input cost* and *natural resource* contributions to total growth; and (iii) using a more complete accounting of farm machinery inputs. None the less, it is likely that the machinery input series is still underestimating actual growth in fixed capital (see Butzer *et al.*, Chapter 15, this volume). Another potential shortcoming is that fertilizer use trends might be a poor proxy for growth in total material inputs. It would be especially helpful if consistent series on animal feed inputs could be developed (possibly from the FAO commodity balance sheets). Any under-accounting of growth in capital, material or other inputs implies an over-accounting of the growth in TFP. Despite these data shortcomings, where comparisons are possible the TFP indexes developed here generally show a good fit

Table 16.6. Difference-in-difference model of technology capital and TFP growth.

Dependent variable: Change in TFP growth rate between 1971–1990 and 1991–2009
Independent variables: Change in **II** and **TM** capitals between 1970/75 and 1990/95

		Coefficients (t-ratios in parentheses)					
Model	Obs.	II		TM		R-squared	Adj. R-sq
All countries	87	0.458 (2.00)	**	0.162 (0.78)	ns	0.111	0.090
LAC	22	0.764 (2.11)	**	0.498 (1.41)	ns	0.393	0.332
Asia	28	0.480 (0.79)	*	−0.176 (−0.41)	ns	0.034	−0.040
SSA	37	0.226 (0.66)	ns	0.328 (0.79)	ns	0.05	−0.004

*,**,***, significant at 10%, 5% and 1% level, respectively. ns, not significant. LAC, Latin America & Caribbean; Asia includes developing countries in East, South and West Asia. SSA, sub-Saharan Africa. *Source*: Author's estimates of Eqn16.8 in text.

with TFP indexes constructed from more detailed, national-level data.

The empirical analysis examined global agricultural growth during 1961–2009. The major empirical finding is that, on the basis of these measures, there does not seem to be a slowdown in sector-wide global agricultural productivity growth. If anything, the growth rate in global agricultural TFP accelerated, in no small part because of rapid productivity gains achieved by developing countries, led by Brazil and China, and more recently because of a recovery of agricultural growth in the countries of the former Soviet Union and Eastern Europe. The results do, however, show clear evidence of a slowdown in the growth in agricultural investment: the global agricultural resource base is still expanding but at a much slower rate than in the past. These two trends – accelerating TFP growth and decelerating input growth – have largely offset each other to keep the real output of global agriculture growing at more than 2% per year since the 1970s. Agricultural producers have substituted productivity for natural and material resources as the primary means of raising agricultural supply. This finding has important implications for the appropriate supply-side policy response to the recent rise in real agricultural prices and the future potential to raise agricultural supply.

One implication is that we should be sanguine about the prospects for global agriculture to respond to the recent commodity price rises by increasing supply in the short run. If TFP were slowing down, it would probably take several years for policy responses to influence this trend. The principal policy lever to increase TFP growth is to increase spending on agricultural research, but there are long time lags between research investments and productivity growth. But the main trend identified here is a slowdown in the rate of growth in agricultural inputs. This is at least in part a consequence of a long period of declining real prices facing producers, who then found better opportunities for their capital outside of agriculture. It was also in part a consequence of the institutional changes in the countries of the former Soviet bloc that precipitated a rapid exit of resources from agriculture in that region. The incentives afforded by the current high commodity prices and a resumption of agricultural growth in the former Soviet republics and Eastern Europe should positively affect the rate of agricultural capital formation at the global level. So long as TFP growth continues at its recent historical pace, this should lead to an increased rate of real output growth in global agriculture in a relatively short period of time.

The evidence presented in this chapter suggests that there has been a convergence in agricultural productivity growth across major world regions, with TFP growth in developed, developing and transition country regions all growing at about 2% per year on average since the turn of the century. This is in marked contrast with previous decades, in which productivity growth in developed countries was markedly higher than elsewhere (a result also demonstrated by Hayami and Ruttan (1985) and Craig et al. (1997), who found developing countries were falling further behind developed countries in agricultural land and labour productivity). None the less, it remains true that many countries have not been able to achieve or sustain productivity growth in agriculture and as a consequence suffer from high levels of poverty and food insecurity. This has not contributed to a *slowdown* in global agricultural TFP growth because their growth rates were never high to begin with. But this certainly has led to agriculture performing below its potential and has kept these countries poor. The largest group of countries in this low-growth category is in sub-Saharan Africa, but also included are several countries in Latin America (notably Bolivia, Panama, Paraguay and several Caribbean states) and in the Asia–Pacific region.

Finally, there is evidence that agricultural productivity growth has been uneven across commodities. Our ability to assess productivity growth at the commodity level is, however, limited mainly to examining harvest yield trends because labour and capital inputs tend to be shared across multiple commodities in the production process. Thus, the slowing of growth in cereal yield (World Bank, 2007; Alston *et al.*, 2009) does raise concerns that there is underinvestment (or low returns) to research directed at these commodities. But even here the picture is uneven, for decomposing cereal yield trends reveal that the slowdown affected primarily wheat and rice yields, with corn yield growth continuing to perform well after 1990. It is possible that the relatively strong performance in corn yield growth is due to the historically higher level of investment in R&D for this crop because

of the strong private-sector interest in breeding for hybrid corn (Fuglie *et al.*, 1996). In any case, the implication for R&D policy is quite different than if a sector-wide productivity slowdown were occurring. Rather than comprehensive changes to agricultural R&D or investment policies, the uneven performance within the agricultural sector suggests a more selective approach that requires a clear understanding of the causes of low productivity growth in particular commodities and countries.

Notes

[1] Strictly speaking, input prices are held constant when estimating total input growth, so any increase in cost comes from using more quantity of the input and not from changes in its price. If input and/or output prices actually change between any two periods over which TFP growth is estimated, this would affect the distribution of the economic gains in TFP but not the measure of TFP growth itself. For example, if output prices fell between the two periods, some of the gains in TFP would be passed on to consumers in the form of lower food prices. If fertilizer prices increased between two periods, some of the gains in TFP would be distributed as higher payments for fertilizers. In competitive equilibrium, any TFP benefits that are retained by the farm sector will be capitalized into the price of sector-specific inputs, namely, land, so as to maintain the zero profit (total cost = total revenue) condition.

[2] Comprehensive statistics on national agricultural extension services are lacking, but I have compiled what information is available from Judd *et al.* (1991) with updates from Swanson *et al.* (1990). The average years of schooling for adult males in the labour force are from Barro and Lee (2001). These are for the labour force as a whole and might overstate average schooling levels of agricultural labour.

[3] Actually, II and TM capacities are measured as an average of observed data from the 1970–1975 and 1990–1995 periods. Because of the spotty nature of the data, it is only possible to derive consistent measures of these indicators for a large number of countries by taking observations over a period of nearby years. For convenience, I refer to these measures as '1970' and '1990' capacities.

[4] Note that the series for Jordan includes a break in the smoothed series between 1967 and 1968. Prior to 1968, FAO's agricultural data for Jordan includes production from the West Bank. But following the Six Day War when the West Bank came under Israeli control, agricultural production from the West Bank is excluded from Jordan's output. Jordan seems to be an exception in the FAO data in that it does not represent a continuous geographical area for the years in which it is included.

[5] Fan and Zhang (1997) also used sown area in their study of agricultural productivity in China. Both the FAO and ERS series on arable land in China show huge discontinuities in the 1970s or 1980s caused by statistical changes to reporting methods. None the less, the sown area series probably overstates growth in cropland somewhat because it includes increases in cropping intensity resulting from expansion of irrigation and other factors.

[6] Some adjustments to these data should be noted. The FAO figure for the number of power thresher-harvesters in use in Indonesia actually includes both pedal and power threshing machines. I include only power thresher-harvesters from Indonesian national data. China reports total 'power' employed in agriculture in terms of kilowatts, but this probably includes some post-harvest processing machinery such as grain mills and oilseed crushers in addition to on-farm machinery. I only include tractors (four-wheel and two-wheel) and power thresher-harvesters in estimating total farm machinery horse power for China.

References

Agricultural Science and Technology Indicators. On-line Database. Agricultural Science and Technology Indicators Project, International Food Policy Research Institute, Washington, DC. Available at http:// www.asti.cgiar.org/ (Accessed October 2008).

Alston, J., Beddow, J. and Pardey, P. (2009) Agricultural research, productivity, and food prices in the long run. *Science* 325, 1209–1210.

Asian Development Bank. On-line Statistical Database System. Manila, The Philippines. Available at https://sdbs.adb.org/sdbs/index.jsp (Accessed August 2011).

Avila, A. and Evenson, R. (2010) Total factor productivity growth in agriculture: The role of technology capital. In: Pingali, P. and Evenson, R. (eds) *Handbook of Agricultural Economics,* volume 4. Elsevier, Amsterdam, pp. 3769–3822.

Ball, V.E. (1985) Output, input and productivity measurement in U.S. agriculture, 1948–79. *American Journal of Agricultural Economics* 67, 475–486.

Ball, V.E., Butault, J., Mesonada, C. and Mora, R. (2010) Productivity and international competitiveness of agriculture in the European Union and the United States. *Agricultural Economics* 41, 611–627.

Barro, R. and Lee, J.-W. (2001) International data on educational attainment: Updates and implications. *Oxford Economic Papers* 53, 54–63.

Binswanger, H., Deininger, K. and Feder, G. (1995) Power, distortions, revolt and reform in agricultural land relations. In: Chenery, H. and Srinivasan, T.N. (eds) *Handbook of Development Economics,* volume 3. Elsevier, Amsterdam, pp. 2659–2772.

CISSTAT. CIS CD-ROM. Interstate Statistical Committee of the Commonwealth of Independent States. Available from http://www.euros.ch/cistop.html

Coelli, T. and Rao, D.S.P. (2005) Total factor productivity growth in agriculture: A Malmquist index analysis of 93 countries, 1980–2000. *Agricultural Economics* 32, 115–134.

Council of Agriculture. Statistics. Executive Yuan, Republic of China. Available at http://eng.coa.gov.tw/list.php?catid=8821 (Accessed September 2011).

Craig, B., Pardey, P. and Roseboom, J. (1997) International productivity patterns: Accounting for input quality, infrastructure, and research. *American Journal of Agricultural Economics* 79, 1064–1076.

Cungu, A. and Swinnen, J. (2003) Transition and total factor productivity in agriculture, 1992–1999. Working Paper 2003/2, Research Group on Food Policy, Transition and Development, Katholieke Universiteit Leuven, Belgium.

Economic Research Service (a) Agricultural productivity in the United States, Agricultural research and productivity briefing room. U.S. Department of Agriculture, Washington, DC. http://www.ers.usda.gov/Data/AgProductivity/ (Accessed September 2011).

Economic Research Service (b) China agricultural and economic data, China briefing room. US Department of Agriculture, Washington, DC. http://www.ers.usda.gov/data/china/ (Accessed September 2011).

EUROSTAT. Database, Statistics, Employment and unemployment (LFS). European Commission. Available at http://epp.eurostat.ec.europa.eu/ (Accessed September 2011).

Evenson, R. and Fuglie, K. (2010) Technological capital: The price of admission to the growth club. *Journal of Productivity Analysis* 33, 173–190.

Evenson, R. and Kislev, Y. (1975) *Agricultural Research and Productivity.* Yale University Press, New Haven, CT.

Evenson, R., Pray, C. and Rosegrant, M. (1999) Agricultural research and productivity growth in India. Research Report Number 109, International Food Policy Research Institute, Washington, DC.

Fan, S. and Zhang, X. (1997) How fast have China's agricultural production and productivity really been growing? EPTD Discussion Paper, International Food Policy Research Institute, Washington, DC.

Fan, S. and Zhang, X. (2002) Production and productivity growth in Chinese agriculture: New national and regional measures. *Economic Development and Cultural Change* 50, 819–838.

FAO. FAOSTAT Database. Food and Agriculture Organization of the United Nations, Rome. Available at http://faostat.fao.org/ (Accessed September 2011).

Fernandez-Cornejo, J. and Shumway, C. (1997) Research and productivity in Mexican agriculture. *American Journal of Agricultural Economics* 79, 738–753.

Fuglie, K. (2008) Is a slowdown in agricultural productivity growth contributing to the rise in commodity prices? *Agricultural Economics* 39, supplement, 431–441.

Fuglie, K. (2010a) Sources of growth in Indonesian agriculture. *Journal of Productivity Analysis* 33, 225–240.

Fuglie, K. (2010b) Total factor productivity in the global agricultural economy: Evidence from FAO data. In: Alston, J., Babcock, B. and Pardey, P. (eds) *The Shifting Patterns of Agricultural Production and Productivity Worldwide.* Midwest Agribusiness Trade and Research Information Center, Iowa State University, Ames, IA. pp. 63–95.

Fuglie, K. (2011) Agricultural productivity in sub-Saharan Africa. In: Lee, D. (ed.) *The Food and Financial Crisis in Africa.* CAB International, Wallingford, Oxon, UK, in press.

Fuglie, K., Ballenger, N., Day, K., Klotz, C., Ollinger, M., Reilly, J., Vasavada, U. and Yee, J. (1996) Agricultural research and development: Public and private investments under alternative markets and institutions. Agricultural Economics Report 735, Economic Research Service, US Department of Agriculture, Washington, DC.

Hayami, Y. and Ruttan, V.W. (1971, 1985) *Agricultural Development: An International Perspective*. 1st edn, 1971; 2nd edn, 1985. Johns Hopkins University Press, Baltimore, MD.

Hayami, Y., Ruttan, V.W. and Southworth, H. (eds) (1979) *Agricultural Growth in Japan, Taiwan, Korea and the Philippines*. University Press of Hawaii, Honolulu, Hawaii.

International Fertilizer Association (IFA). Statistics, IFADATA. Available at http://www.fertilizer.org/ifa/ifadata/search (Accessed September 2011).

IBGE (Instituto Brasileiro de Geografia e Estatística) Agricultural Census Databases, Brazilian Institute for Geography and Statistics (IPGE) Ministry of Planning, Brasilia, Brazil. http://www.ibge.gov.br/ (Accessed July 2008).

Judd, M., Boyce, J. and Evenson, R. (1991) Investment in agricultural research and extension programs: A quantitative assessment. In: Evenson, R. and Pray, C. (eds) *Research and Productivity in Asian Agriculture*. Cornell University Press, Ithaca, NY, 7–46.

Kwon, O.S. (2010) Agricultural R&D and total factor productivity of Korean agriculture. *Korean Journal of Agricultural Economics* 51, 2, 67–88 (in Korean).

LABORSTA. Database, Bureau of Statistics, International Labor Organization, Geneva, Switzerland. Available at http://laborsta.ilo.org/ (Accessed September 2011).

Lerman, Z., Kislev, Y., Biton, D. and Kriss, A. (2003) Agricultural output and productivity in the former Soviet republics. *Economic Development and Cultural Change* 51, 999–1018.

Lusigi, A. and Thirtle, D. (1997) Total factor productivity and the effects of R&D in African agriculture. *Journal of International Development* 9, 529–538.

Pardey, P., Roseboom, J. and Anderson, J. (eds) (1991) *Agricultural Research Policy: International Quantitative Perspectives*. Cambridge University Press, Cambridge.

Peterson, W. (1987) International land quality indexes. Staff Paper P87-10, April. Department of Applied Economics, University of Minnesota, St. Paul, MN.

Rao, D.S.P. (1993) Intercountry comparisons of agricultural output and productivity. FAO Economic and Social Development Paper, United Nations Food and Agriculture Organization, Rome.

Ravn, M. and Uhlig, H. (2002) On adjusting the Hodrick-Prescott Filter for the frequency of observations. *Review of Economics and Statistics* 84, 371–376.

Rozelle, S. and Swinnen, J. (2004) Success and failure of reform: Insights from the transition of agriculture. *Journal of Economic Literature* 42, 404–456.

Schimmelpfennig, D., Thirtle, C., van Zyl, J., Arnade, C. and Khatri, Y. (2000) Short and long-run returns to agricultural R&D in South Africa, or will the real rate of return please stand up? *Agricultural Economics* 23, 1–15.

Shend, J. (1993) Agricultural statistics of the former USSR republics and the Baltic States. Statistical Bulletin No. 863, Economic Research Service, US Department of Agriculture, Washington, DC.

Swanson, B., Farner, B. and Bahal, R. (1990) The current status of agricultural extension worldwide, in *Agricultural Education and Extension Service. Report of the Consultation on Agricultural Extension*, FAO, Rome, 43–76.

Swinnen, J., Dries, L. and Macours, K. (2005) Transition and agricultural labor. *Agricultural Economics* 32: 15–34.

Thirtle, C., Piesse, J. and Schimmelpfennig, D. (2008) Modeling the length and shape of the R&D lag: An application to UK agricultural productivity. *Agricultural Economics* 39, 73–85.

UNESCO. Institute for Statistics, Montreal, Canada. Available at http://stats.uis.unesco.org/unesco/tableviewer/document.aspx?FileId=50 (Accessed July 2008).

Van der Meer, C. and Yamada, S. (1990) *Japanese Agriculture: A Comparative Economic Analysis*. Routledge, London.

Wiebe, K., Soule, M., Narrod, C. and Breneman, V. (2003) Resource quality and agricultural productivity: A multi-country comparison. In: Wiebe, K. (ed.) *Land Quality, Agricultural Productivity and Food Security: Biophysical Processes and Economic Choices at Local, Regional and Global Levels*. Edward Elgar, Northampton, MA, pp. 147–165.

World Bank (2007) *World Development Report 2008*. World Bank, Washington, DC.

Appendix 16.1

Table A16.1. Agricultural input cost shares.

Source study	Input	Input cost shares					Input shares applied to
		1961–1970	1971–1980	1981–1990	1991–2000	2001–2010	
Industrialized countries							
South Africa	Labour	0.232	0.210	0.166	0.161	0.161	South Africa
	Land	0.129	0.143	0.169	0.144	0.144	
Schimmelpfennig et al. (2000)	Livestock	0.252	0.230	0.237	0.239	0.239	
	Fixed capital	0.141	0.138	0.154	0.182	0.182	
	Materials	0.246	0.279	0.275	0.274	0.274	
USA	Labour	0.235	0.184	0.171	0.221	0.226	USA
	Land	0.203	0.225	0.188	0.176	0.152	
Economic Research Service	Livestock	0.291	0.301	0.281	0.250	0.257	
(a), based on Ball (1985)	Fixed capital	0.128	0.134	0.180	0.129	0.131	
	Materials	0.143	0.156	0.180	0.224	0.234	
Canada	Labour	0.345	0.406	0.303	0.431	0.349	Canada
	Land	0.035	0.023	0.022	0.016	0.016	
Cahill and Rich (Chapter 3,	Livestock	0.251	0.213	0.234	0.204	0.222	
this volume)	Fixed capital	0.146	0.147	0.162	0.087	0.085	
	Materials	0.223	0.211	0.279	0.262	0.328	
Australia	Labour	0.176	0.176	0.093	0.089	0.098	Australia and New Zealand
Zhao et al. (Chapter 4, this volume);	Land	0.348	0.348	0.600	0.653	0.539	
with decomposition of total capital	Livestock	0.052	0.052	0.019	0.011	0.016	
stock from Butzer et al. (2012)	Fixed capital	0.222	0.222	0.156	0.114	0.162	
	Materials	0.200	0.200	0.131	0.133	0.186	
Japan	Labour	0.384	0.335	0.309	0.308	0.307	Japan
	Land	0.322	0.291	0.279	0.286	0.278	
Van der Meer and Yamada (1999)	Livestock	0.128	0.123	0.134	0.131	0.130	
	Fixed capital	0.075	0.136	0.157	0.153	0.162	
	Materials	0.091	0.114	0.121	0.122	0.122	
Korea–Taiwan	Labour	0.372	0.558	0.349	0.208	0.156	South Korea and Taiwan
1961–1970 is average for Korea and	Land	0.419	0.227	0.392	0.506	0.519	
Taiwan from Hayami et al. (1979);	Livestock	0.067	0.004	0.009	0.010	0.012	
1970+ from Kwon (2010) using	Fixed capital	0.013	0.016	0.040	0.080	0.122	
Korea data	Materials	0.129	0.194	0.210	0.196	0.191	

Study	Input						Region
UK	Labour	0.327	0.164	0.136	0.137	0.137	UK
	Land	0.084	0.126	0.179	0.216	0.216	
Thirtle et al. (2008)	Livestock	0.251	0.333	0.284	0.235	0.235	
	Fixed capital	0.183	0.199	0.202	0.204	0.204	
	Materials	0.155	0.178	0.199	0.209	0.209	
Europe, Northern except UK	Labour	0.334	0.334	0.244	0.235	0.220	Northern Europe except United Kingdom
	Land	0.040	0.040	0.074	0.079	0.069	
Ball et al. (2010); capital	Livestock	0.261	0.020	0.024	0.017	0.013	
decomposition from Butzer et al.	Fixed capital	0.073	0.073	0.104	0.134	0.134	
(2012)	Materials	0.292	0.533	0.554	0.535	0.564	
Europe, Southern	Labour	0.577	0.577	0.450	0.404	0.469	Southern Europe
	Land	0.085	0.085	0.124	0.154	0.096	
Ball et al. (2010); capital	Livestock	0.016	0.016	0.018	0.014	0.010	
decomposition from Butzer	Fixed capital	0.059	0.059	0.076	0.114	0.105	
et al. (2012)	Materials	0.263	0.263	0.331	0.313	0.319	
Developing countries and regions							
Sub-Saharan Africa	Labour	0.248	0.248	0.248	0.248	0.248	Sub-Saharan Africa
	Land	0.315	0.315	0.315	0.315	0.315	
Fuglie (2011)	Livestock	0.357	0.357	0.357	0.357	0.357	
	Fixed capital	0.024	0.024	0.024	0.024	0.024	
	Materials	0.055	0.055	0.055	0.055	0.055	
Mexico	Labour	0.256	0.239	0.119	0.115	0.115	Central America & Caribbean
Fernandez-Cornejo and Shumway	Land	0.489	0.344	0.179	0.225	0.225	
(1997)	Livestock	0.118	0.221	0.371	0.353	0.353	
	Fixed capital	0.089	0.162	0.315	0.263	0.263	
	Materials	0.048	0.035	0.017	0.045	0.045	
Brazil	Labour	0.434	0.434	0.443	0.415	0.373	South America, North Africa and West Asia
Estimated provided by Nicholas	Land	0.342	0.342	0.159	0.115	0.083	
Rada, unpublished, calculated	Livestock	0.126	0.126	0.168	0.181	0.129	
from Brazilian Agricultural Census'	Fixed capital	0.071	0.071	0.110	0.177	0.161	
1970, 1985, 1996, 2006 (IBGE)	Materials	0.027	0.027	0.120	0.112	0.255	
China	Labour	0.443	0.396	0.413	0.333	0.333	China, Mongolia, and North Korea
	Land	0.261	0.208	0.177	0.255	0.255	
Fan and Zhang (2002)	Livestock	0.228	0.247	0.230	0.206	0.206	
	Fixed capital	0.021	0.070	0.087	0.074	0.074	
	Materials	0.048	0.078	0.093	0.132	0.132	

Continued

Table A16.1. Continued.

Source study	Input	Input cost shares					Input shares applied to
		1961–1970	1971–1980	1981–1990	1991–2000	2001–2010	
India	Labour	0.406	0.419	0.564	*0.564*	*0.564*	South Asia
	Land	0.314	0.210	0.173	*0.173*	*0.173*	
Evenson, Pray and Rosegrant (1999)	Livestock	0.263	0.319	0.173	*0.173*	*0.173*	
	Fixed capital	0.003	0.010	0.024	*0.024*	*0.024*	
	Materials	0.014	0.042	0.066	*0.066*	*0.066*	
Indonesia	Labour	0.370	0.538	0.476	0.388	0.392	Southeast Asia
Fuglie (2010a)	Land	0.219	0.195	0.188	0.306	0.329	and develop-
	Livestock	0.360	0.199	0.278	0.251	0.217	ing countries
	Fixed capital	0.018	0.020	0.004	0.010	0.015	in Oceania
	Materials	0.033	0.048	0.054	0.045	0.046	
Transition countries and regions							
USSR, European	Labour	0.104	0.104	0.104	0.190	0.190	European states
	Land	0.257	0.257	0.257	0.230	0.230	of the former
Lerman et al. (2003) for 1965–1990.	Livestock	0.453	0.453	0.453	0.420	0.420	Soviet Union
Cungu and Swinnen (2003) for 1992+	Fixed capital	0.043	0.043	0.043	0.090	0.090	
	Materials	0.143	0.143	0.143	0.070	0.070	
USSR, Asia	Labour	0.194	0.194	0.194	0.190	0.190	Asian states of
	Land	0.210	0.210	0.210	0.230	0.230	the former
Lerman et al. (2003) for 1965–1990.	Livestock	0.104	0.104	0.104	0.420	0.420	Soviet Union
Cungu and Swinnen (2003) for 1992+	Fixed capital	0.113	0.113	0.113	0.090	0.090	
	Materials	0.379	0.379	0.379	0.070	0.070	

In some cases, studies report cost shares of animal feed or other farm-supplied inputs. This has been included with the cost share of livestock capital. Cost shares in italics are extrapolations using estimates from the nearest period available. When studies did not break out fixed capital from livestock capital, I used average capital component shares for high-income or middle & low-income countries reported by Butzer *et al.* in Chapter 15 of this volume (see Table 15.3). Source: Compiled by author from sources listed. Eldon Ball, Shenggen Fan, Jorge Fernandez-Cornejo, Oh-Sang Kwon, Nicholas Rada, David Schimmelpfennig and Colin Thirtle kindly provided additional, unpublished data.

Table A16.2. Agricultural output and productivity growth by country.

Country	Region	Agricultural output						Agricultural TFP				
		Avg 2006–2009	1961–1970	1971–1980	1981–1990	1991–2000	2001–2009	1961–1970	1971–1980	1981–1990	1991–2000	2001–2009
		Million $										
							Average annual growth (%)					
Sub-Saharan Africa												
Cameroon	Central	3.71	3.89	1.51	2.05	3.20	3.27	-0.06	-1.58	0.79	1.28	2.32
C. African Rep.	Central	0.87	2.99	2.16	2.07	3.80	2.06	-1.55	-0.35	1.46	1.78	-0.04
Congo	Central	0.34	2.57	1.11	1.39	3.26	3.44	-0.78	0.18	0.01	1.39	3.06
Congo, DR	Central	3.67	1.76	1.58	3.07	-2.65	-0.05	-1.09	-0.29	0.69	-0.31	-1.29
Gabon	Central	0.25	1.67	3.42	2.39	1.83	1.30	-0.31	-1.58	-0.75	1.64	0.19
Burundi	Eastern	1.07	2.23	0.83	3.02	-1.38	-1.28	-1.40	-1.33	0.53	0.28	-4.19
Kenya	Eastern	6.10	2.81	3.85	4.34	1.23	3.43	-0.29	1.72	0.71	0.66	1.98
Rwanda	Eastern	1.62	4.69	4.09	1.47	0.63	4.00	0.24	2.53	-0.41	0.51	-2.10
Tanzania	Eastern	6.53	3.18	3.43	2.24	1.87	4.14	-0.50	0.82	0.54	0.38	1.03
Uganda	Eastern	5.45	5.44	-1.63	2.81	2.70	0.84	2.55	-0.02	1.78	-0.06	-1.90
Ethiopia, former	Horn	8.00	2.03	1.36	0.66	3.10	4.82	-1.09	1.23	-1.17	-0.12	1.38
Somalia	Horn	1.59	3.69	2.49	0.91	1.94	0.84	0.40	1.30	-0.32	1.55	0.41
Sudan	Horn	8.06	2.66	3.05	0.79	4.85	1.41	-1.12	1.07	0.54	1.94	0.04
Burkina Faso	Sahel	2.20	3.09	2.10	6.40	4.08	2.67	-0.88	-0.85	1.76	1.03	-2.16
Chad	Sahel	1.42	0.83	1.27	2.87	4.00	1.99	-1.89	0.88	1.02	0.33	-0.13
Gambia	Sahel	0.12	2.54	-2.88	-0.09	3.48	1.90	-1.49	-4.23	-1.84	1.25	-2.03
Mali	Sahel	2.61	3.00	3.45	3.00	3.24	4.95	-1.53	1.95	1.82	1.37	2.39
Mauritania	Sahel	0.45	1.60	1.53	1.74	1.89	1.85	-0.95	0.53	-0.52	0.39	0.57
Niger	Sahel	2.81	2.86	3.82	0.78	5.31	6.16	-2.07	-0.21	0.52	2.33	3.31
Senegal	Sahel	1.20	-0.25	0.89	2.39	1.82	4.04	-3.22	-0.14	0.96	-0.41	2.11
Angola	Southern	2.14	3.04	-4.64	1.14	4.76	6.82	-2.01	-4.76	-0.40	3.94	3.00
Botswana	Southern	0.25	3.52	-0.02	0.82	-1.28	3.33	2.07	-2.06	0.38	-4.37	2.52
Lesotho	Southern	0.12	1.69	0.86	0.61	1.45	-0.52	-0.31	0.77	-1.30	0.17	0.29
Madagascar	Southern	3.14	2.78	1.23	1.64	0.49	3.09	-0.52	-0.76	0.86	-0.19	0.85
Malawi	Southern	2.52	3.97	3.48	1.24	6.43	5.35	0.17	0.57	-0.24	5.17	1.32
Mauritius	Southern	0.25	1.64	0.31	0.99	0.25	-0.29	1.07	0.45	-0.31	-0.27	-0.38
Mozambique	Southern	1.99	3.00	-1.89	-0.98	6.77	1.60	0.25	-2.96	1.14	2.70	-0.03
Namibia	Southern	0.44	3.69	-1.82	0.85	-0.55	0.99	2.56	-1.92	-0.19	-2.29	0.75

Continued

Table A16.2. Continued.

Country	Region	Avg 2006–2009	Agricultural output					Agricultural TFP				
			1961–1970	1971–1980	1981–1990	1991–2000	2001–2009	1961–1970	1971–1980	1981–1990	1991–2000	2001–2009
		Million $	Average annual growth (%)									
Sub-Saharan Africa												
Réunion	Southern	0.17	0.50	2.55	2.46	1.71	0.71	0.01	1.35	3.01	2.48	0.74
Swaziland	Southern	0.28	4.43	3.67	2.23	-0.63	1.43	3.20	2.23	0.32	-0.19	1.52
Zambia	Southern	1.16	3.47	2.27	4.51	1.48	4.13	0.67	1.24	0.77	1.22	2.17
Zimbabwe	Southern	1.53	4.09	1.46	2.37	3.30	-2.58	1.00	0.98	0.67	1.12	-1.65
Benin	Western	1.88	2.48	2.20	4.98	6.07	2.34	-1.48	1.82	2.51	1.91	2.93
Côte d'Ivoire	Western	5.96	4.81	4.64	3.55	3.82	1.59	0.19	-0.05	0.53	2.21	0.76
Ghana	Western	5.67	2.55	-2.64	5.08	5.23	4.00	-0.79	-3.55	3.99	1.79	1.21
Guinea	Western	1.90	1.96	1.43	2.46	3.33	3.23	0.10	0.60	1.58	-1.65	0.05
Guinea-Bissau	Western	0.25	-3.15	3.21	3.32	3.68	1.88	-2.72	-0.35	3.67	-0.24	0.36
Liberia	Western	0.41	4.30	1.80	-0.85	6.13	0.89	-0.36	-0.50	-0.73	3.10	-1.65
Sierra Leone	Western	0.63	2.92	1.35	1.76	-1.34	6.06	-0.75	-0.43	-0.08	1.29	2.24
Togo	Western	0.75	2.38	1.06	3.52	3.92	1.34	-1.28	-1.74	-2.16	0.97	-0.26
Nigeria (FAO data)	Nigeria	35.19	3.30	-0.85	6.34	4.22	1.96	-0.97	-2.36	2.51	3.11	0.48
Nigeria (alt. date)	Nigeria	30.02	3.02	-0.14	4.70	3.85	2.24	-1.32	-2.21	0.33	2.19	0.22
Latin American & Caribbean (LAC)												
Cuba	Caribbean	2.88	2.97	2.99	0.88	-1.77	-2.81	-1.51	1.19	-0.02	-0.73	-2.48
Dominican Republic	Caribbean	2.37	1.40	2.24	0.87	0.33	3.29	-0.30	0.70	-0.60	1.15	2.09
Haiti	Caribbean	1.00	2.01	1.48	-0.58	0.21	1.50	0.24	0.59	-0.66	-1.01	0.99
Jamaica	Caribbean	0.54	0.85	0.17	0.73	2.01	-0.06	1.44	-1.14	-0.15	2.10	3.05
Lesser Antilles	Caribbean	0.33	0.02	-0.57	1.41	-0.70	-2.99	-0.31	-1.08	1.70	-1.13	-0.90
Puerto Rico (USA)	Caribbean	0.31	-2.48	-0.46	0.52	-3.01	-0.35	-0.71	1.80	0.82	-2.07	0.88
Trinidad and Tobago	Caribbean	0.17	2.07	-1.78	-0.81	0.46	1.40	1.30	-2.19	-0.95	0.05	1.42
Belize	C. America	0.17	8.40	4.70	2.60	4.80	0.92	5.03	2.86	-0.44	4.62	-0.20
Costa Rica	C. America	2.75	6.56	2.98	4.21	3.28	3.03	4.77	1.13	3.86	4.13	2.47
El Salvador	C. America	1.15	2.19	2.92	-0.23	1.33	2.58	0.63	1.23	-1.27	1.54	2.12

Country	Region											
Guatemala	C. America	3.72	4.44	3.52	2.24	3.91	4.39	2.50	1.95	1.49	2.86	2.73
Honduras	C. America	1.92	6.51	1.95	1.13	1.50	4.29	2.98	-0.15	-0.51	1.34	1.94
Mexico	C. America	35.22	4.45	4.09	1.34	2.98	1.82	2.65	2.17	-1.98	3.19	2.19
Nicaragua	C. America	1.30	5.97	1.94	-2.94	4.49	3.69	2.44	-1.78	-4.26	2.69	2.88
Panama	C. America	0.93	5.44	1.98	1.02	0.80	1.47	2.57	0.46	0.55	-0.78	1.31
Bolivia	Andean	3.00	4.10	3.02	3.10	4.11	2.96	1.81	0.71	1.70	2.82	-0.47
Colombia	Andean	13.91	2.45	4.11	2.54	1.56	3.02	1.27	2.56	1.43	2.07	2.99
Ecuador	Andean	6.71	2.10	1.12	4.07	3.89	3.34	0.00	-0.18	2.27	1.18	3.55
Peru	Andean	7.75	2.73	0.17	2.16	5.86	4.37	0.85	-0.99	-0.02	3.11	3.46
Venezuela	Andean	6.47	5.09	3.63	2.70	2.65	2.15	3.69	2.57	-0.90	3.20	1.85
Brazil	Northeast	126.64	3.57	3.88	3.44	3.65	4.45	0.19	0.53	3.02	2.61	4.04
Guyana	Northeast	0.32	1.10	1.20	-2.61	4.80	-0.66	-0.04	0.36	-1.64	4.45	-0.60
Suriname	Northeast	0.11	7.59	4.40	0.40	-3.42	3.24	5.09	3.77	0.69	-4.63	1.01
Argentina	S. Cone	41.36	1.80	3.01	0.48	3.24	2.68	0.12	3.13	-0.97	1.45	1.22
Chile	S. Cone	7.75	1.80	2.76	3.44	3.48	2.17	1.70	2.20	1.09	1.71	2.58
Paraguay	S. Cone	4.24	3.20	4.66	4.98	1.79	3.49	0.98	0.63	1.59	-2.35	-1.24
Uruguay	S. Cone	3.60	1.07	0.37	0.54	2.89	4.53	0.87	0.28	0.60	2.03	3.30
West Asia and North Africa (WANA)												
Algeria	North Africa	5.27	-0.97	-0.50	4.67	2.32	4.20	-1.29	-0.93	3.07	0.72	4.12
Egypt	North Africa	21.55	3.16	1.99	4.14	4.57	3.57	1.30	1.41	2.71	2.82	2.76
Libya	North Africa	1.11	8.26	6.23	2.24	3.26	0.95	8.00	3.48	3.60	4.46	3.02
Morocco	North Africa	7.43	4.56	0.97	6.02	1.52	3.82	3.07	-0.71	4.14	0.58	4.11
Tunisia	North Africa	3.66	1.56	2.28	3.84	2.02	2.97	0.75	1.46	3.51	0.38	1.34
Iran	West Asia	24.85	3.93	3.95	4.73	3.86	2.41	2.42	2.65	1.41	2.40	0.73
Iraq	West Asia	2.72	3.90	2.17	2.51	1.47	-1.98	0.85	2.85	1.45	0.39	-0.23
Israel	West Asia	2.70	6.18	3.34	0.58	2.26	1.94	5.65	2.74	0.95	2.41	2.57
Jordan	West Asia	1.03	-6.60	2.86	6.42	2.01	3.81	-8.84	3.94	3.80	2.12	5.87
Kuwait	West Asia	0.20	5.26	7.32	5.19	11.34	3.39	-0.74	2.04	0.08	7.05	-0.23
Lebanon	West Asia	1.26	3.52	0.94	5.76	0.32	0.94	3.44	2.01	8.83	-1.43	3.83
Oman	West Asia	0.32	1.98	7.90	3.49	4.92	1.71	-1.29	2.40	-2.64	3.92	-2.25
Saudi Arabia	West Asia	3.55	3.09	5.65	11.22	0.93	2.80	0.06	1.68	6.35	2.12	5.12
Syria	West Asia	6.58	1.41	7.44	1.22	4.26	0.93	-0.19	6.15	-2.45	2.65	-0.12
Turkey	West Asia	33.97	2.82	3.23	2.54	1.69	1.58	0.75	1.54	0.99	1.50	1.78
United Arab Emirates	West Asia	0.70	4.65	9.97	7.48	13.91	-1.74	2.71	3.93	-0.51	8.20	-4.73
Yemen	West Asia	1.57	-0.40	3.63	3.54	3.66	4.97	-2.94	1.31	1.44	1.72	2.24

Continued

Table A16.2. Continued.

Country	Region	Agricultural output						Agricultural TFP				
		Avg 2006–2009	1961–1970	1971–1980	1981–1990	1991–2000	2001–2009	1961–1970	1971–1980	1981–1990	1991–2000	2001–2009
		Million $				Average annual growth (%)						
Asia & Oceania, developing (LDC)												
China	Northeast	487.20	4.87	3.30	4.53	5.28	3.41	0.93	0.60	1.69	4.16	2.83
Korea, DPR	Northeast	3.76	2.25	4.52	3.22	−3.13	0.84	0.34	1.30	1.47	0.50	1.34
Mongolia	Northeast	0.71	0.59	1.95	1.10	0.01	1.76	0.08	0.12	0.27	0.89	0.58
Cambodia	Southeast	3.08	2.67	−7.04	6.14	4.73	8.13	−0.93	−4.19	2.69	3.06	5.85
Indonesia	Southeast	53.20	2.71	3.34	4.64	1.87	4.86	1.75	1.40	0.59	0.99	3.68
Laos	Southeast	1.52	5.66	1.22	2.98	5.24	4.69	0.61	−0.88	0.96	2.74	2.21
Malaysia	Southeast	13.64	5.41	4.40	4.61	2.50	4.00	3.57	2.56	3.29	1.88	3.81
Myanmar	Southeast	18.14	1.40	4.24	0.34	4.90	7.40	−1.68	2.14	−0.32	2.60	5.97
Philippines	Southeast	20.12	2.64	5.08	1.62	2.43	3.46	−0.18	3.57	0.11	0.80	2.70
Thailand	Southeast	28.79	3.43	4.97	2.63	1.99	2.78	0.44	2.44	0.44	2.79	2.37
Timor Leste	Southeast	0.13	2.42	−1.84	1.80	−0.83	3.06	0.50	0.17	−0.64	−1.79	0.88
Viet Nam	Southeast	26.22	0.45	2.93	4.01	5.91	4.22	−0.68	1.62	1.05	3.08	2.44
Afghanistan	South	3.02	3.12	1.59	−2.41	3.79	2.08	1.30	0.58	−0.12	2.73	−1.83
Bangladesh	South	19.36	2.15	1.97	2.07	2.95	4.32	−0.30	0.39	−0.51	2.12	3.31
Bhutan	South	0.17	2.66	2.71	1.50	2.07	4.12	0.40	−0.22	−0.41	0.55	1.13
India	South	205.34	1.68	2.75	3.35	2.52	3.27	0.49	1.00	1.33	1.11	2.08
Nepal	South	4.63	1.46	1.86	4.70	2.82	2.61	−0.19	−1.22	2.34	0.19	2.49
Pakistan	South	35.87	4.26	2.78	4.78	3.24	3.34	1.90	0.16	3.21	1.19	0.59
Sri Lanka	South	2.61	2.67	3.26	−0.59	1.07	2.33	0.93	2.30	−1.64	1.32	1.17
Fiji	Oceania, LDC	0.22	2.45	2.74	1.25	−0.74	−0.89	0.17	0.17	−1.21	−1.49	−0.36
Papua New Guinea	Oceania, LDC	2.57	2.97	2.33	2.12	2.46	2.51	−1.08	0.59	−0.06	0.52	1.00
Polynesia	Oceania, LDC	0.12	0.15	0.44	−1.94	0.94	1.73	−1.25	−2.54	−2.21	0.19	2.71
Solomon Islands	Oceania, LDC	0.12	1.93	4.89	−0.05	2.22	5.10	−1.79	2.37	−0.70	0.97	3.63
Transition countries												
Albania	E. Europe	1.07	3.81	4.03	0.21	3.03	2.40	−1.28	0.63	−1.31	3.84	3.81
Bulgaria	E. Europe	2.87	3.91	1.35	−0.88	−3.75	−1.97	1.69	0.22	0.82	0.44	0.49
Czechoslovakia, for.	E. Europe	5.91	2.68	1.25	1.08	−3.00	−0.62	1.95	0.33	2.21	0.95	1.56
Hungary	E. Europe	6.29	2.71	2.81	−0.24	−1.45	−0.52	0.67	2.54	1.62	0.14	1.99

Poland	E. Europe	19.99	2.13	0.29	0.96	-1.10	0.32	-0.56	-0.67	2.17	1.12	0.06
Romania	E. Europe	9.56	3.36	3.96	-1.89	-0.60	-0.32	-0.53	1.63	-1.60	1.44	-0.20
Yugoslavia, former	E. Europe	9.21	2.64	2.48	-0.53	-0.68	1.03	1.68	1.87	0.63	1.22	2.09
Estonia	FSU, Baltic	0.55	3.32	1.98	-0.64	-6.89	2.31	1.40	0.19	-0.69	1.29	4.70
Latvia	FSU, Baltic	0.88	2.89	1.02	0.81	-9.20	3.19	1.51	-0.46	0.22	0.48	3.20
Lithuania	FSU, Baltic	1.87	4.08	0.45	1.94	-4.20	1.56	2.73	-0.83	1.35	0.68	0.95
Armenia	FSU, CAC	1.05	0.20	4.85	-1.47	-0.61	7.36	-4.27	2.10	0.59	1.67	5.14
Azerbaijan	FSU, CAC	2.27	4.40	6.93	-0.92	-2.98	4.64	1.58	2.80	-1.10	-0.71	3.02
Georgia	FSU, CAC	0.80	3.22	4.29	-1.59	-0.21	-5.29	-0.59	2.72	-0.64	0.57	-2.97
Kyrgyzstan	FSU, CAC	1.82	3.78	2.00	2.73	1.36	0.87	-0.76	-0.22	1.34	3.74	0.53
Tajikistan	FSU, CAC	1.30	4.52	3.90	0.92	-4.27	5.62	0.81	2.06	0.47	-0.96	2.86
Turkmenistan	FSU, CAC	2.95	4.00	3.78	4.01	2.21	4.72	-0.48	0.62	1.21	0.69	1.01
Uzbekistan	FSU, CAC	10.02	3.21	5.07	0.68	0.73	5.38	-0.50	2.18	-1.38	1.05	3.39
Belarus	FSU, E Euro.	7.13	2.92	1.11	1.45	-3.72	4.38	0.10	-0.28	0.99	0.19	4.74
Kazakhstan	FSU, E Euro	7.92	5.67	1.35	1.69	-7.22	4.21	3.83	-0.45	0.47	3.36	2.41
Moldova	FSU, E Euro	1.51	3.66	1.88	0.11	-6.63	-0.55	0.78	-0.04	0.32	0.52	2.71
Russian Federation	FSU, E Euro	50.61	3.08	0.41	1.40	-4.99	2.25	0.88	-1.35	0.85	1.42	4.29
Ukraine	FSU, E Euro	22.87	2.65	1.13	1.38	-6.04	2.91	0.41	-0.18	1.12	-0.07	5.35
Developed countries (DC)												
South Africa	Africa, DC	11.73	3.13	2.52	1.21	1.55	2.12	0.34	1.15	2.71	2.79	3.01
Japan	Asia, DC	18.45	2.99	1.40	0.51	-1.06	-0.36	2.42	2.17	1.11	1.51	2.43
Korea, Republic	Asia, DC	10.27	4.48	5.81	2.91	2.47	0.64	1.83	4.28	2.81	4.04	2.86
Singapore	Asia, DC	0.18	6.26	3.97	-0.32	-2.18	-0.12	6.71	7.37	13.15	1.00	-5.19
Taiwan (China)	Asia, DC	4.59	3.71	1.48	2.13	1.17	-1.88	2.51	0.73	1.68	3.03	0.51
Australia	Oceania, DC	23.45	2.95	1.84	1.67	3.55	-0.76	0.63	1.65	1.27	2.85	0.55
New Zealand	Oceania, DC	10.07	2.77	1.29	1.01	2.27	1.19	1.47	1.39	1.84	3.20	3.14
Canada	North America	28.00	2.80	2.28	1.25	2.48	1.96	1.41	-0.36	2.67	2.55	2.14
USA	North America	229.48	1.99	2.30	0.61	1.90	1.35	1.21	1.80	1.21	2.17	2.26
Austria	Europe, NW	4.62	1.22	1.18	-0.01	0.65	0.14	1.00	1.90	1.30	2.52	4.39
Belgium-Luxembourg	Europe, NW	6.32	2.08	0.36	1.66	1.67	-1.02	1.99	1.78	1.96	2.77	0.39
Denmark	Europe, NW	6.95	-0.25	1.77	1.20	0.99	0.56	-0.23	2.43	2.31	3.76	2.71

Continued

Table A16.2. Continued.

Country	Region	Agricultural output						Agricultural TFP				
		Avg 2006–2009	1961–1970	1971–1980	1981–1990	1991–2000	2001–2009	1961–1970	1971–1980	1981–1990	1991–2000	2001–2009
Developed countries (DC)		Million $										
						Average annual growth (%)						
Finland	Europe, NW	2.26	0.95	0.81	–0.60	–0.27	0.41	0.05	2.62	1.50	–0.50	2.44
France	Europe, NW	41.28	1.51	1.17	0.43	0.74	–0.78	0.42	1.85	1.51	2.23	1.99
Germany	Europe, NW	35.34	1.87	0.92	0.25	0.19	0.47	2.12	1.05	2.05	2.21	2.98
Iceland	Europe, NW	0.10	–0.13	1.79	–2.12	0.80	1.77	–1.28	2.01	–0.98	2.20	1.75
Ireland	Europe, NW	4.59	1.95	3.19	1.51	0.80	–0.87	–0.22	2.38	1.44	0.93	1.17
Netherlands	Europe, NW	12.40	3.19	3.29	1.41	–0.35	0.84	2.32	1.59	1.49	1.37	2.41
Norway	Europe, NW	1.44	0.49	1.28	0.14	–0.20	0.30	0.92	0.91	1.18	0.56	2.37
Sweden	Europe, NW	3.10	–0.68	1.62	–1.00	0.53	–0.74	–0.31	2.20	1.56	1.82	0.84
Switzerland	Europe, NW	2.80	1.15	1.51	0.15	–0.79	0.52	0.43	1.06	0.06	1.74	2.02
UK	Europe, NW	17.83	1.45	1.53	0.55	–0.41	–0.06	1.97	1.30	0.58	0.18	1.00
Cyprus	Europe, South	0.38	7.21	–1.75	1.83	1.12	–2.99	5.24	1.61	2.95	2.00	–1.24
Greece	Europe, South	8.04	2.72	2.95	0.75	0.79	–2.29	3.13	4.04	1.72	1.36	–0.59
Italy	Europe, South	32.01	2.03	1.49	–0.26	0.32	–0.29	3.93	3.37	0.71	2.59	2.09
Portugal	Europe, South	4.17	0.23	–0.44	2.04	0.66	–0.24	1.19	–2.38	3.50	1.52	1.85

Output is gross agricultural output measuring in constant 2005 international dollars (FAO).TFP growth is the difference between the rate of growth in gross output and total input, where the growth in total input is aggregate growth in agricultural land, labour, livestock herds, farm machinery and fertilizer use (see text for further explanation). DC, developed countries; LDC, less developed countries; FSU, former Soviet Union; CAC, Central Asia and Transcaucasia. Regions are defined by the author. The average annual growth rate in series Y is found by regressing the natural log of Y against time, i.e. the parameter B in ln(Y) = A + Bt.

Index